Third Edition
The Economic Analysis of Industrial Projects

Ted G. Eschenbach
University of Alaska Anchorage

Neal A. Lewis
University of Bridgeport

Joseph C. Hartman
University of Massachusetts Lowell

Lynn E. Bussey

New York Oxford
OXFORD UNIVERSITY PRESS

Oxford University Press is a department of the University of Oxford.
It furthers the University's objective of excellence in research,
scholarship, and education by publishing worldwide.

Oxford New York
Auckland   Cape Town   Dar es Salaam   Hong Kong   Karachi
Kuala Lumpur   Madrid   Melbourne   Mexico City   Nairobi
New Delhi   Shanghai   Taipei   Toronto

With offices in
Argentina   Austria   Brazil   Chile   Czech Republic   France   Greece
Guatemala   Hungary   Italy   Japan   Poland   Portugal   Singapore
South Korea   Switzerland   Thailand   Turkey   Ukraine   Vietnam

Copyright © 2015 by Oxford University Press
Previously published by Prentice Hall 1992, 1978.

> For titles covered by Section 112 of the US Higher Education
> Opportunity Act, please visit www.oup.com/us/he for the latest
> information about pricing and alternate formats.

Published by Oxford University Press
198 Madison Avenue, New York, New York 10016
http://www.oup.com

Oxford is a registered trademark of Oxford University Press

All rights reserved. No part of this publication may be reproduced,
stored in a retrieval system, or transmitted, in any form or by any means,
electronic, mechanical, photocopying, recording, or otherwise,
without the prior permission of Oxford University Press.

Chapter opening photo credits, from left to right: cohdra, MorgueFile; © Janpietruszka, Dreamstime.com; © Christian Lagereek, Dreamstime.com; © K2photoprojects, Dreamstime.com; © Markku Vitikainen, Dreamstime.com.

Library of Congress Cataloging-in-Publication Data
Bussey, Lynn E. (Lynn Edward), 1920–1989.
The economic analysis of industrial projects.—Third edition / Ted G. Eschenbach,
Neal A. Lewis, Joseph C. Hartman, Lynn E. Bussey.
    pages cm
 ISBN 978-0-19-517874-6
1. Capital investments. 2. Corporations—Finance.
I. Eschenbach, Ted. II. Lewis, Neal A. III. Hartman, Joseph C., 1970– IV. Title.
 HG4028.C4B83 2015
    658.15'2—dc23
                            2014033190

This text is dedicated:

To Lynn Bussey & Don Newnan
for their contributions to engineering economy
& trusting me to walk on the paths they had built.
By Ted Eschenbach

To Joan
whose support makes this second career possible.
By Neal Lewis

To My Parents
for encouraging the engineer in me.
By Joseph Hartman

To Lynn E. Bussey, Ph.D.
for his determination in creating a needed text.
By Lewis Bussey and Linda Willard

Contents in Brief

Preface to the Third Edition

PART ONE
Basic Concepts

1. The Firm—Economic Exchanges and Objectives — 1
2. Interest, Interest Factors, and Equivalence — 15
3. Estimating Costs and Benefits — 55
4. Depreciation: Techniques and Strategies — 89
5. Corporate Tax Considerations — 115
6. The Financing Function — 141

PART TWO
Deterministic Investment Analysis

7. Economic Measures — 173
8. Replacement Analysis — 221
9. Methods of Selection Among Multiple Projects — 251

PART THREE
Investment Analysis Under Risk and Uncertainty

10. Optimization in Project Selection — 289
11. Utility Theory — 329
12. Stochastic Cash Flows — 351
13. Decision Making Under Risk — 371
14. Real Options Analysis — 401
15. Capacity Expansion and Planning — 453
16. Project Selection Using Capital Asset Pricing Theory — 471

Appendix   Compound Interest Tables — 491

Index — 505

CONTENTS

Preface to the Third Edition

**PART ONE
Basic Concepts**

1 The Firm—Economic Exchanges and Objectives   1
   1.1 Introduction   1
   1.2 Economic Exchange—The Input-Output Basis of the Firm   3
   1.3 Functions of the Firm: Financing, Investing, Producing   4
   1.4 Objectives of the Firm   6
   1.5 Sources and Uses of Funds   9
   1.6 Summary   11
   References   12
   Problems   12

2 Interest, Interest Factors, and Equivalence   15
   2.1 What Is Interest?   15
      2.1.1 Perfect capital market assumptions   15
      2.1.2 The consumption basis of single-period exchange   16
      2.1.3 Multi-period exchange   19
      2.1.4 Fundamental interest equation   21
      2.1.5 The equilibrium market price concept of interest rates   21
   2.2 Notation and Cash Flow Diagrams   23
   2.3 Tabulated Compound Interest Factors   25
      2.3.1 Factors relating P and F 26
      2.3.2 Factors relating A and F 27
      2.3.3 Factors relating P and A 29
      2.3.4 Arithmetic gradient conversion factors   29
   2.4 Examples of Time Value of Money Calculations   33
   2.5 Geometric Gradients   36
   2.6 Nominal and Effective Interest Rates   38
   2.7 Continuous Interest Factors   41
   2.8 Extended Engineering Economy Factors and Spreadsheets and Calculators   42
      2.8.1 Advantages of extended engineering economy factors   42
      2.8.2 Notation for extended engineering economy factors   42
      2.8.3 Spreadsheet annuity functions   43
      2.8.4 Time value of money (TVM) calculators   44
   2.9 Spreadsheets and Cash Flow Tables   44
      2.9.1 Advantages of spreadsheets for economic analysis   44
      2.9.2 Effective and efficient spreadsheet construction   45

2.10 Economic Interpretation of Equivalent Annual Amount   46
 2.11 Summary   47
 References   48
 Problems   48

3 Estimating Costs and Benefits   55
Lead Coauthor Heather Nachtmann

 3.1 Introduction   55
 3.2 Cash Flow Estimates   55
 3.3 Life Cycle Estimation   56
 3.4 Classification of Estimates   57
 3.5 Estimation Data   59
 3.6 Basic Estimation Techniques—Indexes and Per Unit   60
  3.6.1 Indexes   60
  3.6.2 Unit Technique   61
 3.7 Factor Technique   62
 3.8 Cost Estimation Relationships   63
  3.8.1 Development Process   65
  3.8.2 Capacity Functions   69
  3.8.3 Learning Curves   70
 3.9 Growth Curves   75
 3.10 Estimating Product Costs   77
  3.10.1 Direct costs   77
  3.10.2 Indirect costs   78
 3.11 Sensitivity Analysis   82
 3.12 Summary   82
 References   83
 Problems   85

4 Depreciation: Techniques and Strategies   89

 4.1 Introduction   89
 4.2 Depreciation Strategies   89
 4.3 Definitions   90
  4.3.1 Depreciable property   90
  4.3.2 Basis of property   91
  4.3.3 Recovery period   91
  4.3.4 Salvage value   91
  4.3.5 Symbols and notation   91
 4.4 Basis and Book Value Determination   92
  4.4.1 Definition of initial basis and book value   92
  4.4.2 Special first-year write-offs   92
  4.4.3 Like-for-like replacement   93
 4.5 Methods of Depreciation   93
  4.5.1 Introduction   93
  4.5.2 The straight-line method   94
  4.5.3 The declining balance method   95

Contents   vii

   4.5.4 The sum-of-the-years' digits (SOYD) method 97
   4.5.5 Switching 98
   4.5.6 Units of production 99
   4.5.7 Reasons for accelerated depreciation 99
   4.5.8 Modified Accelerated Cost Recovery System (MACRS) 99
   4.5.9 Job Creation and Worker Assistance Act 101
   4.5.10 Comparing book values with different depreciation methods 102
  4.6 The Present Value of the Cash Flow Due to Depreciation 103
   4.6.1 Straight-line method 103
   4.6.2 Declining balance method 103
   4.6.3 Sum-of-years' digits method 104
   4.6.4 Modified accelerated cost recovery system 105
  4.7 Simple Depreciation Strategies 105
   4.7.1 Accelerated depreciation is better 105
   4.7.2 Declining balance method versus the straight-line method 106
   4.7.3 Declining balance method versus sum-of-year's digits method 106
  4.8 Complications Involving Depreciation Strategies 107
  4.9 Summary of Conclusions: Depreciation 109
  4.10 Depletion of Resources 109
   4.10.1 Entitlement to depletion 109
   4.10.2 Methods for computing depletion deductions 109
   4.10.3 The depletion deduction 110
   4.10.4 Typical percentage depletion rates 110
  4.11 Amortization of Prepaid Expenses and Intangible Property 111
  References 111
  Problems 112

5 Corporate Tax Considerations 115

  5.1 Introduction 115
  5.2 Ordinary Income Tax Liability 117
  5.3 Federal Income Tax Rates 120
   5.3.1 Investment tax credit 121
  5.4 Generalized Cash Flows from Operations 122
  5.5 Tax Liability When Selling Fixed Assets 124
   5.5.1 What are Section 1231 assets? 124
   5.5.2 Tax treatment of 1231 assets 124
  5.6 Typical Calculations for After-Tax Cash Flows 126
  5.7 After-Tax Replacement Analysis 132
  5.8 Value-added Tax 133
  References 133
  Problems 134

6 The Financing Function 141

  6.1 Introduction 141
  6.2 Costs of Capital for Specific Financing Sources 142
  6.3 Cost of Debt Capital 142

viii   Contents

      6.3.1   Short-term capital costs   143
      6.3.2   Capital costs for bonds   143
6.4 Cost of Preferred Stock   145
6.5 Cost of Equity Capital (Common Stock)   146
      6.5.1   Dividend valuation model   146
      6.5.2   The Gordon-Shapiro growth model   147
      6.5.3   The Solomon growth model   149
      6.5.4   Note on book value of stock   150
      6.5.5   Capital asset pricing model (CAPM)   151
      6.5.6   Cost of retained earnings   153
      6.5.7   Treasury stock   154
6.6 Weighted Average Cost of Capital   154
6.7 Marginal Cost of Capital   157
      6.7.1   Market values imply a marginal cost approach   157
      6.7.2   Marginal cost-marginal revenue approach   157
      6.7.3   A discounted cash flow approach   158
      6.7.4   Mathematical approach to marginal cost of capital   159
6.8 Numerical Example of the Marginal Weighted Average Cost of Capital   162
      6.8.1   Calculation of the present weighted average cost of capital   162
      6.8.2   The future weighted average cost of capital after provision for new capital   163
      6.8.3   The marginal cost of capital   166
6.9 MARR and Risk   166
6.10 WACC and the Pecking Order Model   167
6.11 Summary   167
References   168
Problems   169

PART TWO
Deterministic Investment Analysis

7 Economic Measures   173
7.1 Introduction   173
7.2 Assumptions for Unconstrained Selection   174
7.3 Some Measures of Investment Worth (Acceptance Criteria)   175
7.4 The Payback Period   175
      7.4.1   Payback rate of return   177
      7.4.2   Discounted payback   179
7.5 Criteria Using Discounted Cash Flows   180
7.6 The Net Present Value Criterion   181
      7.6.1   Production–consumption opportunities of the firm   182
      7.6.2   The present value criterion for project selection   186
      7.6.3   Multi-period analysis   187
      7.6.4   Characteristics of net present value   187
7.7 The Benefit-Cost Ratio Criteria   190
7.8 Internal Rate of Return   191

7.8.1 Defining the internal rate of return   191
7.8.2 The fundamental meaning of internal rate of return   192
7.8.3 Conventional and nonconventional investments (and loans)   196
7.8.4 Conventional investments and internal rate of return   197
7.9 Nonconventional Investment   198
7.9.1 Nonconventional investment defined   198
7.9.2 Conventional, pure investments   199
7.9.3 Analyzing nonconventional investments   200
7.9.4 Numerical examples   200
7.10 Roots for the PW Equation   204
7.10.1 Using the root space for P, A, and F 204
7.10.2 Defining the root space for P, A, and F 206
7.10.3 Practical implications of the root space for P, A, and F 208
7.11 Internal Rate of Return and the Lorie-Savage Problem   208
7.11.1 Multiple positive roots for rate of return   208
7.11.2 Return on invested capital   209
7.11.3 Present worth and the Lorie-Savage problem   212
7.12 Subscription/Membership Problem   213
7.13 Summary   214
References   215
Problems   217

8 Replacement Analysis   221

8.1 Introduction   221
8.2 Infinite Horizon Stationary Replacement Policies   222
8.2.1 Stationary costs (no technological change)   223
8.2.2 Technological change and stationary results   228
8.3 Nonstationary Replacement Policies   230
8.3.1 Age-based state space approach   230
8.3.2 Length of service state space approach   232
8.3.3 Applying dynamic programming to an infinite horizon problem   236
8.3.4 Solving with linear programming   237
8.4 After-Tax Replacement Analysis   238
8.5 Parallel Replacement Analysis   241
8.6 Summary and Further Topics   245
References   245
Problems   248

9 Methods of Selection Among Multiple Projects   251

9.1 Introduction   251
9.2 Project Dependence   251
9.3 Capital Rationing   252
9.4 Comparison Methodologies   254
9.5 The Reinvestment Rate Problem   256
9.6 The Reinvestment Assumption Underlying Net Present Value   258

9.7 The Reinvestment Assumption Underlying the Internal Rate of Return: Fisher's Intersection 262
9.8 Incremental Rates of Return 264
    9.8.1 Incremental rate of return applied to the constrained project selection problem 266
    9.8.2 Inclusion of constraints 270
9.9 The Weingartner Formulation 271
    9.9.1 Objective function 271
    9.9.2 Constraints 272
    9.9.3 The completed Weingartner model 274
    9.9.4 Constrained project selection using Solver 275
9.10 Constrained Project Selection by Ranking on IRR 276
    9.10.1 The opportunity cost of foregone investments 276
    9.10.2 Perfect market assumptions 277
    9.10.3 Internally imposed budget constraint 278
    9.10.4 Contrasting IRR and WACC assumptions 279
    9.10.5 Summary of ranking on IRR 280
9.11 Summary 281
References 281
Problems 283

PART THREE
Investment Analysis Under Risk and Uncertainty

10 Optimization in Project Selection 289
(Extended Deterministic Formulations)
10.1 Introduction 289
10.2 Invalidation of the Separation Theorem 290
10.3 Alternative Models of the Selection Problem 291
    10.3.1 Weingartner's horizon models 291
    10.3.2 The Bernhard generalized horizon model 293
    10.3.3 Notation 294
    10.3.4 Objective function 294
    10.3.5 Constraints 295
    10.3.6 Problems in the measurement of terminal wealth 296
    10.3.7 Additional restrictions 297
    10.3.8 The Kuhn-Tucker conditions 298
    10.3.9 Properties of ρ_t^* 301
    10.3.10 Special cases 301
10.4 Project Selection by Goal Programming Methods 302
    10.4.1 Goal programming format 303
    10.4.2 An example of formulating and solving a goal programming problem 304
    10.4.3 Project selection by goal programming 310
10.5 Summary 316
Appendix 10.A Compilation of Project Selection Problem 317

References   320
Problems   323

11 Utility Theory   329
11.1 Introduction   329
    11.1.1 Definitions of Probability   330
11.2 Choices Under Uncertainty: The St. Petersburg Paradox   331
11.3 The Bernoulli Principle: Expected Utility   333
    11.3.1 The Bernoulli solution   333
    11.3.2 Preference theory: the Neumann-Morgenstern hypothesis   335
    11.3.3 The axiomatic basis of expected utility   335
11.4 Procuring a Neumann–Morgenstern Utility Function   338
    11.4.1 The standard lottery method   338
    11.4.2 Empirical determinations of utility functions   339
11.5 Risk Aversion and Utility Functions   340
    11.5.1 Risk aversion as a function of wealth   340
    11.5.2 Other risk-avoiding utility functions   341
    11.5.3 Linear utility functions: Expected monetary value   341
    11.5.4 Complex utility functions: Risk seekers and insurance buyers   342
    11.5.5. Reconciling firm's utility and behavior by employees and managers   344
11.6 Summary   344
References   345
Problems   347

12 Stochastic Cash Flows   351
12.1 Introduction   351
12.2 Single Risky Projects—Random Cash Flows   352
    12.2.1 Estimates of cash flows   352
    12.2.2 Expectation and variance of project net present value   354
    12.2.3 Autocorrelations among cash flows (same project)   356
    12.2.4 Probability statements about net present value   357
12.3 Multiple Risky Projects and Constraints   359
    12.3.1 Variance of cross-correlated cash flow streams   360
    12.3.2 The candidate set of projects   362
    12.3.3 Multiple project selection by maximizing expected net present value   363
12.4 Accounting for Uncertain Future States   363
12.5 Summary   364
References   365
Problems   366

13 Decision Making Under Risk   371
13.1 Introduction   371
13.2 Decision Networks   371
13.3 Decision Trees   372
13.4 Sequential Decision Trees   377

13.5 Decision Trees and Outcome Variability   380
    13.5.1 Stochastic decision trees   382
    13.5.2 Applications   383
13.6 Expected Value of Perfect Information   383
13.7 Simulation   384
13.8 Summary   393
References   393
Problems   394

14 Real Options Analysis   401

14.1 Introduction   401
14.2 Financial Options   402
14.3 Real Options   403
    14.3.1 Historical development   403
    14.3.2 The real option model   405
    14.3.3 Interest rates   405
    14.3.4 Time   406
    14.3.5 Present value of future cash flows   406
14.4 Real Option Volatility   407
    14.4.1 Actionable volatility   407
    14.4.2 Logarithmic cash flow method   408
    14.4.3 Stock proxy method   408
    14.4.4 Management estimates method   409
    14.4.5 Logarithmic present value returns method (CA method)   409
    14.4.6 Standard deviation of cash flows   410
    14.4.7 Internal Rate of Return   410
    14.4.8. Actionable volatility revisited   411
14.5 Binomial Lattices   411
14.6 The Deferral Option: Dementia Drug Example   412
    14.6.1 Definition and NPV calculation   412
    14.6.2 Volatility   415
    14.6.3 Black-Scholes results   418
    14.6.4 Binomial lattices   418
14.7 The Deferral Option: Oil Well Example   420
    14.7.1 NPV   421
    14.7.2 Delay option formulation   421
    14.7.3 Black-Scholes results   426
    14.7.4 Binomial lattices   426
14.8 The Abandonment Option   428
14.9 Compound Options   431
    14.9.1 Multi-stage options modeling   431
    14.9.2 Multi-stage option example   432
    14.9.3 Closed form solution   437
    14.9.4 Volatility issues in multi-stage modeling   437
14.10 Current Issues with Real Options   437
14.11 Summary   438
Appendix 14.A Derivation of the Black-Scholes Equation   439

References   444
Problems   446

15 Capacity Expansion and Planning   453
15.1 Introduction   453
15.2 Expansion Analysis   454
    15.2.1 Dynamic deterministic evaluation   454
    15.2.2 Dynamic probabilistic evaluation   456
15.3 Capacity Planning Strategies   458
    15.3.1 Maximizing market share strategy   462
    15.3.2 Maximizing utilization of capacity strategy   466
15.4 Summary   468
References   469
Problems   469

16 Project Selection Using Capital Asset Pricing Theory   471
16.1 Introduction   471
16.2 Portfolio Theory   473
    16.2.1 Securities and portfolios   473
    16.2.2 Mean and variance of a portfolio   474
    16.2.3 Dominance among securities and portfolios   475
    16.2.4 Efficient portfolios   475
    16.2.5 The risk in a portfolio   478
16.3 Security Market Line and Capital Asset Pricing Model (CAPM)   479
    16.3.1 Combinations of risky and riskless assets   479
    16.3.2 The security market line   480
    16.3.3 The capital asset pricing model (CAPM)   481
16.4 Firm's Security Market Line and Project Acceptance   481
    16.4.1 Projects and the capital asset pricing model (CAPM)   481
    16.4.2 Risk/return trade-offs and the firm's security market line   482
16.5 The Firm's Portfolio of Projects   483
    16.5.1 Why do firms use project portfolios?   483
    16.5.2 Can security portfolio theory be extended to project portfolios?   484
    16.5.3 Reasonable inferences from security portfolio theory to project portfolios   485
    16.5.4 Can the capital asset pricing model for securities be extended to projects?   486
16.6 Summary   487
References   487
Problems   488

Appendix of Engineering Economy Factors   491

Index   505

Preface to the Third Edition

In two previous editions *The Economic Analysis of Industrial Projects* was one of the leading advanced texts in engineering economy, with comprehensive coverage of the theoretical foundations of engineering economics. It is in this spirit that we present this third edition, which has been completely revised and expanded. Every existing chapter has been meticulously rewritten, and five new chapters have been added. We continue to build on the theoretical foundations of economic analysis to provide the best possible methods for applying this theory to practice.

This text is suitable for an introductory course, but it is more rigorous and comprehensive than many instructors would want for that class. Most instructors will find it more suitable for a more advanced course. The text is targeted for use by undergraduate or graduate students doing advanced study in engineering economics or project selection, the practitioner who needs more than an introductory text, and academics looking for the theory behind economic analysis methods. A working knowledge of statistics and calculus is assumed, as well as previous experience with the time value of money from introductory engineering economics or finance.

Also available from Oxford University Press is *Cases in Engineering Economy, 2nd edition*. These 54 cases and 4 chapters are designed to allow student application of the material in this text. There is an instructor's manual with solutions to all problems. There is an extensive library of other supplements in engineering economy from Oxford University Press. Thus teaching and learning are easier to accomplish and more complete.

The text begins with basics in engineering economy, but includes theoretical foundations that are often not included in introductory texts. Later chapters contain material that is rarely found elsewhere—and then only in the research literature. This is particularly true in the areas of real options, implications of double roots, replacement analysis, and project portfolios. The emphasis on original references has been continued, and in this digital age, many of the original articles in finance and engineering economy are more available than ever before.

The coauthor team for this edition was assembled to write the best possible and most authoritative text in engineering economy. Since writing the second edition of this text, Ted Eschenbach has learned from coauthoring 14 other engineering economy texts and numerous articles and presentations. New coauthors have been added to ensure expertise on advanced topics. Neal Lewis has won two best article awards for his work on real options, and awards for research on other topics. Joe Hartman has won four best article awards from *The Engineering Economist* as well as serving as editor-in-chief of *TEE,* as well as being a leading researcher on replacement analysis. Heather Nachtmann has won two best article awards and is an authority on cost and benefit estimating.

Together, they have completely rewritten this text. The changes in this edition include:

    1. Every chapter has been updated and revised. Many of the chapters have been reorganized and significantly expanded in scope.
    2. Five new chapters have been added, including
        Chapter 3   Estimating Costs and Benefits

Chapter 8   Replacement Analysis
Chapter 13   Decision Making Under Risk
Chapter 14   Real Options Analysis
Chapter 15   Capacity Expansion and Planning

3. Outdated end-of-chapter problems have been revised or deleted, and new problems have been added.
4. References to recent work have been added.

This edition bears more resemblance to a new text than an updated edition; the individual changes are simply too numerous to list. However, we continue to follow the focus that was begun by Lynn Bussey in the first edition on how best to choose which projects in which to invest.

We would like to acknowledge some of those people whose work made this book possible. First, to Lynn Bussey and his family, who wrote the first edition on which the current text is built. Heather Nachtmann led the work on estimating costs and benefits (Chapter 3). We thank our reviewers, Richard Bernhard, Stanley Bullington, Robert Creese, Camille DeYong, David Enke, Gordon Hazen, W.J. (Biff) Kennedy, James Luxhoj, William Peterson, Surendra Singh, and Changchun Zeng for offering constructive feedback that improved the book. We would also like to thank our students at the University of Alaska Anchorage, University of Bridgeport, Lehigh University, University of Florida, and the University of Arkansas for their part in class testing many parts of the text. We thank the team at Oxford University Press, including Nancy Blaine, Keith Faivre, Patrick Lynch, Christine Mahon, and Nathaniel Rosenthalis for their support in producing this book. And finally, we wish to thank our families for putting up with us during the many weeks, months, and years that were required to complete this work.

Ted Eschenbach   Neal Lewis   Joe Hartman

PART ONE
Basic Concepts

1
The Firm—Economic Exchanges and Objectives

1.1 INTRODUCTION

Three fundamental decisions must be continually made within the industrial firm. The first is to decide on the firm's mission—what service or product, which market, and how objectives are to be accomplished. The second decision concerns who to hire and retain—which people, which skills, how many, and so forth. The third decision allocates the other available resources of the firm, primarily money, for which projects, to which departments, in what amounts, and under what conditions.

This book details how the third type of decision is made: how money should be allocated within the firm to improve the ultimate worth of the shareholders' holdings. We say *should* because investment research is not merely descriptive of what is or was done. Investment analysis is goal-oriented.

The three fundamental decisions are not made in isolation; they are mutually dependent. The concept of *organization* implies an interrelated *system* of people, structures, machines, tools, materials, methods, money, and other inputs. So the three decisions concerning the organizational objectives, the people and their skills, and the allocation of money are very much related.

Nevertheless, we have tried to isolate the capital aspect of the resource allocation decision. Why? First, the subject is generally factual, its data are generally numerical, and its processes are generally amenable to mathematical formulation. Second, it is a repetitive decision in most firms with even the decision's format remaining somewhat constant over time (that is, virtually the same facts are elicited, they are typically analyzed in the same manner, and generally the same persons, or their successors, make the decisions).

The third, and most important, reason for focusing on the resource allocation decision is that these decisions fix the organization's focus for a long time. Capital expenditure decisions are made about *capital* items—what plants to build, where to locate the plants, what processes to use, what equipment to select, what product to make, how much capacity to install, and so forth. In short, the decision's consequences can extend far into the future. Furthermore, although capital expenditures can usually be modified, they are seldom reversible. There is usually no secondhand market for many industrial capital goods—only the scrap dealer. The alternative to a "bad" resource allocation decision is to scrap the project, recognize the loss, and start again.

Because of these two factors—longevity and irreversibility—the resource allocation decision often commits the firm to a fixed technology and largely determines future operating expenses as well. The initial outlay may be the least important consequence of the decision, even though the outlay's immediacy may make it the focus of discussion. This book is about the economic analysis of industrial projects. A capital investment project may be defined as involving the outlay of cash in exchange for an anticipated return of future benefits. Returns or benefits may be monetary, or they may be nonmonetary. It is the exchange of present expenditure for future benefits that distinguishes the economic analysis of projects from typical decisions among feasible alternatives.

Most engineering economy texts focus on the comparison of mutually exclusive alternatives, such as the choice between engineering design alternatives. This is based on the heritage of engineering economy (Fish, 1915; Grant, 1930; Lesser, 1969), derived principally from Wellington's (1887) early railroad location economics and the economics of early transcontinental telephone lines (Grant, 1930). An economic criterion, such as minimum annual cost or maximum present value, is invoked to delineate the preferred choice. Generally the implicit intent is to select and execute at least one of the feasible alternatives.

Most texts also include more limited coverage of the economic analysis of projects, but we believe this subject needs a more detailed and complete presentation. Quite simply, it is often impossible to execute all worthy (that is, cost-saving or profit-improving) projects. Even though all candidate projects may be technically feasible and economically attractive, limited resources require that some must be selected and some rejected. The real question is whether or not a project should be accepted at all. This is very different from a choice between executable alternatives and it is based on a different decision criterion (Bernhard, 1976; Horowitz, 1976). An emphasis on a strategic perspective, which is now common in theory and practice, pushes the horizons out even further. Now the question is, "What is best for the firm, overall?" not "Which of these alternatives should we do?"

This book deals with the investment decision in the context of the industrial firm. The first topics are basic—the nature of the firm, its objectives, the time value of money, and taxes. The exposition proceeds to deterministic project evaluation and selection methods, in which all of the data is known with certainty. Finally, this assumption is relaxed to examine the effects of an uncertain future.

In this chapter we want to build an integrated economic concept—the *ideal firm*. Even though the ideal firm does not exist, the concept enables us to interpret and understand the economic behavior of actual businesses. In addition, it often supports surprisingly accurate decisions and predictions.

Our concept of the ideal firm will begin with an input-output analysis of economic exchange—the fundamental mechanism for observing the firm's actions. We progress to the firm's economic functions: financing, investing, and producing. We look into the objectives of the firm, and the sources and uses of its funds. Finally, we see that the firm pursues its objectives by examining investment opportunities, or *projects,* on the basis of their financial inputs and outputs.

1.2 ECONOMIC EXCHANGE—THE INPUT-OUTPUT BASIS OF THE FIRM

Economic analysis of financing, project evaluation and selection (investment), replacement, and divestment defines the economic behavior of the *firm*. A firm may be an individual, a sole proprietorship, a partnership, a corporation, or any other social organization that, *as an entity, engages in economic exchanges for consideration or for items of economic value*. In each case, the firm acts as *an entity* when it engages in economic exchange.

Our focus is on the exchanges of goods and services between firms. For example, a home-appliance manufacturing company makes economic exchanges to acquire the *inputs* of steel, electrical parts, labor, and machinery. Further economic exchanges take place when the *outputs* (refrigerators, stoves, washers, and dryers) are sold.

All firms acquire inputs of goods, services, money, and credit to provide outputs to other economic entities. Thus, exchanges of goods and services are the basic activities common to all firms. Although each party to an exchange may think that their utility has increased, the economist takes the position that the objects or the considerations exchanged were equivalent in value when exchanged.

Modern complex economies rely on an artificial medium of exchange, called money. Money is basically cash and promises. Cash, in most nations, is merely the state's social promise to provide at some future date or upon demand an economic equivalent. Similarly, bonds, notes, accounts payable, and other kinds of future intents are simply promises of firms to provide an economic equivalent. The economist views cash and promises to pay together as credit. Fundamentally, any complex economy operates on faith in its sources of credit.

The value of the economic goods and services exchanged in both directions (input and output) is usually quantified in dollars, euros, pounds, yuan, or other local units. A firm may therefore exchange:

> Economic goods and services,
> Cash,
> Promises to be satisfied in the future.

There are two equivalent halves to every exchange. One half, which the firm receives, is an input. The other half, which the firm gives up, is an output. Accountants refer to the considerations or goods that are received (or to be received) as debits and the considerations or goods that are given up (or to be given up) as credits. So when a design engineering firm receives cash for services provided, the cash received is a debit to the firm's sales. Note in this example that an economic good (engineering services) was exchanged for a promise (cash, which is the promise of the State). Accountants record both halves of the exchange—one as debits, the other as credits.

The input-output nature of the firm can be shown graphically, as in Figure 1.1. Suppliers provide goods and services to the firm (inputs) in exchange for cash (output). Similarly, the firm provides goods and services to customers (output) in exchange for cash. Investors provide cash to the firm in exchange for equity in the firm. Similar exchanges occur with lenders and their investments.

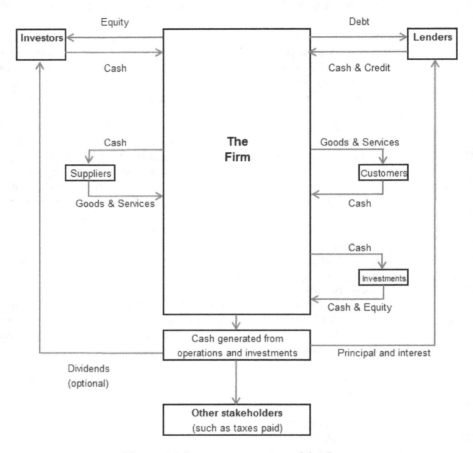

Figure 1.1 Input-output nature of the firm

1.3 FUNCTIONS OF THE FIRM: FINANCING, INVESTING, PRODUCING

Three fundamental exchanges define the firm's basic functions: the firm secures the capital it needs (financing), it employs its available capital (investing), and it generates financial returns on its invested funds (producing). The relationship of these exchanges and the firm are illustrated in Figure1.2. The firm acquires capital funds from *investors* or *lenders,* chooses one or more *investments* for the capital works to produce returns from the investments over time, and then periodically pays *returns on capital* to the sources of capital. While this process in Figure1.2 may appear sequential, in reality firms carry out these functions simultaneously and almost continuously.

Securing capital from investors and lenders and paying returns to them is the firm's *financing function.* These capital funds are of two types: equity funds and borrowed funds. Equity funds are obtained by selling shares of stock or by letting new partners buy-in. These investors exchange funds for shares in the firm's ownership in the expectation of receiving returns on their invested capital. The returns generally take two forms: payment of *dividends* and the increased value of the firm (and the owners' shares) over time. In general, not only do owners of equity shares stand to gain if the firm is profitable, but they also stand to lose a part or all of their investment if the firm fails. So the burden of uncertainty and risk is borne principally by the equity shareholders.

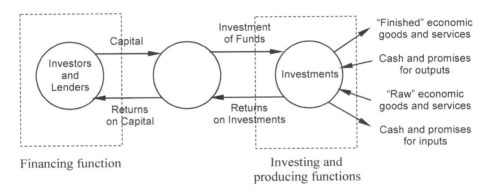

Figure 1.2 The firm's functional investment, financing, and producing cycle

Capital is also secured from *lenders*. Lenders furnish money to the firm without ownership being obtained. Banks, insurance companies, investment companies, and bondholders retain title to the funds. The firm has the right to use, but not own, a given sum of money (or other financial promises) for a stated period of time. In exchange it pays interest—the return on lent capital. Conceptually, leasing is equivalent, except that the firm "borrows" money and invests it with the lender in return for the use of equipment, buildings, or land owned by the lender. Capital may also be secured by retaining a portion of the *returns* on *capital* that the firm generates. Net income may either be kept within the firm as *retained earnings* or can be given to shareholders in the form of *dividends*. It is conceptually proper to say that all net earnings of the firm belong totally to the equity investors. So the retention of earnings by the firm is, in effect, keeping something that rightfully belongs to the equity investors. An equivalent process to retaining earnings would be to pay out all the firm's net earnings to the equity investors and then promptly acquire new capital in the same amount from the same investors. We say that retained earnings are a form of new capital for the firm, and retaining earnings is considered to be a part of the firm's financing function.

The firm has a finite number of opportunities for investing its funds at any point in time. Each such investment opportunity is termed an *investment project* or, simply, *a project*. At the same time there may be *production activities* that occur from previous projects that have already been executed. The distinction between a project and a production activity is simple: a project is *a future* opportunity to generate a return on *a potential* investment, while a production activity is *presently* generating returns on *actual* investments. The investment and production functions of the firm are interrelated. The set of unselected, available opportunities for investment is called the *candidate set of projects*, while the previously selected and executed set of projects is called the *production activities set* or *executed set of projects*. The executed set accounts for the firm's ongoing, day-to-day activities; the candidate set of projects is the opportunity set. These new projects will generate future sales, expenses, and net incomes and opportunities for abandoning (divesting) unprofitable activities and projects.

The firm generally obtains capital to invest in future projects. The firm expects the returns on its investments to exceed the amounts of capital obtained and invested. Only by returning the capital to the investors and lenders can the firm assure a continuing source of new capital. The firm's managers are stewards with custody of and legal responsibility for the firm's physical and financial assets and also the moral and ethical responsibility to use the investors' capital properly and efficiently. Such moral and ethical responsibility is called *fiduciary* responsibility. Fiduciary integrity and stewardship in management requires that decisions regarding investment opportunities (projects) be made not only from the standpoint of the firm, so as to benefit the firm's investors, but also so that the firm's investors may be benefited as greatly as possible.

The decisions concerning investment projects and productive activities include, in the most general sense, (1) whether to accept any, some, or all of the candidate set of new projects and (2) whether to continue with any, some, or all of the firm's existing productive activities. These decisions are made from the firm's point of view, by managers who stand as fiduciary representatives of the firm's investors and lenders.

In some cases, such as small firms owned and managed by members of the same family group, ownership and management are vested in the same people. Then, the fiduciary responsibility problems resulting from separation of ownership and management do not exist. Publicly held firms tend to employ professional managers who own relatively little of the firm's shares. One area of finance research is *agency theory*, which concerns how to ensure that management acts so as to ensure the maximum benefit to the shareholders. Potential conflicts in goals focus on management "perks," different time horizons, and differing risk evaluations; the solutions focus on incentive compensation systems (Xu and Birge, 2008). There are also examples where management acted contrary to shareholder goals, such as Enron, Global Crossing, WorldCom, AIG, and others.

When the owner-investor and managerial functions are separated, the owner-investors are not necessarily a cohesive group. Often, their collective action is expressed through stock prices. Basically, if the price of a firm's shares falls, the buyers and sellers are agreeing that the firm's investment and production performance or prospects have deteriorated. Conversely, if the market price of a firm's shares increases, then potential buyers are expressing confidence in the firm's future returns on capital. In either case, events external to the firm may be the driving force.

This leads us to consider the relationship between the firm's investment function, which is illustrated in Figure 1.3, and the firm's objectives, the subject of the next section. The firm examines the proposed project's required investment, anticipated cash flows, and intangible factors. The decision is then made to execute or to abandon the proposed project. This entire investment procedure is the subject of project evaluation and selection.

1.4 OBJECTIVES OF THE FIRM

What is, or should be, the firm's objectives in pursuing some kind of economic gain? There are several distinct approaches, each leading to a different kind of objective.

The first approach to determining the firm's objectives is given by organization theory. As an example, we cite first an organizational concept due to Barnard (1938). Essentially, Barnard develops the idea that people form organizations to coordinate physical activities, to overcome environmental or physical (biological) limitations, and to accomplish a common purpose.

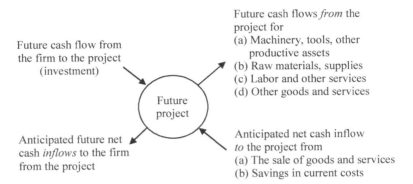

Figure 1.3 Net cash flows in the investment function of the firm

In Barnard's concept, for an organization to exist in the short run there must be (1) an agreed upon or commonly accepted purpose, (2) willingness and ability to provide economic goods and services, and (3) organizational communication. For long-run survival of the organization, Barnard requires that the value produced by the organization must exceed the costs incurred by the organization. For the simplified exchange model of Figure1.1, the equivalent of Barnard's requirement is that the receivers of the cash (or other outputs) from the firm value these more highly than the economic goods and services that they provide to the firm.

Along a somewhat parallel vein, Simon (1957) also conceives of the firm's principal objective as survival—to survive comfortably and securely as a good-size organization—by satisficing rather than optimizing. Baumol (1959) presents a compromise position for oligopoly cases (dominance in a market by relatively few producing firms). However, the human behavior approach leads us to the conclusion that the firm is not trying to optimize any quantifiable variable, and does not explain the desire for real firms to earn a *profit,* which is quantifiable.

The second approach toward identifying the firm's objectives comes from simplistic or naive economic theory. The naive view has the rational entrepreneur maximizing his net profit—that excess over the production costs. The entrepreneur provides organizing, decision-making, and ultimate risk-taking functions and is therefore entitled to a profit. So a firm should maximize its net profit to realize the optimum return on the entrepreneurial services.

There are several major problems with this naive theory. First, the ultimate risk belongs not to the firm itself but to its owners. Second, *ownership* is often separated from *control*; control of the business is frequently given to professional managers who do not risk their own fortunes but may risk their jobs. How does one then separate net profit as a reward between risk-taking and managerial control? Third, a theoretical deficiency exists in the naive theory. This deficiency was pointed out by Knight (1921), who demonstrated that profit is equivalent to a difference between incomes in an economy in equilibrium and in disequilibrium rather than as compensation for risk-taking. In Knight's analysis, profits and losses are borne by *all* members of an economy and not by a special entrepreneurial class.

The third major approach toward identifying the firm's objectives comes from

Knight (1921), who has the firm maximizing net *revenue* (the accountant's *gross profit*). This is a corollary of Knight's opposition to the naive theory. In Knight's system, the firm maximizes *net income* on all productive services lumped together.

The fourth principal approach toward identifying the firm's economic objective is from Bronfenbrenner (1960). Bronfenbrenner has the firm generate profit as compensation for the risks that are due to those having no *contractual* claim to the firm's income. There are people who accept part or all of their compensation from what is left of the firm's income after contractual claims are paid. These *residual claimants* include the common and preferred stockholders (who hold contingent claims to income), the salaried partner (whose salary is a contractual claim, but whose share of partnership net income is only a residual claim), and the executives who receive a bonus (salary is a contractual claim, but bonus is a residual claim).

Many modern financial authorities state that the firm's objective should be to maximize the *current value of the firm* to its shareholders. The firm's value is represented by the market value of the firm's shares, which in turn is a reflection of the firm's financing, investment, and dividend decisions.

Maximizing any kind of net income differs from maximizing the firm's value over time. The latter objective is more desirable. While maximizing net income is a valid short term goal, the maximization of shareholder wealth (the market value of the firm), over time, is the appropriate objective for a firm (Solomon, 1963). The value of net income is only of value where it can be translated into cash, and ultimately returned to those who have invested in the longer term success of the enterprise. In this book, we shall assume that this objective is measured in *present value* terms. Future wealth is directly related to this present value and to the future cash flows of accepted projects.

The value of a firm is the present value of the future cash flows generated by the firm. This linkage between the firm's financial performance and the strategic selection of projects has been well expressed by Porter (1980, 1985, and 1989) in his work on competitive strategy.

Finally, we would like to briefly mention two alternative views of the firm's goals. The first is often attributed to the Japanese (Kagano, 1985), whose economic success in the 1980s provides at least some justification. This approach is not stated financially but rather in terms of human and organizational development and long-term goals and survival. In financial terms, the shareholders are entitled to a fair return. However, equal recognition is given to the rights of the employees and managers to satisfying work, stable employment, and fair wages, and even to the rights of vendors to fair treatment and of customers to quality products.

Another approach has been stated by Werner (2009). In the old economic model, the primary inputs were land, labor, and capital. In the days of Adam Smith, there was an abundance of available land, labor was plentiful, but capital was limited. The focus of the firm was to produce the most with the least amount of required capital. In today's world, available land is no longer abundant, skilled labor may be in short supply in the location where it is needed, and capital is no longer limited (only difficult to justify). We still need to use scarce resources in the most efficient manner, but those constraints have changed while most business models have not.

Werner also notes that today's businesses have the ability to do immense damage to the environment and society to a far greater extent than in the past. The Deepwater Horizon oil platform explosion in 2010 is an example. Solutions to environmental and social problems are best handled within the business, not by government intervention. Business goals can no longer be solely financial if society wishes to derive the maximum benefit from business activity as it has in the past. Business needs to find a balance between the purely financial goals and the need to serve society by being a good citizen.

While these approaches frame the question of the firm's objectives quite differently than in purely financial terms, it is clear that many issues regarding strategic objectives and numerical goals influence project selection models and practices.

1.5 SOURCES AND USES OF FUNDS

To evaluate a candidate project, its year by year cash flows are modeled similarly to Figure 1.3. This requires that all items that create the net cash flows be identified for each year. A simple investment example will be described.

Projects usually start with an investment, which is the estimated cash flow(s) moving *from* the firm *to* the project. Thus, the firm invests *in* the project. Investment-type cash flows are negative from the firm's perspective. A net cash flow of "$-1,000,000" would indicate that $1 million flows from the firm to the project (at a specified time).

A project can require more than one increment of investment by the firm. For example, a project costing $772,000 might have its initial cash flows distributed as follows:

Year	Net Cash Flow, $
0	−188,000 (design)
1	−584,000 (construction)
2	−10,000 (startup and first year sales)
3	+416,500 (second year of operation)
4	+520,000 (third year of operation)

Initial investment costs are usually determined by skilled estimators and engineers who are experts in an industry's process technology and construction methodology. Large projects are often analyzed by consulting engineering firms, while smaller projects are often analyzed by in-house experts. Design and cost engineers are of particular value since their training and experience usually qualify them in their employer's technology. They need only to add basic economic principles to equip them for making estimates, not only for initial investments but also for the net cash inflows generated by the projects. We cover the topic of cost estimation in Chapter 3. Obviously, for a project to be worthwhile, it should return to the firm over the life of the project a series of net cash inflows much greater than the initial investments.

There are a variety of sources that a firm may use to fund internal projects. The most popular sources of funds include retained earnings, debt, common equity, and occasionally preferred equity. Debt can be as simple as a short term loan from the bank, or as complex as long term bonds. The source of funds may determine the cost of money for the project (which is explored in more detail in Chapter 6).

Even if the source of funding is internal (such as retained earnings), the cost of funds is not free. Funds could always be invested in another project (there is an opportunity cost) or they could be placed in an interest bearing bank account. Periodic cash inflows (from the project to the firm) are assumed to be generated by the project's production function. In discrete cash flow analysis, these cash inflows are usually assumed to occur at specific equal time intervals, for example, at the end of each operating year. The end of period convention is observed in the text unless otherwise stated. In continuous cash flow analysis, the cash flows are taken as occurring continuously but often expressed as occurring annually. Cash flows from a project to the firm are considered to be positively valued (+). The discrete case is somewhat easier to describe and illustrate.

Example 1.1 Cash flows
Consider a typical project being considered by the ABC Company that is expected to develop a cash flow for the year 20XX as illustrated in Figure 1.4. The statement indicates that the project is expected to develop a total cash inflow to the project of $1,043,500, the sum of $933,500, $50,000, and $60,000. In the same period, $167,000 is to be retained in the project as an increase in its working capital, $140,000 is a cash inflow from the project to repay the principal on a former loan, and $320,000 is a cash outflow for new equipment purchased. The *balancing* amount of $416,500 is the net cash inflow to the firm (the ABC Company) from the project. If this balancing cash flow were negative, the net cash flow would be from the firm to the project.

ABC Company, Inc.
CASH FLOW STATEMENT—TRANSFORMER PROJECT
for the year ended December 31, 20XX

SOURCE OF FUNDS		
Operating Revenues (Sales)		$3,000,000
Less:		
Cash operating costs (except interest paid)	$1,640,000	
Interest paid on debt	60,000	
Income tax paid	366,500	2,066,500
Net Funds Generated for Operations		$ 933,500
Add:		
Sale of excess equipment and plant (net)		50,000
New borrowed capital (loan)		60,000
Total Cash Inflow to Project:		$1,043,500
APPLICATION OF FUNDS		
Increase in working capital of the project during the period		$ 167,000
Principal payment on debt		140,000
Capital expenditures for new equipment		320,000
Net Cash Return (Cash Inflow to ABC Company)		416,500
Total Cash Outflow from Project		$1,043,500

Figure 1.4 Funds flow statement in year 20XX for a project

Combining the investment cash flows *to* a project and the return cash flows *from* a project results in a series of net cash flows distributed by time periods. This series is often called the *net cash flow stream,* and it is the basis for deciding whether to engage in the project. A typical *net* cash flow stream for a project might be illustrated as follows:

End-of-Year	0	1	2	3	4	5	6
Net Cash Flow	$−188,000	−584,000	−10,000	+416,500	+520,000	540,000	500,000

The cash flow diagram for this series is shown in Figure 1.5.

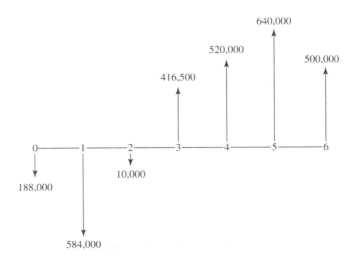

Figure 1.5 Cash flow diagram

Even though the cash flows may actually take place throughout the year, each flow is typically assumed to occur instantaneously at the year's end as a single discrete cash flow. This simplifies calculations when interest is considered, which will be discussed in Chapter 2.

1.6 SUMMARY

In this chapter we have developed some important economic concepts about the entrepreneurial firm. First, the firm was defined as an entity that engages in economic exchanges for the purpose of making a gain. The economic exchanges were described as a series of inputs and outputs, from which an ultimate net gain was realized for the owners. This final gain was then analyzed using the concept of *opportunity cost,* by which we divided the gain into a return on input investment, a return on input labor, and an economic profit.

Second, the three functions (financing, investment, and production) of the entrepreneurial firm were described and illustrated. *Financing* includes exchanges by which the firm acquires funds from owners and lenders in exchange for future promises to pay returns on capital (dividends and/or interest). *Investing* includes anticipated exchanges by which the firm expects to invest funds in projects, in anticipation of receiving returns from each executed project. *Production* includes economic exchanges pertaining to the operation of a group of projects after they have been selected and placed in operation by the firm.

The objectives of the firm were examined in detail, from both sociological and economic standpoints. For purposes of economic analysis, the firm's objective should be maximization of the shareholder's wealth over time.

Finally, the concept of *funds flow,* or generation of cash flows by a project, was considered in detail to demonstrate how production activities generate net cash inflows to the firm from executed projects. The *net cash flow stream* is the project's basic economic measure, and it is used to determine the project's profitability to the firm.

REFERENCES

BARNARD, CHESTER I., *The Functions of the Executive* (Harvard University Press, 1938).

BAUMOL, WILLIAM J., *Business Behavior, Value and Growth* (Macmillan, 1959), Chapters 4, 6–8.

BERNHARD, RICHARD, "Comments on: Horowitz, Ira, 'Engineering Economy: An Economists Perspective,'" *AIIE Transactions*, **8**(4) (December 1976), pp. 438–442.

BRONFENBRENNER, MARTIN, "A Reformulation of Naive Profit Theory," *Southern Economic Journal,* **26**(4) (April 1960).

FISH, JOHN CHARLES LOUNSBURY, *Engineering Economics* (McGraw-Hill, 1915).

GRANT, EUGENE L., *Principles of Engineering Economy* (Ronald, 1930).

HOROWITZ, IRA, "Engineering Economy: An Economist's Perspective," *AIIE Transactions,* **8**(4) (December 1976), pp. 430–437.

KAGONO, TADAO, IKUJIRO NONAKA, KIYONORI SAKA KIBARA, and AKIHIRO OKUMURA, *Strategic vs. Evolutionary Management: A U.S.-Japan Comparison of Strategy and Organization* (Elsevier Science, 1985).

KNIGHT, FRANK H., *Risk, Uncertainty, and Profit* (Houghton-Mifflin, 1921).

LESSER, ARTHUR, JR., "Engineering Economy in the United States in Retrospect—An Analysis," *The Engineering Economist,* **14**(2) (Winter 1969), pp. 109–116.

PORTER, MICHAEL E., *Competitive Strategy: Techniques for Analyzing Industries and Competitors* (Free Press, 1980).

PORTER, MICHAEL E., *Competitive Advantage: Creating and Sustaining Superior Performance* (Free Press, 1985).

PORTER, MICHAEL E., *The Competitive Advantage of Nations and Their Firms* (Free Press, 1989).

SIMON, HERBERT, *Models of Man* (Wiley, 1957), Chapter 10.

SOLOMON, EZRA, *The Theory of Financial Management* (Columbia University Press, 1963).

WELLINGTON, ARTHUR MELLEN, *The Economic Theory of the Location of Railways* (Wiley, 1887).

WERNER, FRANK, *The Amazing Journey of Adam Smith* (CreateSpace, 2009).

XU, XIAODONG and JOHN R. BIRGE, "Operational Decisions, Capital Structure, and Managerial Compensation: A News Vendor Perspective," *The Engineering Economist,* **53**(3) (July–September 2008), pp. 173–196.

PROBLEMS

1-1. From an economic standpoint, what is a firm?

1-2. Define the following terms: (a) economic good, (b) service, (c) input, (d) output, (e) commodity, (f) exchange, (g) money, (h) cash, (i) credit.

1-3. Describe an economic exchange.

1-4. What are the fundamental types of exchange that business firms engage in?

1-5. What is meant by "to invest"?

1-6. From the standpoint of project analysis, what is an investment?

1-7. What is the financing function of the firm?

1-8. What relates the financing and the investing functions of the firm?

1-9. What is the difference between an owner and a lender?

1-10. How is an investment project defined?

1-11. What is meant by the *fiduciary* responsibility of the management of a firm? Why is this responsibility carefully defined and protected under the laws of most states?

1-12. Describe three possible goals of an entrepreneurial firm.

1-13. Why is the goal of maximization of *net income* not the same as maximization of the *value of the firm* over time?

1-14. What is the principal difference (in words) between *net profit after taxes* and *net cash flow* for a firm in a particular year?

1-15. What is the reason for concentrating on the expected *cash flows* in evaluating a proposed project rather than on its anticipated *net income?*

1-16. Why is maximizing market share an intuitively appealing goal for the firm in financial terms?

1-17. What is the intuitive appeal of maximizing growth or firm size? Why is this not the best goal?

1-18. Why is agency theory crucial in examining the objective of publicly held firms?

1-19. A new R&D project is being proposed by an experienced research team. Funding must be approved by the group's chief scientist, the lab's manager, the vice-president of R&D, and the firm's executive committee. Each of the five levels will have strategic, tactical, and personal goals; and these may sometimes be expressed in financial terms. Identify two possible goals for each level—one should be aligned with maximizing the firm's value to the stockholders and the other should be in conflict with maximizing the firm's value to the stockholders.

1-20. Construct an annual cash flow table for a project whose start-up costs total $250,000, and this is evenly split between now and the end of the first year. The project will cost $50,000 per year in years two through six and the revenues are $125,000 per year over the same period. The equipment will have a salvage value of $30,000 and the project requires $45,000 in working capital.

1-21. Why is the concept of social responsibility for industrial firms controversial?

1-22. What are some advantages and disadvantages of a firm paying a dividend?

1-23. Net earnings can be used in two ways. Briefly describe these.

2
Interest, Interest Factors, and Equivalence

This chapter introduces the concept of interest and the time value of money. Interest factors and their notation are defined, as are simple interest, compound interest, and equivalence. Discrete compound interest factors are discussed, and nominal and effective interest rates are defined. We move onto continuous interest and the use of factors, financial calculators, and spreadsheets for solving problems.

2.1 WHAT IS INTEREST?

Most people start with a simple notion of *interest*, as money paid by a borrower or received by a lender for the use of borrowed money. Interest payments are transacted this way, but we will develop a broader conceptual model of interest.

Interest rates are, in general, established by imperfect markets. Nevertheless, we can gain considerable insight by examining *interest theory* under what are called *perfect capital market* conditions, as initially developed by Irving Fisher (1954) and later extended by J. Hirshleifer (1958, 1988).

2.1.1 Perfect capital market assumptions

1. *Financial markets are perfectly competitive.* In a perfect market, no individual or firm is large enough to affect prices by their market actions. More specifically, every borrower and lender trades in such small amounts that no one borrower or lender has any appreciable effect on interest rates.
2. *There are no transaction costs.* Thus, all transactions occur at market prices (interest rates) with no transaction taxes and no market costs for bankruptcies, brokers, middlemen, mergers, etc.
3. *Information is complete, costless, and available to all.* Market prices (interest rates) and all other market factors are fully known to all; otherwise, some would borrow (or lend) at rates other than the market rate violating the fourth assumption. Moreover, *complete* information implies perfect knowledge about future events and outcomes—for oneself, for others, and for the market.
4. *All individuals and firms are able to borrow and lend on the same terms.* In other words, there is only one interest rate, it is known with certainty, and all borrow or lend at that rate.

While no real financial market fully meets these assumptions, many securities markets with open trading come sufficiently close so that perfect market interest models are good approximations of reality.

2.1.2 The consumption basis of single-period exchange.

Let us begin by examining the consumption behavior of an individual, whose income is fixed, in a perfect capital market. To concentrate on essentials, assume that time may be expressed simply as *now* and *later*, so that only one time period exists between the two instants. *Now* is $t = 0$, and *later* is $t = 1$. *Note:* Later may be a day, a year, or a decade.

Suppose that the perfect capital market assumptions (Section 2.1.1) apply, and suppose also that an individual is planning a consumption pattern over this single time period—that is, some consumption now and some later. Consumption, borrowing, and lending will occur at time $t = 0$ and again at $t = 1$.

We assume that the individual's sole problem is to choose from alternative combinations of total consumption. The only knowledge we assume about the individual's preferences is that the person prefers more consumption to less at any given time. For example, consuming $110 worth of goods now and $100 worth later is preferred to consuming $100 now and $100 later.

In this system consumption is limited by two factors: (1) income and (2) the available terms for transferring income from time $t = 0$ to $t = 1$, and vice versa. For convenience, let us assume that prices are stable and that consumption can be measured in dollars. In a world of certainty the income would be known exactly (for example, $100 now and $100 later). The second factor (transfer of income from $t = 0$ to $t = 1$) is governed by a capital market. This market trades this year's income for next year's. The operation of this capital market is assumed to be governed by the perfect capital market conditions (Section 2.1.1).

The capital market obeys the forces of supply and demand just as markets for fruits and vegetables. Unlike other markets, however, Fisher (1954) tells us that in the capital market people tend to consume their income immediately and not postpone the purchase of goods and services.

People *save*, Fisher says, only because there is an inducement to do so. The inducement is that *tomorrow's consumption is made more attractive (more valuable) than today's*. The premium for saving today—and thus transferring consumption to tomorrow—is the *interest rate* established by the capital market. For example, one might forego spending $1 of this year's income in exchange for being able to spend an additional $1.07 next year. Alternatively, one might borrow by foregoing $1.07 of next year's income at $t = 1$ in order to increase present ($t = 0$) consumption by $1.

Transaction costs or other market imperfections that might cause differences in the interest rates for borrowing and lending are not permitted in a perfect capital market. Here, we speak only of the terms or *rate of interest* by which present consumption is traded for future consumption, or vice versa. The rate of interest is expressed as a fraction of the present amount. For the foregoing example

$$\text{Rate of interest} = \frac{\$1.07 - \$1.00}{\$1.00} = 0.07$$

This discussion can be illustrated by an example and a simple graphical model. Consider Mr. A, whose income *stream* consists of $Y_0 = \$100$ at time $t = 0$(now) and $Y_1 = \$100$ at time $t = 1$ (next year). We assume that Mr. A's entire *wealth* is this income stream. Thus, Mr. A's initial income position is represented by point Y in Figure 2.1. Point Y is called the *endowment position* or *endowment point*, and represents the anticipated current and future incomes endowed on Mr. A before he begins trading.

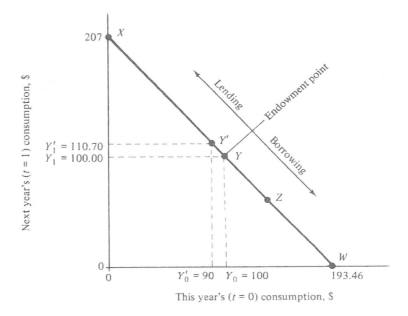

Figure 2.1 One time period borrowing and lending

Without a capital market, Mr. A's consumption pattern is completely dictated by his fixed income stream—that is, he could spend $100 now and $100 next year. In fact, Mr. A can transfer income and consumption from one time to another, anywhere along the market line *XW*, as he chooses. For example, consider point W. He could borrow against next year's income, receiving $100/(1 + 0.07) = $93.46 now, and thereby increase his present consumption to $193.46. His next year's endowment income ($100) would just repay his borrowing, however, so his consumption would be zero at time t = 1.

However, if he lends or borrows only a portion of his endowment income, he can consume any combination on the line *X-W*. For example, if Mr. A reduces his $t = 0$ consumption (see Figure 2.1) to $Y_0' = \$90$, then he can *lend* his present savings of $10 in the capital market and receive back $10.70 next year (increasing Y_1' to $110.70). Similarly, he might increase his present consumption by *borrowing* from future income, which would correspond to some point, say Z, on the market line.

Which consumption pattern—that is, which point on the market line—should Mr. A choose? In a perfect market where borrowing and lending occur at the same interest rate, he should choose the point on the market line he likes best. Mr. A would not choose an interior consumption point (inside *0XW0*) since such a choice gives away some of his consumption opportunity. Thus, rational consumption combinations lie only on the market line.

The market line can be described by its slope and an intercept. The slope is based on the market rate of exchange to trade current for future consumption, that is, the rate of interest. Specifically, the slope of the market line is $-(1 + i)$. Since the market rate of interest applies to all borrowers and lenders in a one-period model $(t = 0, 1)$, any individual consumption pattern can be fully described by the intercept of a market line passing through the endowment position. The convention is to use the horizontal intercept—the amount that could be consumed this year by foregoing all claims to future income. This is the *present value* of the endowed income stream or the individual's wealth. In Figure 2.1, Mr. A's wealth is $193.46 (point W).

How is present value, or wealth (W), determined in general? It is simply the *market value* of the individual's endowment of present (Y_0) and future (Y_1) consumption claims (income). Then the present market value, W, is:

$$W = Y_0 + \frac{Y_1}{1+i} \qquad (2.1)$$

All income streams that have the same present value, *regardless of initial endowment* (for example, Y_a, Y_b, Y_c, ..., in Figure 2.2), lie on the same present value line, for example, on *XW*. Different initial income endowments (for example, *Y'*) with different present values, such as *W'* in Figure 2.2, lie on different present value lines (for example, *X' W'*). The present value lines through both *XW* and *X'W'* have a common slope of $-(1+i)$ and are therefore parallel.

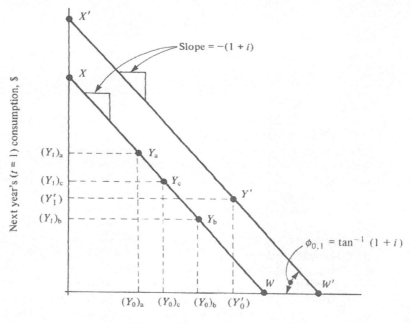

Figure 2.2 Market model with different endowments

The terms *present value (PV)*, *present worth (PW)*, and *net present value (NPV)* are three terms that have the same meaning and are used interchangeably; they are synonyms. All three terms are found in the literature, but some authors tend to prefer one term over another. All three will be found in this book, and have the same meaning except when used as Microsoft Excel functions. Excel uses the terms PV and NPV as specific financial functions, and in this context they are not interchangeable.

2.1.3 Multi-period exchange. The single-period capital exchange model can be extended to several periods. For the two-period case, consider Figure 2.3.

The individual has an income stream, represented by the vector **OY**, consisting of income amounts $\mathbf{Y} = (Y_0, Y_1, Y_2)$ to be received at times $t = (0, 1, 2)$, respectively. The market rate of interest in effect between $t = 0$ and $t = 1$ is i_1, and the rate between $t = 1$ and $t = 2$ (period 2) is i_2. Under perfect market conditions borrowing and lending can take place freely at those rates. The consumption opportunities available under these conditions (assuming no give away of income) are defined by a triangle in space, **WXZ** in Figure 2.3. The point **Y** of the endowment income vector, **OY**, and i_1, and i_2 determine this triangle. By analogy with the single-period model, the angles $\varphi_{0,1}$ and $\varphi_{1,2}$ are defined as

$$\varphi_{0,1} = \tan^{-1}(1 + i_1), \qquad \varphi_{1,2} = \tan^{-1}(1 + i_2).$$

Point **W**, the point at which the triangle intersects the $t = 0$ axis, is the maximum amount available for consumption at time zero if nothing is consumed at times $t = 1$ and $t = 2$. So, the magnitude of W is the *present value* of the income stream (Y_0, Y_1, Y_2). Thus,

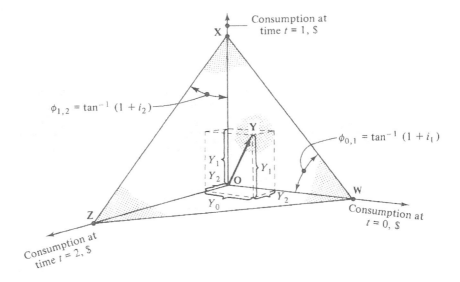

Figure 2.3 Two-period intertemporal exchange

$$W = Y_0 + \frac{Y_1}{\tan\phi_{0,1}} + \frac{Y_2}{(\tan\phi_{0,1})(\tan\phi_{1,2})}$$

$$= Y_0 + \frac{Y_1}{1 + i_1} + \frac{Y_2}{(1 + i_1)(1 + i_2)}$$

This line of reasoning and development extends directly to the N-period case. The present value, or wealth, of the individual is then given by:

$$W = Y_0 + \frac{Y_1}{1+i} + \frac{Y_2}{(1+i_1)(1+i_2)} + \cdots + \frac{Y_N}{(1+i_1)(1+i_2)\cdots(1+i_N)}$$

$$= \sum_{t=0}^{N} \frac{Y_t}{\prod_{j=0}^{t}(1+i_j)} \quad (2.2)$$

where Y is the income stream at time t and i_j is the assumed known interest rate in effect during period j, that is, from $t = j - 1$ to $t = j$.

In general, we can state that under perfect market conditions any two income streams with the same present value are exactly equivalent since they provide the same consumption choices. If one income stream has a greater present value than another, it provides increased consumption in all combinations relative to the smaller.

Two further observations concern Equation (2.2). If it is assumed that the current market interest rate will not change in the future, such that all i_j equal i, then Equation (2.2) for present value becomes:

$$W = Y_0 + \frac{Y_1}{1+i} + \frac{Y_2}{(1+i)^2} + \cdots + \frac{Y_N}{(1+i)^N} = \sum_{t=0}^{N} Y_t(1+i)^{-t} \quad (2.3)$$

Examples 2.1 and 2.2 define and detail the distinction between simple and compound interest.

Example 2.1 TR Enterprises simple interest calculation

TR Enterprises is a start-up firm developing an innovative green technology for use in third world countries. The firm has borrowed $100,000 from a philanthropist, who is charging 9% simple interest to be repaid after 10 years. The amount borrowed, $100,000, is the *principal*, and the present amount, P. The annual interest is 9% of $100,000 or $9,000 each year for a total of $90,000 over 10 years. The *rate of interest*, r, is the amount of interest accrued (earned) by a unit of principal over a unit of time. Thus, in this example, 100,000 units of principal earned 90,000 units of interest in 10 years, so the rate of interest was

$$r = \frac{90,000}{(100,000)(10)} = 0.09$$

or *9% per annum or year.*

This equivalence problem used simple interest. The interest (I) owed to the bank for the use of P dollars is calculated as

$$I = F - P = PNr$$

because the interest earned is directly proportional to the principal involved (P), the interest rate (r), and the number of periods (N) the principal was used. Thus, the future amount, F, is equivalent to the present amount, P, using simple interest:

$$F = P + I = P + PNr = P(1 + Nr)$$

or, conversely, the *present value* (of the loan) is made equivalent to a larger *future value* (loan + interest) by subtracting interest:

$$P = F - I = \frac{F}{1 + Nr}$$

Example 2.2 TR Enterprises compound interest calculation

The terms of the TR Enterprises loan would normally be based on compound interest. Suppose that each year the interest is simply added to the amount owed with a final payment at the end of year 10 of all interest. Thus, during the *second* year, the bank lent the principal ($P = \$100{,}000$) and the interest due at the beginning of the second year (\$9,000). So interest for the second year also includes 9% of the \$9,000 interest accrued at the end of year 1. Interest treated in this manner is *compounded,* since interest is earned on interest. Compounding the loan has the following results:

End of Period	Amount Owed at Beginning of Period	Interest Charge for Period	Amount Owed at End of Period
0	-0-	-0-	100,000
1	100,000	9,000	109,000
2	109,000	9,810	118,810

Thus, $F = \$118{,}810$ would be due the bank at the end of the second period under compounding, whereas with simple interest $F = \$118{,}000$. The difference of \$810 equals 9% interest on \$9,000. The value at the end of 10 years can be calculated by repeating this calculation or by using Equation (2.4), which is presented next.

2.1.4 Fundamental interest equation. Equation (2.3) can be simplified by assuming there is a single present value at time 0 and a single future value at the end of year N. Using $P \equiv$ present value and $F \equiv$ future value, the result is:

$$F = P(1 + i)^N \qquad (2.4)$$

Applying Equation (2.4) to Example 2.2 to find the amount owed at the end of year 10, $F = \$100{,}000(1.09)^{10} = \$236{,}736$.

2.1.5 The equilibrium market price concept of interest rates. The previous sections dealt with interest rates and consumption behavior. Here we examine the productive function supplied by the capital market.

Let us begin by reviewing some concepts about production-type organizations. For convenience, the *generalized factor inputs* are (1) land and natural resources,[1] (2) labor,

---

[1] Natural resource inputs of all kinds are often called simply *land* by economists.

and (3) capital.[2] *Natural resources* (land) are provided by Nature. The amount paid, or the return, for the use of land is *economic rent*. *Human labor* is determined by biological and social determinants. The *return* to the laborer or corporation executive is called *wages*—the amount paid for the labor consumed. These two factors taken together—natural resources and labor—are the *primary factors of production,* since their supply is largely determined outside the economic system itself.

Capital, the third input, is produced *within the economic system.* It is an *intermediate* factor, as it is both an output of and an input to the productive process. For example, an injection molding machine is *a real-capital* item, or a capital good, and it is the output of a manufacturing process. The injection molding machine can be *rented out* (leased or sold) where it becomes an input in another productive process—perhaps producing ballpoint pens.

Why, and how, are these intermediate factors of production, capital goods, produced? Quite simply, all capital goods are produced by a process where:
1. Some individuals refrain from total consumption of goods and services produced within the economic system, thereby generating savings.
2. Other individuals employ the resulting savings and use tools, building, equipment, etc. to produce other consumption and capital goods.

Answering why is more difficult. In general, the economist recognizes a technological fact of life: Society can realize more future consumption, in total and in the long run, by using indirect production methods. This requires capital goods and concentrates productive effort in factories. As Fisher (1954) indicates, people tend to consume immediately, rather than postponing consumption until tomorrow. Thus, people postpone consumption, create savings, and invest those savings in the production of intermediate capital goods, because of the expectation of receiving more tomorrow than they save today.

The users of the intermediate goods pay *rent,* which is the *interest* paid to the savers. Thus, from a productive standpoint, interest is the return on an investment in intermediate capital goods. In Fisher's words (1954, pp. 14–15):

> The *value* of capital [goods] must be computed from the value of its estimated future net income, not *vice versa.* This statement may at first seem puzzling, for we usually think of causes and effects running forward not backward in time. It would seem that income is derived from capital; and in a sense, this is true. Income [that is, services] *is* derived from capital *goods.* But the *value* of the income is not derived from the *value of* the capital goods. On the contrary, the value of the capital [goods] is derived from the value of the income.... These relations are shown in the following scheme in which the arrows represent the order of sequence– (1) from capital goods to their future services; that is, income; (2) from these services to their value; and (3) from their value back to capital value:

---

[2] As Hirshleifer (1988, pp. 366–368) points out, however, these three classifications are arbitrary and traditional only, probably originating among early English economists in the 18th and 19th centuries when there was an obvious three-class society in England (the landowning aristocracy, the working peasant, and the capital-lending middle class). In economic terms, these separate factor inputs are not functionally defensible.

Capital goods → flow of services (income)
                        ↓
Capital value ← income value

> Not until we know how much income an item of capital will probably bring us can we set any valuation on that capital at all. It is true that the wheat crop depends on the *land which yields it*. But the *value of* the crop does not depend upon the value of the land. On the contrary, the *value of the land depends upon the expected value of its crops*. [Italics supplied]

From a productive standpoint, the interest rate is viewed as a return rate, the ratio of the income value produced to the capital value of the intermediate goods. From the consumer's standpoint, the interest rate is the inducement to save or to postpone consumption. Thus, the capital market has two opposing forces operating to establish *the* interest rate: (1) the consumer's demand for interest payments as an inducement to save, and (2) the producer's limited ability to generate income value from the intermediate capital goods employed. At equilibrium a market price for the rate of interest is established. Note that in the economic evaluation of projects, capital values are set for the productive assets. From Fisher's analysis above, these capital values are established from the income value of the products and/or services produced. Because these capital and income values occur at different times, it is simply an exchange problem.

2.2 NOTATION AND CASH FLOW DIAGRAMS

Calculations of present value, such as those required by Equation (2.4), are generally made easier with *interest factors*. These factors help students understand the time value of money, and are matched to common cash flow patterns of timing and amount. Formulas are available but tabulated values, Time Value of Money (TVM) calculators (often called financial calculators), or spreadsheets are often used instead.

The following notation, which follows the ANSI (1988) standard, is used for interest calculations (unless specifically noted otherwise):

- r = nominal interest rate per period.
- i = effective interest rate per interest period.
- N = number of periods.
- P = present sum of money, or the *equivalent* present value of one or more future cash flows. P occurs at the end of period 0 which is the start of period 1.
- F = future sum of money or the *equivalent* future value of one or more cash flows. F occurs at the *end* of period N.
- A = end-of-period cash flow (or equivalent end-of-period amount) in *a uniform series* continuing for a specified number of periods, N. A is one of a series of *equal* amounts, say $A_1 = A_2 = A_3 = \ldots = A$, *each* of which occurs at the end of a period in an uninterrupted series from $t = 1$ to N.
- G = uniform period-by-period change in cash flow (the arithmetic gradient). G is assumed to occur at the *end* of periods 1 to N with the first nonzero cash flow at $t = 2$.

g = uniform rate of change of cash flows (the geometric gradient) that occur at the end of periods 1 to N with the first change in cash flows at $t = 2$.

Equivalence relationships and calculations are easier to visualize with *cash flow diagrams*. Figure 2.4 uses cash flow diagrams to clearly show the assumptions used in the definitions of P, A, F, and G above.

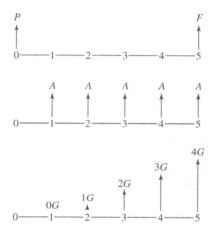

Figure 2.4 Cash flow diagram for definition of P, A, F, and G

These diagrams show how much money was spent or received, and when. Example 2.3 and Figure 2.4 illustrate this. The cash flow diagram uses several conventions:

1. The horizontal line is *a uniform time scale* on which time moves from left to right. Time periods, are numbered sequentially 1, 2, 3, ... with each *discrete* value of time t marking the end of one period and the beginning of the next, that is, $t = 0, 1, 2, ..., N$.
2. The arrows signify cash flows or money amounts. Normally, arrows directed in a negative (downward) direction indicate disbursements (cash out flows), and those directed in a positive (upward) direction indicate cash inflows or receipts.
3. In project evaluation work, the cash flow diagram is conventionally taken from the firm's standpoint. The firm invests in or lends money to the project and later receives the returns or savings from the project.

Example 2.3 Example cash flow diagram
Joan, Karen, and Chris Enterprises is doing the preliminary evaluation of moving from their current leased space to facilities designed to support their operations. This will cost $22M initially. In the first year net savings will be $3M, and these net savings will increase by $0.5M annually as their volume of operations grows. The plan projects another move to larger facilities at the end of year 6. They expect a salvage value of $4M for equipment that can be sold or reused in the new facility.
Figure 2.5 is the cash flow diagram that summarizes the above situation.

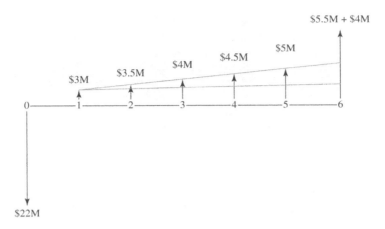

Figure 2.5 Cash flow diagram for Joan, Karen, and Chris Enterprises

2.3 TABULATED COMPOUND INTEREST FACTORS

Equivalence calculations can rely on compound interest factors. Eight factors are commonly defined and tabulated. They are designed to convert from one to another of present value (P), uniform series (A), future value (F), and arithmetic gradient series (G). The geometric gradient (g) is a formula based factor that is described in Section 2.5. The factors assume end-of-period cash flows (end of period 0 for P), and they assume the compounding and interest rate periods are of unit length (effective interest rates are used). Using the ANSI standard notation these factors are

$(P/F, i\%, N)$ $(F/P, i\%, N)$
$(P/A, i\%, N)$ $(A/P, i\%, N)$
$(A/F, i\%, N)$ $(F/A, i\%, N)$
$(P/G, i\%, N)$ $(A/G, i\%, N)$

A number of observations about these factors are appropriate. First, the top six factors for P, A, and F factors occur in three pairs of reciprocals. On the other hand, the common arithmetic gradient tables convert from, but not to, the arithmetic gradient series. Geometric gradients do not usually appear in factor tables as they require three input variables.

Second, the letters that identify the factor provide a useful "clearing the fraction" mnemonic. For example, if you start with a time zero cash flow (P) and multiply by (F/P, $i\%$, N); then, the P's cancel and an equivalent F results.

Third, the interest rate carries a % sign. It is highly recommended that interest rates always be stated as a percentage (9%) or as a decimal equivalent (.09) but never stated as an integer (9). Using integers tends to cause the error of interchanging the values of $i\%$ and N. The only time that an integer should be used is when entering an interest value into a TVM calculator.

Fourth, these are the end-of-period factors, and there are other cash flow assumptions that can be and have been made. For example, rather than assuming that the cash flow occurs instantaneously at the end of the period, it could be assumed to flow continuously throughout the period. For example, rather than $365 at the end of the year, it could be $0.00069/minute for a year. This type of continuous or distributed flow would be a more realistic representation of most engineering costs and benefits; however, the common use of spreadsheets and financial calculators requires discrete cash flows, so that today the continuous cash flow assumption is only of theoretical interest. The notation for continuous flow assumptions uses horizontal bars over the continuous or distributed $\overline{A}$ or $\overline{F}$. Earlier editions of this book included interest rate factors derived with continuous cash flows.

Fifth, the time period need not be a year, although the A for the uniform series comes from the word annuity. For example, the same factors are used where i (the effective interest rate) is stated *as percent per annum (or year) compounded annually, percent per quarter compounded quarterly, percent per month compounded monthly,* or, in general, *percent per period compounded per period.* These factors require that *the compounding period equal the interest period.* If the interest is stated another way, then one must first convert to an effective interest rate, i (see Section 2.5), before using the interest factors.

We will now derive these factors.

2.3.1 Factors relating P and F. Single-payment compound amount factor (finding F, given P).

In Example 2.2 interest accumulated at 9% per annum on the $100,000 principal for 10 years for an equivalent future amount of $236,736. In general terms, this relationship between a future and a present amount is called the *single-payment compound amount factor*.

If P dollars exist at a point in time and i% is the effective interest rate, then P will grow as described in Table 2.1.

Table 2.1 Compound amounts

End of Period (t)	Amount at Beginning of Period	Interest (or Growth) During Period	Compound Amount at End of Period	
1	P	Pi	$F_1 = P + Pi$	$= P(1+i)$
2	$P(1+i)$	$P(1+i)i$	$F_2 = P(1+i) + P(1+i)i$	$= P(1+i)^2$
3	$P(1+i)^2$	$P(1+i)^2 i$	$F_3 = P(1+i)^2 + P(1+i)^2 i$	$= P(1+i)^3$
N	$P(1+i)^{N-1}$	$P(1+i)^{N-1} i$	$F_N = P(1+i)^{N-1} + P(1+i)^{N-1} i$	$= P(1+i)^N$

This last relationship between P and F_N, that is, $(1+i)^N$, is the *single-payment compound amount factor*, which is also our fundamental interest equation:

$$F = P(1+i)^N \quad (2.5)$$

or, from the ANSI notation introduced earlier, which is also that used for the column headings of Appendix A:

$$F = P(F/P, i\%, N) \quad (2.6)$$

For Example 2.2, we have the following equation:

2.3 Tabulated Compound Interest Factors

$$F = P(F/P, i\%, N)$$
$$= 100{,}000(F/P, 9\%, 10)$$
$$= 100{,}000\ (2.3674) = \$236{,}740.$$

Note that there is round-off error introduced by the more limited accuracy of the tables.

Single payment present value factor (finding P, given F). Equation (2.5) may be rearranged to solve for an unknown P, assuming F is given:

$$P = F\left(\frac{1}{1+i}\right)^N \quad (2.7)$$

Using the *single-payment present value factor* (also tabulated in Appendix A) this becomes:

$$P = F(P/F, i\%, N). \quad (2.8)$$

For example, we could find the *present* equivalent at $i = 9\%$ of a future amount $F = \$236{,}736$ that occurs 10 years from now.

$$P = 236{,}736\ (P/F, 9\%, 10)$$
$$= 236{,}736\ (0.4224) = \$99{,}972.$$

This does not equal 100,000 due to the limited accuracy of the tabulated value.

2.3.2 Factors relating A and F. *Uniform-series compound amount factor (finding F, given A).* Uniform amounts, each an A, occur at the ends of periods 1 through N. These form a uniform series, often called an *annuity*, which is illustrated in Figure 2.6. Three important facts about cash flow timing should be noted:

1. The uniform series values, A, occur at the end of each period.
2. The equivalent *present* value, P, occurs one interest period *before* the first A of the uniform amounts.
3. The equivalent future value, F, occurs *at the same time* as the last A.

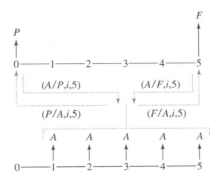

Figure 2.6 Equivalent periodic amount cash flow diagram

The future value, F, at the end of the Nth period equals the sum of the future values of the N flows of A each. Thus, to find F in Figure 2.6 we would sum the future values of each A:

t	F_t
1	$A(1 + i)^{N-1}$
2	$A(1 + i)^{N-2}$
$N-1$	$A(1+i)^1$
N	$A(1+ i)^0$

$$F = A[1 + (1 + i) + (1 + i)^2 + \cdots + (1 + i)^{N-1}]. \tag{2.9}$$

This equation is a geometric series with common ratio $(1 + i)$. The sum of the first N terms of this series can be found by multiplying the equation by the common ratio:

$$F(1 + i) = A[(1 + i) + (1 + i)^2 + (1 + i)^3 + \cdots + (1 + i)^N] \tag{2.10}$$

and then subtracting Equation (2.9) from Equation (2.10) term by term to get Equation (2.11):

$$F(1 + i) - F = A[(1 + i)^N - 1] \tag{2.11}$$

and upon solving for F we have:

$$F = A\left[\frac{(1 + i)^N - 1}{i}\right] \tag{2.12}$$

The term in brackets is the *uniform-series compound amount factor,* which is tabulated in Appendix A:

$$(F/A, i\%, N) = \left[\frac{(1 + i)^N - 1}{i}\right]$$

As an example, suppose that equal cash flows of $1,000 occur at the end of five consecutive years. If interest is 10% per year compounded annually, what is the equivalent future amount of these cash flows at the end of the fifth year? The solution is

$$F = A(F/A, 10\%, 5) = \$1,000(6.1051) = \$6,105.10.$$

Uniform-series sinking fund factor (finding A, given F). Taking Equation (2.12) and solving for A, we have:

$$A = F\left[\frac{i}{(1 + i)^N - 1}\right] \tag{2.13}$$

and the quantity in brackets is called the *sinking fund factor,* or $(A/F, i\%, N)$.
This reciprocal of the uniform series compound amount factor takes its name from

historical sinking funds. Before income taxes and inflation, it was common for a firm to deposit a portion of its profits each month in a bank account, so that a fund of sufficient size (F) would exist at the end of N years to replace machinery. This was called *a sinking fund*, and the required deposit (A) was determined using the sinking fund factor. As an example, what amount in periods 1 through 5 is equivalent to a future amount $F = \$6,105$ occurring at the end of the fifth period, if the effective interest rate $i = 10\%$? The solution is

$$A = F(A/F, 10\%, 5) = \$6,105(0.1638) = \$1,000.$$

2.3.3 Factors relating P and A. Uniform-series present value factor (finding P, given A). To relate a series of uniform end-of-period amounts, A, to their present value, P, we combine two factors already developed. From Equation (2.4), $F = P(1+ i)^N$, and substituting F's value in Equation (2.12) we have:

$$F = P(1+i)^N = A\left[\frac{(1+i)^N - 1}{i}\right]$$

Dividing both sides by $(1 + i)^N$ produces:

$$P = A\left[\frac{(1+i)^N - 1}{i(1+i)^N}\right] \quad (2.14)$$

the *uniform-series present value factor*, $(P/A, i\%, N)$.

For example, to find the present equivalent of five equal cash flows of \$1,000 each occurring at the end of years 1 to 5, with effective interest $i = 10\%$ per year compounded annually:

$$P = A(P/A, 10\%, 5) = \$1,000(3.7908) = \$3,790.80.$$

Uniform-series capital recovery factor (finding A, given P). This reciprocal of the series present value factor finds the required cash amount to recover a capital investment; hence, its name. The *uniform-series capital recovery* factor, $(A/P, i\%, N)$, is obtained by taking the reciprocal of Equation (2.14):

$$A = P\left[\frac{i(1+i)^N}{(1+i)^N - 1}\right] \quad (2.15)$$

This factor, as shown in the derivation of Equation (2.2), equals $(A/F, i\%, N) + i\%$.

For example, to find the value of the uniform series of five equal cash flows that is equivalent to a present value $P = \$3,790.80$ with effective interest $i = 10\%$ per period compounded each period, we use:

$$A = P(A/P, 10\%, 5) = \$3,790.80(0.2638) = \$1,000.$$

2.3.4 Arithmetic gradient conversion factors. Sometimes a series of cash flows increase or decrease by a fixed *amount* each period. This forms an arithmetic gradient. For example, maintenance and repair expenses often increase with the age of an asset.

Figure 2.7 shows a series of end-of-period amounts, 0, G, 2G, 3G, ... , that differ by a constant gradient, G.

The timing of the arithmetic gradient amounts is

End of Period	Amount
1	0
2	G
N–1	(N–2)G
N	(N – 1)G

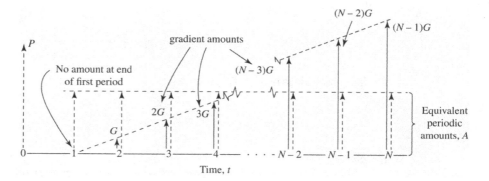

Figure 2.8 Cash flow diagram for gradient conversion factor

The 0 cash flow for period 1 exists because the gradient is a change from the base level. This base level for the first period is part of a uniform series. So the first real cash flow in a gradient series occurs at the end of period 2. We wish to find (P/G, i%, N) and (A/G, i%, N) factors. We begin with the equivalent present amount:

$$P = \frac{G}{(1+i)^2} + \frac{2G}{(1+i)^3} + \frac{3G}{(1+i)^4} + \cdots + \frac{(N-2)G}{(1+i)^{N-1}} + \frac{(N-1)G}{(1+i)^N}$$

Now, multiply Equation (2.15) by $(1+i)^N$, giving

$$P(1+i)^N = G(1+i)^{N-2} + 2G(1+i)^{N-3} + 3G(1+i)^{N-4} + \ldots + (N-2)G(1+i) + (N-1)G \quad (2.16)$$

Note that this sequence contains $N - 1$ terms and can be recomposed by stating as a series of equations where each G has a coefficient of 1. This is shown in Figure 2.8, where each column corresponds to a term in Equation (2.16).

2.3 Tabulated Compound Interest Factors

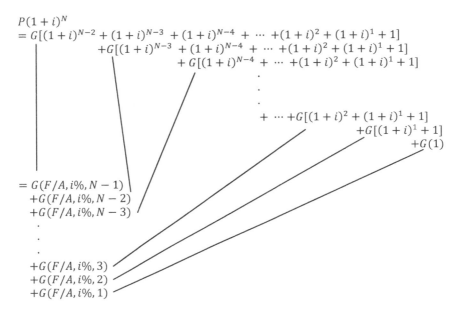

Figure 2.8 Reconstructing the arithmetic gradient formula

Restating the right hand side using the formula for the (F/A) factors results in:

$$= G\left[\frac{(1+i)^{N-1}-1}{i} + \frac{(1+i)^{N-2}-1}{i} + \cdots + \frac{(1+i)^2-1}{i} + \frac{(1+i)-1}{i}\right]$$

Collecting the $N-1$ terms of -1 and moving $1/i$ outside the brackets results in:

$$= \frac{G}{i}[(1+i)^{N-1} + (1+i)^{N-2} + \cdots + (1+i)^2 + (1+i) - (N-1);]$$

and if we move the $-N$ term (on the right side) outside the brackets and combine with the G/i, then

$$P(1+i)^N = \frac{G}{i}[(1+i)^{N-1} + (1+i)^{N-2} + \cdots + (1+i)^2 + (1+i) + 1] - \frac{NG}{i}. \quad (2.17)$$

The terms inside the brackets in Equation (2.17) define the $(F/A, i\%, N)$ factor; hence, Equation (2.17) becomes:

$$F = P(1+i)^N = \frac{G}{i}\left[\frac{(1+i)^N - 1}{i}\right] - \frac{NG}{i}$$

or

$$P = G\left[\frac{(1+i)^N - 1}{i^2(1+i)^N} - \frac{N}{i(1+i)^N}\right]$$

$$= G(P/G, i\%, N). \tag{2.18}$$

Note that an alternate algebraic derivation can be found in Newnan, Eschenbach, and Lavelle (2014). To find an equivalent uniform end-of-period amount, we multiply Equation 2.18 on both sides by the A/F factor:

$$A = F(A/F, i\%, N)$$

$$= \left\{\frac{G}{i}\left[\frac{((1+i)^N - 1)}{i}\right] - \frac{NG}{i}\right\}\left[\frac{i}{(1+i)^N - 1}\right]$$

$$= \frac{G}{i} - \frac{NG}{(1+i)^N - 1} \tag{2.19}$$

$$= G\left[\frac{1}{i} - \frac{N}{(1+i)^N - 1}\right].$$

The resulting factor is the *uniform-series gradient conversion factor*, $(A/G, i\%, N)$.

As discussed, the gradient factors assume a *zero* cash flow at the beginning of the gradient. If the initial flow is not zero, the cash flows can be decomposed using $A_1 =$ the initial *nonzero* flow at $t = 1$. Then the resulting equivalence is simply

$$P = A_1(P/A, i\%, N) + G(P/G, i\%, N) \tag{2.20}$$

$$A = A_1 + G(A/G, i\%, N) \tag{2.21}$$

For a decreasing gradient series, G is negative.

Example 2.4 Gradient
Suppose that the expense estimates had been:

End of Period, t	Expense
1	$3,000
2	5,000
3	7,000
4	9,000

If the effective interest rate is still 15%, what is the equivalent end-of-period amount of these expenses?

Solution
The initial amount is $A_1 =$ $3,000 and the gradient is $G =$ $2,000. Applying Equation (2.20), we obtain

$$A = A_1 + G(A/G, i\%, N) = \$3{,}000 + \$2{,}000(A/G, 15\%, 4)$$

$$= \$3{,}000 + \$2{,}000(1.3263) = \$5{,}653.$$

Example 2.5 Decreasing gradient
Suppose the amounts in Example 2.4 were reversed:

End of Period, t	Expense
1	$9,000
2	7,000
3	5,000
4	3,000

At 15% effective interest rate, what is the equivalent end-of-period amount?

Solution
Here, $A_1 = \$9{,}000$ and the gradient is negative ($G = -\$2{,}000$ per period). Applying Equation (2.20) yields

$$A = \$9{,}000 + (-\$2{,}000)(A/G, 15\%, 4)$$

$$= \$9{,}000 - \$2{,}000(1.3263) = \$6{,}348.$$

Note that the equivalent amount is larger in Example 2.5 since the larger cash flows occur at the beginning rather than the end of the series.

2.4 EXAMPLES OF TIME VALUE OF MONEY CALCULATIONS

Most real problems require the use of multiple factors to calculate equivalent economic values. Example 2.6 illustrates how to segment cash flows and to apply multiple factors as needed. In particular, notice how a second P/F factor is applied for uniform and gradient series that do not start in period 1. Example 2.7 illustrates an overhaul that occurs in the middle of a machine's life. Example 2.8 illustrates a common pattern called a *deferred annuity*. The latter two examples focus on the key cash flow, but similar patterns are part of many problems.

Example 2.6 Combining multiple factors
Find the present value of these cash flows, at an effective interest rate $i = 15\%$:

End of Year	Cash Flow in Year
0	−200
1	900
2	1,000

34 Chapter 2, Interest, Interest Factors, and Equivalence

EOY	Cash Flow
3	1,000
4	1,000
5	1,200
6	1,300
7	1,400
8	1,600

Solution

The first step is drawing the cash flow diagram (Figure 2.9), and the 2nd step is constructing the *anatomical* solution shown in Figure 2.10.

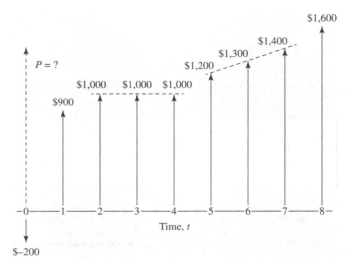

Figure 2.9 Cash flow diagram

End of Year

```
0:  P = $-200        +$782.64              +$1,985.30          + $1,684.92 + $523.04
         (P/F,15%,1)                 (P/F,15%,1)
1:  $900( 0.8696 ) =         $2,283( 0.8696 ) =
2:  $1,000
               (P/A,15%,3)
3:  $1,000   = $1,000( 2.283 ) =
4:  $1,000                                          (P/F,15%,4)
                                      $2,946.69( 0.5718 ) =
5:  $1,200
                    (A/G,15%,3) (P/A,15%,3)
6:  $1,300  = [$1,200 + $100( 0.9071 )]( 2.283 ) =
7:  $1,400
         (P/F,15%,8)
8:  $1,600( 0.3269 ) =
```

Figure 2.10 Anatomical diagram of solution

We must break up the cash flows into sections and take advantage of the patterns that are present. Years 2 through 4 contain uniform cash flows, and years 5 through 7 contain an arithmetic gradient. We can solve the problem by breaking it up into parts.

For year 0, $P = -200$
For year 1, $P = 900(P/F, 15\%, 1) = 782.64$
For years 2–4, $P = 1000(P/A, 15\%, 3)(P/F, 15\%, 1) = 1,985.30$
For years 5–7, $P = \{[1200 + 100(A/G, 15\%, 3)](P/A, 15\%, 3)\}(P/F, 15\%, 4)$
$\qquad = 1,684.92$
For year 8, $P = 1,600(P/F, 15\%, 8) = 523.04$
Totaling these, in any currency,

$$P = -200 + 782.64 + 1,985.30 + 1,684.92 + 523.04 = \underline{4,775.90.}$$

Having established the equivalent present value of this nonuniform series, we may now establish (a) a future equivalent amount (F) at any time or (b) an equivalent periodic amount (A) for the 8-year period. Thus, for example, the future equivalent at the end of year 3 is
$\quad F_3 = P(F/P, 15\%, 3) = 4,775.90(1.5209) = 7,264.$
Likewise, the equivalent annual amount over the 8-year period is
$\quad A = P(A/P, 15\%, 8) = 4,775.90(0.2229) = 1,065$ per year.

Example 2.7 Equivalent annual cost of an overhaul
The calculation of an equivalent annual (periodic) amount (A) always proceeds from either a single present equivalent (P), a single future equivalent (F), or a single gradient series (G) with the same number of periods. This example illustrates this with the problem of finding the equivalent annual cost at 8% interest over 5 years of an overhaul that costs $3000 at the end of year 3.

Solution
To find a uniform annual amount at the ends of years 1 through 5, first a P at time 0 or an F at the end of year 5 must be calculated. First the present value approach.
$\quad P = \$3,000(P/F, 8\%, 3) = \$3,000(0.7938) = \$2,381.5$ then the equivalent annual amount for 5 years:
$\quad A = P(A/P, 8\%, 5) = \$2,381.5(0.2505) = \$596.5.$
For the future value approach, $F = \$3,000(F/P, 8\%, 2) = \$3,000(1.664) = \$3,499$ then the equivalent annual amount for 5 years (ends of years 1 through 5).
$\quad A = F(A/F, 8\%, 5) = \$3,499(0.1705) = \$596.6.$
If the A/P or A/F 5-year factor is applied directly to the $3000 at the end of year 3, then the 5 uniform periods are at the ends of years 4 to 8 or years –1, 0, 1, 2, and 3.

However, the calculation of a single P or F can be an intermediate step, so that we can calculate a second series, A_2, that is equivalent to a first series, A_1 where the two uniform series are not of the same length. Example 2.8 illustrates this in finding the equivalent annual cost of repairs when a warranty covers the cost for an initial term.

Example 2.8 A deferred annuity
Northwest Machines has just purchased a milling machine which comes with a two-year warranty. Once the warranty expires repair costs are expected to average $5500 per year. Find the equivalent annual cost (EAC) for the repair costs over the 10-year life of the milling machine. The interest rate is 9%.

Solution
This is a deferred annuity since repair costs are incurred only in years 3 through 10. Even this short 2-year warranty lowers the EAC for repairs to 28% below the $5500 per year. Note that the easiest way to solve this is to find the future value for the 8 years of repair costs, and then to spread that total over all 10 years.
$$EAC = \$5500(F/A,9\%,8)(A/F,9\%,10) = 5500 \times 11.028 \times 0.06582 = \$3992$$

2.5 GEOMETRIC GRADIENTS

A geometric gradient changes cash flows at a constant rate rather than by a constant amount. Consider an example where the annual maintenance costs increase from a base of $1000 by $200 per year. If this increase had been stated as 20% per year, the cash flows would have been $1000, $1200, $1440, $1728, and $2073.6. The first 2 years are the same, but the 5th year differs by $273.6, and this difference is greater for larger rates of change and longer time periods.

This geometric gradient, g, or constant rate of change, is compound growth and is similar to the basic interest rate, i. In a practical sense, geometric gradients are often used to model inflation and changes in the volume of operations. A rate of change is one of the most common ways to predict the cost or revenue of a future activity. Common examples include the labor cost to run an assembly line and the revenue from a product with a growing market.

Thus, each cash flow series may be affected by the following four classes of geometric gradients:

1. Inflation or deflation in the price of the item.
2. Changes in the volume of the item that is used or sold.
3. Inflation or deflation in the economy as measured by the value of a dollar.
4. Discount rate for the time value of money.

Every project will have multiple cash flow series within it. These might include a series for revenue, one for energy costs, another for labor costs, one for loan repayments, and yet another for income taxes. While each of these should use the same discount rate, the other sources of geometric gradients are likely to be different.

It is often easier to use a spreadsheet package and model the cash flows on a period by period basis (see Section 2.9). However, it is sometimes practical to use formulas that have been developed for a single geometric gradient in addition to the interest rate. These formulas are useful either for a volume change or for a *differential* inflation rate.

In many cases it is easier and more accurate to estimate the difference between the inflation rates for the item and for the value of the currency used. This differential inflation rate is often more stable than the two inflation rates on which it is based.

Moreover, for most items the differential inflation rate is about 0%, and *inflation can be ignored for that item.* Thus, for many problems and for even more cash flows inflation is ignored.

Inflation rates for the economy and individual items are difficult to estimate accurately. However, several classes of items have known inflation rates of 0%. Consequently, their differential inflation rate is negative if inflation is decreasing the value of the currency, such as the dollar. For example, most loans and bonds have fixed terms. Therefore, even if inflation changes the value of the payments, the payments do not change, and there is a 0% inflation rate for those payments. The interest rate for a bond or loan is set to compensate for an expected level of future inflation, but any inflation in prices decreases the value of future payments and is thus a negative differential inflation. Depreciation and its associated income tax deductions (see Chapters 4 and 5) also have a 0% inflation rate.

To develop the formula for a single geometric gradient we begin by stating the assumptions. First, the cash flow has a continuing basis so that it occurs at the end of every period through period N. Second, the cash flow at the end of the first period is c. Third, the geometric gradient and the time value of money are both constant through period N. Fourth, the geometric gradient is first applied at the end of the second period. This assumption is consistent with our model of the arithmetic gradient. With these assumptions the formula for present value becomes:

$$P = c[1/(1+i) + (1+g)^1/(1+i)^2 + \cdots + (1+g)^{N-1}/(1+i)^N]$$
$$= c[(1+g)/(1+i) + (1+g)^2/(1+i)^2 + \cdots + (1+g)^N/(1+i)^N]/(1+g)$$

$$P = \frac{c}{1+g} \sum_{t=1}^{N} \frac{(1+g)^t}{(1+i)^t}$$

Now, so long as $g \neq i$, this can be expressed in ANSI standard notation (ANSI, 1988) as follows:

$$(P/A, g, i, N) = \frac{1 - (1+g)^N (1+i)^{-N}}{(i-g)} \quad (2.22)$$

If $g = i$, then $(P/A, g, i, N) = N/(1+g)$ or $N/(1+i)$. However, for the cases where $g \neq i$, it may be more useful to develop formulas that rely on a single interest rate, so that the tables of factors can be used. This is done by defining an equivalent discount rate, x. Two formulas are required since only tables for positive discount rates are typically available. The first is for $g<i$, and the second is for $g>i$.

If $g<i$, then:

$$P = [c/(1+g)](P/A, x, N) \quad \text{where } (1+x) = (1+i)/(1+g). \quad (2.23)$$

If $g >i$ then the x as calculated above would be negative. This might occur where the volume was increasing at a high rate. Since we have no tables for negative interest rates, we want to invert the definition of $1 + x$. This is like the definition of *(F/A, i, N)* so we want to reduce the limits of the summation by one. We do this by removing $(1+g)/(1+i)$ from the summation:

$$P = \frac{c}{1+i} \sum_{t=0}^{N-1} \frac{(1+g)^t}{(1+i)^t}$$

Now we let $(1+x) = (1+g)/(1+i)$:

$$P = \frac{c}{1+i} \sum_{t=0}^{N-1} (1+x)^t = \frac{c}{1+i}(F/A, x, N) \qquad (2.24)$$

So when the geometric gradient is larger than the discount rate, a future value factor is used to calculate the present value. Also note that if $g<i$ we divide c by $1+g$, but if $g>i$ we divide c by $1+i$.

Example 2.9 O&M expenses with a geometric gradient
Consider the example stated above, where maintenance begins at $1000 per year, and increases annually. If the increase is 20% per year, the cash flows would be $1000, $1200, $1440, $1728, and $2073.6.

Solution
Using the ANSI formula, we have the following:

$$P = \$1{,}000(P/A, g, i, N) = \$1{,}000(P/A, 20\%, 10\%, 5)$$
$$= \$1{,}000(1 - 1.2^5 \times 1.1^{-5})/(0.1 - 0.2) = \$1{,}000(-0.5451/-0.1)$$
$$= \$5{,}450.51.$$

Using present value factors, we have the following:

$$P = \$1{,}000/1.1 + \$1{,}200/1.1^2 + \$1{,}440/1.1^3 + \$1{,}728/1.1^4 + \$2{,}073.6/1.1^5$$
$$= \$5{,}450.51.$$

Using the future value formulation for the equivalent discount rate since $g = 20\%$ exceeds $i = 10\%$, we have $(1+x) = 1.2/1.1$ or $x = 9.091\%$. Then the formula becomes

$$P = [\$1{,}000/(1+i)](F/A, x, N) = (\$1{,}000/1.1)(F/A, 9.091\%, 5)$$
$$= \$909.09(5.9956) = \$5{,}450.51.$$

As should be expected, all three approaches give the same answer.

2.6 NOMINAL AND EFFECTIVE INTEREST RATES

In most equivalence applications, the interest period matches the compounding period. It is possible, however, for the compounding to occur more frequently. Some applications are listed in Table 2.1, where the nominal interest period is taken to be annual.

If an interest rate is compounded more frequently than its nominal interest period, then the *effective* interest rate is increased. For example, if the *nominal interest rate* is 8% per annum but compounding is semiannual, then the *effective interest rate* is 8.16%. Consider $100 as the principal; then the interest during the first 6 months is:

$$I_1 = \$100\left(\frac{0.08}{2}\right) = \$4.00$$

and the total principal at the beginning of the second 6-month period is:

$$P + I_1 = \$100 + \$4 = \$104.$$

The interest during the second 6 months is:

$$I_2 = \$104\left(\frac{0.08}{2}\right) = \$4.16$$

so that the principal plus interest at year's end is:

$$P + I_1 + I_2 = \$100 + \$4.00 + \$4.16 = \$108.16$$

Table 2.2 Example of several interest compounding frequencies

Frequency of Compounding	Example
Annual	Economy studies, long-term loans
Semiannual	Most bonds
Quarterly	Many certificates of deposit
Monthly	Credit Union deposits; mortgage loans
Weekly or daily	Bank savings accounts (day-in day-out interest); savings and loan accounts

and the *effective rate of* interest is simply:

$$i = \frac{I_1 + I_2}{P} = \frac{\$8.16}{\$100} = 8.16\%$$

We note that i has units *of percent per year compounded yearly*. The effective rate, i, when compounded per period, has the same effect as does the nominal rate compounded more often.

In this book, effective interest rates are designated by i and are interpreted as i *percent per period compounded per period*. Nominal interest rates are designated by r and are interpreted as r *percent per period* (no compounding). Usually $i = r$; that is, the nominal and effective rates are equal because the compounding and interest periods are both a year.

For noncomparable compounding and interest periods, the effective interest rate follows by analogy from the numerical example, where r is now the nominal *rate per year:*

Semiannual compounding: $\quad i = \left(1 + \frac{r}{2}\right)^2 - 1$

Quarterly compounding: $\quad i = \left(1 + \frac{r}{4}\right)^4 - 1$

Weekly compounding: $i = \left(1 + \dfrac{r}{52}\right)^{52} - 1$

or in general for M periods per year:

$$i = \left(1 + \frac{r}{M}\right)^M - 1 \tag{2.25}$$

In the limit, interest may be compounded an infinite number of times per year, that is, continuously. The effective interest rate may be derived as follows. From the fact that

$$\left(1 + \frac{r}{M}\right)^M - 1 = \left[\left(1 + \frac{r}{M}\right)^{M/r}\right]^r - 1$$

and

$$\lim_{M \to \infty} \left(1 + \frac{r}{M}\right)^{M/r} = e$$

then

$$i = \lim_{M \to \infty} \left[\left(1 + \frac{r}{M}\right)^{M/r}\right]^r = e^r - 1 \tag{2.26}$$

So, when interest at nominal rate r is compounded continuously, the effective rate i is $e^r - 1$.

Only those interest rates placed on a comparable basis may be compared. For example, is it better to receive interest at a rate of 17% per year compounded annually or at 16% per year compounded monthly? The nominal rates cannot be compared, but the effective rates can be. The effective rate of 17% per year compounded annually is of course 17%; while for 16% compounded monthly the *effective* annual rate is:

$$i = [1 + (0.16/12)]^{12} - 1 = 0.1722 \text{ or } 17.22\%.$$

The compound interest factors derived in Section 2.2 are all based on an effective interest rate—an i expressed as *percent per period compounded per period*. Thus, the tables in Appendix A may be used if the interest and compounding periods correspond; for example, if i is

Percent per annum compounded annually
Percent per quarter compounded quarterly
Percent per week compounded weekly
:
Percent per period compounded per period

Whenever the interest and the compounding periods do not correspond, then the conversion in Equation (2.25) must be made first.

2.7 CONTINUOUS INTEREST FACTORS

In certain cases, interest may be considered to compound continuously, as in Equation (2.26). This concept of continuous generation of interest leads to another family of interest factors, similar to the discrete compounding factors of Section 2.2. These *discrete payment continuous compounding* factors involve various combinations of e^{rN}. These factors must not be confused with the *fund flows* factors, which assume cash flows are distributed in time instead of discrete amounts at points in time. As discussed in Section 2.2, these factors are omitted from this book because of a low and decreasing level of use.

The factors that are based on e^{rN} have an important theoretical advantage over the discrete-event factors (for example, they can be functionally integrated, while the discrete factors cannot). Virtually all of the research in finance assumes continuous interest. Most work in capital budgeting, and virtually all problems solved in practice use discrete cash flows. Real options analysis covered in Chapter 15, uses both discrete and continuous interest.

These continuous compounding formulas may be used when payments occur in discrete points in time, but interest is compounded continuously. In such cases, we use a nominal interest rate, r. As in discrete compounding, the period N must match the interest rate period.

We can substitute i with e^{r-1} to develop a series of continuous compounding factors:

$F = P(1 + i)^N$ becomes $F = P(1 + e^{r-1})^N$ or $F = Pe^{rN}$

similarly:

$P = F/(1 + i)^N$ becomes $P = F/(1 + e^{r-1})^N$ or $P = Fe^{-rN}$

A summary of continuous compounding formulas with discrete payments is shown in Table 2.3 (Park & Sharp-Bette, 1990).

Table 2.3 Continuous compounding formulas with discrete payments

Present Value	$(P/F, r\%, N)$	$P = F(e^{-rN})$
Compound Amount	$(F/P, r\%, N)$	$F = P(e^{rN})$
Series Present Value	$(P/A, r\%, N)$	$P = A \left[\dfrac{e^{rN} - 1}{e^{rN}(e^r - 1)} \right]$
Series Compound Amount	$(F/A, r\%, N)$	$F = A \left[\dfrac{e^{rN} - 1}{e^r - 1} \right]$
Capital Recovery	$(A/P, r\%, N)$	$A = P \left[\dfrac{e^{rN}(e^r - 1)}{e^{rN} - 1} \right]$
Sinking Fund	$(A/F, r\%, N)$	$A = F \left[\dfrac{e^r - 1}{e^{rN} - 1} \right]$

Arithmetic gradient Present Value	$(P/G, r\%, N)$	$P = G\left[\dfrac{e^{rN} - 1 - N(e^r - 1)}{e^{rN}(e^r - 1)^2}\right]$
Geometric gradient Present Value (Park and Sharp-Bette, 1990)	$(P/A, g, r\%, N)$	$P = \begin{cases} F_1\left[\dfrac{1-e^{(g-r)N}}{e^r - e^g}\right] \\ \dfrac{NF_1}{e^r} \quad (\text{if } g = e^r - 1) \end{cases}$

2.8 EXTENDED ENGINEERING ECONOMY FACTORS AND SPREADSHEETS AND CALCULATORS

2.8.1 Advantages of extended engineering economy interest factors. The extended engineering economy factors are available as spreadsheet functions and as built-in functions on financial calculators. These extended factors will solve for the fifth variable given values for any four variables chosen from i, N, A, P, and F. In comparison with using the tabulated factors, these extended factors have the following advantages.

- Time is saved by entering fewer numbers and by the increased TVM capabilities of the functions.
- Because the tabulated factor values do not have to be entered, there is a smaller chance of transposed digits or wrong row lookup or similar errors.
- If all three of P, A, and F are present, a single set of calculator values is needed rather than two tabulated factors, as shown in Example 2.10.
- There is no need to interpolate for interest rates or lives that are not tabulated, such as a life of 120 monthly periods, a mortgage at 5.85%, a car loan at 8.95%, or a credit card at 14.85%.
- There is no need for iterative solutions to solve for i or N.
- For an A that represents beginning-of-period cash flows, such as a lease, there is a TYPE spreadsheet argument and a BEGIN/END financial calculator choice.

2.8.2 Notation for extended engineering economy factors. The notation we will use is based on the most common choices for the calculators, on the standard notation for engineering economy, and on the notation for the very similar spreadsheet functions. The typical equivalents are:

- i, I, RATE
- N, n, NPER
- A, PMT
- P, PV, PW
- F, FV, FW

Finally, the order of the items in each function follows the order of the items in typical spreadsheet annuity functions $\{i, N, PMT, PV, \text{and } FV\}$. Note that most financial calculators order the buttons $\{N, i, PV, PMT, \text{and } FV\}$. However, each function removes the variable that is being solved for from this list. So our notation is:

- $PV(i, N, A, F)$ or $PV(i, N, A, \quad F)$
- $PMT(i, N, P, F)$ or $PMT(i, N, \quad P, F)$
- $FV(i, N, A, P)$ or $FV(i, N, A, P \quad)$

2.8 Extended Engineering Economy Factors, Spreadsheets and Calculators

- $i(N, A, P, F)$ or $i(\ N, A, P, F)$
- $N(i, A, P, F)$ or $N(i, A, P, F)$

When implemented on financial calculators, these extended engineering economy factors are solving Equation 2.27. This equation is written so that i is entered as a percentage, so 8 is entered for 8%. Equation (2.28) rewrites this in standard factor notation. When implemented on a spreadsheet, the extended engineering economy factors assume that i is entered as 0.08 or 8%, not as an integer:

$$PMT\left[\frac{1-(1+i/100)^{-N}}{i/100}\right] + FV(1+i/100)^{-N} + PV = 0 \quad (2.27)$$

$$A(P/A,i,N) + F(P/F,i,N) + P = 0 \quad (2.28)$$

Since the factors are positive, one of *PMT, FV,* and *PV* (or *A, F,* and *P*) must be different in sign from the other two. If two positive cash flows are entered to solve for the third cash flow, then that third value must be negative to solve the equation. Equation (2.27) is the foundation for the sign convention that is used in the TVM calculators *and* the spreadsheet annuity functions, PV, FV, and PMT.

2.8.3 Spreadsheet annuity functions. The extended engineering economy factors are implemented in spreadsheets as the five functions listed in Table 2.4. Because these functions are solving the equivalent of Equation 2.28, the PMT, PV, and FV functions reverse the sign of the cash flows (they are solving for the third cash flow value that will produce a present value of 0 for all cash flows. That sign convention is why those three functions are given an initial minus sign in Table 2.4. When solving for i or N or when using RATE and NPER, the nonzero cash flows of P, A, and F must differ in sign.

Table 2.4 Spreadsheet annuity functions for Excel
PV(i, N, A, F, Type)
PMT(i, N, P, F, Type)
FV(i, N, A, P, Type)
NPER(i, A, P, F, Type)
RATE(N, A, P, F, Type, guess)

The type argument in the spreadsheet functions is equivalent to the BEGIN/END setting of financial calculators. The choice is between whether the uniform periodic amount, A, is assumed to be an end-of-period or a beginning-of-period value. The Type argument does *not* affect P which is always assumed to be at time 0 nor F, which is always assumed to be at the end of period N. The GUESS argument supplies a starting point finding the unknown interest rate. The formats do not need to be memorized; they are available on a dropdown menu when the function is typed. Alternatively, the f_x button may be clicked and the information added.

Example 2.10 PV of a simple project
Assume a project has a first cost of $140,000, net savings of $22,000 annually, a final salvage value of $20,000, and a life of 10 years. If the firm's interest rate is 8% what is the project's present value?

Known Quantity	Cell Address
$i = 8\%$,	A1
$N = 10$ periods	A2
$A = 22{,}000$	A3
$F = 20{,}000$	A4

The PV function may be used to find the present value of the project's future cash flows, $-PV(i, N, A, F)$. The default for Type is zero, and it does not need to be included. Using cell addresses, $-PV(\$A\$1, \$A\$2, \$A\$3, \$A\$4)$ will provide an answer of \$156,886. Subtracting the \$140,000 first cost, provides the final answer of \$16,886.

2.8.4 Time value of money (TVM) calculators. There are three classes of TVM calculators. First, there are many *financial* or *business* calculators that have buttons labeled N, i, PV, PMT, and FV or equivalents. The button for the uniform series cash flow, A, is labeled PMT, because one of the most important uses of these calculators is for calculating loan payments (PMT). These calculators also typically have the capability to find the NPV of complex cash flow patterns, spreadsheets are a better tool for more complex patterns (see Section 2.9 Spreadsheets and Cash Flow Tables).

There are also *graphing* calculators that have the TVM calculations in a menu. Finally, there are programmable scientific calculators where the TVM equation (Equation 2.27) can be entered to create the menu that is built into the graphing calculators. The Hewlett Packard 33s and 35s calculators are currently allowed for use on the Fundamentals of Engineering (FE) exam, while the financial and graphing calculators are not (Eschenbach and Lewis, 2011).

Some calculators are shipped from the factory with a setting of 12 months per year, so that N will be entered as years for loans with monthly payments. It is recommended that this be changed to 1 payment per "year or period" and left on this setting. Then N is entered as the number of periods or payments, and there is no confusion for problems with different length periods.

2.9 SPREADSHEETS AND CASH FLOW TABLES

2.9.1 Advantages of spreadsheets for economic analysis. Spreadsheets are the principal tool of the practicing engineer for economic analysis. In fact, the original use of early spreadsheets was for financial analysis, and spreadsheets are credited with helping the dramatic early growth of personal computers. The primary advantages of spreadsheets include:
1. It is easy to perform scenario analysis, comparing alternative plans
2. Available graphics aid in presenting the information
3. The electronic format makes it easy to share work with others, and for several people to work on a common spreadsheet
4. Realistic models can be used to analyze complex problems

Real projects often involve forecasting future costs, revenues, sales volumes, and other factors. Forecasts often include projections of growth—with growth rates that may change over time. Many projects have construction or research phases that last for years, sales volumes may follow an S-curve of growth and then decline, and prices may change at different rates that vary over time. There may be renovation, overhaul, or expansion expenses. Finding a complex project's economic value is best done through a cash flow table in a spreadsheet.

2.9.2 Effective and efficient spreadsheet construction. Example 2.11 illustrates the construction and use of a cash flow table with a geometric gradient. The *data block* above the cash flow table in Figure 2.11 clearly states the assumed values for each variable. These values can easily be updated as better data is received, and the use of a data block as the single point of entry for every variable ensures that the entire spreadsheet is updated whenever a value is changed. All variables should be defined in the *data block*, and no values should be entered directly into formulas.

The formulas shown in Figure 2.11 illustrate the use of relative and absolute addresses. For example, the formula in cell B13 refers to the annual change in the rate with the absolute address $A$6, and the previous year's rate of change in net revenue with the relative address B12. Thus, when this formula is copied to fill the cash flow table, the absolute addresses never change and the relative address always refers to the cell above. For the final year's cash flow, this formula is modified by adding in a reference to the salvage value included in the data block.

Economic analysis of cash flow tables relies on two functions that calculate equivalent economic values for a range of cells that corresponds to each period's net cash flow. These functions assume: (1) cash flows are at the end of periods 1 to N, (2) each time period is the same length, and (3) the same interest rate applies to each time period. Because these two functions work with a range or block of cells, they are described as the spreadsheet *block* functions, in contrast to the spreadsheet *annuity* functions presented in Section 2.8.3.

- NPV(i, $CF_1:CF_N$)
- IRR($CF_0:CF_N$, *guess*)

The NPV (net present value or present value) function assumes that the cash flow at time 0 is not part of the specified block of cells. Thus the total PW = CF_0 + NPV(i, range). In contrast, the internal rate of return (IRR) which is the interest rate at which the cash flows have a PW of 0 includes the cash flow at time 0 in the specified range.

Example 2.11 Economic value with a changing geometric gradient

The assembly line to manufacture a new product is expected to cost $800,000 to build. The salvage value in 10 years is expected to be $120,000. The new product is expected to produce net revenues of $150,000 the first year. In year 2 the net revenue will increase by 6%, but in each following year the rate of change will decrease by 2%. Thus after year 5 the net revenues will be shrinking each year. Compute the project's internal rate of return and the present value at the firm's interest rate of 12%.

Chapter 2, Interest, Interest Factors, and Equivalence

	A	B	C	D	E
1	$800,000	First cost			
2	$120,000	Salvage value			
3		10	Horizon		
4	$150,000	Initial net revenue			
5		6%	Initial gradient in revenue		
6		-2%	Annual change in gradient		
7		12%	Interest rate		
8					
9	Year	Revenue rate of change	Cash flow		
10	0		-$800,000		
11	1		150,000		
12	2	6%	159,000	=C11*(1+B12)	
13	3	4%	165,360		=B12+$A$6
14	4	2%	168,667		
15	5	0%	168,667		
16	6	-2%	165,294		
17	7	-4%	158,682		
18	8	-6%	149,161		
19	9	-8%	137,228		
20	10	-10%	243,505	=C19*(1+B20)+A2	
21					
22			NPV	$124,934	=NPV(A7,C11:C20)+C10
23			IRR	15.6%	=IRR(C10:C20)

Figure 2.11 Spreadsheet with changing geometric gradient

Note that the assumed cash flows follow a common pattern of first increasing and then decreasing over a product's life. Computing each year's cash flow is easily done based on the values for the previous year. Then the NPV computes the PW of the cash flows in years 1 to 10, which are combined with the initial first cost for a total of $124,934.

2.10 ECONOMIC INTERPRETATION OF EQUIVALENT ANNUAL AMOUNT

Early in the history of engineering economy, some analysts argued that industrial project economic studies could be satisfactorily made by using simple interest rates. They advocated use of straight-line depreciation plus average (simple) interest to approximate the equivalent annual amount, A. At least as early as 1938, Grant (1938, pp. 83–92) showed this method to be only an approximation that is potentially useful only when the life N and the interest rate i are both small. The availability of spreadsheets and calculators have made this approximation obsolete.

Instead, we can show that a theoretically correct calculation of the equivalent annual amount also provides a useful economic interpretation. Consider a project with depreciable assets having the following characteristics:

P = initial cost of the assets at time $t = 0$.

F = salvage value of assets at time $t = N$.
N = life of the depreciable assets.
i = the effective interest rate (percent per period, compounded per period).

Note that $(P - F)$ *is* the depreciable amount, to be recovered over N periods. When we apply Equations (2.13) and (2.15) the net equivalent annual *cost of capital recovery* is simply:

$$A = P(A/P, i\%, N) - F(A/F, i\%, N). \quad (2.29)$$

Our objective is to recast Equation (2.26) to clarify its economic meaning. To do so, add and subtract $F(A/P, i\%, N)$ to the right side of Equation (2.29), obtaining:

$$A = (P - F)(A/P, i\%, N) + F[(A/P, i\%, N) - (A/F, i\%, N)]$$

and then reduce as follows:

$$A = (P - F)\left(\frac{A}{P}, i\%, N\right) + F\left[\frac{i(1+i)^N}{(1+i)^N - 1} - \frac{i}{(1+i)^N - 1}\right]$$
$$= (P - F)(A/P, i\%, N) + Fi\left[\frac{(1+i)^N - 1}{(1+i)^N - 1}\right]$$
$$= (P - F)(A/P, i\%, N) + Fi. \quad (2.30)$$

This formula for calculating the *annual equivalent cost of capital recovery* is particularly convenient when $F = P$. An economic interpretation of Equation (2.30) is that the annual equivalent cost of capital recovery, A, is composed of two factors:

1. The term $(P - F)(A/P, i\%, N)$, which is the annual equivalent of the *depreciable* value of the asset.
2. The term Fi, which is the annualized return required on the salvage value.

The first term, $(P - F)(A/P, i\%, N)$, provides for the recovery of the invested capital, $(P - F)$, *at interest*, over the asset's life. This can be shown to be equivalent to annual depreciation plus interest on the *unrecovered* depreciable balance. See Hartman (2007, pp. 646–647) for a continued discussion of the topic.

Unfortunately, inflation complicates the issue. If the initial first cost, the final salvage value, and the equivalent annual amount are not stated in constant value dollars, then interpreting their meaning is unclear. In particular, if the equivalent annual value is not in constant value dollars then each successive annual amount really represent less buying power—and uniformity only in nominal amounts.

2.11 SUMMARY

This chapter has covered the development and use of the basic and extended engineering economy factors. While the mathematics are straightforward, a clear understanding of these factors and of time value of money principles is the required foundation for this text. For theoretical research the formulas are likely to be used, for industrial practice spreadsheets are the tool of choice, and the Fundamentals of Engineering Exam will often be taken while relying on tabulated factors—even though programmable calculators can be used to function as a TVM calculator.

REFERENCES

ESCHENBACH, TED G. and NEAL A. LEWIS, "The Roles of Tabulated Factors, Financial Calculators, and Spreadsheets in Engineering Economy Teaching," *The Engineering Economist*, **56**(4) (October-December 2011), pp. 283–294.

FISHER, IRVING, *The Theory of Interest* (Macmillan, 1930; reprinted, Kelley & Millman, 1954).

GRANT, EUGENE L., *Principles of Engineering Economy,* revised ed. (Ronald, 1938).

HARTMAN, JOSEPH C., *Engineering Economy and the Decision-Making Process*, (Pearson Prentice-Hall, 2007).

HIRSHLEIFER, J., "On the Theory of Optimal Investment Decision," *Journal of Political Economy,* **66**(4) (August 1958), pp. 329–352.

HIRSHLEIFER, J., *Price Theory and Applications,* 4th ed. (Prentice-Hall, 1988).

NEWNAN, DONALD, TED ESCHENBACH, and JEROME LAVELLE, *Engineering Economic Analysis*, 12th ed., Oxford University Press (2014).

PARK, CHAN S. and GUNTER P. SHARP-BETTE, *Advanced Engineering Economics*, (Wiley, 1990).

"Report of the Engineering Economy Subcommittee (Z94.5) of the ANSI Z94 Standards Committee on Industrial Engineering Terminology," *The Engineering Economist,* **33**(2)(Winter 1988), pp. 145–151.

PROBLEMS

2-1. How much money would have to be invested today to provide a balance of $10,000 at the end of 10 years if the investment earns 5% simple interest per year?

2-2. For what period of time will $1,000 have to be invested to equal a future amount of $1500 if the investment earns 6% simple interest per year?

2-3. What is the simple rate of interest per annum if $1,000 invested today amounts to $1,250 at the end of 4 years?

2-4. If $10,000 is borrowed now for 10 years and if the interest is to be paid at the end of each year at the rate of 8% per year simple interest on the entire loan, what is the total amount of repayment during the 10-year period, including the principal repayment at the end of the tenth year?

2-5. If $10,000 is borrowed now for 10 years and if the interest is paid at the end of the 10th year at the rate of 8% per year compounded annually, what is the total amount of repayment at the end of the 10th year, including the principal repayment?

2-6. What is the total amount of repayment in problem 2-5 if the rate of interest is 8% per year compounded quarterly?

2-7. What is the effective interest rate per year for
(a) 8% per year compounded annually?
(b) 8% per year compounded semiannually?
(c) 8% per year compounded quarterly?
(d) 2% per quarter compounded quarterly?

2-8. What is the present value of these future payments?
(a) $5000 in 10 years at 15% per year compounded annually?
(b) $1000 in 10 years at 10% per year compounded semiannually?
(c) $2000 in 5 years at 8% per year compounded quarterly?

2-9. At 8% per year compounded annually, what is the annual end-of-the-year payment that will provide a future amount of $10,000 in 10 years.

2-10. How much money at 8% per year compounded annually can be loaned today on the agreement that $10,000 will be paid 6 years from now?

2-11. At 10% per year compounded annually, what single payment 10 years from now is equivalent to a payment of $2000 at the end of 3 years?

2-12. A lending institution offers to lend $300 now on a contract in which the borrower must pay back $10 at the end of each week for the next 35 weeks. What is the effective interest rate expressed in percent per year compounded annually?

2-13. What present amount is equivalent to $5000 in 10 years from now using an interest rate of 6% per year compounded semiannually?

2-14. Using an interest rate of 8% per year compounded annually, convert the following sequence of end-of-year amounts to
(a) Present value amount
(b) Annual equivalent amount
(c) Equivalent amount at the end of year 5
(d) Future sum at the end of the sequence

EOY	Amount	EOY	Amount	EOY	Amount
0	-	5	5,000	10	6,000
1	1,000	6	5,000	11	5,000
2	1,000	7	5,000	12	5,000
3	-	8	6,000	13	10,000
4	-	9	6,000		

2-15. Using an interest rate of 8% per year compounded annually, convert the following series of end of year amounts to
(a) Present value amount (P).
(b) Annual equivalent amount (A).
(c) Equivalent amount at the beginning of year 8.
(d) Future sum at the end of the sequence (F).

EOY	Amount	EOY	Amount	EOY	Amount
0	1,000	4	2,000	8	5,000
1	2,000	5	2,000	9	8,000
2	2,000	6	3,000	10	9,000
3	0	7	4,000	11	10,000

2-16. At an interest rate of 8% per year compounded annually, convert the following cash flows to
(a) Present value amount (P).
(b) Annual equivalent amount (A).
(c) Future sum at the end of the sequence (F).

EOY	Amount	EOY	Amount	EOY	Amount
0	−6000	4	2450	8	2850
1	2150	5	2550	9	2950
2	2250	6	2650	10	9050
3	2350	7	2750		

2-17. Suppose that 9 years ago you started putting $200 per month into a bank at the beginning of each month at 6% per year compounded monthly. You made 71 continuous payments and then stopped but left the accumulated deposit in the bank. One year from now you plan to buy a house, paying $1000 per month for 10 years with payments to begin at the end of the first month. For how many months can you make the housing payments using the money in the bank?

2-18. At an interest rate of 8% per year compounded annually, convert the following cash flows to
(a) Present value amount (P).
(b) Annual equivalent amount (A).
(c) Future sum at the end of the sequence (F).

EOY	Amount	EOY	Amount	EOY	Amount
0	-	4	1200	8	1800
1	1800	5	1800	9	1800
2	1800	6	1800		
3	1800	7	1800		

2-19. At an interest rate of 8% per year compounded annually, convert the following cash flows to
(a) Present value amount (P).
(b) Annual equivalent amount (A).
(c) Future sum at the end of the sequence (F).

EOY	Amount	EOY	Amount	EOY	Amount
0	-	4	1200	8	2000
1	1000	5	1400	9	2200
2	1000	6	1600	10	2400
3	1000	7	1800		

2-20. At an interest rate of 8% per year compounded annually, convert the following cash flows to
(a) Present value amount (P).
(b) Annual equivalent amount (A).
(c) Future sum at the end of the sequence (F).

EOY	Amount	EOY	Amount	EOY	Amount
0	-	4	1800	8	1500
1	1800	5	1500	9	1500
2	1800	6	1500	10	1500
3	1800	7	1500	11	1500

2-21. At an interest rate of 8% per year compounded annually, convert the following cash flows to
(a) Present value amount (P).
(b) Annual equivalent amount (A).
(c) Future sum at the end of the sequence (F).

EOY	Amount	EOY	Amount	EOY	Amount
0	-	4	3500	8	3500
1	3500	5	3500	9	4900
2	3500	6	4900	10	3500
3	4900	7	3500	11	3500
				12	4900

2-22. Using an interest rate of 8% per year compounded annually and two gradient series, convert the following series of end of year amounts to
(a) Present value amount (P).
(b) Annual equivalent amount (A).
(c) Future sum at the end of the sequence (F).

EOY	Amount	EOY	Amount	EOY	Amount
0	-	4	1400	8	800
1	2000	5	1200	9	700
2	1800	6	1000	10	600
3	1600	7	900		

2-23. A lease is available on a $300,000 piece of equipment. The lease payments are $75,000 per year for 5 years. If the equipment were purchased, it would have a salvage value of $100,000 after 5 years. It does not matter whether the equipment is leased or purchased. In either case operating costs will be paid by the user. Agreeing to the lease is financially equivalent to borrowing the money at what interest rate? *(Note:* Lease payments are made at the beginning of the year.)

2-24. A $100,000 serial maturity bond is to be paid back in 10 annual payments. The first payment is $20,000, the second is $19,000, the third is $18,000, etc. The final payment is $11,000. What is the interest rate on the bond?

2-25. The net revenue stream from a new product begins at $15,000 at the end of year 1. Revenues are expected to increase at 8% per year over the 10-year life of the product. If the interest rate is 11%, what is the present value of the 10 years of net revenue?

2-26. If the interest rate in problem 2-27 is 6%, then what is the present value?

2-27. Prove that formulas (2.22) and (2.23) are equivalent.

2-28. Which revenue stream has a larger present value at an interest rate of 6%? Stream A has a receipt of $20,000 in year 1 and a gradient of $1000 per year for 20 years. Stream B begins with a receipt of $17,500, and it has a gradient of 5% for 20 years. How much is the difference in the present values?

2-29. A government-sponsored student loan requires that only half the loan be repaid so long as the student returns "home" for five years to practice as a professional. The maximum loan amount available is $40,000 to be "repaid" in five payments of $4000 at the end of years 1 to 5. What is the interest rate on this loan?

2-30. On the same graph include the following:
(a) $(P/A, 0.12, N)$ for $N = 1$ to 40,
(b) $(P/A, 0.04, N)$ for $N = 1$ to 40, and
(c) $(P/A, 0.04, 0.12, N)$ for $N = 1$ to 40.

2-31. What is the present value at 8% of a project whose cash flows are defined as follows? The first cost is $1 million, the life is 15 years, the salvage value in 15 years is $200,000, annual net revenues are $300,000 per year, and the working capital for operations is $200,000. (Note: Working capital is required to cover the "cash float" in management operations. Working capital is returned intact when the project is completed.)

2-32. On the same graph include the following:
(a) $(A/P, 0.12, N)$ for $N = 1$ to 40
(b) $(A/P, 0.04, N)$ for $N = 1$ to 40
(c) $(A/G, 0.04, N)$ for $N = 1$ to 40

2-33. A credit card charges 1.5% interest per month. What is the nominal annual interest rate? If this is compounded monthly, what is the effective interest rate?

2-34. An auto loan is for 6.9%, compounded monthly. What is the effective interest rate?

2-35. A 5-year certificate of deposit has a nominal interest rate of 3.9%. If $1000 is deposited, what is the value at maturity if interest is compounded
(a) annually?
(b) semiannually?
(d) quarterly?
(d) monthly?
(e) continuously?

2-36. You go to a payday loan company to borrow enough money to get you through the next 2 weeks. They are willing to loan you $500. There will be a $25 up-front fee, and you will pay them $525 at the end of 2 weeks. What is the effective interest rate of the loan?

2-37. A new bank in town is offering continuous compounding on savings accounts. If their interest rate is 4.5%, what is the value 1 year from now of a deposit of $1000?

2-38. A new car costs $24,000. In addition to the cost of the car, taxes, license, and fees that total $2000 must be paid. $12,000 is paid in cash, but the rest must be financed at an annual interest rate of 6.9% for 4 years. What is the monthly payment on the loan?

2-39. $1000.00 is owed on a credit card. The credit card company charges 24% interest (annual basis) and wants a minimum payment of $28.77 each month.
(a) If the minimum payment of $28.77 is paid each month, how long will it be until the balanced is paid? No other purchases will be made.
(b) How much will the credit card company be paid, including principal and interest, when the loan is paid off?

2-40. If the nominal rate on savings is 3.5%, what is the effective interest rate when interest is compounded
(a) Annually?
(b) Semiannually?
(c) Quarterly?
(d) Monthly?
(e) Weekly?
(f) Continuously?

3

Estimating Costs and Benefits
Lead Coauthor Heather Nachtmann

3.1 INTRODUCTION

Estimating project cash flows is the first step in economic analysis of projects. This chapter discusses the importance of estimation accuracy and addresses the tradeoffs involved with generating cost and benefit estimates. The classification of estimates within a realm of the project life cycle is discussed along with the importance of data within the estimation process. Basic estimation methods including indexes and the unit and factor techniques are presented and demonstrated. A general overview of cost and benefit estimating relationships is presented along with two common development methodologies and an examination of special cases such as capacity functions and growth curves.

Because forecasts are uncertain, we suggest that the material in this chapter be considered in conjunction with the sensitivity analysis material in Chapter 4 of *Cases in Engineering Economy*, 2nd edition, which is included on the CD packaged with this text.

3.2 CASH FLOW ESTIMATES

Cash flow estimates are frequently based on the estimator's knowledge, experience, analysis, and subjective judgment. Unavailability and uncertainty of relevant information create difficulties in developing accurate cash flow estimates. Direct cost information, such as labor and material costs, tend to be relatively easy to calculate due to the availability and accuracy of associated information. Indirect costs and revenue streams, including sales volumes, administrative costs, salvage values, and new product revenues, are inherently more uncertain and therefore more difficult to estimate. The quality of the economic analysis of an industrial project can only be as good as the quality of its cash flow estimates.

A good estimate is realistic (represents what can actually happen), independent (unaffected by other estimates), and unbiased (void of preconceived notions). Difficulty in producing good cash flow estimates is increased in companies with diverse processes and products, competitive operating environments, and tight design and quality specifications.

The accuracy of a cash flow estimate is directly related to the estimation model and the availability of information, time, and resources. All organizations are limited by time and resource constraints that affect the accuracy of their cash flow estimations.

Westney (1997, p. 17) provides a classic Estimating Paradox that describes the tradeoffs between estimation accuracy and resource consumption:

> the more accurate the estimate, the more information required; the more information required, the more time is required to produce the estimate; the more time required to produce the estimate, the more resources are required to develop the estimate; the more resources required, the more money it will cost to produce the estimate; the more money spent, the more pressure to reduce resources, time, information, and accuracy.

Clearly, it is necessary to trade off the resource requirements of developing good estimates and the desired accuracy. The possibility of error exists in all estimates to some degree. Estimates that are developed with little or no historical information and are based on guesswork will have a high degree of uncertainty. Estimates are presumed to be more certain when they are based on accurate data and reliable expertise.

The relationship between estimation and pricing is important to understand. It is imperative that organizations have accurate information regarding their costs and benefits in order to competitively price their goods or services. However, it is pertinent that the estimate of a cost should never be influenced by its selling price. Selling price is often a function of cost, but the reverse should never occur. Allowing selling price to influence cost estimate values can lead to underestimated and inaccurate cost information generated in hopes of meeting the current sales constraints.

3.3 LIFE CYCLE ESTIMATION

Most engineering projects begin with the costs to create the product, build the structure, and so forth. There are even professional societies such as the Project Management Institute that focus on the initial phases of a project:
1) Concept: project concept is identified and defined,
2) Definition: project concept is verified and a plan for implementation is developed,
3) Execution: implementation plan is carried out, and
4) Closeout: project is completed and finished product is transferred to the owner.

There are three common estimating methods that are suitable for use during these initial project life cycle phases:
1) Analogy method—estimates inferred from the costs of similar previously conducted projects through analytical or judgmental means
2) Engineering build-up method—estimates are developed by summing individual project component costs from a detailed component listing, and the
3) Parametric method—estimates based on functional relationship between the project's costs and important parameters (characteristics or attributes).

Table 3.1 summarizes the suitability of these methods during the initial project life cycle phases.

A more complete life-cycle view of engineered products and infrastructure looks at the reason why the projects are undertaken—what benefits are being delivered. This includes sales, demand, and use over a much longer period. Bridges, highways, and dams tend to see ever-increasing levels of use, but new products often involve the S-curve shown in Figure 3.1. Sales tend to begin slowly, grow as more people adopt the product, reach a maximum as the market becomes saturated, and eventually decline. Many products have remained on the market for an extended period of time by introducing improvements, repeating the cycle many times. Good economic analysis relies on correctly estimating the kind and length of the life cycle and the current stage of that life cycle.

TABLE 3.1 Initial project life cycle estimation methods

Estimation Method	Initial Project Life Cycle Phase			
	Concept	Definition	Execution	Closeout
Analogy	Good	Good	Fair	Fair
Engineering buildup	Poor	Fair	Good	Good
Parametric	Good	Good	Fair	Fair

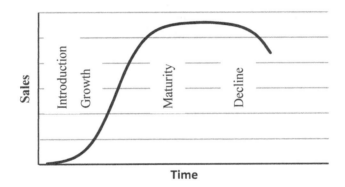

Figure 3.1 Generic product life cycle

3.4 CLASSIFICATION OF ESTIMATES

AACE International (formerly the Association for the Advancement of Cost Engineering) prescribes the Cost Estimation Classification System as a recommended set of guidelines for applying estimate classification principles to project *cost* estimates (AACE International Recommended Practice No. 18R-97, 2013). While Table 3.2 summarizes definitions, methods, and results for *cost* estimating in the process industries, it also provides a foundation for other applications. It provides rough guidance for the relationships between the maturity level of the project definition, the methodologies used, the achievable accuracy, and the costs incurred.

The focus of Table 3.2 is for the construction of certain types of facilities where there is a long history of constructing somewhat similar facilities. Not surprisingly, the achievable level of accuracy is often much lower for R&D projects, new product or software development, and very much lower for estimating sales or benefits that are further into the future and subject to more uncertainties.

Figure 3.2 represents a generalized view of another important aspect of economic analysis and decision-making for projects. Decisions that are made early in a life cycle may have little direct cost, but they often largely determine what costs will be incurred. For example, the analysis to choose a project size or a product feature may take days or weeks to make and months to implement. While not emphasized in Figure 3.2, the revenues and benefits that will be received for years are often also largely determined by early decisions. The most important implication is that it is often *NOT* cost effective to focus on reducing early costs, as that far too often causes larger costs later.

TABLE 3.2 AACE Cost estimate classification matrix for the process industries (AACE International Recommended Practice No. 18R-97, November 2011)

	Primary Characteristic	*Secondary Characteristic*		
Estimate Class & End Usage	**Level of Project Definition** as % of complete	**Methodology**	**Expected Accuracy Range** [a]	**Preparation Effort**[b] relative to least cost index of 1
Class 5 concept screening	0% to 2%	capacity factored, parametric models, judgment, or analogy	−20% to −50%; +30% to +100%	1
Class 4 study or feasibility	1% to 15%	equipment factored or parametric models	−15% to −30%; +20% to +50%	2 to 4
Class 3 budget authorization or control	10% to 40%	semidetailed unit costs with assembly level line items	−10% to −20%; +10% to +30%	3 to 10
Class 2 control or bid/tender	30% to 75%	detailed unit cost with forced detailed takeoff	−5% to −15%; +5% to +20%	4 to 20
Class 1 check estimate or bid/tender	65% to 100%	detailed unit cost with detailed take-off	−3% to −10%; +3% to +15%	5 to 100

Notes: [a] The +/− value represents typical percentage variation of actual costs from the cost estimate after application of contingency (typically at a 50% level of confidence) for given scope.
[b] This column is from an earlier version of this standard. If the range index value of "1" represents 0.005% of project costs, then an index value of 100 represents 0.5%.

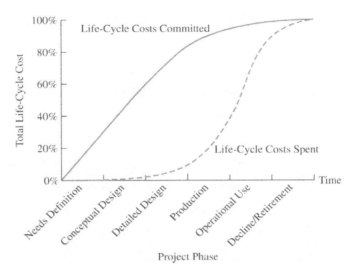

Figure 3.2 Cumulative life-cycle costs committed and dollars spent
(Newnan, Lavelle, Eschenbach, 2014)

3.5 ESTIMATION DATA

The data used to develop cash flow estimates must be accurate, robust, and collectable. Accurate data is representative of the organization's history and absent of abnormalities. As estimate development requires prediction of the future, it is important that the data is robust enough to reasonably represent all possible events and outcomes. Collection of all data must be feasible and conducive to collection within a reasonable expenditure of time and resources. It is imperative to remember that an estimate is only as good as the data from which it is generated. The "garbage in, garbage out" principle strongly applies here, and its recognition prevents inaccurate estimate development from poor quality data.

Data collection and management practices range from individual to large-scale systematic. Individual estimation is an informal data collection and management approach employed when estimate development is infrequent or low in volume. Individual estimation involves specific data collection for a particular estimate. Data sources are contacted or observed each time an estimate is made. Systematic data collection and management is employed to support more frequent estimating needs. Typically, standard cost data is developed based on historical and current information and housed in a formal data management system. The setup and operating costs and time required for the data management system must be balanced by the resulting cost and time savings from systematic cost estimation.

There are many internal and external data sources available to the estimator. The estimator must identify which sources can provide the required data in a reasonable amount of time and effort. Select internal sources and the type of data that they typically provide are listed in Table 3.3. Table 3.4 provides similar information for a selection of external sources.

TABLE 3.3 Internal data sources

Internal Sources	Available Data
Accounting	Historical product costs
Engineering	Design specifications, bills of materials
Purchasing	Material costs, suppliers
Human resources	Wages, fringe rates
Marketing	Competitor information, market demand
Production	Manufacturing routing, work standard time data
Quality control	Inspection procedures
Shipping	Transportation requirements

TABLE 3.4 External data sources

External Sources	Available Data
Bureau of Labor Statistics	Earning statistics, Consumer Price Index, Wholesale Price Index
Office of Business Economics	Gross National Product
Public firms	Income published in annual report
Banks	Business cycles, cost of money, local economics
Universities	Published research and case studies
Trade associations	Dissemination of trade specific information

3.6 BASIC ESTIMATION TECHNIQUES—INDEXES AND PER UNIT

Basic estimation techniques can be applied in situations where minimal information is available and the time required to develop a more detailed estimate is not justified. Limited historical data is adjusted to represent current or future cash flow estimates. These preliminary estimates are often referred to as order of magnitude or conceptual estimates.

3.6.1. Indexes. Indexes are used to develop current and future cash flows estimates from historical data by accounting for variation in aspects such as technological advancement, geographical location, and inflation over time. An index is a dimensionless number that shows how a cash flow of interest has changed over time with respect to a base year. Rough estimates can be developed by updating historical cash flows from a specified base year by applying the appropriate indexes to determine their current or future values. The index estimation technique requires application of Equation (3.1).

3.6 Basic Estimation Techniques

$$C_x = C_k \left(\frac{I_x}{I_k}\right) \tag{3.1}$$

where C_x = estimated cash flow in year k
C_k = actual cash flow in base year x
I_x = index value of year x for which the cash flow is to be estimated ($x > k$)
I_k = index value of base year k for which the cash flow is known

Example 3.1 Index
During a lunch meeting, a group of cardiac doctors was discussing the lack of a state-of-the-art care facility for heart disorders. One of the doctors suggested that it would be helpful to have a hospital specializing in such disorders. Intrigued by the idea, the doctors decided to investigate the cost of developing such a hospital. Since this was just an idea, they did not want to invest a large amount of time or money in the development of a cost estimate. The doctors found a facility in Little Rock, Arkansas, that was similar to the one they envisioned. This facility had been built eight years earlier. The facility cost approximately $4.5 million dollars to build. Since the doctors only need a rough estimate to decide if they should investigate their idea further, they decided to use a cost index. Using the known cash flow of the Little Rock facility's building cost and the provided indexes, calculate the estimated cost of constructing the new heart disorders hospital in year 20. The hospital in Little Rock was built in year 12.

Year	Index	Year	Index
1	50.0	11	107.9
2	54.0	12	116.6
3	58.3	13	125.9
4	63.0	14	136.0
5	68.0	15	146.9
6	73.5	16	158.6
7	79.3	17	171.3
8	85.7	18	185.0
9	92.5	19	199.8
10	100.0	20	215.8

Solution
Using Equation (3.1), the known cash flow at time $k = 12$, the values from the table, and the estimated cash flow for time $x = 20$ is:

$$C_{20} = C_{12}\left(\frac{I_{20}}{I_{12}}\right) = \$4,500,000 \times \left(\frac{215.8}{116.6}\right) = \$8,328,473$$

3.6.2. Unit Technique. The unit technique develops estimates simply by multiplying the value of a particular unit of interest by a per-unit rate. These rates can be developed from historical data and used to predict future costs and revenues. Some examples include construction cost per square foot, revenue per truckload, fuel cost per mile, and inventory cost per product. The unit technique equation is given in Equation (3.2).

$$C = Q \times R \qquad (3.2)$$

where C = estimated cash flow
Q = number of units
R = rate per unit

Example 3.2 Unit technique
The doctors in Example 3.1 are also investigating the possibility of purchasing a preexisting building instead of constructing a new hospital. An average price per square foot for buildings similar to the one required by the doctors is $125. The doctors have found five buildings that fit their design requirements and are roughly the same size: (1) 52,000 ft^2, (2) 49,500 ft^2, (3) 51,400 ft^2, (4) 49,800 ft^2, and (5) 52,500 ft^2. Given the square footage of each building, determine what the doctors should expect to pay for each building.

Solution
Using Equation (3.2), the square feet per building and the price per square footage unit, the estimated cost per building is:

Building 1: C = 52,000 ft^2 x $125.00/ft^2 = $6,500,000
Building 2: C = 49,500 ft^2 x $125.00/ft^2 = $6,187,500
Building 3: C = 51,400 ft^2 x $125.00/ft^2 = $6,425,000
Building 4: C = 49,800 ft^2 x $125.00/ft^2 = $6,225,000
Building 5: C = 52,500 ft^2 x $125.00/ft^2 = $6,562,500

3.7 FACTOR TECHNIQUE

The factor technique is a simple extension of the unit technique that allows direct estimates and multiple per-unit estimates to be combined into a single estimate. While the unit technique estimates the entire cash flow using a single per-unit rate, the factor technique allows the cash flow to be broken down into smaller components. Each component can be separately estimated as shown in Equation (3.3).

$$C = \sum_{i=1}^{n} c_i + \sum_{j=1}^{m} q_j r_j \qquad (3.3)$$

where C = total cost flow
c_i = cash flow of component i that is directly estimated
q_j = number of units of component j
r_j = per unit rate of component j
n = number of components directly estimated
m = number of components estimated on a per unit basis

Example 3.3 Factor technique

The doctors have decided to build a new facility instead of purchasing an existing one. Before beginning construction, the doctors would like to estimate the major costs involved in the setup and supply of the hospital. There are several major components that can be estimated including the cost of the building itself and the cost of the administrative setup. These two costs are considered direct estimates. Two other components, hospital room costs and operating room costs, are dependent upon the actual number of rooms chosen. Given the various cost estimates for the major hospital components, use the factor technique to provide an estimate of total costs.

Cost Component	Cost per Unit	Units Required
Building	$8,328,473	n/a
Administrative Setup	$540,000	n/a
Hospital Room	$25,000	75
Operating Room	$385,000	4

Using Equation (3.3), calculate the total directly estimated cost. Then calculate the estimated total per unit cost and compute the total estimated cost.

Solution

Directly Estimated Costs
$$C_{Direct} = \$8,328,473 + \$540,000 = \$8,868,473$$

Per Unit Estimated Costs
$$C_{PerUnit} = \$25,000 \times 75 + \$385,000 \times 4 = \$3,415,000$$

Total Estimated Costs
$$C_{Total} = \$8,868,473 + \$3,415,000 = \$12,283,473$$

3.8 COST ESTIMATION RELATIONSHIPS

A cost estimation relationship (CER) is a mathematical equation that relates cost to one or more physical or performance variables associated with the item being estimated. Here, the single dependent variable is cost, while there may be multiple independent physical or performance variables. CERs are based on historical information and typically derived from statistical regression analysis. These statistically based CERs were first studied and documented by the RAND Corporation (Asher, 1956; Noah, 1962; Teng, 1966). This early work focused on predicting the cost of military equipment at early design phases. CER development has relatively low resource requirements when compared to detailed estimation, and was soon applied to budget planning and high-level systems analysis (Sato, 2012; Lai and Barkan, 2009; Jorgensen, 2007; Williams, 2003; Harbuck, 2002).

The use of CERs in practice has become commonplace, driven predominately by the prevalence of personal computers in the workplace. CERs are most frequently used in the early phases of project life when there is insufficient specific information to develop a detailed estimate. CERs are also employed when the resources required to develop a detailed estimate cannot be economically justified or when a fast, approximate estimate is required. It is critical that good statistical relationships exist between the cost and the predictive variables, also known as cost drivers, and that the proper model that best describes this relationship is used. The common CER equation forms are linear, power curves, exponential curves, and logarithmic curves as shown in Figure 3.3. The linear CER, as expected, assumes a linear relationship exists between the dependent and independent variable. Variable costs typically are not driven in a linear manner as opposed to the economies of scale phenomenon where the variable cost per unit decreases as the unit quantity increases. The power CER can take economies of scale into account through the assumption that a relationship between the dependent variable and independent variable and cost exists such that a percent change in the independent variable causes a constant percent change in dependent variable. Another real world cost phenomenon is represented by the exponential CER, which assumes that a unit change in the independent variable causes a relatively constant percent change in dependent variable. The logarithmic CER is simply the inverse of the exponential CER. The equations for the singular variable forms of these CERs are given in Table 3.5, where Y = predicted estimate, X = independent variable, and a and b are constants derived from least squares regression. The power and exponential CERs can be linearly transformed as shown in Table 3.5 to allow the application of least squares regression.

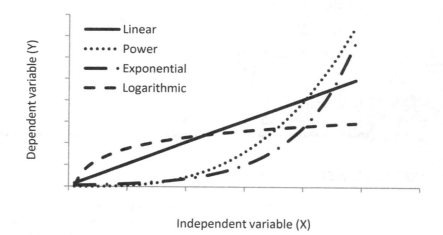

Figure 3.3 Common CER forms

TABLE 3.5 Common CER equations

CER	Equation	Linear Form
Linear	$Y = a + bX$	$Y = a + bX$
Power	$Y = aX^b$	$\ln Y = \ln a + b \ln X$
Exponential	$Y = ae^{bX}$	$\ln Y = \ln a + bX$
Logarithmic	$Y = a + b \ln X$	$Y = a + b \ln X$

3.8.1 Development Process

Least squares regression. The most common method for developing a CER is the method of least squares regression. A least squares regression fits a curve (a line in the linear case) between the data, where the sum of squared deviations of all data points from the curve is minimized. Specifically, the method seeks to fit this curve so that the vertical distances between the curve and the points above and below are equal. The method of least squares requires paired historical data that relate the dependent variable (cost) to one or more physical or performance independent variables. When the dependent variable is predicted by a single independent variable case, this is referred to as simple regression; the multiple independent variable case is referred to as multiple regression. The determination of which CER best fits the data is performed by measuring the goodness of fit of the various forms and selecting the form with the best fit. The parameters for a CER developed from simple linear regression can be estimated using Equations (3.4) and (3.5).

$$y = a + bx$$

where a = y intercept
b = slope

$$b = \frac{\sum_{i=1}^{n} x_i y_i - \bar{x} \sum_{i=1}^{n} y_i}{\sum_{i=1}^{n} x_i^2 - \bar{x} \sum_{i=1}^{n} x_i} \quad (3.4)$$

$$a = \bar{y} - b\bar{x} \quad (3.5)$$

Two common measures of goodness of fit are the correlation coefficient (r) and coefficient of determination (r^2). The correlation coefficient measures the closeness of the fit of the regression line to the data points and results in values between -1 and 1. An $r = 0$ indicates that no relationship exists between the variables, while an $r = 1$ or -1 indicates a perfectly positive or negative relationship respectively. A positive r designates that both variables increase at the same time, while a negative r denotes a situation where one variable decreases as the other increases. The coefficient of determination measures the portion of total variation that is explained by the regression line, and represents how well the CER is expected to predict items with a similar relationship. The coefficient of determination falls between 0 and 1. A value of $r^2 = 0$ indicates that none of the variation in the dependent variable (cost) is represented by the CER, while $r^2 = 1$ indicates that all of this variation is represented. Clearly it is desirable to have r^2 values as high as possible. The correlation coefficient can be computed using Equations (3.6) through (3.9).

$$r = \frac{S_{xy}}{\sqrt{S_{xx} \times S_{yy}}} \quad (3.6)$$

Chapter 3, Estimating Costs and Benefits

$$S_{xy} = \sum_{i=1}^{n} x_i y_i - \frac{\left(\sum_{i=1}^{n} x_i\right)\left(\sum_{i=1}^{n} y_i\right)}{n} \quad (3.7)$$

$$S_{xx} = \sum_{i=1}^{n} x_i^2 - \frac{\left(\sum_{i=1}^{n} x_i\right)^2}{n} \quad (3.8)$$

$$S_{yy} = \sum_{i=1}^{n} y_i^2 - \frac{\left(\sum_{i=1}^{n} y_i\right)^2}{n} \quad (3.9)$$

Multiple regression has been presented in the context of *cost* estimating relationships, but it is also a very common tool for estimating future revenues and benefits. This is often done where time is *the* variable or one of the variables. This can be done by either estimating a rate of growth or by projecting historical demand data into the future with a linear or nonlinear model. Example 3.4 illustrates this.

Example 3.4 Regression

The manager of a firm is analyzing the link between the number of major customers for a division and total revenue. After plotting the data, the manager can see that there appears to be a strong linear trend. Therefore, he wishes to create a linear CER to estimate the relationship between the two variables.

Team Data

Period	1	2	3	4	5	6	7	8	9	10
Major customers	1	2	4	6	7	9	12	15	17	20
Total revenue	4	14	16	18	21	21	26	34	63	67

Period	11	12	13	14	15	16	17	18	19	20
Major customers	27	32	34	36	37	41	45	47	48	50
Total revenue	72	81	82	118	110	111	127	133	139	150

Solution

The linear CER is developed using these values and Equations (3.4) and (3.5).

$$b = \frac{50104 - 24.5 \times 1407}{17538 - 24.5 \times 490} = 2.825$$

$$a = 70.35 - 2.825 \times 24.5 = 1.1375$$

$$Y = 1.1375 + 2.825X$$

3.8 Cost Estimation Relationships

Period	X	Y	X×Y	X^2	Y^2
1	1	4	4	1	16
2	2	14	28	4	196
3	4	16	64	16	256
4	6	18	108	36	324
5	7	21	147	49	441
6	9	21	189	81	441
7	12	26	312	144	676
8	15	34	510	225	1156
9	17	63	1071	289	3969
10	20	67	1340	400	4489
11	27	72	1944	729	5184
12	32	81	2592	1024	6561
13	34	82	2788	1156	6724
14	36	118	4248	1296	13924
15	37	110	4070	1369	12100
16	41	111	4551	1681	12321
17	45	127	5715	2025	16129
18	47	133	6251	2209	17689
19	48	139	6672	2304	19321
20	50	150	7500	2500	22500
Total	490	1407	50104	17538	144417
Average	24.5	70.35			

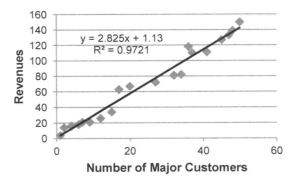

Problem 3.4 Regression line

To determine how good the linear fit of the CER is, the manager computes its coefficient of determination using Equations (3.6)–(3.9).

$$S_{xy} = 50104 - \frac{(490)(1407)}{20} = 15632.5$$

$$S_{xx} = 17538 - \frac{(490)^2}{20} = 5533$$

$$S_{yy} = 144417 - \frac{(1407)^2}{20} = 45434.6$$

$$r = \frac{15632.5}{\sqrt{5533 \times 45434.6}} = 0.986$$

$$r^2 = 0.972$$

An r^2 value of 0.972 indicates that the model accounts for almost all of the variation found in the regression line and should provide valid estimates.

Theoretically, CERs can be developed from many independent variables; however in practice, CERs are generally built with few independent variables. This is likely because the relationships between fewer variables are easier to explain and understand. Also, multiple variables require increased data collection. As previously mentioned, multiple regression can be used to develop a CER with multiple independent variables (that is, $Y = a + b_1X_1 + b_2X_2 + b_3X_3$ ... and $Y = aX_1^{b1}X_2^{b2}X_3^{b3}$ ... for the linear and power CER cases). It is important to identify the limits of the independent variable to ensure that prediction of the dependent variable is not occurring outside these limits. The developed CER is only strictly valid within bounds of the data used to develop the CER, and extrapolation beyond these bounds is not generally recommended. Confidence intervals about the regression line can be developed to provide the reliability of the estimate of the dependent variable from the fitted line, which can provide additional insight under conditions of high risk or uncertainty.

Neural networks. Typically, least squares regression has been used to take a parametric approach to developing CERs. Smith and Mason (1997) provide a detailed overview and survey of the relevant literature in a more recently developed approach of neural networks. These authors apply neural networks to a previously published real world problem, and compare their results to the results of comparable regression models. They conclude that neural networks may be a possible substitute for regression when the model commitment step (functional form selection, interaction selection, and data transformation) of the regression modeling is unsuccessful.

Neural networks are algorithms designed to emulate how biological neurons learn. These purely data driven algorithms are used to develop CERs when the explicit form of the variable relationship (i.e. linear, power, etc.) is unknown. A neural network uses a set of processing elements (or nodes) loosely analogous to neurons in the brain. These nodes are interconnected in a network. The neural network identifies a pattern of connections between the nodes, and uses a training algorithm to determine weights on these connections. Through iterative training, the algorithm transitions from a random state to a final model or CER.

The primary limitations of the neural network approach are that it requires large amounts of historical data and there is a greater learning curve associated with setting the parameters of the algorithm. The benefits of this approach include the abilities to capture data nonlinearities, discontinuities, and interactions and to accept a very large number of cost drivers. One concern of developing CERs with neural networks is that the approach may create credibility concerns with management and customer, as it is less well-known and somewhat of a black box approach.

General overviews of neural networks in cost estimation are provided by Setyawati et al. (2002) and Bode (1998). Neural networks have been successfully implemented in the development of CERs across many industries including engineering project costing (Odeyinka et al., 2013; Liu, 2010; Setyawati et al., 2003; Al-Tabtabai and Alex, 2000), manufacturing production estimation (Wang, 2007), and air travel and cargo demand forecasting (Chen et al., 2012). Comparisons of neural networks to regression models for cost estimation include deBarcelos Tronto (2007) and Smith and Mason (1997).

3.8.2 Capacity Functions. A widely used CER model is the capacity function or power-sizing model. The capacity function uses a capacity parameter [also known as Lang factors (Lang, 1948)] to estimate a facility's cost. As shown in Equation (3.10), the model assumes that facility cost varies as some power of the change in capacity. Table 3.6 is an abbreviated list of these exponents specified by facility type, which are applied in Example 3.5.

$$C_x = C_k \left(\frac{S_x}{S_k} \right)^n \qquad (3.10)$$

where C_x = estimated facility cost of capacity S_x n = capacity exponent (typically < 1)
C_k = known facility cost of capacity S_k

TABLE 3.6 Capacity exponents [From Table 2.1, Humphreys and Wellman, (1996)]

Facility	Exponent (n)
LP-gas recovery in refineries	0.70
Polymerization, small plants	0.73
Polymerization, large plants	0.91
Steam generation, large, 200 psi	0.61
Steam generation, large, 1000 psi	0.81
Power generation, 2,000–20,000 kW	0.88
Power generation, oil field, 20–200 kW	0.50
Sulfur from H_2S	0.64
Oxygen plants	0.65
Styrene plant	0.65
Ammonia, and nitric acid or urea	0.98
Chlorine plants, electrolytic	0.75
Refineries, small	0.57
Refineries, large	0.67
Hydrogen sulfide removal	0.55

Example 3.5 Lang factors
A contractor is submitting a proposal for construction of a new refinery. The refinery is a small one in a remote location, so it will also need a small amount of independent power generation. Past data shows that a similar refinery with a capacity of 1.5 million barrels per year cost $15 million dollars four years ago. A 35 kW generator cost approximately $125,000 dollars two years ago. Cost indexes exist for the refinery. Four years ago, the cost index for the refinery was 123, while the current index is 156. Likewise, indexes exist for the generator. Two years ago, the cost index for the generator was 131, while it is currently 152. The contracted refinery must have a capacity of 2.25 million barrels per year and will require 50 kWs of power. How much should the contractor allocate to cover the project costs?

Solution
To estimate the cost of the new refinery, the contractor determines the cost of the similar facility built four years ago. The resulting cost is then converted to present dollars using the provided cost index.

$$C_x = \$15,000,000 \times \left(\frac{2.25M}{1.5M}\right)^{0.57} = \$18,900,064$$

$$I_x = \$18,900,064 \left(\frac{156}{123}\right) = \$23,970,813$$

To estimate the cost of the generator, the cost of a similar generator two years ago is computed and converted to present dollars using the provided index.

$$C_x = \$125,000 \times \left(\frac{50kW}{35kW}\right)^{0.50} = \$149,404$$

$$I_x = \$149,404 \left(\frac{152}{131}\right) = \$173,354$$

The total cost (C) for the project is therefore estimated to be:

$$C = \$23,970,813 + \$173,354 = \$24,144,167$$

3.8.3 Learning Curves. The effect of learning on labor cost was first studied by Wright (1936) within the aerospace industry, and was formally combined with cost estimation in the mid-fifties (Asher, 1956). The basic premise behind learning (or product improvement) curves is that the time required to complete a task decreases as the task is repeated. As learning occurs, the labor requirements necessary to complete a unit of production will decrease by a constant percentage each time the production quantity is doubled. This constant percentage is referred to as the rate of learning, and is used as a descriptive characteristic of each learning curve. The extent and rate of time reduction is influenced by many factors including the production process, the stability of the product design, the time duration between production runs, and employee attitudes.

Learning curves are defined by their slope, which is defined as 100% minus the rate of learning. For example, a learning curve with a slope of 85% describes a production situation where the labor requirements necessary to complete a unit of production will decrease by 15% each time the production quantity is doubled. Two common forms of learning curves exist, unit curve and cumulative average curve. With identical underlying equations, the definitive difference between these two curves is how the dependent variable is defined. The dependent variables for the unit and cumulative average curves are clearly defined in Equations (3.11) and (3.12) respectively.

Unit Curve

$$Y_x = KX^\theta \tag{3.11}$$

where Y_x = cost required to produce the Xth unit
K = theoretical cost of the first production unit
X = sequential number of the unit for which the cost is to be computed
θ = exponent relating to the rate of learning

Cumulative Average Curve

$$Y_x = KX^\theta \tag{3.12}$$

where Y_x = average cost of the first X units
K = theoretical cost of the first production unit
X = sequential number of the last unit for which average cost is to be computed
θ = exponent relating to the rate of learning

Both learning curves are exponential functions, and thus can be transformed to the familiar linear equation form by taking the natural logarithm of both sides of the equation. This yields Equation 3.13 which allows the rate of learning (θ) to be estimated from historical production data through least squares regression. Note that learning curve values for the unit curve will indicate faster learning than for the cumulative average curve, since for the same process the average will decrease much more slowly—especially for larger volumes.

$$\ln Y = K + \theta \ln X \tag{3.13}$$

Example 3.6 Cumulative learning curve
An engineering firm has an estimating department to support its analysis of potential projects. As new estimators come on board, the personnel must learn and familiarize themselves with the software packages and data bases. The firm has carefully tracked the performance and output of an average new-hire from their first day to ten weeks of full-time employment. It should be noted that the estimator has no other responsibilities outside of estimating. Using the data collected, construct the cumulative average learning curve for the new-hire estimator.

Week	Cumulative Hours	Cumulative Units
1	40	0
2	75	0
3	110	1
4	145	1
5	195	2
6	260	3
7	302	4
8	360	6
9	399	8
10	440	11

Solution

Using the data collected by the firm, identify each time that the cumulative production increases. Note that while this data could also support a unit cost learning curve for the first 4 units, beginning in week 8 only cumulative or average cost data is available.

Week	Cumulative Hours	Cumulative Units
1	40	0
2	75	0
3	110	1
4	145	1
5	195	2
6	260	3
7	302	4
8	360	6
9	399	8
10	440	11

Then take each week with an increase in cumulative units and divide the cumulative production hours by the cumulative number of units produced. The cumulative number of units produced becomes the X value, while the cumulative hours divided by the cumulative units produced becomes the Y value. The natural logarithm of each X and Y value are then determined.

Week	X	Cumulative Hours	Y	ln X	ln Y
3	1	110	110.00	0.0000	4.7005
5	2	195	97.50	0.6931	4.5799
6	3	260	86.67	1.0986	4.4621
7	4	302	75.50	1.3863	4.3241
8	6	360	60.00	1.7918	4.0943
9	8	399	49.88	2.0794	3.9095
10	11	440	40.00	2.3979	3.6889

Find the linear regression of the curve using the form using Equations 3.4 and 3.5 where $\ln Y$ is the predicted variable and $\ln X$ is the independent variable:

$$\ln Y = 4.8312 - 0.4297 \ln X$$

Taking the value of $\theta = -0.4297$ and substituting it into the following equation produces the rate of learning for this curve:

$$P = 2^n = 2^{-0.4297} = 74.24\%.$$

Example 3.6 demonstrates the development of a cumulative average curve from historical production and cost data. The cumulative average curve was selected over the unit curve for this example because analysts are more likely to obtain cumulative total hours, as this information is easier to collect. However, as long as historical data is available, this same process can be applied to develop a unit curve.

In the absence of data, the rate of learning can be estimated through educated guess or approximation. The rate of learning exponents for common learning curve percentages are provided in Table 3.7.

TABLE 3.7 Common Learning Curve Exponents

Curve Percentages	θ
65	−0.624
70	−0.515
75	−0.415
80	−0.322
85	−0.234
90	−0.152
95	−0.074

Companies that use learning curves typically maintain records of historical learning rate percentages and often use these historical rates to predict learning rates for new products. Learning curves commonly range from around 70% to 100%. A 100% learning curve indicates no learning occurs and each additional unit takes as much time as the first unit to produce. A 50% cumulative learning curve would indicate that no additional time or cost would be required for additional units beyond the first unit given that the cumulative average time or the unit time would decrease by 50% each time output doubled.

Unit Curves. This section contains some additional equations that are useful when estimating costs from a unit learning curve. Equation (3.14) provides the total cost required to build a group of consecutively manufactured units (T_g).

$$T_g \cong \frac{K}{\theta+1}\left[(X_l+0.5)^{\theta+1} - (X_f-0.5)^{\theta+1}\right] \quad (3.14)$$

where K = theoretical cost of the first production unit
X_f = sequential number of the first production unit of the group

X_1 = sequential number of the last production unit of the group
θ = exponent relating to the rate of learning

Equation (3.14) reduces to Equation (3.15) when the consecutive group begins with the first unit, ($X_f = 1$). Therefore, Equation 3.15 can be used to calculate the total cost (T) required to build a cumulative total of X_1 units.

$$T \cong \frac{K}{\theta + 1}\left[(X_l + 0.5)^{\theta+1} - (0.5)^{\theta+1}\right] \tag{3.15}$$

where K = theoretical cost of the first production unit
X_1 = sequential number of the last production unit of the group
θ = exponent relating to the rate of learning

Equations (3.14) and (3.15) are generally accurate within a 5% percent range when the production quantity of the group is greater than 10 (Malstrom, 1981). The cumulative average cost (A) to manufacture a total of X_l units can be found using Equation (3.16).

$$A = \frac{T}{X_l} \tag{3.16}$$

Example 3.7 Unit curves
A new product is estimated to have a cost for the first unit of $80,000 and a unit learning curve rate of 65%. Given this learning curve, what is the average cost for the first 15 units?

Solution
Using Equation (3.15) and Table 3.7, substitute in the numerical values provided.

$$T = \frac{(\$80,000)}{-0.624+1}\left[(15+0.5)^{-0.624+1} - (0.5)^{-0.624+1}\right] = \$432,352$$

$$A = \frac{\$423,252}{15} = \$28,823$$

Cumulative average curves. This section contains formulae useful when working with cumulative average curves. The total cost (T) to manufacture a cumulative total of X_1 units can be computed from Equation (3.17).

$$T = KX_l^{\theta+1} \tag{3.17}$$

where K = theoretical cost of the first production unit
X_1 = sequential number of the last production unit of the group
θ = exponent relating to the rate of learning

The total cost (T_g) to manufacture a group of consecutively manufactured units is calculated from Equation (3.20).

$$T_g = KX_l^{\theta+1} - K(X_f - 1)^{\theta+1} \tag{3.18}$$

where K = theoretical cost of the first production unit
X_f = sequential number of the first production unit of the group
X_l = sequential number of the last production unit of the group
θ = exponent relating to the rate of learning

The unit cost (U) required to manufacture the Xth unit can also be computed from Equation (3.19) when $X \geq 10$.

$$U = (1+\theta)KX^\theta \tag{3.19}$$

3.9 GROWTH CURVES

Many of the topics covered in this chapter thus far are typically used to estimate costs. Analysis of industrial projects also heavily depends on revenue estimation. Growth curves, which were originally used to model behavior of biological growth, are often used to model sales and revenue over time. Growth curves, also referred to as S-curves after their S-like shape, assume that sales are initially low with a rapid increase after market establishment and then eventually taper off due to market saturation and technical advancement. Two common forms of growth curves are the Pearl and Gompertz curves.

The Pearl curve is attributed to its namesake, Raymond Pearl (1925), whose original work as a biologist and demographer led to the popularity of the logistic growth curve. However, the original work can be traced back to the French scientist, Pierre Verhulst (1845; 1847). While similar in general shape, as shown in Figures 3.4 and 3.5, the Pearl curve is symmetric about its inflection point, whereas the Gompertz curve does not have this property (Gompertz, 1825).

The equations for the Pearl and Gompertz curves are given in Equations (3.20) and (3.21) respectively. The parameters of these curves determine where the curve will be located on the time axis and the steepness of the rising portion, as shown in Figures 3.4 and 3.5. Note how the base case Pearl curve is similar to the generic product life cycle curve in Figure 3.1.

$$S = \frac{L}{1 + ae^{-bt}} \tag{3.20}$$

where
S = revenue estimated by the Pearl curve
L = upper limit of the growth of S
t = time
a, b = parameters determined from historical data

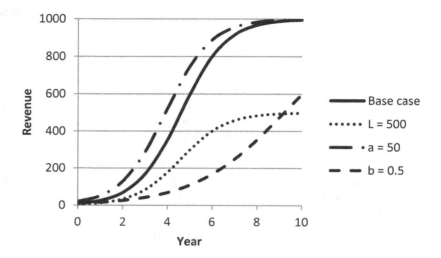

Figure 3.4 Pearl curve example (Base case: $L = 1000$, $a = 100$, $b = 1$)

$$S = Le^{-ce^{-dt}} \tag{3.21}$$

where S = revenue estimated by the Gompertz curve
  L = upper limit of the growth of S
  t = time
  c, d = parameters determined from historical data

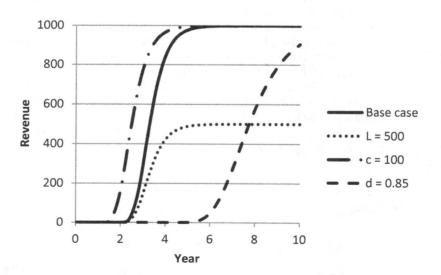

Figure 3.5 Gompertz curve example (Base case: $L = 1000$, $c = 500$, $d = 2$)

There are other techniques that are widely used, including moving averages, exponential smoothing, and double exponential smoothing (Holt's method). These are typically used for single point forecasts, and thus are not relevant for economic justification of long-term projects. Seasonal effects are used for one-year models, and not typically for longer term projects. These topics are well covered in other texts, and are not included here.

3.10 ESTIMATING PRODUCT COSTS

Globalization and deregulation of markets, rapid technological changes, and advances in information technology have made tremendous changes in the operations of companies throughout the world. Companies in today's industrial environment find themselves with higher proportions of overhead costs, shorter product life cycles, more complex product lines, and increased sales and distribution expenses. These developments create the need and capability to have a better awareness and understanding of the product costs being generated by companies.

Product costs are classified as direct or indirect costs. Direct costs are costs incurred during the manufacturing process of a product or the delivery process of a service that are tracked and readily assigned to a cost driver. A cost driver is any output measure that causes cost. The most common direct costs are direct labor and direct material. When a cost is difficult to track or to identify a cost driver for, this cost is classified as an indirect (or overhead) cost. Typical indirect costs include management salaries, utilities, software packages, insurance, and depreciation. Due to the ease of tracking and cost driver identification, direct costs are less challenging to estimate than indirect costs.

3.10.1 Direct costs. The three common direct costs are direct labor, direct material, and manufacturing (direct) overhead. Direct labor cost is the cost of all hands-on effort required to manufacture a product. Typical direct labor activities include fabrication, assembly, testing, inspection, and rework. Direct material cost is the cost of all components and raw materials included in the end product. Overhead includes items which are required but do not appear in the product, such as manufacturing management, direct engineering support, utilities, lubricants, and cleaning supplies. Each of these items can be directly traced to the manufacture of a particular product.

A typical process for estimating direct labor cost involves five steps: (1) obtain the manufacturing operation sequence, (2) specify all work elements in each operation, (3) estimate times for all work elements, (4) calculate the total direct labor hours for each labor rate, and (5) compute direct labor cost (C_{DL}) using Equation (3.22).

$$C_{DL} = \sum_{i=1}^{r} (H_{DLi})(R_i) \qquad (3.22)$$

where H_{DLi} = number of direct labor hours for each labor rate i
R_i = direct labor rate in dollars
r = number of distinct direct labor rates i

Direct material costs can be conveniently charged directly to products using information provided by supplier catalogs, quotes, and material handbooks. Direct material cost estimation involves the following steps: (1) obtain the bill of materials, (2) determine the quantity of each material used in the end product, (3) compute the direct material cost (C_{DM}) using Equation (3.23).

$$C_{DM} = \sum_{j=1}^{m}(Q_{Mj})(C_{Mj}) \qquad (3.23)$$

where Q_{Mj} = number of units of material j
C_{Mj} = unit cost of material j in dollars
m = number of distinct materials j

3.10.2 Indirect costs. Indirect costs include all costs incurred during the time period involving the management and sales of a product other than the direct costs. A few examples of indirect costs are nonmanufacturing facility rent and utilities, insurance, management salaries, and nonmanufacturing equipment depreciation. Traditionally, volume based cost (VBC) accounting systems allocate indirect costs to products based on a direct labor rate. This may provide distorted costing information when products are diverse in size, complexity, material requirements, and/or setup procedures (Cooper and Kaplan, 1998). In 1987, Johnson and Kaplan condemned standard costing systems as inadequate, irrelevant, and incorrect. They challenged cost analysts to develop more flexible cost management systems. Soon after, Cooper (1988a, 1988b, 1989a, 1989b) proposed activity based costing to meet this challenge.

Traditional volume based costing. Traditional volume based costing (VBC) systems allocate indirect costs with a single volume based cost driver such as direct labor costs or hours. This process, where all indirect costs are grouped into a single cost pool and divided by a single cost driver, can lead to distorted product cost information. The traditional VBC system only takes direct labor requirements into account when allocating overhead costs to products and does not consider other important indirect characteristics. VBC systems have been found to provide inadequate information regarding profit effects of modifying services, product lines, production volume, capacity, or outsourcing (Walker, 1999).

The VBC process has six steps as follows: (1) calculate the total cost pool costs (C_i) for cost pool i using Equation (3.24), (2) compute the total cost driver level (D_i) for each cost pool i, (3) calculate the VBC overhead rates (R_i) for cost pool i using Equation (3.25), (4) calculate the VBC overhead cost (OH_k) for each product k using Equation (3.26), (5) calculate the VBC total product cost (PC_k) for each product k using Equation (3.27), and (6) compute the VBC profit (loss) (PL_k) from the profit (loss) analysis for product k using Equation (3.28).

$$C_i = \sum_{j=1}^{n} X_{ij} \qquad (3.24)$$

where X_{ij} = overhead cost element j for cost pool i
n = total number of overhead cost elements
i = 1

$$R_i = \frac{C_i}{D_i} \qquad (3.25)$$

where D_i = total cost driver level for cost pool i

$$OH_k = \sum_{i=1}^{m} (R_i)(CDC_{ik}) \qquad (3.26)$$

where CDC_{ik} = product k consumption of cost driver for cost pool i
  m = total number of cost pools

$$PC_k = DL_k + DM_k + OH_k \qquad (3.27)$$

where DL_k = direct labor cost for product k
  DM_k = direct material cost for product k

$$PL_k = P_k - PC_k \qquad (3.28)$$

where P_k = selling price for product k

Example 3.8 Volume based costing
An outdoor equipment manufacturer is currently using a VBC system to estimate the costs associated with producing backpacks and tents. Using direct labor hours as the overhead cost driver, compute the volume based product cost for these two products and analyze their profitability. The average hourly labor cost is $10, and the manufacturer spent $25,428 on overhead costs to produce and sell these products.

Product	Direct Labor Hours per Unit	Material Cost per Unit	Sales Volume	Selling Price per Unit
Backpack	3	$12	250	$150
Tent	2	$45	75	$200

Solution
Equation (3.25) is used to compute the overhead rate per direct labor hour.

$$R_{VBC} = \frac{\$25,428}{(3hrs \times 250)+(2hrs \times 75)} = \frac{\$25,428}{900hrs} = \$28.25/hr$$

Each product's overhead cost per unit is found using Equation (3.26).

$$OH_{Backpack} = \$28.25 * 3hrs = \$84.75$$
$$OH_{Tent} = \$28.25 * 2hrs = \$56.50$$

While the total product cost calculated using Equation (3.27).

$$PC_{Backpack} = (3 hrs \times \$10) + \$12 + \$84.75 = \$126.75$$
$$PC_{Tent} = (2 hrs \times \$10) + \$45 + \$56.50 = \$121.50$$

The profitability of each product is calculated from Equation (3.28).

$$PL_{Backpack} = \$150 - \$126.75 = \$23.25$$
$$PL_{Tent} = \$200 - \$121.50 = \$78.50$$

Both products are shown to be profitable under the VBC system.

Activity Based Costing. Activity based costing (ABC) was described as "the most significant managerial accounting development within the last fifty years" (Harrison and Sullivan, 1996, p. 55). ABC was developed to provide more accurate indirect cost information based on the premise that overhead costs should not be allocated solely on a volume basis. ABC analysis identifies business activities that are performed during the production process. Indirect costs associated with these activities are identified and grouped into multiple cost pools based on similar cost drivers. Activity based costing systems better measure the way different jobs, products, and customers use company resources (Horngren, Foster, and Datar, 1997). Collecting sufficient information to develop an ABC system is a formidable task (Bruesewitz and Talbott, 1997), and companies tend to underestimate how large the task of gathering this information is (Ness and Cucuzza, 1995). Because of the extensive data needs, ABC systems often rely on estimated costs since installing elaborate measuring and monitoring devices to learn exactly the quantity and cost of indirect resources is rarely justified (Kaplan, 1992). Kaplan (1992, p. 59) states that "estimates based on interviews, employee judgments, and available operating data are usually sufficiently accurate for the managerial use of the information from an ABC model." In addition, ABC systems have been combined with known cost estimation techniques to provide quantitative means for developing ABC input parameters (Nachtmann and Needy, 2001; Nachtmann and Needy, 2003). A review of activity based costing literature is provided by Gosselin (2006).

The ABC development process consists of six steps: (1) calculate the total cost pool cost (C_i) for each cost pool i using Equation (3.29), (2–6) see Steps 2–6 of the VBC process. The number of cost pools is the primary difference between the VBC and ABC development processes, where the VBC system is restricted to $i = 1$.

$$C_i = \sum_{j=1}^{n} X_{ij} \qquad (3.29)$$

where X_{ij} = overhead cost element j for cost pool i
n = total number of overhead cost elements

3.10 Estimating Product Costs

Example 3.9 Activity based costing

The outdoor equipment manufacturer in Example 3.8 wants to use these two products to test an ABC system. The manufacturer decides to collect the additional data required to develop an ABC system.

Cost Pools	i	Overhead Cost	Cost Driver	Total Cost Driver Level	Backpack Cost Driver Level	Tent Cost Driver Level
Utilities	1	$5,241	Facility Space (ft²)	1400	450	950
Stitching Machine	2	$14,235	Hours Used	200	55	145
Packaging	3	$2,500	Box Size (in³)	700	100	600
Shipping	4	$3,452	# of Orders	70	45	25
Total		$25,428				

Solution

Using each cost driver, the overhead rate for each cost pool is calculated using Equation (3.25).

$$R_{Utilities} = \left(\frac{5241}{1400}\right) = \$3.74/ft^2$$

$$R_{Stitching\ Machine} = \left(\frac{14,235}{200}\right) = \$71.18/hr$$

$$R_{Packaging} = \left(\frac{2500}{700}\right) = \$3.57/in^3$$

$$R_{Shipping} = \left(\frac{3452}{70}\right) = \$49.31/order$$

Using the overhead rates associated with each cost pool and Equation (3.26), the product's overhead cost per unit is computed.

$$OH_{Backpack} = \frac{(\$3.74 \times 450 + \$71.18 \times 55 + \$3.57 \times 100 + \$49.31 \times 45)}{250} = \$32.70$$

$$OH_{Tent} = \frac{(\$3.74 \times 950 + \$71.18 \times 145 + \$3.57 \times 600 + \$49.31 \times 25)}{75} = \$229.98$$

Now the product cost of each product is found with Equation (3.27).

$$PC_{Backpack} = (3hrs \times \$10) + \$12 + \$32.70 = \$74.70$$

$$PC_{Tent} = (2hrs \times \$10) + \$45 + \$229.98 = \$294.98$$

The profitability of each product can then be computed from Equation (3.28).

$$PL_{Backpack} = \$150.00 - \$74.70 = \$75.30$$
$$PL_{Tent} = \$200.00 - \$294.98 = \$-94.98$$

Under the ABC system, the profit margins change drastically. The backpacks, which were only marginally profitable under the VBC system, are found to be very profitable. The tents, which were shown to be quite profitable under VBC, are found to be unprofitable under ABC. The reason for this lies largely in the allocation of overhead. The stitching machine, which was allocated on a basis of direct labor hours, is now allocated on the basis of how much it is actually used in the production of each item. The backpacks only use the stitching machine roughly a quarter of the time. When this is reflected accurately, the true profitability of each item appears. Therefore, the price of the tents should be increased, alternative methods of manufacturing should be found, or the product line should be discontinued altogether.

3.11 SENSITIVITY ANALYSIS

Because the cash flow estimates that are the focus of this chapter are not precisely known, a good economic analysis normally includes a sensitivity analysis of the estimates and the economic measures presented in later chapters. Chapter 4 of *Cases in Engineering Economy 2nd*, which is included on the CD packaged with this text, covers the basics of sensitivity analysis. Eschenbach and McKeague (1989) and Eschenbach (1992 and 2006) provide more examples and details.

More advanced techniques for sensitivity analysis often include explicit consideration of uncertainty. This is distinct from the modeling of uncertainty that is presented in later chapters although sometimes the stochastic sensitivity analysis also uses this information (Eschenbach and Gimpel, 1990; Garvey, 2000; and Saltelli, Chan, and Scott, 2000).

3.12 SUMMARY

Accurate estimates of costs, volumes, revenues, and cash flows are critically important to evaluating industrial projects. Estimates can be grouped into classes, ranging from rough to highly precise, each requiring different amounts of time and cost. A number of techniques have been discussed, including the use of indices and factor techniques. Different estimating models were also discussed, including the use of least squares, capacity functions, and growth curves. Product costs were also addressed, including the use of activity based costing. Sensitivity analysis is often needed because of the inherent uncertainty of cash flow estimates.

REFERENCES

AACE INTERNATIONAL RECOMMENDED PRACTICE NO. 18R-97, "Cost Estimation Classification System," (accessed 2013, last updated Nov. 29, 2011).

AL-TABTABAI, HASHEM, and ALEX P. ALEX, "Modeling the Cost of Political Risk in International Construction Projects," *Project Management Journal*, **31**(3) (September 2000), pp. 4–14.

ASHER, HAROLD, "Cost-Quantity Relationships in the Airframe Industry," Report 291 (Rand Corporation, 1956).

BODE, JURGEN, "Neural Networks for Cost Estimation," *Cost Engineering*, **40**(1) (January 1998), pp. 25–30.

BRUESEWITZ, STEPHEN, and JOHN TALBOTT, "Implementing ABC in a Complex Organization," *CMA Magazine*, **71**(6)(July-August 1997), pp. 16–19.

CHEN, SHU-CHUAN, SHIH-YAO KUO, KUO-WEI CHANG, and YI-TING WANG, "Improving the Forecasting Accuracy of Air Passenger and Air Cargo Demand: The Application of Back-Propagation Neural Networks, *Transportation Planning and Technology*, **35**(3) (2012), pp. 373–392.

COOPER, ROBIN, "The Rise of Activity-based Costing - Part One: What is an Activity-based Cost System?" *Journal of Cost Management*, **2**(2) (Summer 1988a), pp. 45–54.

COOPER, ROBIN, "The Rise of Activity-based Costing - Part Two: When Do I Need an Activity-based Cost System?" *Journal of Cost Management*, **2**(3) (Fall 1988b), pp. 41–48.

COOPER, ROBIN, "The Rise of Activity-based Costing - Part Three: How Many Cost Drivers Do You Need and How Do You Select Them?" *Journal of Cost Management*, **2**(4) (Winter 1989a), pp. 34–46.

COOPER, ROBIN, "The Rise of Activity-based Costing - Part Four: What Do Activity-based Cost Systems Look Like?" *Journal of Cost Management*, **3**(1) (Spring 1989b), pp. 38–49.

COOPER, ROBIN, and ROBERT S. KAPLAN, "The Promise and Peril of Integrated Cost Systems," *Harvard Business Review*, **76**(4) (July-August 1998), pp. 109–119.

DE BARCELOS TRONTO, I.F., "Comparison of Artificial Neural Network and Regression Models in Software Effort Estimation," *International Joint Conference on Neural Networks*, (2007), pp. 771–776.

ESCHENBACH, TED, "Technical Note: Constructing Tornado Diagrams with Spreadsheets," *The Engineering Economist*, **51**(2) (2006), pp. 195–204.

ESCHENBACH, TED, "Spiderplots vs. Tornado Diagrams for Sensitivity Analysis," *Interfaces*, **22**(6) (November-December 1992), The Institute for Management Sciences, pp. 40–46.

ESCHENBACH, TED, and Robert J. Gimpel, "Stochastic Sensitivity Analysis," *The Engineering Economist*, **35**(4) (Summer 1990), pp. 305–321.

ESCHENBACH, TED, and LISA S. MCKEAGUE, "Exposition on Using Graphs for Sensitivity Analysis," *The Engineering Economist*, **34**(4) (Summer 1989), pp. 315–333.

GARVEY, PAUL R., *Probability Methods for Cost Uncertainty Analysis: A Systems Engineering Perspective*, (Marcel Dekker, 2000).

GOMPERTZ, B., "On the Nature of the Function Expressive of the Law of Human Mortality, and on a New Mode of Determining the Value of Life Contingencies," *Phil. Trans. Roy. Soc. London,* **115** (1825), pp. 513–583.

GOSSELIN, MAURICE, "A Review of Activity-Based Costing: Technique, Implementation, and Consequences," *Handbook of Management Accounting*, **2** (2006), pp. 641-671.

HARBUCK, ROBERT H., "Using Models in Parametric Estimating for Transportation Projects," *AACE International Transactions*, (2002), pp. EST.05.1-EST.05.9.

HARRISON, DAVID S., and WILLIAM G. SULLIVAN, "Activity-based Accounting for Improved Product Costing," *Engineering Valuation and Cost Analysis*, **1**(1) (1996), pp. 55–64.

HORNGREN, CHARLES T., GEORGE FOSTER, and SRIVANT M. DATAR, *Cost Accounting: A Managerial Emphasis,* 9th ed. (Prentice Hall, 1997).

HUMPHREYS, KENNETH K., AND PAUL WELLMAN, *Basic Cost Engineering*, 3rd ed. (Marcel Dekker, 1996).

JORGENSEN, M., "A Systematic Review of Software Development Cost Estimation," *IEEE Transactions on Software Engineering*, **33**(1) (2007), pp.33–53.

JOHNSON, H. THOMAS, AND ROBERT S. KAPLAN, *Relevance Lost: The Rise and Fall of Management Accounting* (Harvard Business School Press, 1987), pp. 224.

KAPLAN, ROBERT S., "In Defense of Activity-Based Cost Management," *Management Accounting*, **74**(5), (1992), pp. 58–63.

LAI, Y., AND CHRISTOPHER P. L. BARKAN, "Enhanced Parametric Railway Capacity Evaluation Tool," *Transportation Research Record: Journal of the Transportation Research Board*, **2117** (2009), pp. 33–40.

LANG, HANS J., "Simplified Approach to Preliminary Cost Estimates," *Chemical Engineering*, **55** (June 1948), pp. 112–113.

LIU, HEPING, "Cost Estimation and Sensitivity Analysis on Cost Factors: A Case Study on Taylor Kriging, Regression and Artificial Neural Networks," *The Engineering Economist*, **55**(3) (2010), pp. 201–224.

MALSTROM, E., *What Every Engineer Should Know About Manufacturing Cost Estimating*, (Marcel Dekker, Inc., 1981).

NACHTMANN, HEATHER, and KIM LASCOLA NEEDY, "Methods for Handling Uncertainty in Activity Based Costing," *The Engineering Economist*, **48**(3) (2003), pp. 259–282.

NACHTMANN, HEATHER, and KIM LASCOLA NEEDY, "Fuzzy Activity Based Costing: A Methodology for Handling Uncertainty in Activity Based Costing Systems," *The Engineering Economist*, **46**(4) (2001), pp. 245–273.

NESS, JOSEPH A., and THOMAS G. CUCUZZA, "Tapping the Full Potential of ABC," *Harvard Business Review*, **73**(4)(July-August 1995), pp. 130–138.

NEWNAN, DONALD G., JEROME P. LAVELLE, and TED G. ESCHENBACH, *Engineering Economic Analysis 12th*, Oxford University Press (2014).

NOAH, J. W., "Identifying and Estimating R&D Costs," RM-3067-PR (Rand Corporation, 1962).

ODEYINKA, HENRY A., JOHN LOWE, and AMMAR P. KAKA, "Artificial Neural Network Cost Flow Risk Assessment Model," *Construction Management and Economics*, **31**(5) (2013), pp. 423–439.

PEARL, RAYMOND, *The Biology of Population Growth* (Alfred A. Knopf, 1925).

SALTELLI, ANDREA, KAREN CHAN, AND E. MARIAN SCOTT, *Sensitivity Analysis*, (John Wiley and Sons, 2000).

SATO, Y., "Optimal Budget Planning for Investment in Safety Measures of a Chemical Company," *International Journal of Production Economics*, (2012).

SETYAWATI, BINA R., SIDHARTA SAHIRMAN, and ROBERT C. CREESE, "Neural Networks for Cost Estimation," *AACE International Transactions*, (2002), pp. 13.1–13.9.

SETYAWATI, BINA R., ROBERT C. CREESE, AND SIDHARTA SAHIRMAN, "Neural Networks for Cost Estimation (Part 2)," *AACE International Transactions*, (2003), pp. EST 14.1-EST 14.10.

SMITH, ALICE E., and ANTHONY K. MASON, "Cost Estimation Predictive Modeling: Regression versus Neural Networks," *The Engineering Economist*, **42**(2) (1997), pp. 137–161.

TENG, C., "An Estimating Relationship for Fighter/Interceptor Avionic System Procurement Cost," RM-4851-PR (Rand Corporation, 1966).

VERHULST, PIERRE F., " Recherches Mathématiques sur la loi D'accroissement de la Population," *Nouv. mém. de l'Academie Royale des Sci. et Belles-Lettres de Bruxelles*, **18**(1–41) (1845).

VERHULST, PIERRE F., "Deuxième Mémoire sur la loi D'accroissement de la Population," *Mém. de l'Academie Royale des Sci., des Lettres et des Beaux-Arts de Belgique*, **20**(1–32) (1847).

WANG, QING, "Artificial Neural Networks as Cost Engineering Methods in a Collaborative Manufacturing Environment," *International Journal of Production Economics*, **109**(1–2) (2007), pp. 53–64.

WESTNEY, RICHARD E., *The Engineer's Cost Handbook*, (Marcel Dekker Inc., 1997).

WILLIAMS, TREFOR P., "Predicting Final Cost for Competitively Bid Construction Projects using Regression Models," *International Journal of Project Management*, **21**(8) (2003), pp. 593–600.

WRIGHT, T. P., "Factors Affecting the Cost of Airplanes," *Journal of Aeronautical Sciences*, **3** (February 1936), pp. 122–128.

PROBLEMS

3-1. Describe the characteristics of a good cash flow estimate and explain why good cash flow estimates can be challenging to achieve in real world industrial projects.

3-2. Westney's Estimating Paradox describes the tradeoffs between estimation accuracy and resource consumption. Explain these tradeoffs in the context of industrial project selection and management.

3-3. Describe why it is important to distinguish between cost estimation and pricing.

3-4. Explain the project life cycle and how it relates to cost estimation and estimation method selection.

3-5. Describe the cost estimation classification system and how it can be utilized in practice.

3-6. Describe the importance of quality data in cash flow estimation.

3-7. Compare and contrast the basic cost estimation techniques discussed in this chapter. Find examples of inflation and geographical indices and use unit costs to project the cost of a 40,000 square foot facility to be built, for example, in Anchorage, Alaska, in 2013 based on the cost of a 50,000 square foot facility built in Arkansas in 2008.

3-8. What is a cost estimating relationship (CER) and how can a CER be applied in industrial project analysis?

3-9. Regression and neural networks are two methods commonly used to develop CERs. Discuss the advantages and disadvantages of both methods.

3-10. Provide a real world cost/driver example for the following common CER forms; linear, power, exponential. For example, labor cost (cost) and the number of labor hours worked (driver) typically has a linear CER. Also, provide a rationale for each of your examples.

3-11. Use a best-fit regression approach to develop a CER to predict the building project cost of gas production facility as a function of gas production output. The Lang factor is associated with which CER form? Discuss how the coefficient of determination influenced your CER model selection.

Gas Production Output (MBTU/h)	Building Project Cost ($M)
4300	1080
4600	1350
5000	1466
5200	1586
6000	1667
6500	1732
7800	1802
7900	1825
8400	1863
9000	1890

3-12. Why is it important to consider the effect of learning when developing cost estimates?

3-13. The following data describes a new assembly project. Utilize this data to answer parts a-d below.
- A learning rate of 85% is associated with a cumulative average curve.
- 500 units are to be built.
- A set up time of 3 hours has been estimated.
- Standard data predicts the run time to be 0.5 hours per piece.

(a) At the completion of production, it is known that 35 hours in total were required to complete the last 100 units. If the 85% learning rate applies, how many hours were expended to build the first unit?
(b) How many units were produced before the standard time was reached?
(c) How many hours were spent to build the 200th unit?
(d) How many hours were spent to build units #50 through #100 inclusive?

3-14. Calculate and draw Pearl and Gompertz growth curves over 12 years. Let $L = 1500$ and the numbers for years 1 and 2 be 6 and 60 respectively. Contrast the two curves. Is there any mathematical basis for believing one is more accurate? If not, how would you choose which to use.

3-15. Compare and contrast volume based costing and activity based costing. What was the motivation behind the development of activity based costing?

3-16. A manufacturing company produces two models—a deluxe and a standard. The deluxe model can be distinguished from the standard model in that it uses more expensive materials and requires more craftsmanship. Presently, the company uses the number of direct labor hours to allocate overhead to each product. The company has expended 15,000 total hours of direct labor and pays an average hourly rate of $25. Management is concerned that product costing may be distorted using the present method of overhead allocation and wants to investigate. As a result, they are considering the possibility of implementing an ABC system. Preliminary analysis divides all overhead spent into three major cost pools: design, manufacturing, and delivery. An appropriate cost driver has been identified for each cost pool. The following data has been collected.

Product	Direct Labor Hours (hours/unit)	Material Cost ($/unit)	Sales Volume (units)	Selling Price ($/unit)
Standard	3	100	4000	450
Deluxe	6	300	500	800

Overhead Cost Pool	Amount ($)	Cost Driver
Design	300,000	Number of Drawings
Manufacturing	150,000	Production Orders
Delivery	75,000	Customer Orders
Total	525,000	

Cost Driver	Standard	Deluxe	Total
Number of Drawings	10	30	40
Production Orders	80	10	90
Customer Orders	400	250	650

(a) Determine the product cost of the standard and deluxe models using the present traditional cost accounting method.
(b) Determine the product cost of the standard and deluxe models using an ABC approach.
(c) What conclusions can you draw based on your analysis in parts a and b and the company's profit or loss? What is your recommended course of action?

4

Depreciation: Techniques and Strategies

4.1 INTRODUCTION

A firm's future cash flows from a project depend on three items that are *not* cash flows—because they are part of determining the taxes paid by the firm.
1. *Depreciation* of productive assets.
2. *Depletion* of natural resources.
3. *Amortization* of prepaid expenses.

Depreciation of a robot, depletion of a mineral deposit, and amortization of a purchased license or patent are part of determining the economic value of a firm's assets and thus the firm. However, when evaluating a project, our focus is on the resulting cash flows for property and income taxes.

This chapter summarizes the depreciation and depletion methodologies with an introduction to their application in different countries. While the information is as accurate and authoritative as possible, no attempt is being made here to render legal, accounting, or tax advice. For such assistance, consult a competent professional.

4.2 DEPRECIATION STRATEGIES

The economic evaluation of alternative depreciation strategies has a long history; see, for example, Schoomer (1966), Kaufman and Gitman (1988), Berg and Moore (1989), and Fleischer et al. (1990). More recent examples include Hurley and Johnson (1996), Berg et al. (2001), and Hartman (2002). These evaluations focus on the present worth of the income tax deductions. We show that this generally implies taking the deduction as soon as legally permitted.

The following paragraph describes why a firm might prefer a smaller depreciation deduction so that it will be more profitable, even though it would pay more in income tax. However, in many cases firms maintain two versions of their accounts—one for computing taxes and another for describing the firm's performance. If two sets of accounts are maintained then the depreciation approach that maximizes the present worth of depreciation deductions and thus minimizes the present cost of taxes can be used. At the same time another depreciation approach can be used for other purposes.

Recall that depreciation is a deduction which reduces taxes, but it also decreases profits as it is an expense. It might be important to the firm's managers or owners that the firm shows *a profit* (or a larger one) in a given year. Smaller depreciation deductions or depletion allowances increase income taxes (bad) by increasing the firm's profits (good) at the cost of decreasing future profits (bad) and decreasing future taxes (good). Similarly, a firm may want to appear strong or stable during mergers and acquisitions or during negotiations with a prospective lender or investor. The reasons for not maximizing the present worth are varied and behavioral, and thus beyond the scope of our analysis.

It can also be shown that a firm that is not profitable may actually benefit from taking slower deductions. For example, a high tech start-up firm with normal early year losses might use the slowest depreciation schedule to delay deductions until the firm is profitable. The value of this strategy is reduced by carry-forward and carry-back provisions for losses.

The amount of tax savings generated through depreciation in jurisdictions that use income and property taxes can be very large. For example, studies indicate that depreciation, amortization, and depletion deductions are considerably more than *twice* the annual additions to retained earnings for many US corporations. Thus, it should not be surprising that considerable attention is paid to international comparisons of accounting rules and their impact on US firms facing global competition. (See, for example, Fleischer and Leung, 1988; Ashbaugh and Pincus, 2001; and Hartman et al., 2007.)

The basic depreciation methods include straight-line, declining balance, and sum-of-the-years' digits. These methods and others differ not so much in the amount of total depreciation allowed, but in the fraction that may be taken early in the asset's life. In general, depreciation should be taken as early as possible, so the tax savings happen as early as possible. Then the funds thus generated can be invested longer.

Tax agencies typically recognize both item and class accounting procedures. In the former, depreciation is computed separately for each depreciable capital asset. The individual results are then added to find the total for the year. Alternatively, each capital asset is put into one of several asset classes. Depreciation is then computed separately for each asset class. This chapter's results apply in either case, but we present this in terms of item accounting since it is easier to follow.

This analysis applies to property held for the production of income, not for capital gains. Furthermore, we initially restrict ourselves to the usual corporate situation in which the incremental tax rate is constant over time and depreciation is fully covered by pretax earnings. However, we do address these issues as they may impact the optimal depreciation choice.

This exposition is divided into five major parts: (1) a description of the various procedures that are commonly used for depreciation accounting, their characteristics, and the legal restrictions on their use; (2) a derivation of the discounted tax savings resulting from use of the straight-line and accelerated depreciation methods; (3) certain conditions under which one method of depreciation is preferable to another; (4) a summary of the results in a form suitable for day-to-day use; and (5) an analysis of depletion.

4.3 DEFINITIONS

While the following definitions use the term depreciation, there are nations which use the term capital allowances instead. The basic concepts remain the same.

4.3.1 Depreciable property. Any property, personal or real, tangible or intangible, that is used in a business to produce income and that has a limited and estimable life exceeding 1 year (property with a useful life of 1 year or less is expensed). *Note:* Personal property is a legal term for any property that is not realty (land, buildings, roads, or bridges).

4.3.2 Basis of property. The basis of any purchased depreciable item is its purchase cost, including installation and delivery costs, plus the cost of any capital additions. Determining the basis for assets obtained in other ways, such as building it or through a trade, can be more complicated.

4.3.3 Recovery period. This defines the number of periods over which depreciation deductions are taken. The recovery period may be set by statute, as in the Modified Accelerated Cost Recovery System (MACRS) that is used in the U.S. More often it equals the asset's useful life.

To apply the depreciation method over the recovery period, the *placed-in-service date* must be known. This coincides when the asset actually began its function as a business asset. Just as the recovery period may define a noninteger number of years for depreciation recovery, the placed-in-service date dictates whether a full, half, or quarter of a year is claimed as depreciation in the first year of service. As these rules vary by country, we assume full-year depreciation in this chapter's examples and problems.

4.3.4 Salvage value. The salvage value is the taxpayer's estimate of the asset's market value at the end of its useful life. Salvage value enters explicitly into calculations of depreciation with the straight-line and the sum-of-the-year's digits methods, it is a limit for declining balance methods, and it is assumed to be zero for the MACRS method.

4.3.5 Symbols and notation. We use the following notation:

N = asset's depreciable life (years)
B = asset's original installed cost or *basis*
L = asset's estimated salvage value after N years
i = effective interest rate
T = effective income tax rate
BV_t = book value or undepreciated balance remaining at the end of period t
D_t = depreciation taken (claimed) during period t
t = subscript referring to a given period ($t = 0, 1, 2, ..., N$)
P = present value of a tax savings due to depreciation
P_0 for straight-line
P_α for declining balance
P_s for sum-of-years' digits
P_M for MACRS
α/N = depreciation rate used for the declining balance method; $\alpha = 1.5$ for used property, $\alpha = 2$ for new property
M = some period (number of years) where $M < N$
π = combined depreciation and discount factors for any depreciation method (SL, DB, or SYD)
w = right side of inequality (4.13)
X = right side of inequality (4.14)

4.4 BASIS AND BOOK VALUE DETERMINATION

4.4.1 Definition of initial basis and book value. Determining the asset's relevant basis, B, is the first step in calculating the depreciation charges over time. The amount that can be depreciated includes the purchase price and all necessary costs to place the asset in service—including taxes, delivery, and installation costs. This total may then be reduced according to reductions noted in Section 4.4.2. An asset's book value equals its initial basis less any accumulated depreciation. It equals:

$$BV_t = B - \sum_{j=1}^{t} D_j \qquad (4.1)$$

While this is used as a measure of an asset's economic value, it may or may not be correlated with market value. The book value is an important part of determining the tax consequences of selling or replacing an asset.

4.4.2 Special first-year write-offs. A common tool to help promote investment by firms is to provide faster depreciation, such as additional write-offs for the first year of an asset's purchase. These special first-year write-offs usually (but not always) reduce the basis for computation of depreciation in later years.

There are two common provisions for faster depreciation that modify the basis and book value for later computations with the methods presented in Section 4.5.

- First year allowances (FYA). For example, in South Africa the rates for manufacturing and R&D equipment equal 40% for the first year, and 20% for the next three years. This is straight-line depreciation with an extra 20% FYA.
- Small cost investments. For example, in Japan assets costing less than ¥100,000 can be immediately expensed. In some years this limit has been tripled to stimulate economic investments by firms.

In the United States, tax codes (2001–2005) and (2008–now) have allowed for an extra first year depreciation of 30%, 50%, or even 100%. See for example, the Economic Stimulus Act, which has been extended and modified in each of the last several years. The extra depreciation has been intended to be a short-term economic stimulus measure.

Section 179 is the US provision for "small" cost investments which has increased from $5000 by the Economic Recovery Act of 1981, when it was established to $560,000 in 2012. The Section 179 deduction allows smaller companies to immediately expense some of their investment costs in the year of expenditure, as opposed to depreciating them over time. The current IRS tax code allows firms that purchase, finance, or lease less than $2 million in new or used equipment during the year to deduct the full price of qualifying equipment up to a maximum of $139,000. The Section 179 benefit is designed for small and medium size businesses, and rules change from year to year.

Qualifying property must be purchased for active use in a trade or business, and not property held merely for the production of income (such as real estate). The property must be tangible (not patents, copyrights, franchises, designs, etc.) with a life of 3 years or more, although software is allowed.

The Section 179 deduction can be applied to new equipment purchases of up to $560,000. For organizations that place more than $139,000 (as of 2012) of qualifying property in service, there is a 50% reduction, allowing 50% of the costs in excess of $139,000 to be deducted. For example, a small business places $280,000 of qualifying property in service. The deduction is:

$$\$139{,}000 + (0.50)(\$280{,}000 - \$139{,}000) = \$209{,}500$$

The Section 179 deduction reduces the initial basis for computing depreciation, and taxable income is reduced by any depreciation. This mutual dependence complicates the calculation of the Section 179 deduction, if the income limit is the limiting factor. In this case, the sum of depreciation and the Section 179 deduction is limited to the taxable income before depreciation. For this case Equation (4.2) describes the maximum Section 179 deduction that can be claimed:

$$\text{Section 179} = (\text{Tax Income before depreciation} - BV_0) / (1 - \text{MACRS\%}_1) \qquad (4.2)$$

The Section 179 tax deduction changes nearly every year, so it is important to use the appropriate tax code before applying such deductions.

4.4.3 Like-for-like replacement. When an asset is salvaged and a new asset is acquired to take over functions of the salvaged asset (replacement), the initial basis of the newly acquired asset is determined by the purchase cost of the new asset plus any residual value of the salvaged asset. Specifically, the initial basis B is:

$$B = P - (SV - BV_N) \qquad (4.3)$$

or the initial purchase price less any difference between the salvage value of the replaced asset and its book value at the time of sale. Note that this adjustment occurs before any Section 179 deduction can be calculated. This definition of the initial basis of the new asset greatly complicates the analysis of when to replace an asset (see Hartman and Hartman, 2001). This provision is often ignored in economic analyses—through ignorance of it or through a conscious decision to simplify the analysis.

4.5 METHODS OF DEPRECIATION

4.5.1 Introduction. Which depreciation methods are allowable varies by national jurisdiction, which revision of a tax code, and even what kind of property is being depreciated. For example, in the United States until 1981, the basic techniques of straight-line, declining balance, and sum-of-the-years digits were used for calculating depreciation—individually and with switching between them. In 1981, the accelerated cost recovery system, ACRS, was adopted. ACRS specified fixed percentages that were "close" to constant after an initial half-year deduction, but the lives were shorter than had been permitted before. The Tax Reform Act of 1986 adopted the Modified Accelerated Cost Recovery System, or MACRS, which is still in use today.

The MACRS schedules provide combinations of declining balance and straight-line techniques, one for each life class. These schedules are very similar to the optimal depreciation strategy when more flexibility was permitted.

Even for US centric studies, we cannot limit our discussion to depreciation permitted under the current tax code. First, straight-line and declining balance must be understood as they provide the foundation for development of the MACRS schedules. Second, in retirement and replacement problems, abbreviating the depreciation schedule of an aging asset has tax consequences and this depreciation schedule is determined by the allowable depreciation strategy chosen *at the time the asset was purchased.* Third, it is unknown when and how the tax code will be revised next. Fourth, many firms operate under other national jurisdictions that allow one or more of these depreciation methods (as summarized in Table 4.1).

TABLE 4.1 Permitted depreciation methods for equipment by country

Country	Straight Line	Sum-of-Years' Digits	Declining Balance
Australia	✓		✓
Brazil	✓		
Chile	✓		
China	✓		
Costa Rica	✓	✓	
Egypt	✓		
France			
Germany	✓		✓
India			✓
Indonesia	✓		✓
Japan	✓		✓
Malaysia	✓		
Mexico	✓		
New Zealand			✓
Panama	✓	✓	✓
Philippines	✓	✓	✓
Russia	✓		✓
Saudi Arabia	✓		
Singapore	✓		
South Africa	✓		
South Korea	✓		✓
Spain	✓	✓	✓
Sweden	✓		✓
Taiwan	✓		✓
Thailand	✓		
Turkey	✓		✓
United Kingdom			✓
United States	Modified Accelerated Cost Recovery System		

4.5.2 The straight-line method. This is the simplest depreciation method as it has the same deduction in each period over the recovery period. For the straight-line method, the *rate* of depreciation is constant and equal to the reciprocal of the life. The deduction amount is also constant over the life, and it equals:

4.5 Methods of Depreciation

$$D_t = \frac{B-L}{N} \qquad (4.4)$$

where 1/N is called the *straight-line rate*. If the book value, BV_t, is graphed as in Figure 4.1, it is a straight line; hence, the method's name.

Straight-line depreciation is the standard against which other *accelerated* methods have been measured. That is, other methods take a larger share of the depreciation in their early years; so, the depreciation is accelerated. As noted earlier, these other methods are generally more profitable than straight-line due to their accelerated nature. (*Note:* There is a technique called sinking fund depreciation (Grant et al., 1982) that depreciates items more slowly than straight-line. However, it is permitted so rarely, and used even less, that it is not discussed here.)

Historically and internationally, the straight-line method has had the broadest applicability. It is the primary method used in the United States for determining depreciation of tangible and intangible assets for financial reporting, even when other methods are used to determine taxes. Also, before 1986 it could be used on any depreciable property and was the only method that could be used on intangible property. A modified version (alternate MACRS) is still applicable today and must be used in a number of situations.

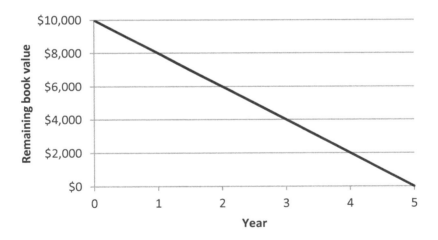

Figure 4.1 Book value for straight-line depreciation assuming no salvage value

4.5.3 The declining balance method. Depreciation in each year is computed by multiplying the depreciation rate, which is constant throughout the asset's life, by the asset's current basis. The basis is then reduced by the amount of the depreciation, and the procedure is repeated. This cycle continues until either the asset's useful life is expended or the current basis equals the salvage value. As this is an accelerated method, the initial depreciation amount is larger than straight-line when placed in service.

Define the rate of depreciation as α/N, with $2/N$ (double declining balance, or DDB, method) and $1.5/N$ (150% declining balance method) being traditionally common and currently used in MACRS. We now calculate the book value, BV_t, after the tth year and the amount of depreciation, D_t, allowable during the tth year (Myers 1958):

$$BV_1 = B - B\left(\frac{\alpha}{N}\right) = B\left(1 - \frac{\alpha}{N}\right),$$

$$BV_2 = BV_1 - BV_1\left(\frac{\alpha}{N}\right) = B\left(1 - \frac{\alpha}{N}\right)^2,$$

$$\vdots$$

$$BV_t = BV_{t-1} - BV_{t-1}\left(\frac{\alpha}{N}\right) = B\left(1 - \frac{\alpha}{N}\right)^t,$$

and:

$$D_i = BV_{t-1}\left(\frac{\alpha}{N}\right) = B\left(\frac{\alpha}{N}\right)\left(1 - \frac{\alpha}{N}\right)^{t-1} \tag{4.5}$$

The asset's salvage value is implicit in the declining balance method (Myers 1960), and it depends on a and N, as shown in Figure 4.2. For a given basis and α, the salvage value increases monotonically with N, approaching the limit:

$$\lim_{N \to \infty} BV_N = \lim_{N \to \infty} B\left(1 - \frac{\alpha}{N}\right)^N = Be^{-\alpha}$$

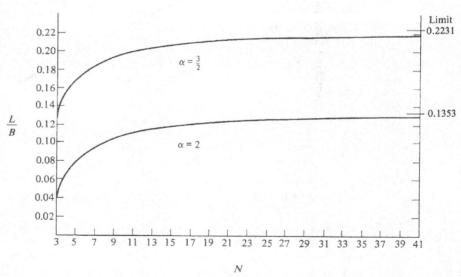

Figure 4.2 The salvage value implicit in the declining balance method

The implied salvage value will only coincidentally match an item's true salvage value. Under class accounting of depreciation, this is not important. However, under item accounting it is common to shift from declining balance to straight-line depreciation to match the item's salvage value.

Example 4.1 Double declining balance depreciation
A depreciable asset has an initial basis of $10,000 and an estimated useful life of 5 years. For the double declining balance method, $\alpha = 2$, and the depreciation rate is 2/5 or 0.4.

Table 4.2 Example of double declining depreciation

t	α/N	D_t	BV_t	
0	0.4		$10,000	
1	0.4	$4,000	$6,000	
2	0.4	$2,400	$3,600	$= \$10{,}000\,(1-0.4)^2$
3	0.4	$1,440	$2,160	
4	0.4	$864	$1,296	
5	0.4	$518	$778	$(0.4)(\$1296) = \$518$

4.5.4 The sum-of-the-years' digits (SOYD) method. This accelerated depreciation method calculates the current year's depreciation deduction as the original basis minus the salvage value, multiplied by a declining fraction. The fraction's denominator is constant and equals the sum of the digits from 1 to N, the asset's depreciable life. For example, for $N = 5$, the denominator equals $1 + 2 + 3 + 4 + 5 = 15$. The numerator for the first year is N, and it declines by 1 each year. This equals the number of useful years of life left at the start of the year. For example, with $N = 5$, the fractions become 5/15, 4/15, 3/15, 2/15, 1/15. Thus, the depreciation rate declines with each year as an arithmetic gradient.

In general, with an N year recovery period, the denominator equals $N(N + 1)/2$ or $SOYD$ and the numerator equals $N - t + 1$. So:

$$D_t = (B-L)\left[\frac{2(N-t+1)}{N(N+1)}\right] = (B-L)\frac{(N-t+1)}{SOYD} \qquad (4.6)$$

The book value remaining at any time t can easily be found by subtracting the cumulative depreciation to time t from the original basis, B, or:

$$BV_t = B - \sum_{j=1}^{t} D_j = B - \frac{B-L}{SOYD}\sum_{j=1}^{t}(N-j+1)$$

$$= \frac{B-L}{SOYD}\left[(Nt+t) - \sum_{j=1}^{t} -j\right]$$

$$= \frac{B-L}{SOYD}\left[(Nt+t) - \frac{t(t+1)}{2}\right]$$

$$= (B-L)\frac{(2N+1-t)t}{N(N+1)}$$

or:

$$BV_t = L + (B-L)\left(\frac{N-t}{N}\right)\left(\frac{N-t+1}{N+1}\right) \qquad (4.7)$$

While the depreciation rate changes each period, the change in the rate is constant at 1/SOYD each period. Thus, an arithmetic gradient factor can be used in time value of money calculations.

Example 4.2 Sum-of-year's digits depreciation
Consider again Example 4.1 with an asset whose basis $B = \$10,000$ and life $N = 5$ years. Assuming a salvage value $L = \$778$ (the same as the DDB method), then the sum-of-years' digits method yields

$$D_1 = (\$10,000 - \$778)(5/15) = \$3074$$
$$D_2 = (\$10,000 - \$778)(4/15) = \$2459$$
$$D_3 = (\$10,000 - \$778)(3/15) = \$1844$$
$$D_4 = (\$10,000 - \$778)(2/15) = \$1230$$
$$D_5 = (\$10,000 - \$778)(1/15) = \underline{\$\ 615}$$
$$\sum_{t=1}^{5} D_t = \$9,222 = B - L$$
$$BV_1 = \$778 + (\$10,000 - \$778)(4/5)(5/6) = \$6926$$
$$BV_2 = \$778 + (\$10,000 - \$778)(3/5)(4/6) = \$4467$$
$$BV_3 = \$778 + (\$10,000 - \$778)(2/5)(3/6) = \$2622$$
$$BV_4 = \$778 + (\$10,000 - \$778)(1/5)(2/6) = \$1393$$
$$BV_5 = \$778 + (\$10,000 - \$778)(0)(1/6) = \$778$$

Note that the final salvage value was assumed to match the DDB model, and that the depreciation amounts for SOYD are smaller than for DDB in the early years but larger in later years.

4.5.5 Switching. Switching merely allows one to combine any of the three previous methods. As discussed later, and shown in Schoomer (1966), switching from declining balance to straight-line depreciation maximizes the present worth of depreciation deductions. For MACRS the switch is from double or 150% declining balance at the optimal point, or when the straight-line depreciation amount exceeds that from declining balance, as shown in Example 4.3.

Example 4.3 Switching from DDB to straight line
Consider the asset in Example 4.1 again with basis $B = \$10,000$ and life $N = 5$ years. However, now its salvage value, L, is defined to be $200. For simple, straight-line depreciation the annual depreciation charge $= (B - L)/N = \$9800/5 = \$1960$. At least initially, the double-declining balance produces a larger write-off, as shown in Table 4.3. However, for each year we evaluate a potential switch to straight-line by dividing the current book value less salvage value by the number of years remaining. In year 4, when the switch is made, this is $(\$2160 - \$200)/2$ years $= \$980$ per year.

TABLE 4.3 Example: switch from declining balance to straight line (underlined values are the D_t values for the combined method.)

t	DDB D_t	SL D_t	BV_t
0			$10,000
1	<u>$4,000</u>	$1,960	$ 6,000
2	<u>$2,400</u>	$1,450	$ 3,600
3	<u>$1,440</u>	$1,133	$ 2,160
4	$ 864	<u>$ 980</u>	$ 1,180
5		<u>$ 980</u>	$ 200

4.5.6 Units of production. The units-of-production technique allows a firm to determine depreciation deductions according to the usage of an asset in a year according to its expected lifetime usage. For example, the life of heavy construction equipment may best be measured by operating hours and high uncertain annual usage levels could make the conversion of these total operating hours into an expected lifetime problematic at best. Thus, equipment with an expected life of 20,000 hours and usage of 3120 hours would generate a 15.6% depreciation charge on its initial basis minus expected salvage value.

4.5.7 Reasons for accelerated depreciation. Accelerated depreciation methods are used primarily to determine the income tax owed by the firm. When using accelerated depreciation methods, taxable income is determined to be smaller over the near term, decreasing taxes during that time. When firms assemble their financial statements, straight line depreciation is often used to determine the book value of assets and the level of profit that will be shown on income statements. This is true even when firms use accelerated methods for determining income taxes.

Many countries allow accelerated depreciation (see Table 4.1), and local tax law is a factor in determining where manufacturing operations will take place. Locations having low taxes are often more attractive as manufacturing sites, allowing manufacturing plants to be written off over shorter periods of time.

While our use of the term depreciation is not the same as the loss of value that a capital item will experience, the fact is that accelerated depreciation more closely matches the actual loss in value of those capital items. Thus, accelerated depreciation allows book value to more closely follow actual market value for many capital items.

4.5.8 Modified accelerated cost recovery system (MACRS). This system is unique to the United States. The Tax Reform Act of 1986 modified the ACRS system established in 1981 by adding classes for lengths of depreciable lives, by lengthening the write-off period for some kinds of assets, by increasing the initial declining balance rate from 150 to 200% for most categories, and by mandating an initial half-year convention for all personal property. MACRS ignores estimated salvage values, assuming a salvage value of 0.

The class definitions shown in Table 4.4 are the starting point for MACRS depreciation calculations: In what class is a particular asset? Specific asset classes, class lives, and recovery periods for the United States are listed in IRS Publication 946.

Table 4.4 Class examples for MACRS

3-Year	Tractors for over-the-road use, qualified rent-to-own property
5-Year	Automobiles*, buses, trucks, computers, office machinery, and R&D equipment
7-Year	Office furniture, agricultural machinery
10-Year	Marine vessels, qualified small electric meter and qualified smart electric grid systems, placed in service after Oct. 3, 2008.
15-Year	Roads, shrubbery, and municipal wastewater treatment facilities, high voltage electric transmission property
20-Year	Farm buildings and most municipal sewers
25-Year	Water utility property
27.5-Year	Residential real property
39-Year	Nonresidential real property (31.5 years if placed in service before May 13, 1993)

* Passenger automobiles subject to limits

The percentages shown in Table 4.5 are for personal property and are based on initial declining balance with optimal switching to straight-line. For the shorter life classes of 3, 5, 7, and 10 years, the rate is 200%; for the 15- and 20-year classes, the rate is 150%. For the real property classes (27.5 and 39 years), a straight-line method is used. However, for these classes, a mid-month convention is used along with tables matched to the month the asset is placed in service. Similar optional tables exist for mid-quarter assumptions if a large fraction of the personal property is placed in service either early or late in the year.

Table 4.5 MACRS percentage tables

	Recovery Period					
Year	3-Year	5-Year	7-Year	10-Year	15-Year	20-Year
1	33.33%	20.00%	14.29%	10.00%	5.00%	3.750%
2	44.45	32.00	24.49	18.00	9.50	7.219
3	14.81	19.20	17.49	14.40	8.55	6.677
4	7.41	11.52	12.49	11.52	7.70	6.177
5		11.52	8.93	9.22	6.93	5.713
6		5.76	8.92	7.37	6.23	5.285
7			8.93	6.55	5.90	4.888
8			4.46	6.55	5.90	4.522
9				6.56	5.91	4.462
10				6.55	5.90	4.461
11				3.28	5.91	4.462
12					5.90	4.461
13					5.91	4.462
14					5.90	4.461
15					5.91	4.462
16					2.95	4.461
17						4.462
18						4.461
19						4.462
20						4.461
21						2.231

4.5 Methods of Depreciation

The derivation of the MACRS percentages and their application are explained in Table 4.6 and Example 4.4, which applies MACRS to our earlier example.

MACRS permits an alternative calculation using the straight-line method. The life on real property is 39 years, for autos and computers it is 5 years, for personal property without a class life it is 7 years, and for other property, 10 pages of classes govern. As shown in Table 4.7, many of these lives are longer for the alternate MACRS. Note that assets generally used outside of the United States must be depreciated using alternate MACRS.

Example 4.4 MACRS depreciation

We return to Example 4.1 with an asset whose basis is B = \$10,000 and depreciable life $N = 5$ years. For DDB, the depreciation rate is 40% per year (2.0/5 years). With an initial half-year convention, the initial declining balance rate shown in Table 4.6 is 20%. The underlined percentages become the fraction of the initial basis that is tabulated in Table 4.6.

For our example the depreciation charges use these percentages applied to the \$10,000 initial basis. This results in the following:

Year	1	2	3	4	5	6
Depreciation	\$2000	\$3200	\$1920	\$1152	\$1152	\$576

Since the MACRS ignores salvage value, whenever this asset is disposed of the book value is used to compute a gain or loss (see Section 5.6). Note that only half of the depreciation is allotted in the final year if it is salvaged early (before year 6 in this example).

Table 4.6 Derivation of 5-year MACRS percentages

Year	Initial Book Value	Declining Balance Rate		Straight-Line Rate	
1	1	0.2		0.1	
2	0.8	0.32		0.178	(4.5 years)
3	0.48	0.192		0.137	(3.5 years)
4	0.288	0.1152	or	0.1152	(2.5 years)
5	0.1728			0.1152	
6	0.0576			0.0576	

4.5.9 Job Creation and Worker Assistance Act. The following is an example of a tax program to promote investment in capital assets. In 2002, an economic stimulus package was passed that changed depreciation allowances for assets purchased between September 11, 2001, and January 1, 2005. Specifically, the act allotted for an "additional" 30% of depreciation to be taken in the first year for assets that fall into MACRS property classes of 20 years or less (exceptions were made for water utility property, software, and leasehold property).

Table 4.7 Recovery periods for MACRS and Alternate MACRS

Asset Type	MACRS	Alternate MACRS
Furniture	7	10
Airplanes	5	6
Buses	5	9
Tractor units	3	4
Trailers	5	6
Marine vessels	10	18

Hartman (2002) showed that this "bonus" depreciation could increase the present worth of tax savings from depreciation by nearly 20% of the original investment times the tax rate, depending on the discount rate. As may be expected, the savings are greatest for longer lived assets.

Table 4.8 presents the "adjusted" MACRS percentages for the six affected asset classes as a result of the increased first year of depreciation. These are derived by reducing the initial basis of the asset by 30% and then applying the correct declining balance switching to straight-line strategy and applying the half-year convention. Similar tables can be produced for any level of first year "bonus" depreciation.

Table 4.8 Adjusted MACRS percentages due to Job Creation and Worker Assistance Act of 2002

Year	New 3-yr	New 5-yr	New 7-yr	New 10-yr	New 15-yr	New 20-yr
1	53.33	44.00	40.00	37.00	33.50	32.63
2	31.11	22.40	17.14	12.60	6.65	5.05
3	10.37	13.44	12.24	10.08	5.99	4.67
4	5.19	8.06	8.75	8.06	5.39	4.32
5		8.06	6.25	6.45	4.85	4.00
6		4.03	6.25	5.16	4.36	3.70
7			6.25	4.59	4.13	3.42
8			3.12	4.59	4.13	3.17
9				4.59	4.13	3.12
10				4.59	4.13	3.12
11				2.29	4.13	3.12

4.5.10 Comparing book values with different depreciation methods. The resulting book values for these different methods are summarized in Figure 4.3. All include a cost basis of $10,000, a salvage value of $0, and a 5-year depreciable life, similar to the previous examples. The MACRS graph reflects standard MACRS depreciation from Table 4.5. Note that the double declining balance method is shown without switching to straight line; without a switch, the curve is asymptotic with the x axis. The MACRS method includes optimal switching to straight line in year 4.

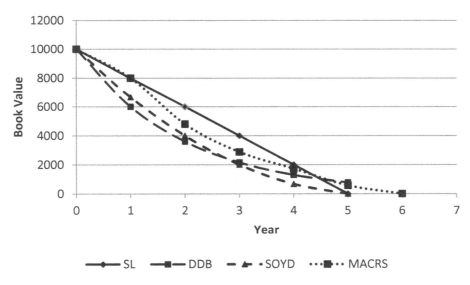

Figure 4.3 Book values over time for various depreciation methods

4.6 THE PRESENT VALUE OF THE CASH FLOW DUE TO DEPRECIATION

We now compute the present value of tax savings due to depreciation deductions. Since the Section 179 deduction and the investment tax credit are deducted at the end of the first year, if legal, their present value clearly equals the total deduction divided by $(1 + i)$.

4.6.1 Straight-line method. The annual deduction is $(B - L)/N$ for each of the N years of the asset's depreciable life. Assuming a constant tax rate, T, savings of $T(B - L)/N$ are provided each year, with a present value of:

$$P_0 = T[(B-L)/N](P/A, i\%, N) \qquad (4.8)$$

4.6.2 Declining balance method. Equation (4.5) defines periodic depreciation for declining balance, such that the present value of tax savings over the recovery period is:

$$P_\alpha = TB\frac{\alpha}{N}\sum_{t=1}^{N}\left(1-\frac{\alpha}{N}\right)^{t-1}(1+i)^{-t} \qquad (4.9)$$

To simplify this expression, first multiply the right side by $[1 - (\alpha/N)]/[1 - (\alpha/N)]$, to obtain:

$$P_\alpha = TB\frac{\alpha/N}{1-\alpha/N}\sum_{t=1}^{N}\left(1-\frac{\alpha}{N}\right)^{t}(1+i)^{-t}$$

Second, let:

$$C = TB\frac{\alpha/N}{1-(\alpha/N)}$$

and:

$$\frac{1}{1+x} = \frac{1-\alpha/N}{1-i} \text{ or } x = \frac{1+i}{1-\alpha/N} - 1.$$

Since the shortest depreciable life is 3 years and $(1 - \alpha/N)$ is greater than 0 and less than 1, $x > i$. Thus, substituting the x variable results in:

$$P_\alpha = C\sum_{t=1}^{N} \frac{1}{(1+x)^t} = C(P/A, x\%, N). \tag{4.10}$$

Example 4.5 Present value of tax savings, DDB
Again from Example 4.1, $\alpha = 2$, $N = 5$, and $B = \$10,000$. Letting $i = 10\%$ and $T = 34\%$, then:

$$x = \frac{1+i}{1-\alpha/N} - 1 = \frac{1.1}{1-0.4} - 1 = 0.833.$$

The present value of the tax savings resulting from the double declining balance depreciation deductions is:

$P_a = [(0.34)(10,000)(2/5)/(1 - 2/5)](P/A, 83.3\%, 5) = 2266.7(1.14206) = \$2588.67.$

4.6.3 Sum-of-year's digits method. Equation (4.6) defines the periodic depreciation for the SOYD method. This series can be decomposed into a uniform periodic amount of:

$$A = \frac{(B-L)N}{\text{SOYD}}$$

and a periodic arithmetic gradient of:

$$G = \frac{B-L}{\text{SOYD}}$$

If we assume that the tax savings TD_t occur at the ends of periods $t = 1, 2, \ldots, N$, then the present value is:

$$P_s = (T(B-L)/\text{SOYD}) [N(P/A, i\%, N) - (P/G, i\%, N)] \tag{4.11}$$

Example 4.6 Present value of tax savings, SOYD
Referring to Example 4.2, the present value of tax savings resulting from the SOYD method may be calculated from Equation (4.11) as follows:

$$P_s = \frac{0.34(10,000 - 778)}{15} [5(P/A, 0.1, 5) - (P/G, 0.1, 5)]$$

$$= (209.03)[(5)(3.7908) - (6.8617)] = \$2527.7$$

4.6.4 Modified accelerated cost recovery system. Because the MACRS write-off technique is based on declining balance with a switch to straight-line, it would be possible to derive formulas similar to those above. However, the variety of write-off periods ensures that the formulas would be complex. It seems easier to use the brute force technique of applying single-period present worth factors to the listed percentages of Table 4.5. As will be detailed in Chapter 5, the salvage value will be taxed as recaptured depreciation since the book value is depreciated to $0 under MACRS. If this is ignored, then the present value of the MACRS depreciation schedule is overstated.

Example 4.7 illustrates the technique, but it does not represent a fair comparison of MACRS with other techniques. This series of examples has used a 5-year write-off period for each technique, but one of the major consequences of MACRS was the use of shorter write-off periods.

Example 4.7 Present value of tax savings, MACRS
Referring to Example 4.4 (with the salvage shifted to year 6), the present value of the tax savings resulting from the MACRS method may be calculated as follows:

$$P = BT \sum_t \%t(P/F, i, t) - TL(P/F, i, Nlast) \tag{4.10}$$

$= [(10{,}000)(0.34)][0.2(P/F, 0.1, 1) + 0.32(P/F, 0.1, 2) + 0.192(P/F, 0.1, 3)$
$\quad + 0.1152(P/F, 0.1, 4) + 0.1152(P/F, 0.1, 5) + 0.0576(P/F, 0.1, 6)]$
$\quad - [(0.34)(778)](P/F, 0.1, 6)$
$= \$2479.77$

4.7 SIMPLE DEPRECIATION STRATEGIES

Throughout the remainder of this chapter, we shall select as optimal that depreciation policy or strategy that *maximizes the present value of the resulting tax savings* since this is a cash inflow to the firm. Assuming a constant tax rate T and no losses over time, this strategy is equivalent to minimizing the present worth of an asset's book value over time. In other words, maximizing the tax deduction of depreciation lowers the book value as rapidly as possible or minimizes the present worth of the book values.

4.7.1 Accelerated depreciation is better. The principle of minimizing the book value's present worth is equivalent to maximizing the cumulative fraction depreciated. Graphs of book value can be drawn and compared, and then the lower curve can be chosen. The following results are trivial mathematically but very important in practice:

Method A	is preferred to	Method B
Double declining balance (DDB)		150% declining balance
DDB with SL switch		Straight-line (SL)
DDB with SL switch		Double declining balance
Sum of the year's digits		Straight-line
MACRS		Alternative MACRS

Two other results follow from this principle. First, the recovery period should be chosen to be as short as possible. Second, any program of extra first year depreciation or Section 179 expensing should be used whenever possible and if it can only be applied to some assets; it should be applied to those assets with the longest lives.

4.7.2 Declining balance method versus the straight-line method. Declining balance without a switch to straight-line is rarely, if ever, used. Its implied salvage value is often unrealistic, as well as economically unattractive. However, the comparison of declining balance with straight-line is a foundation for other more realistic comparisons. We shall find that high rates of return and salvage values that are large relative to the basis favor the declining balance method. The former is not surprising; a high rate of return on the tax savings produced should increase the value of accelerating depreciation. The latter is true because the salvage value that is implicit in the declining balance method (see Figure 4.2) may be quite large with respect to the asset's true salvage value. So considerably more depreciation in total may be available with the straight-line method. Obviously, this effect decreases as the actual salvage value of the asset increases.

Setting

$$P_\alpha > P_0$$

gives by substitution:

$$B\frac{\alpha}{N}\frac{1-\{[1-(\alpha/N)]/(1+i)\}^N}{(\alpha/N)+i} > (B-L)\frac{1}{Ni}[1-(1+i)^{-N}], \quad (4.12)$$

which yields, after rearrangement:

$$\frac{L}{B} > 1 - \frac{\alpha i\{(1+i)^N - [1-(\alpha/N)]^N\}}{[(\alpha/N)+i][(1+i)^N - 1]} \quad (4.13)$$

as the condition under which declining balance is preferable to straight-line. We shall designate the right side of the inequality (4.13) by w. Then declining balance is preferable to straight-line if the ratio of the salvage value to the original basis (L/B) is greater than w, whereas straight-line is preferable if the ratio L/B is less than w. Curves of w are given in Figure 4.4 as a function of i for several values of N and for $\alpha = 3/2$. We are not concerned here with values of α greater than 3/2 since whenever these values could be used, sum-of-the-years' digits could also be used, and the latter would be used in preference to straight-line. We shall consider the latter comparison in Section 4.7.3.

4.7.3 Declining balance method versus sum-of-year's digits method. Under what conditions is declining balance superior to sum-of-the-years' digits? Which method is superior is determined by the rate of return and by the ratio of the asset's salvage value to its original basis. Again, large values of the rate of return and of the ratio of salvage value to basis favor the declining balance method.

Setting

$$P_\alpha > P_s$$

gives, by substitution of Equations (4.7) and (4.8):

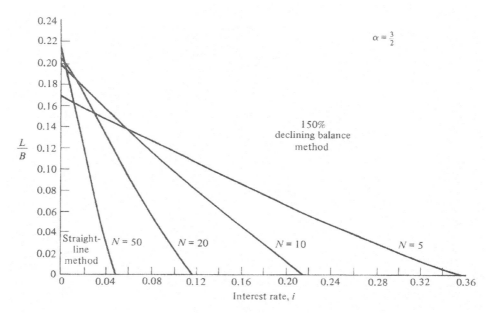

Figure 4.4 Straight-line versus 150% declining balance

$$B\frac{\alpha}{N}\frac{1-\{[1-(\alpha/N)]/(1+i)\}^N}{(\alpha/N)+i} > (B-L)\frac{2}{i(N+1)}\left\{1-\frac{1}{Ni}[1-(1+i)^{-N}]\right\},$$

which yields, after rearrangement:

$$\frac{L}{B} > 1 - \frac{i^2\alpha(N+1)\{(1+i)^N - [1-(\alpha/N)]^N\}}{2[(\alpha/N)+i][(Ni-1)(1+i)^N + 1]} \tag{4.14}$$

as the condition under which declining balance is preferable to the sum-of-the-years' digits.

We define the variable X as the right side of inequality (4.14). As before, declining balance is preferable if L/B is greater than X, while sum-of-the-years' digits is preferable if L/B is less than X.

Curves of X are given as a function of i for several values of N and for $\alpha = 2$ in Figure 4.5. Obviously, we are not concerned with values of α less than 2 since a taxpayer will use as large a value of α as possible, and if the sum-of-the-years' digits method was allowable, so was the double declining balance method.

4.8 COMPLICATIONS INVOLVING DEPRECIATION STRATEGIES

The introduction of MACRS reduced the number of depreciation choices for assets. However, one can still choose between an accelerated method (double or 150% declining balance switching to straight line) and use of straight-line (AMACRS) for most asset

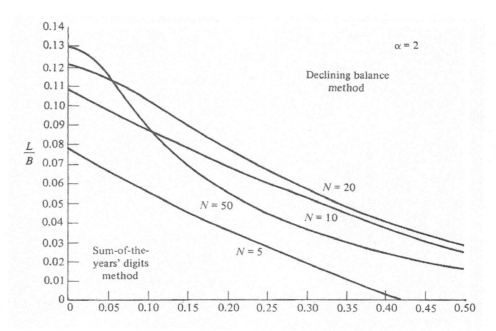

Figure 4.5 Sum-of-the-years' digits versus double declining balance

classes. The choice of the accelerated method is optimal when maximizing the present worth of the tax savings under the assumptions of a constant tax rate and profitability. Optimality is not clear when the tax rate changes with time or the entity sustains losses.

Fleischer et al. (1990) compared the use of MACRS and AMACRS over different horizons under conditions of changing tax rates, interest rates, and the year of disposal. It was shown that straight-line would be preferred in the case of increasing tax rates, especially with low interest rates and short service lives (as the savings from straight-line occur in the future with respect to accelerated methods). Hurley and Johnson (1996) developed a lower bound on the present value of the tax shield when the tax rate was assumed to be stochastic.

Berg et al. (2001) showed that in light of losses, straight-line methods may be superior because the large, early deductions from accelerated methods could be lost. This is dependent on the number of periods in which a loss can be carried forward and how long the company is unprofitable. As noted by Kulp and Hartman (2011), losses can be carried forward for 20 years in the United States, making it unlikely that straight-line methods would be optimal. However, in light of losses, the choice of the depreciation method becomes less important because by the time the tax savings can be utilized, the assets may already be fully depreciated (meaning the timing would be the same for either method).

4.9 SUMMARY OF CONCLUSIONS: DEPRECIATION

This chapter has presented decision rules for maximizing the present value of the income tax savings resulting from depreciation deductions, assuming a constant marginal tax rate and a firm's profitability over the asset's useful life.

1. Any available extra first year deduction or Section 179 expensing should be used wherever possible.
2. Always use as small a value of N as possible.
3. When using the declining balance method, use as large a value of α as possible.
4. Sum-of-the-years' digits is always preferable to straight-line.
5. Declining balance with optimal switching is always preferable to straight-line.
6. MACRS is preferred to the alternate MACRS.

4.10 DEPLETION OF RESOURCES

Depletion, like depreciation, operates to reduce income tax cash outflows. Depletion allowances reduce the basis of *natural resources,* such as mines, wells, and timberlands. The depletion deduction is intended to compensate the taxpayer for *capital* consumed in severance and production, and for *loss* of the resource.

4.10.1 Entitlement to depletion. Annual depletion deductions depend on the owner's economic interest. Generally, every firm, partnership, corporation, etc. *receiving income from the exhaustion of a natural resource is* entitled to deduct depletion allowance as an expense on its tax return *according to the capital interest in the property.*

The right to a depletion deduction depends on (1) the economic interest in the resource and (2) the actual production of the resource. If either factor is absent, no depletion deduction is allowable.

4.10.2 Methods for computing depletion deductions. The basic method for calculating depletion, applicable to all types of depletable assets, is known as *cost depletion.* A second method, known as *percentage depletion,* applies to mineral deposits and, with limits, to domestic oil/gas wells, but not to timber.

Cost depletion is the cost of the mineral interest (or timber, gravel, etc.), divided by the estimated recoverable reserves, and multiplied by the number of units sold during the taxable year. More specifically, cost depletion is found by first estimating the number of tons, barrels, board feet, and so forth that can be produced from the resource deposit. Then the number of producible units is divided into the property's cost (or that part of the total cost allocable to the resources deposit). The allocation is necessary, for example, for timber, where only the value of the tree and not the cost of the land is depleted. The quotient is the *cost depletion per unit* of resource. This amount is multiplied by the number of units *sold* or, for timber, felled, during the year. The product of the number of units times the cost depletion per unit establishes the year's depletion deduction.

Example 4.8 Cost depletion
The Empire Mining Company bought a mine for $2,650,000. The mine was estimated to contain 600,000 tons of recoverable ore. The land's estimated value, *excluding the ore*

deposit, was $300,000. During the first year, 65,000 tons of ore were extracted, and 60,000 tons were sold. Cost depletion for the year is calculated as follows:

$$\text{Cost depletion per unit} = \frac{\text{Net basis of resource}}{\text{estimated total reserve}}$$

$$= \frac{\$2,650,000 - \$300,000}{600,000 \text{ tons}}$$

$$= \$3.92 \text{ per ton.}$$

Cost depletion = ($3.92/ton)(60,000 tons) = $235,000.

Percentage depletion is limited to those who are otherwise entitled to *cost depletion*. It does not apply to timber resources, or to petroleum and natural gas deposits. Percentage depletion is calculated by multiplying a percentage factor (dictated by law) times the *gross income from the property*. Percentage depletion is also limited to a maximum of 50% of the property's net income calculated without the depletion deduction.

Example 4.9 Percentage depletion
The owner/operator of a sulfur well sells $1,000,000 of crude sulfur during the taxable year. Operating expenses (labor, power, taxes, etc.) are $600,000, leaving a taxable income of $400,000 (before depletion). The percentage depletion is allowed at a 22% rate, resulting in a tentative depletion deduction of (0.22)($1,000,000) = $220,000. However, the allowable percentage depletion is limited to (0.5)($1,000,000 − $600,000) = $200,000. So the depletion is $200,000 for the year.

4.10.3 The depletion deduction. For all depletable assets except timber, the taxpayer must (and wants to) use the depletion method that results in the greater depletion deduction. Thus, depletion is calculated by both the cost and percentage methods, and the greater value is used. For timber, only the cost method may be used.

Because the maximum of cost or percentage depletion must be used, the resource's cost basis (except timber) can be depleted to a negative value. Once the basis is reduced to zero, and if the resource continues thereafter to produce, then percentage depletion is still allowable (to the extent of 50% of the property's income). This drives the property's book value below zero. Rather than show negative values for producing resources on their balance sheets, most firms show the excess percentage depletion as a credit to their surplus or retained earnings.

4.10.4 Typical percentage depletion rates. Under the Internal Revenue Code [Section 613(b)], percentage depletion is allowed at specified rates on specified natural resources. The following are some typical rates applied to *gross income from the property* for the indicated resources (from IRS Publication 535):
    a. 22%   Sulfur, uranium, asbestos, mica, and ores of lead, zinc, and nickel, regulated natural gas, natural gas sold under a fixed contract, and any geothermal deposit determined to be a natural gas well
    b. 15%   US deposits of gold, silver, copper, iron ore, oil shale

    c. 14%   Borax, granite, limestone, marble, potash, slate, soapstone
    d. 10%   Coal, lignite, perlite, sodium chloride
    e. 7.5%   Clay for sewer pipe and tile or for lightweight aggregates
    f. 5%   Gravel, sand, stone, clay for roofing tile and flower pots

4.11 AMORTIZATION OF PREPAID EXPENSES AND INTANGIBLE PROPERTY

Prepayment of future expenses is generally recovered by expense deductions in future years. Such a process is called *amortization,* and the expense deduction is called an *amortization expense.* The cost of starting a business or of organizing a corporation or partnership must be amortized and cannot be expensed.

Some representative categories *of prepaid items* that can (or must) be amortized over the life of the item, rather than deducted in total in the year of acquisition include:

a. Market surveys, salaries during training *of* employees and trainers, travel to secure distributors, and consulting fees.
b. Premium paid for the purchase of a bond in excess of its face value (amortization elective).
c. Expenses incurred in acquiring a lease (mandatory amortization).

In general, the method of amortization is straight-line, with zero salvage value. The time period is a minimum of 60 months.

Intangible property can be depreciated under straight-line, *if* its useful life can be determined. Patents and copyrights have a life that is granted by the government. Similarly, an agreement not to compete may have a contractual life. On the other hand, a business's *goodwill* has no defined limit to its useful life and it may not be depreciated. Franchises, designs, and customer lists may be depreciable if useful lives and values separate from goodwill can be estimated. Repairs and replacements that do not increase a property's value usefulness or life can be expensed; otherwise, these must be depreciated.

REFERENCES

ASHBAUGH, HOLLIS, and MORTON PINCUS, "Domestic Accounting Standards, International Accounting Standards, and the Predictability of Earnings," *Journal of Accounting Research*, **39**(3) (December 2001), pp. 417–434.

BERG, MENACHEM, ANJA DE WAEGENAERE, and JACCO L. WIELHOUWER, "Optimal Tax Depreciation with Uncertain Future Cash-Flows", *European Journal of Operational Research*, **132**(1) (July 2001), pp. 197–209.

BERG, MENACHEM, and GIORA MOORE, "The Choice of Depreciation Method Under Uncertainty," *Decision Sciences*, **20**(4) (December 1989), pp. 463–654.

FLEISCHER, GERALD A., and LAWRENCE C. LEUNG, "Depreciation and Tax Policies in China and the Four Little Dragons," *IIE Integrated Systems Conference Proceedings* (1988), pp. 314–320.

FLEISCHER, G. A., A. K. MASON, and L. C. LEUNG, "Optimal Depreciation Policy Under the Tax Reform Act of 1986," *IIE Transactions,* **22**(4) (December 1990), pp. 330–339.

Grant, Eugene Lodewick, William Grant Ireson, and Richard S. Leavenworth, *Principles of Engineering Economy,* 7th (Wiley, 1982), pp. 202–203. [Not in 8th edition.]

Hartman, Joseph C., "New Depreciation Rules and their Impact on Capital Investment from the Job Creation and Worker Assistance Act of 2002," *The Engineering Economist,* **47**(3) (Fall, 2002), pp. 354–367.

Hartman, Joseph C., and Raymond V. Hartman, "After-Tax Economic Replacement Analysis," *The Engineering Economist,* **46**(3) (2001), pp. 181–204.

Hartman, Joseph C., S. L. Liedtka, and L. V. Snyder, "The Impact of US Tax Depreciation Law on Asset Location and Ownership Decisions," *Computers and Operations Research,* **34**(12) (2007), pp. 3560–3568.

Hurley, W. J., and L. D. Johnson, "Capital Investment under Uncertainty: Calculating the Present Value of the Deprecation Tax Shield when the Tax Rate is Stochastic," *The Engineering Economist,* **41**(3) (Spring, 1996), pp. 243–252.

Kaufman, Daniel J., and Lawrence J. Gitman, "The Tax Reform Act of 1986 and Corporate Investment Decisions," *The Engineering Economist,* **33**(2) (Winter, 1988), pp. 95–108.

Kulp, Alison, and Joseph C. Hartman, "Optimal Tax Depreciation with Loss Carry-Forward and Backward Options," European Journal of Operational Research, **208**(2) (2011), pp. 161–169.

Myers, John H., "Useful Formulae for DDB and SYD Depreciation," *The Accounting Review,* **33**(1) (January 1958), pp. 93–95.

Myers, John H., "Influence of Salvage Value upon Choice of Tax Depreciation Methods," *The Accounting Review,* **35**(4) (October 1960), pp. 598–602.

Schoomer, B. Alva, "Optimal Depreciation Strategies for Income Tax Purposes," *Management Science,* **12**(12) (August 1966), pp. B552–B579.

PROBLEMS

4-1. A new machine is purchased at the beginning of the year for $70,000. The machine's expected life is 7 years, at which time it will be scrapped. Compute the book value and depreciation for the first two years using the following depreciation methods: (a) straight-line, (b) double declining balance, (c) sum-of-years' digits, and (d) MACRS assuming a 7-year recovery period.

4-2. A new machine is purchased at the beginning of the year for $9000. Delivery and installation charges are $1000. The expected life of the machine is 8 years at which time it will be sold for $1000. Compute the book value and depreciation for the fourth year using the following depreciation methods: (a) straight-line, (b) 150% declining balance, (c) double declining balance, (d) sum-of-years' digits, and (e) MACRS assuming a 5-year recovery period.

4-3. A machine was purchased for $75,000 4 years ago. When purchased, the machine had an estimated life of 10 years, and a salvage value of $15,000. What is the machine's book value after 4 years of depreciation? Assume that the machine has been depreciated under (a) straight-line, (b) double declining balance, (c) sum-of-years' digits, (d) 7-year MACRS.

4-4. An asset costs $25,000 and is expected to have a life of 8 years with a salvage value equal to $1,000. (a) What is the depreciation charge for the fifth year and the book value at the end of the fifth year using the 150% declining balance method? (b) What is the depreciation charge for the fifth year and what is the book value at the end of the fifth year using the 200% declining balance method?

4-5. A small engineering firm will have only one capital purchase this year, a computer graphics workstation. The computer will cost $5000, which the firm will recover using MACRS. The firm's taxable income currently averages $50,000 per year. (a) What is the recovery period? (b) What is the annual depreciation schedule? (c) If Section 179 is available, what is the annual depreciation schedule?

4-6. An asset costs $75,000 and is expected to have a life of 20 years with a salvage value of 20% of the initial cost. What is the accumulated depreciation for the first 10 years for each of the following depreciation methods: (a) straight-line, (b) double declining balance, (c) sum-of-the-years' digits, and (d) MACRS with a 20-year recovery period?

4-7. An asset costs $10,000 and is expected to have a life of 5 years, at which time it will have no salvage value. (a) What is the book value at the end of the second year and the depreciation charge for the third year using sum-of-the-years' digits depreciation? (b) Using double declining balance depreciation? (c) Using MACRS with a five-year recovery period?

4-8. An asset was purchased for $4000. Shipping and nonrecurring installation costs amounted to $1000. After 6 years the machine will be sold for $1100 after $500 is paid to dismantle it. (a) What is the depreciation and book value at the end of the third year using double declining balance depreciation? (b) Using SOYD? (c) Using MACRS with a five-year recovery period?

4-9. An asset costs $10,000 and has a life of 10 years with no salvage value. When do you switch from double declining balance depreciation to straight-line depreciation to obtain the *fastest* depreciation possible? How much does this change if the salvage value is $1000?

4-10. A machine will cost $10,000 and have a life of 10 years. It will be depreciated using double declining balance with a switch to straight-line. (a) For salvage values from $0 to $5000 graph the optimal year to switch. (b) Repeat the graph assuming that the life is 20 years.

4-11. (a) Develop a spreadsheet to analyze double declining balance depreciation with a switch to straight-line depreciation. The spreadsheet should automatically determine the year to switch. Show the formulas. (b) Use your spreadsheet to find the book value in years 3, 6, and 9 for an asset that follows this depreciation technique. The asset costs $100,000, it has a 10-year life, and its salvage value is $0. (c) How much does the answer change if the salvage value is $2000?

4-12. A small firm must choose between Section 179 followed by straight-line and MACRS for a potential asset (the only one to be chosen this year). The machine's salvage value is $0 and both write-offs will be 5-year recovery periods. If the machine will cost (a) $20,000, (b) $150,000, or (c) $300,000, which recovery choice is better? Does this answer depend on the firm's interest rate?

4-13. An asset costs $10,000 and is expected to have a life of 5 years, at which time it will have a salvage value of $2000. (a) Determine the depreciation charge for each year and the book value at the end of each year for the entire life of the asset using straight-line, double declining balance, sum-of-the-years' digits depreciation, and MACRS with a 5-year recovery period. (b) What is the present value of each depreciation schedule where

$$PV = \sum_{t=1}^{N} D_t(P/F, i, t)$$

and the discount rate is 8%? (c) On the same graph, plot the book values for each depreciation method.

4-14. A new computerized machine tool costs $335,000, the firm's interest rate is 10%, and the salvage value is $40,000. The tool's expected life is 18 years (with periodic upgrades), but the MACRS write-off period is 5 years. Compare the PV of the depreciation deductions with (a) straight-line, (b) double-declining balance, (c) SOYD, and (d) MACRS.

4-15. Redo problem 4-8, graphing the PV's for i equals 5, 10, and 15%.

4-16. A revised investment tax credit has been proposed. For assets with a recovery period of seven or more years the rate is 10%. This investment tax credit is subtracted from the initial basis of the asset. An asset costs $100,000 and has a MACRS recovery period of 10 years. What is the credit/deduction schedule for the asset?

4-17. When considering the *advantage* of MACRS over double declining balance with an optimal switch to straight-line depreciation it is worthwhile to consider the source of the MACRS advantage. Consider some equipment used in the forging industry with a life of 18 years and a 10-year MACRS recovery period. The installed first cost is $3 million, and the salvage after 18 years is $0.5 million. Use the firm's interest rate of 10% to calculate the present worth of each depreciation schedule. (a) What percentage of the DDB value is the MACRS value? (b) What fraction of this advantage is due to the shorter write-off period permitted by MACRS? (c) What fraction is due to depreciating under MACRS to $0 and then recapturing the excess depreciation in year 18?

4-18. Redo problem 4-17 and compare MACRS with straight-line depreciation. Add the following: (d) What fraction is due to the *acceleration* of the depreciation schedule?

4-19. Considering the asset and comparison of problem 4-17, analyze the relative importance of the salvage value, which equals the recaptured depreciation. Graph (1) the advantage MACRS has in dollars and (2) the fraction of the advantage that is due to the salvage value as a function of the salvage value. Vary the salvage value from $0 to $2 million. (To generalize this graph, rescale the y axis as the fraction of installed first cost and the x axis as the fraction salvage value is of installed first cost.)

4-20. A coal company expects to produce 125,000 tons of coal each year for 15 years. The deposit cost $30 million to purchase. The annual gross revenues are expected to be $60 per ton, and the net revenues are expected to be $40 per ton.
(a) Determine the annual depletion on a cost basis
(b) Determine the annual depletion on a percentage basis.

4-21. 20,000 acres of timber is bought for $10M. The firm estimates that the land will be worth $200 per acre after the timber is removed. It is estimated that 50 million board-feet of lumber is available from the property. The land will take 5 years to clear, with about equal amounts of timber being harvested each year. Determine the cost depletion allowance for each year.

5

Corporate Tax Considerations

5.1 INTRODUCTION

This chapter analyzes the role of taxes in project evaluation. Some of the more common taxes are the following:

a. *Property* or *ad valorem* taxes are levied against the *value of property*. These taxes are commonly levied by school districts, municipalities, counties, and states on real estate or business and personal property.

b. *Sales* or *use* taxes are assessed when property is *transferred* from one owner to another. This tax is usually a percentage of the property's value and is usually levied by a state or city.

c. *Excise* taxes are *federal* taxes levied against a commodity's value at a certain stage of manufacture or transfer. Originally (1792), excise taxes were the main source of US federal revenue, and the levies were against nonessential commodities, such as whiskey and molasses. Over the years, excise taxes have been extended to cover gasoline, automobiles, tires, firearms, telephone service, and air transportation. Generally, this tax is paid by the manufacturer or supplier and is added to the price paid by the customer.

d. *Value-added taxes* are applied to the value that a firm adds while making or selling a good or service. As detailed in Section 5.8, the value added can be thought of as the wages and profits that are added to the value of purchased raw materials, parts, equipment, and services when making a product. The value added can also be considered the value of the output minus the value of the inputs. Value-added taxes are widely used around the world, and are applied at each step of the supply chain, including when the final good or service is sold.

The above taxes depend only on the *value* of the item or service at the time it is taxed. Income taxes do not depend on this value, but they are the most prevalent since they are assessed by the federal government, most states, and many municipalities, as well as governments around the world.

e. *Income* taxes are generally calculated as a percentage of the *taxable income* of individuals and corporations. "Taxable" income is gross income minus certain expenses permitted by the tax laws.

This chapter focuses on income taxes. From the firm's perspective, the other taxes are included either in the price of the product or service or as a production expense. The net effect of such taxes is recovered either in gross income or as a reduction of income tax paid.

Projects can be studied either *before-tax* or *after-tax,* by either ignoring or including income taxes. Because of their simplicity, before-tax studies are common in engineering economy analyses. However, modern capital budgeting theory indicates that before-tax studies are theoretically limited to very special cases in the comparison of alternatives (see Chapters 6 and 7). However, in a practical sense it may be better to compare alternatives before taxes, so long as income taxes will affect the alternatives similarly. The greater simplicity of before-tax comparisons facilitates sensitivity analysis, the inclusion of uncertainty and risk, and communicating with stakeholders.

In reality, income taxes affect all business and industrial firms except those owned by the government and certain cooperatives and nonprofit organizations (such as hospitals and charitable organizations). *Except as described above, income taxes should be taken into account in economic studies for the firm.* The reason is simple: Income taxes are a major cash outflow. This is consistent with the objective stated in Chapter 1: The firm ought to maximize the present value of future *cash flows* since this is the equivalent of maximization of future wealth for the share owners—even if the "firm" is owned by a nonprofit cooperative or is a government project owned by the citizens.

The income tax statutes, regulations, and case law are lengthy and complex. Federal laws and regulations are linked with many variations to state and municipal tax laws. Furthermore, the income tax laws are constantly changing. Added to this is the number of tax procedures that are changed simply by executive order or by Internal Revenue Service regulation. So we present some tax principles rather than the complete methodology. In general, the income tax principles, rates, and regulations are those in existence in 2011.

Most of the relevant literature was discussed in the last chapter, since the bulk of the research has focused on depreciation of capital expenditures. However, there has been substantial work on tax policy questions (see, for example, Amoako-Adu and Rashid (1990) on analysis of a tax cut and Kaufman and Gitman (1988) on elimination of the investment tax credit and long-term capital gains). A review of how taxes and depreciation are handled in engineering economy texts was performed by Lundquist (2009).

One difficulty, however, is that many firms pay substantially less than the tabulated rates due to the large variety of tax breaks in the complex code. For example, General Electric paid no US federal income tax in 2010; in fact, it received a tax credit of $3.2 billion on US profits of $5.1 billion (Kocieniewski, 2011). While some tax breaks may not be linked to individual projects, others are linked directly to the many industrial projects that have received tax holidays, reduced rates, and other incentives. The net effect of these is to reduce the importance of taxes in some cases (due to lower effective rates) and increase their importance in others (due to significant differences between how different potential project locations are taxed).

We omit *individual* income taxes, even though there are many businesses and small industries that pay income taxes as if they were individuals (e.g., a partnership, a Limited Liability Corporation, or a "Subchapter S" corporation, a corporation whose shareholders elect under the tax laws to be taxed individually). The major difference between a business taxed as a corporation and one taxed as an "individual" (or group of individuals) is the *rate* of tax paid. Other differences between corporate and personal income taxes are not relevant to the firm's project selection process. What follows focuses on US *corporate* income taxes at the federal level. These comprise the major income tax burden of business and industry. This discussion is also limited to tax provisions that can be linked to individual projects; it omits minimum tax calculations for the firm as a whole and accumulated earnings taxes.

Similarly, capital gains provisions as applied to stocks, art, and real estate are not covered. Our focus is engineering projects, where the assembly lines, bulldozers, and other infrastructure lose value over time—so capital gains are rare. Capital gains are taxes based on the length of time they are held; less than one year is considered short-term capital gains, and greater than one year is considered long-term capital gains. Short term capital gains are usually taxed at the same rate as income, but long term capital gains are generally taxed at a lower rate (currently 15%). The rate in future years depends on tax policy, which is subject to change.

For corporate analysis of projects there are two basic income tax calculations:

1. The *ordinary* income tax liability. Ordinary income has a distinct technical meaning. Ordinary income is defined, in the broad sense, as gross profit minus the deductions allowed by law. Such deductions generally are noncapital expenses required for the production of income.
2. Income tax liability due to the selling of fixed assets. This tax liability arises if an asset, such as a machine or vehicle, is sold at a gain or a loss during the year.

While the rates for income taxes and the specific methods for calculating a particular tax may change, the general principles underlying these basic income tax calculations are relatively constant. Hence, the principles of income tax calculation will be illustrated, but anyone who undertakes or is responsible for economic studies should be familiar with current income tax laws and regulations.

5.2 ORDINARY INCOME TAX LIABILITY

Ordinary income tax is paid on taxable income which equals *gross profit* minus certain specified deductions. Gross profit is the total income that results from the firm's principal trade or business. Customarily, this is total receipts from sales minus the cost of goods sold.

In manufacturing firms, the cost of goods sold includes the entire *factory* cost, that is, any cost item required to convert the raw materials into a finished product. In merchandising firms, cost of goods sold includes the purchase price of the articles bought for resale, less cash and trade discounts, plus delivery charges to get the articles into inventory, all adjusted for beginning and ending inventories. The direct materials for any year equals the value of the initial inventory, plus purchases during the year, minus the closing inventory.

In manufacturing firms, the cost of goods sold is a variable cost, and is deducted from gross receipts to find gross profit. Cost of goods sold is made up of direct materials, direct labor, and manufacturing overhead. The depreciation of manufacturing equipment and buildings is generally accounted for within the manufacturing overhead account. The depreciation of non-manufacturing equipment is considered as a period cost, and is part of operating expenses, not cost of goods sold. As an example, in the 4th quarter of 2009, Johnson & Johnson recorded a depreciation charge of $1.2 billion, of which $113 million was included in the cost of goods sold. On most income statements, depreciation is listed only as a period cost, and the portion accounted for in cost of goods sold can be difficult to find.

Other operating expenses, such as administration, marketing, and interest are costs that are not directly dependent on unit sales. These period costs are deducted from gross profit to find the taxable income, on which the ordinary income tax is based. Calculating taxable income closely follows the itemization used on a conventional income (profit and loss) statement. An example will illustrate the procedure for calculating taxable income.

Example 5.1 Federal and state tax for transformer project

In Chapter 1, Figures 1.4 and 1.5 illustrated the cash flows from the ABC Company's Transformer Project. Figure 5.1 shows the same data with added detail, illustrating how taxable income is calculated for this project from an income statement.

First, gross profit includes gross *sales,* merchandise allowances and returns, cost of goods sold (includes direct material, direct labor, and manufacturing overhead). Second, *cash* operating expenses (except interest paid) are deducted from gross profit. Third, *noncash* expense items (depreciation, depletion, and amortization deductions) are subtracted from gross profit. Fourth, a special cash item, *interest* paid on debt, is deducted from gross profit. The gross profit minus the total of these three classifications of *operating expenses* yields *taxable income.*

Assuming that the following ordinary tax rates apply in this example,

Federal tax rate = 34% of entire $1,000,000 taxable income (see Section 5.3)
State tax rate = 2.65% of taxable income (assumed)

then the ABC Company would calculate its ordinary income tax liability as

Ordinary tax = 0.34($1,000,000) + 0.0265($1,000,000) = $366,500

The corresponding cash flow from implementing the transformer project would then be calculated by (1) deducting the income tax paid (a cash outflow) from taxable income, and (2) adding back the noncash items in the income statement:

Taxable income	$1,000,000
Less ordinary income tax paid	366,500
After-tax income	$ 633,500
Add:	
Depreciation deducted	300,000
Cash flow from operations	$ 933,500

5.2 Ordinary Income Tax Liability

ABC Company, Inc.
INCOME STATEMENT - TRANSFORMER PROJECT
for the year ended December 31, 20XX

GROSS SALES OF TRANSFORMERS			3,100,000
Less Merchandise Returns and Allowances			100,000
NET SALES			3,000,000
Less Cost of Goods Sold:			
Direct Materials		885,000	
Beginning Inventory	686,000		
Purchases	695,000		
Less Ending Inventory	496,000		
Direct Labor		250,000	
Manufacturing Overhead		250,000	
Cost of Goods Sold			1,385,000
INCOME FROM OPERATIONS			1,615,000
Add Interest Received		25,000	
Royalties Received		10,000	
Rents Received		10,000	
Total additions			45,000
GROSS PROFIT			1,660,000
Less Operating Expenses:			
1. Cash Items (such as)			300,000
a. Compensation for personal services not included in direct labor (wages, salraies, bonuses, etc.)			
b. Rentals paid			
c. Repairs (not reconstructions)			
d. Losses due to bad debts			
e. Contributions for charitable purposes (limit: 5% of taxable income)			
f. Casualty losses (fire, wind, theft) not covered by insurance			
g. Advertising and sales promotion expenses			
h. Utilities and communications (electricity, power, heat, telephone, postage)			
i. Record keeping expenses (computer rental, supplies, etc.)			
j. Contributions to qualified employee retirements plans			
2. Noncash Items:			300,000
a. Depreciation deductions			
b. Depletion deductions			
c. Amortization deductions			
3. Special Cash Items			60,000
Interest paid on borrowed money (interest paid on bonds, notes, and all other indebtedness of the firm)			
TAXABLE INCOME			1,000,000

Figure 5.1 Typical itemized calculation of taxable income

Now compare these actual cash flows above, and in Figure 5.1, to those in Figure 1.4:

From Income Statement (Figure 5.1)			From Cash Flow Statement (Figure 1.4)	
Net sales		$3,000,000	Operating revenues Less:	$3,000,000
Less:				
Cost of goods sold	$1,385,000			
Operating expenses	300,000			
Other income	−45,000	1,640,000	Cash operating costs	1,640,000
Interest paid		60,000	Interest paid	60,000
Income tax paid		366,500	Income tax paid	366,500
Cash flow from operations		933,500	Net funds generated	933,500

By adding noncash items back to after-tax income on the operating statement, the same cash flows result as if actual cash outflows were deducted from cash inflows on a cash flow statement.

5.3 FEDERAL INCOME TAX RATES

The federal corporate tax rate structure applies several tax rates to ordinary income. Lower marginal tax rates (15% and 25%) are applied to the first $75,000 of taxable income, as shown in Table 5.1 and Figure 5.2a. Table 5.1 comes from IRS Publication 542 (2006). The tax rates at higher income levels are shown in Figure 5.2b. If state income taxes are a significant factor, the state income tax rate must be included to find the total *effective* ordinary income tax rate. State tax policies and rates vary considerably, and will not be addressed in detail here. Marginal tax rates, which must include the effective federal plus state income tax rates, are important from a project perspective because added revenue (or savings) from a project will often be taxed at the marginal rate.

Table 5.1 US corporate income tax rate schedule

Over—	But not over—	Marginal Tax Rate
0	50,000	15%
50,000	75,000	25%
75,000	100,000	34%
100,000	335,000	39%
335,000	10,000,000	34%
10,000,000	15,000,000	35%
15,000,000	18,333,333	38%
18,333,333		35%

5.3 Federal Income Tax Rates

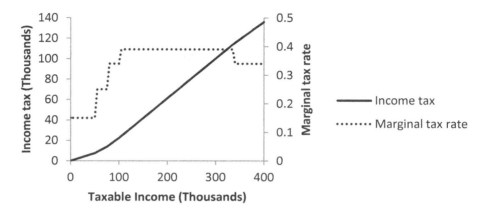

Figure 5.2a Income tax and marginal tax rate at low incomes

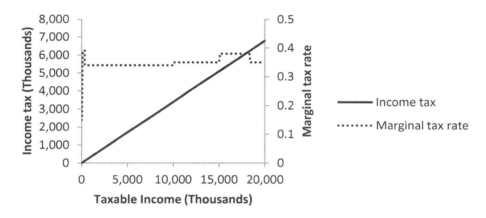

Figure 5.2b Income tax and marginal tax rate at higher incomes

$$T_e = \text{effective ordinary income tax rate} = \sum \left(\frac{\text{all ordinary income taxes}}{\text{taxable income}}\right)$$

5.3.1 Investment tax credit. Investment tax credits are instituted to encourage investment in new plants and equipment. A certain fraction of the invested money was credited against the taxpayer's income tax liability. Tax credits are subtracted from the tax to be paid, rather than deducted from taxable income. In its various reincarnations the investment tax credit has at times been in addition to the allowable depreciation, and at other times it has reduced the asset's depreciable basis.

The credit was significantly reduced in the Tax Reform Act of 1986, allowing only for specialized credits for reforestation, rehabilitation of older buildings, and solar/geothermal/ocean thermal energy projects. These credits are still in existence today but have various rules and limitations. For example, 10% of costs for reforestation can be applied up to $10,000 in a year while 10% of investment costs for certain energy investments can be taken as a credit.

The investment tax credit, like Section 179 expensing, is economically attractive when available. Not only does it reduce taxes for the year property is acquired, as a credit it reduces those taxes on. At the present time, investment tax credits are available only for specific energy related projects.

5.4 GENERALIZED CASH FLOWS FROM OPERATIONS

Example 5.1 and the effective tax rate concept of Section 5.3 can be generalized to find cash flows from a project for any given period, t. Let

- G_t = the project's gross income; this *cash inflow* results from operating the project during period t.
- E_t = the *cash outflows* during t for all deductible expenses, excluding interest paid on project indebtedness.
- D_t = the sum of all *noncash* items chargeable during t, such as depreciation, depletion, and amortization expenses.
- I_t = the *cash interest paid* during t on borrowed funds.
- T_e = the effective *ordinary* income tax rate (federal, state, and other).
- Y_t = the net cash flow from the project during t.

By analogy to Example 5.2.1, the ordinary income tax due is

$$(G_t - E_t - D_t - I_t)T_e \tag{5.1}$$

and the *after-tax* income is taxable income minus the income tax payable, or,

$$\underbrace{(G_t - E_t - D_t - I_t)}_{\text{Taxable income}} - \underbrace{(G_t - E_t - D_t - I_t)T_e}_{\text{Income tax}}$$

or,

$$\text{after-tax income} = (G_t - E_t - D_t - I_t)(1 - T_e) \tag{5.2}$$

The cash flow is after-tax income plus the noncash items, or

$$Y_t = (G_t - E_t - D_t - I_t)(1 - T_e) + D_t \tag{5.3}$$

Equation (5.3) for the project cash flow can also be stated as

$$Y_t = (G_t - E_t - I_t)(1 - T_e) + D_t T_e \tag{5.4}$$

The last term in Eq. (5.4), the effective tax rate times the depreciation deduction, is the equivalent cash contribution of the noncash depreciation deduction. This term—the cash flow equivalent of the depreciation—was used in Chapter 4 to evaluate the effectiveness of the different depreciation methods presented there.

5.4 Generalized Cash Flows from Operations

It is not necessary for taxable income to be greater than zero. In fact, many economic analyses compare alternatives with equal incomes that are not quantified. For example, two different material handling systems may differ only in their costs and not in their contribution to the factory's income. In this case, the G, terms would equal zero and the taxable incomes would be negative (for the project analysis, not for the factory).

Example 5.2 After-tax cash flows

An estimator is forecasting the cash flows for a certain project, as listed in Table 5.3. The effective income tax rate (state and federal) is estimated to be 37% for each year. Find the net cash flows for each year and find the present value at an effective interest rate of 15% per year.

Table 5.3 Data for example 5.2

End of Year, t	Gross Income, G_t	Operating Expenses, E_t	Interest Payments, Deduction, I_t	Depreciation, D_t
0	-0-	-0-	-0-	-0-
1	$100,000	$50,000	$4,000	$20,000
2	140,000	60,000	2,000	16,000
3	180,000	70,000	1,000	12,000
4	160,000	80,000	-0-	10,240
5	150,000	90,000	-0-	8,190

Solution

For the first year, the cash flow by Equation (5.4) is

$$Y_1 = (\$100{,}000 - \$50{,}000 - \$4{,}000)(1 - 0.37) + \$20{,}000(0.37)$$
$$= \$46{,}000(0.63) + \$20{,}000(0.37) = \$36{,}380$$

The cash flows for the second to fifth years are found similarly. For the fourth and fifth years Y_4 and Y_5 are both zero. The resulting cash flows are tabulated in Table 5.3 with the present worth calculations.

Table 5.3 Cash flows for Example 5.2

End of Year, t	Cash Flow, Y_t	Present Value Factor $(P/F, 15\%, t)$	Present Value of Incremental Cash Flow
0	-0-	-0-	-0-
1	$36,380	0.8696	$ 31,636
2	55,060	0.7561	41,631
3	73,110	0.6575	48,070
4	54,189	0.5718	30,985
5	40,830	0.4972	20,301
			$P_0 = 172{,}623$

5.5 TAX LIABILITY WHEN SELLING FIXED ASSETS

In the US tax code, Section 1231 assets are used *to carry on the production process* and their tax liability arises when the asset is sold, lost, destroyed, or otherwise disposed of. Due to depreciation, these assets have a book value different from their initial cost. Only rarely does this book value equal the value received from sale or disposal. The difference is a gain or loss. This gain or loss is treated differently than *capital gains* and *losses.* The differentiation was developed in response to business pressures. For example, suppose a machine was classed as a *capital* asset. If it was later sold at a loss, that loss would be *a capital* loss, which can only reduce prior or future capital gains. But if the machine was classed as a *noncapital* asset, then an *ordinary* loss could be deducted as an expense against gross income in the year of sale. Accordingly, *noncapital* asset treatment for *depreciable* property was included in the Revenue Act of 1938.

5.5.1 What are Section 1231 assets? These assets, if held for more than 1 year, are:
1. Property used in a trade or business on which depreciation is allowable. This includes buildings, structures, equipment, machinery, tools, materials handling devices, etc. used in connection with the production of a product or service.
2. Real property used in trade or business and not held regularly for sale to customers. This includes the land on which a factory is built and also improvements to the land, such as street pavements and landscaping. Costs of road construction, excavating, grading, and dirt (and rock) removal for a specific production-type building are not improvements, but rather costs that are included and depreciated with the building.
3. Coal, domestic iron ore, and timber royalties received in return for mining or harvesting of the mineral or timber.
4. Other items including cut timber on which the taxpayer reported the gain at the time of cutting; unharvested crops sold with the land; and livestock held for draft, breeding, dairy, or sporting purposes.

An industrial firm commonly has assets in depreciable property and real property *used in the trade or business.* These are the Section 1231 assets considered here.

5.5.2 Tax treatment of 1231 assets. When Section 1231 assets are sold, first the property is classed as either depreciable or nondepreciable. For depreciable Section 1231 assets, the tax law requires that excess prior depreciation deductions be "recaptured" as ordinary income in the year of sale. The resulting tax essentially equals the taxes saved in prior years by the excess depreciation.

Figure 5.3 illustrates the four possibilities with an asset whose initial cost, P_0, is $50,000 and whose current book value is $10,000.

1. The asset is sold at its book value, SP_1. There is no gain or loss and no tax consequence.
2. The asset is sold at a selling price, $SP_2 = \$25,000$, which exceeds its book value but not its original cost. The sale price less book value, $SP_2 - BV = \$25,000 - \$10,000 = \$15,000$, is recaptured depreciation, which is taxed as ordinary income.
3. The asset is sold at a selling price, $SP_3 = \$8000$, which is less than its book value.

The loss is $SP_3 - BV = \$8{,}000 - \$10{,}000 = (\$2{,}000)$. The net loss is fully deductible as an ordinary expense.

The asset is sold at a selling price, $SP_4 = \$65{,}000$, which exceeds its original cost. The gain is $SP_4 - BV = \$65{,}000 - \$10{,}000 = \$55{,}000$. The depreciation claimed in prior years is $P_0 - BV = \$50{,}000 - \$10{,}000 = \$40{,}000$. However, \$40,000 of the gain is recaptured depreciation, and \$15,000 is "capital gain on sale." The \$40,000 is taxed as ordinary income and \$15,000 is taxed at capital gain rates.

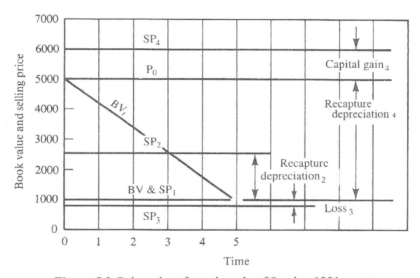

Figure 5.3 Gain or loss from the sale of Section 1231 assets

Thus, the principal tax differences on the sale of a depreciable capital asset and a depreciable Section 1231 asset are (1) the recapture of "excess" depreciation for the Section 1231 asset if a gain results and (2) the ordinary income deduction if a loss results. In either case ordinary income tax rates apply.

As noted in Chapter 4, while there are special rules about the amount of depreciation that can be claimed in the year of disposal—they don't matter. Taxes on recaptured depreciation and loss on sale are calculated at ordinary income tax rates. The final result is the same whether no, 50%, or 100% of the final year depreciation is claimed.

For nondepreciable Section 1231 assets the only change is that comparisons of sale prices are with the initial book value. If a net gain results, it is taxed at the capital gain rate, if it exists. If a net loss results, it is again fully deductible as an ordinary loss at ordinary income tax rates.

The importance of these Section 1231 transactions increased with the Tax Reform Act of 1986. MACRS assumes salvage values of \$0 when depreciating, so there will be more *recaptured depreciation* under this depreciation method.

5.6 TYPICAL CALCULATIONS FOR AFTER-TAX CASH FLOWS

Examples 5.4–5.10 illustrate how to find after-tax cash flows for some typical situations in project evaluation. The following assumptions are included: (1) Income tax cash flows occur at the same year's end as the income (or expense deduction) on which they are based. (2) Unless noted all productive assets are Section 1231 assets, which are disposed of to a disinterested third party at the project's end.

The after-tax cash flow calculations are based on Equation (5.4):

$$Y_t = (G_t - E_t - I_t)(1 - T_e) + D_t T_e \qquad (5.4)$$

This equation is modified as needed to include the Section 1231 tax on gains and losses.

For complex problems, the calculations required by Equation (5.4) can be simplified by using a tabular or spreadsheet format. This tabulation treats expenses as positive numbers, but subtracts them from income figures. Other noncash expenses are included with depreciation expenses in column (7).

(1)	(2)	(3)	(4)	(5)
Year, t	Gross Income, G_t	Cash Expenses (Except Interest), E_t	Interest Expense, I_t	Operational Income = (2) − [(3) + (4)]

(6)	(7)	(8)	(9)	(10)
Ordinary Income Tax =(5)(T_e)	Depreciation Expense, D_t	Tax-Saving Cash Flow = (7)(T_e)	Untaxed or 1231 Cash Flow	Total Cash Flow = Y_t = (5) − (6)+ (8) + (9)

Example 5.3 Straight-line depreciation
Some new machinery will cost $200,000 installed, and it will reduce net annual operating expenses by $50,000 per year for 10 years, when it will have a salvage value of $47,000. Assuming that the tax rate is 34%, calculate (a) annual depreciation based on the straight line method, (b) annual net cash flows, and (c) present value if the discount rate is 10%.

Solution
a. *Depreciation*

$$\text{Depreciation(SL)} = \frac{B_0 - L_{10}}{N} = \frac{\$200,000 - \$47,000}{10 \text{ yr}}$$

$$= \$15,300/yr$$

5.6 Typical Calculations for After-Tax Cash Flows

b. *Net cash flow table*

(2) and (5) Expense Savings Before Tax	(6) Tax	(7) Depreciation Expense	(8) Tax Savings from Depreciation	(9) Section 1231 Cash Flow	(10) Total Cash Flow
-0-	-0-	-0-	-0-	$-200,000	$-200,000
$50,000	$17,000	$15,300	$5,202	-0-	38,202
50,000	17,000	15,300	5,202	47,000	85,202

c. *Present value*

$$P_0 = \$-200{,}000 + 38{,}202\underset{(P/A,\ 10\%,\ 10)}{(6.1446)} + 47{,}000\underset{(P/F,\ 10\%,\ 10)}{(0.3855)} = 52{,}855$$

With a financial calculator, this is $-200{,}000 + PV(i, N, A, F)$
$= -200{,}000 + PV(10\%, 10, 38202, 47000) = 52{,}855$

Example 5.4 Sum-of-years' digits depreciation
Use sum-of-years' digits depreciation to determine
   a. Annual depreciation
   b. Annual net cash flows
   c. Present value if the discount rate is 10%.
Use the data of Example 5.3

Solution
a. *First-year depreciation*
    First-year depreciation (SYD):
$$B = \$200{,}000$$
$$L = \$\ 47{,}000$$
$$\text{Depreciable amount} = B - L = 153{,}000.$$

By Equation (4.6),
$$(D_B)_t = (B - L)\left[\frac{2(N - t + 1)}{N(N + 1)}\right]$$

$$\therefore (D_B)_t = (153{,}000)\left[\frac{2(10 - 1 + 1)}{10(11)}\right] = 27{,}818$$

For years 2–10 this decreases by $(B - L)(1/\text{SOYD})$ each year,
$$\Delta D_B = (\$153{,}000)/55 = \$2782.$$

b. *Net cash flow table*

(1) End of Year, t	(2) and (5) Expense Savings Before Tax	(6) Tax	(7) Depreciation Expense	(8) Tax Savings from Depreciation	(9) Section 1231 Cash Flow	(10) Total Cash Flow
0	-0-	-0-	-0-	-0-	$-200,000	$-200,000
1	$50,000	$17,000	$27,818	$9,458	-0-	42,458
2	50,000	17,000	25,036	8,512	-0-	41,512
3	50,000	17,000	22,255	7,567	-0-	40,567
4	50,000	17,000	19,473	6,621	-0-	39,621
5	50,000	17,000	16,691	5,675	-0-	38,675
6	50,000	17,000	13,909	4,729	-0-	37,729
7	50,000	17,000	11,127	3,783	-0-	36,783
8	50,000	17,000	8,345	2,837	-0-	35,837
9	50,000	17,000	5,564	1,892	-0-	34,892
10	50,000	17,000	2,782	946	47,000	80,496

c. *Present value*
Put the Total cash flow values into a spreadsheet, use
=NPV(rate, value cell 0: value cell 10)
=NPV(0.1,cell −200000:cell 80496)
$P_0 = \$+51,984.62$

Example 5.5 MACRS depreciation

Use MACRS to determine
  a. Annual depreciation
  b. Annual net cash flows
  c. Present value, if the discount rate is 10%
Use the data of Example 5.3 and 5.4
Assume the machinery has a five-year recovery period. Notice that there will be recaptured depreciation equal to the salvage value at $t = 10$. This recaptured depreciation is taxed like a gain on sale.

Solution
a. *Depreciation and book values, first through sixth years*

End of Year, t	MACRS %	MACRS Depreciation %, BV_0	Book Value, BV_t
1	0.2000	40,000	$160,000
2	0.3200	64,000	96,000
3	0.1920	38,400	57,600
4	0.1152	23,040	34,560
5	0.1152	23,040	11,520
6	0.0576	11,520	-0-
7–10	-0-	-0-	-0-

b. *Net cash flow table*

(1) End of Year, t	(2) and (5) Expense Savings Before Tax	(6) Tax	(7) Depreciation Expense	(8) Tax Savings from Depreciation	(9) Section 1231 Cash Flow	(10) Total Cash Flow
0	-0-	-0-	-0-	-0-	$-200,000	$-200,000
1	$50,000	$17,000	$40,000	$13,600	-0-	+46,600
2	50,000	17,000	64,000	21,760	-0-	54,760
3	50,000	17,000	38,400	13,056	-0-	46,056
4	50,000	17,000	23,040	7,834	-0-	40,834
5	50,000	17,000	23,040	7,834	-0-	40,834
6	50,000	17,000	11,520	3,917	-0-	36,917
7	50,000	17,000	-0-	-0-	-0-	33,000
8	50,000	17,000	-0-	-0-	-0-	33,000
9	50,000	17,000	-0-	-0-	-0-	33,000
10	50,000	17,000	-0-	-0-	31,020*	64,020

* Taxed since $BV_{10} = 0$. After-tax cash flow = $47,000 \times (1 - 0.34)$.

c. Present value at 10% equals $67,312.

Examples 5.4, 5.5, and 5.6 illustrate that, first, the *total* amount of depreciation deducted over the 10-year period is the same for each, $200,000 − $47,000 = $153,000. For the MACRS depreciation, $200,000 is depreciated over years 1 through 6, and then $47,000 is *recaptured* in year 10. Second, the equivalent cash inflows from the savings are the same ($33,000 per year). Third, all have total cash flows (at 0% interest) of $229,020. Nevertheless, the net present values of the cash flows differ as follows:

    i.   For SL depreciation, $53,855,
    ii.   For SYD depreciation, $57,355,
    iii. For MACRS depreciation, $67,312.

Obviously, the MACRS method provides the highest net present value of the cash flows, due to the *timing* of the depreciation deductions. Since MACRS is based on double declining balance, it forces a greater proportion of the *total* depreciation into the project's early years. The 6-year write-off period reinforces this, and so the tax *savings* occur earlier.

It may seem illogical to depreciate an asset below its salvage value, as we do with MACRS, but it is advantageous. The tax savings in Example 5.6 in years 1–6 from the "extra" $47,000 in depreciation arrives 4 years earlier than it is recaptured. As detailed in Example 5.6 without this "extra" depreciation, the net present value decreases from $67,312 to $61,116.

Example 5.6 MACRS with salvage value $\neq 0$
We use the same data as in Example 5.5, but we assume that only $200,000 − $47,000 = $153,000 is depreciated.

Solution
a. *Depreciation and book values, first through sixth years*

End of Year, t	MACRS %	MACRS Depreciation ($%_t$)($153,000)	Book Value, BV_t
1	0.2000	30,600	$169,400
2	0.3200	48,960	120,440
3	0.1920	29,376	91,064
4	0.1152	17,626	73,438
5	0.1152	17,626	55,813
6	0.0576	8,813	47,000
7–10	-0-	-0-	47,000

b. *Net cash flow table*

(1) End of Year, t	(2) and (5) Expense Savings Before Tax	(6) Tax	(7) Depreciation Expense	(8) Tax Savings from Depreciation	(9) Section 1231 Cash Flow	(10) Total Cash Flow
0	-0-	-0-	-0-	-0-	$-200,000	$-200,000
1	$50,000	$17,000	$30,600	$10,404	-0-	+43,404
2	50,000	17,000	48,960	16,646	-0-	49,646
3	50,000	17,000	29,376	9,988	-0-	42,988
4	50,000	17,000	17,626	5,993	-0-	38,993
5	50,000	17,000	17,626	5,993	-0-	38,993
6	50,000	17,000	8,813	2,996	-0-	35,996
7	50,000	17,000	-0-	-0-	-0-	33,000
8	50,000	17,000	-0-	-0-	-0-	33,000
9	50,000	17,000	-0-	-0-	-0-	33,000
10	50,000	17,000	-0-	-0-	47,000*	80,000

* Not taxed since BV_{10} = $47,000.

c. Present value at 10% equals $61,116.

Example 5.7 Sale of Section 1231 asset at less than book value
Continuing with Example 5.6, using MACRS depreciation for the producing assets, we assume that the assets will cost $5,000 to dispose of at the end of year 10. This problem also illustrates the tax treatment for an *existing* asset that must be sold at a book loss.

Solution
a. The depreciation is calculated exactly as in Example 5.6 (a).
b. The *net cash flow table* is virtually the same as for Example 5.6 (b), except the tenth year in the Section 1231 asset column represents *a loss* of $5000 rather than a salvage value of $47,000.

The tax *saving* resulting from the *loss* on selling or disposing of the asset is calculated at *ordinary* tax rates (since this is a Section 1231 loss):

$$\text{Loss} = \text{selling price} - \text{book value} = \$-5000 - \$0 = \$-5000,$$

$$\text{Tax } saving = (\text{loss})T_e = \$5000(0.34) = \$1700.$$

Thus, disposing of the asset at a book loss of $5000 at the end of year 10 results in cash loss of only $5000 - $1700 = $3300.

c. The net present value is
$P_0 = \$54{,}080.$

Example 5.8 Immediate expensing
Use the data of Example 5.4, but assume that Section 179 or some other special tax treatment allows immediate expensing of the machinery's $200,000 installed cost. Assuming that the tax rate is 34%, calculate (a) annual net cash flows and (b) present value if the discount rate is 10%.

Solution
a. *Net cash flow table*

(2) and (5) Expense Savings Before Tax	(6) Tax	(7) Recaptured Depreciation Expense	(8) Tax Savings from Depreciation	(9) Section 1231 Cash Flow	(10) Total Cash Flow
$-200,000	$-68,000	-0-	-0-	-0-	$-132,000
$50,000	$17,000	-0-	-0-	-0-	33,000
50,000	17,000	$-47,000	$-15,980	-0-	48,980

b. *Present value*

$$P_0 = \$-132{,}000 + 33{,}000\underset{(P/A,\ 10\%,\ 10)}{(6.1446)} + 47{,}000(1 - 0.34)\underset{(P/F,\ 10\%,\ 10)}{(0.3855)} = \$82{,}730$$

Example 5.9 Borrowed funds
If a project is to be analyzed along with a specific financing proposal, then the tax deductibility of the interest must be considered along with the pattern for loan repayment. This approach should be considered only when the financing source is tied to this specific project. For example, a city, county, or state government may offer special incentive financing to compete with other communities for a major manufacturing plant.

For consistency, we rely on the data of Example 5.6 with the addition of a $160,000 loan to be paid off in eight annual payments of $20,000 with interest at 8% on the unpaid balance.

a. *Net Cash Flow Table*

Year	Expense Savings Before Tax	Tax	Depreciation Expense	Tax Savings from Depreciation	Section 1231 Cash Flow	Principal Cash Flow to/from Bank	Interest Payments	Tax Savings	Total Cash Flow
0	0	0	0	0	$-200,000	$160,000			$-40,000
1	$50,000	$17,000	$40,000	$13,600	0	-20,000	$-12,800	$4,352	18,152
2	50,000	17,000	64,000	21,760	0	-20,000	-11,200	3,808	27,368
3	50,000	17,000	38,400	13,056	0	-20,000	-9,600	3,264	19,720
4	50,000	17,000	23,040	7,834	0	-20,000	-8,000	2,720	15,554
5	50,000	17,000	23,040	7,834	0	-20,000	-6,400	2,176	16,610
6	50,000	17,000	11,520	3,917	0	-20,000	-4,800	1,632	13,749
7	50,000	17,000	0	0	0	-20,000	-3,200	1,088	10,888
8	50,000	17,000	0	0	0	-20,000	-1,600	544	11,944
9	50,000	17,000	0	0	0			0	33,000
10	50,000	17,000	0	0	31,020				64,020

b.   Present value = $92,470.

5.7 AFTER-TAX REPLACEMENT ANALYSIS

There is a significant difference between the retirement of an asset and a like-for-like exchange, or replacement (Hartman and Hartman, 2001). Residual book value must be transferred to the new asset in the case of replacement. The costs of replacement assets are dependent not only on the initial purchase price and salvage value, but also on the residual book value of the asset being replaced. This complicates the after-tax analysis of replacement projects, and care must be taken especially where gains or losses from asset sales are large. This is reexamined in Chapter 8.

5.8 VALUE-ADDED TAX

Value added is considered as anything added before selling an improved product or service. Materials and components are purchased, people are paid wages to improve these materials, profit is added, and the final good or service is sold. Value can be determined from the additive side (wages plus profits) or from the subtractive side (output minus input) (Tait, 1988).

There are four ways to view this value added, all of which should provide an identical result:

1. Tax the wages plus profits, using existing accounting information. This is called the additive-direct method.
2. Tax wages, and separately tax profits. This is called the additive-indirect method.
3. Tax the (output-input), called the subtractive-direct method, and sometimes called the business transfer tax.
4. Identify the tax on the output then subtract the equivalent tax on the input. This is the subtractive-indirect method, and is the method most widely used around the world, including Europe. In this case, the central evidence in determining taxes is the individual invoice, not the company's accounting records.

Value-added taxes can be involved when dealing with capital purchases. There are several options for dealing with capital. Under a consumption VAT, capital purchases are treated as expenses and the tax on the purchase price is deducted as an input. Under an income VAT, the tax paid on the capital purchase is amortized over the expected life of the capital item. Most countries, including the European Union, use the consumption type of tax, treating capital the same as expense for the purposes of value added taxes (Bickley, 2003).

REFERENCES

AMOAKO-ADU, BEN, and M. RASHID, "Corporate Tax Cut and Capital Budgeting," *The Engineering Economist,* **35**(2) (Winter 1990), pp. 115–128.

BICKLEY, JAMES M., *Value Added Tax*, (Nova Science Publishers, 2003).

DEPARTMENT OF THE TREASURY, Internal Revenue Service, Publication 542 Corporations, March 8, 2006.

HARTMAN, JOSEPH C. and RAYMOND V. HARTMAN, "After-Tax Economic Replacement Analysis," *The Engineering Economist,* **46**(3) (2001), pp. 181–204.

KAUFMAN, DANIEL J. JR., and LAURENCE J. GITMAN, "The Tax Reform Act of 1986 and Corporate Investment Decisions," *The Engineering Economist,* 33(2) (Winter 1988), pp. 95–108.

KOCIENIEWSKI, DAVID, "G.E.'s Strategies Let It Avoid Taxes Altogether," *The New York Times*, March 24, 2011.

LUNDQUIST, ROBERT, "The Pedagogy of Taxes and Tax Purpose Depreciation," *Proceedings of the 2009 American Society for Engineering Education National Conference*, Austin, TX, June 2009, CD.

TAIT, ALAN A. *Value Added Tax*, (International Monetary Fund, 1988).

PROBLEMS

5-1. A small company in the warehouse business had operating expenses of $30,000, depreciation of $25,000, and interest expenses of $15,000. The cost of goods sold was $150,000. What is the taxable income (TI), tax, and after-tax cash flow if the company had sales of (a) $235,000? (b) $635,000?

5-2. A small company's annual taxable income is $52,000. A potential project would increase this annual taxable income by $15,000 for the next 10 years. (a) What is the effective tax rate and tax without the project? (b) What is the effective income tax rate and tax with the new project? (c) What is the effective income tax rate applicable to the project's increment of earnings?

5-3. A machine was purchased for $8000, used in production for 3 months, and then sold for $10,000 more than its installation and removal expenses. How much tax will be paid by a firm in the $10M tax bracket?

5-4. A computer system costing $15,000 will be written off using MACRS by a small firm whose tax rate is 15%. If annual operating costs are $5000 for ten years, what are the after-tax cash flows? The firm will donate the used computer system to the local community college after ten years.

5-5. A machine costing $10,000 will be depreciated using double declining balance depreciation with a life of 10 years. Assume a capital gains tax rate of 30% and an ordinary tax rate of 34%. What is the after-tax cash flow due to the salvage in year 10, if the asset is sold for (a) $2000? (b) $0 (that is, scrap)?

5-6. A company purchased a machine 10 years ago for $50,000. It was depreciated using sum-of-the-years' digits depreciation with a life of 20 years and no salvage value. If the machine is sold for $2000, what is the after-tax cash flow due to the salvage in year 10? Assume that the tax rate is 40% and the capital gains tax rate is 15%.

5-7. A large engineering firm is opening an office in a new city. The firm will be purchasing nearly $300,000 worth of furniture, which will be depreciated under MACRS. (a) What is the after-tax cash flow for each of the first 10 years? (b) If the furniture is replaced in year 12, and sold to a secondhand office furniture store for 10% of its initial cost, what tax is paid?

5-8. During the year, machines A and B were sold for $10,000 and $40,000, respectively. Both were depreciated using double declining balance. Machine A was purchased 20 years ago for $100,000 and a depreciable life of 25 years was used. Machine B was purchased 10 years ago for $150,000 and was depreciated using a salvage value of $30,000 and a life of 8 years. If these were the only depreciable assets sold during the year, what after-tax cash flow resulted using an effective tax rate of 34% and a capital gains tax rate of 30%?

5-9. The R&D lab of a large manufacturing organization is considering the purchase of a $1.2 million piece of equipment. It will be replaced in 4 years by a newer model, when the salvage value of the old one should be $250,000. What is the after-tax cash flow due to the equipment in year 4?

5-10. In problem 5-9, what is the equivalent annual after-tax cost of the equipment, if the firm's interest rate is 10%?

5-11. A proposed project will require four machines that will be depreciated using straight-line with no salvage value.

Machine Number	Purchase Cost	Depreciable Life (years)	Actual Life (years)	Selling Price
1	$100,000	6	7	$1,000
2	24,000	6	7	0
3	32,000	8	8	2,000
4	20,000	5	7	1,000

Assume a tax rate of 34%, a capital gains rate of 15%, and that the machines are purchased at the project's start. When machines are sold at the ends of years 7 and year 8, what are the cash flows?

5-12. A proposed project requires the purchase of four machines.

Machine Number	Purchase Cost	Salvage Value	Depreciable Life (years)	Selling Price	Actual Life (years)	Purchased at End of Year
1	$50,000	$5,000	9	$0	9	0
2	60,000	2,000	8	20,000	8	1
3	60,000	0	4	5,000	4	1
4	60,000	0	4	5,000	4	5

Machine 1 will be depreciated using straight-line, but declining balance will be used for the others. Assume a tax rate of 50% and a capital gains tax rate of 30%. What are the year 5 and year 9 cash flows due to the sale of these Section 1231 assets?

5-13. A project requires four machines, which will all be sold at the end of the eighth year.

Machine Number	Depreciation Model	Purchase Cost	Depreciable Life (years)	Salvage Value	Selling Price at t = 8
1	SL	$100,000	10	$0	$10,000
2	MACRS	50,000	5	0	10,000
3	DDB	50,000	10	10,000	20,000
4	SOYD	60,000	11	0	30,000

The machines will all be purchased at the project's start (that is, t = 0). What is the tax paid, and what is the cash flow in year 8 due to the sale and depreciation of these machines? Assume a tax rate of 52% and a capital gains tax rate of 30%.
SL = straight-line depreciation
MACRS = modified accelerated cost recovery
DDB = double declining balance depreciation
SOYD = sum-of-the-years' digits depreciation

5-14. A large company is considering a 10-year project for manufacturing a new product. The required machinery costs $40,000 and has an economic life of 10 years with no appreciable salvage value. The company's effective tax rate is 40%, and no borrowed capital will be used to fund the project if it is undertaken. Project data:

EOY	Income	Deductible Operating Expenses	EOY	Income	Deductible Operating Expenses
1	10K	25K	6	60K	30K
2	20K	20K	7	60K	30K
3	20K	15K	8	40K	25K
4	30K	15K	9	30K	25K
5	50K	25K	10	25K	20K

Determine the net cash flow stream if the following depreciation technique is used: (a) straight-line, (b) SOYD, (c) double declining balance, (d) 7-year MACRS, (e) each method if the machine's salvage value is $10,000.

5-15. A proposed 8-year project includes leasing land and purchasing a small building and production equipment. These Section 1231 assets (building and machines) will be purchased at the beginning of the first year for $600,000 (allocated $200,000 to the building and $400,000 to the machines). Their economic life is 8 years, and salvage values are $50,000 for the building and $100,000 for the machinery. Straight-line depreciation will be used on the building and double declining balance used on the machinery. Equity capital will be used to fund the project. The expected gross income less cash operating expenses are:

End of Year:	1	2	3	4	5	6	7	8
Income−Expense	70K	100K	150K	180K	200K	200K	120K	60K

The company's effective tax rate is 34%, and capital gains are taxed at 15%. Assume that the company can utilize any tax benefits that may result from this project. (a) Determine the project's net cash flows. (b) Assume that the company borrows $300,000 at the beginning of the first year, to be paid back in six equal annual payments beginning at the end of year 1 at 8% interest. Calculate the net cash flow stream. (c) Assume that the land can be bought for $20,000 and that its value will increase to approximately $50,000 in 8 years. [The lease costs were $8000 per year in (a) and (b)]. Working capital totaling $70,000 will be needed at the beginning of year 2, and losses due to inventory shrinkage and bad debts are estimated to be $20,000 over the project life. Determine the net cash flows for each year.

5-16. A manufacturer may introduce a new product, whose income and expenditure forecasts are:

End of Year	Gross Income	Operating Expenses	Investment
1	$0	$750K	$
2	0	500K	3800K
3	1500K	1000K	2000K
4	2000K	1200K	800K
5	3000K	1100K	700K
6	4500K	1600K	300K
7	5500K	1700K	
8	6000K	2000K	
9	5750K	2200K	
10	5500K	1800K	200K
11	5000K	1800K	
12	5000K	1600K	
13	4500K	1400K	
14	4000K	1200K	
15	3000K	1000K	

The $750,000 expense in year 1 is for R&D and since the company is large enough, it can be expensed. The investments are for the following items:

EOY 2	Working capital	$500,000
	Building (20-year life, salvage = $500,000)	2,000,000
	Land for building (to be sold in 20 years for $600K)	300,000
	Machine(12-year life, salvage = $200,000)	1,000,000
		$3,800,000
EOY 3	Additional working capital	$700,000
	Machine (8-year life, salvage = $300,000)	1,300,000
		$2,000,000

The remaining investments are for more working capital. The income cash flows are from the new product's sale. To help fund the new building a loan for $1,000,000 is to be made at the beginning of year 3, which will be repaid in 5 equal principal payments of $200,000 each, plus 8% interest on the unpaid balance. Another loan of $1,500,000 is to be obtained at the beginning of year 4. This loan is to be paid back in 10 equal annual installments of

$150,000 each, plus 6% on the unpaid balance. The company's effective tax rate is 34%, and capital gains are taxed at a rate of 15%. (a) Determine the net cash flow stream if all assets are depreciated by the straight-line method. (b) Determine the net present value of the net cash flow stream if the discount rate is 12%.

5-17. The Sawyer Chemical Company is considering a new process for making phosphoric acid, which they use in manufacturing commercial fertilizer. The new machinery and equipment will have an installed cost of $2,000,000. The new process will result in excess productive capacity, but the excess can be sold to other fertilizer manufacturers, and the cost of in-house acid will be reduced. The total annual savings and excess acid sales together are estimated to be $1,500,000 per year. Operating expenses are estimated to be $500,000 per year. The equipment used in the new process has an estimated life of 6 years, with a salvage value of $200,000. Sawyer has an effective income tax rate of 52% and requires a marginal investment rate of 8%. Assume that capital gains are taxed at 15% and that the new equipment will be sold for its salvage value after 6 years. (a) Using the optimal method of depreciation, what is the net present value of the project? (b) What is the net present value of the project if a $300,000 investment in working capital is also required now? (Assume that the working capital will be recovered at the project's horizon). (c) If the project (cost = $2,000,000) is financed 60% by borrowed funds and 40% by equity funds, what is the net present value of the equity portion? (Assume that the principal of the loan, obtained at the beginning of the project, will be paid back in six equal principal payments at the end of each year and that $50,000 interest will be paid annually. Include the working capital requirement.)

5-18. A revised investment tax credit has been proposed. For assets with a recovery period of seven or more years the rate is 10%. This investment tax credit is subtracted from the initial basis of the asset. An asset costs $100,000 and has a MACRS recovery period of ten years. (In problem 4-16 you calculated the credit/deduction schedule for the asset.) Compare the investment tax credit with the Section 179 write-off. If the tax rate is 34%, which is more attractive (investment tax credit or Section 179) for an asset that costs (a) $100,000, (b) $25,000, and (c) $10,000?

5-19. Two alternative investment tax credits are under consideration. At a tax rate of 34%, which of the following is more attractive? The firm's interest rate is 12%. The first investment tax credit policy allows a 10% credit with a 10 year MACRS recovery period. The second policy allows only a 7% credit, but a 5 year MACRS recovery period is permitted. In both cases, the credit is subtracted from the basis for depreciation.

5-20. For Problem 5-19 calculate (a) a breakeven interest rate and (b) a breakeven tax rate where the firm is indifferent between the two potential investment tax credit policies.

5-21. Under earlier depreciation techniques (straight-line, declining balance, and sum-of-the-years digits) the salvage value was a lower limit on the book value. This can lead to surprising results. Compare the present worth after taxes for the following machine assuming a salvage value of (a) $30,000 and (b) $0. Use straight-line depreciation, an interest rate of 10%, and a tax rate of 39%. The machine costs $50,000 and it has a life of 30 years.

5-22. Redo problem 5-21 using SOYD depreciation.

5-23. In problem 5-21 it is possible to vary the interest rate (10%), the salvage value ($30,000), and the tax rate (39%). What is the value of each that makes the present values of the machine with nonzero and zero salvage values equal?

5-24. In problem 5-22 it is possible to vary the interest rate (10%), the salvage value ($30,000), and the tax rate (39%). What is the value of each that makes the present values of the machine with nonzero and zero salvage values equal?

5-25. A firm must choose whether to build a new manufacturing facility in country A or country B. To analyze the importance of tax policy the firm will briefly assume that the costs of the facility are the same. The first cost will be $8 million, the annual operating costs will be $2 million, the life will be 20 years, and the salvage value will be $0. The firm uses an interest rate of 10%, and both countries require the use of straight-line depreciation. In country A the depreciation period is 20 years, and in B it is 5 years. In country A the tax rate is 40%, and in country B it is 45%. In which country does the facility have a more attractive present value? By how much?

5-26. In problem 5-25, what is the breakeven tax rate for country B that would make the present values the same?

6

The Financing Function

6.1 INTRODUCTION

We have been assuming that the project's interest (discount) rate relating future cash flows to its present value is a known quantity. In reality, the discount rate is not known, but it can be estimated with sufficient precision to support investment decisions. Therefore the financing and investment decisions should be made simultaneously but separately. However, very large projects may require a specific funding strategy, and the financing and investment decisions may be directly linked.

Acceptance or rejection of future investment projects depends in part on how the projects are to be financed. As we saw in Chapter 1, the firm is an *intermediary* which obtains funds from owners and other investors and invests these funds in projects. In short, the *interest* or *discount rate* used in evaluating investment projects is linked to the cost of financing these projects. This financing cost is commonly called the *cost of capital* and is usually expressed as a rate (percent per year).

In this chapter, we discuss how to measure this cost of capital. We begin by recalling from Chapter 1 that an appropriate objective for the firm is maximizing the shareholders' future wealth, which is equivalent to maximizing the firm's present value (in a perfect capital market). Maximizing the firm's present value involves discounting future cash flow streams at some interest rate. Generally, this interest rate is related to the risk of the project, with high risk projects requiring a high discount rate.

The firm needs to invest in future projects whose internal rates of return are greater than the firm's cost of capital (assuming no restrictions on the supply of capital or interdependencies among projects). This will theoretically maximize the market price of the firm's shares.

The cost of capital has at times been controversial because there have been widespread theoretical differences as to how to measure it. Authors such as Solomon (1963) and Durand (1959) maintain that an optimal balance can be struck between the amounts of equity and borrowed funds to minimize the cost of capital; whereas others, notably Modigliani and Miller (1958, 1959), advocate that the cost of capital is constant and independent of the ratio of borrowed funds to equity funds.

More recent work supports the view that the capital structure can be managed to reduce the cost of capital. Some level of borrowing or leverage is advantageous, but extreme levels of borrowing are linked to the risk of bankruptcy (see Leland, 1994).

6.2 COSTS OF CAPITAL FOR SPECIFIC FINANCING SOURCES

The firm's funds are derived from three principal sources (debt, equity, and retained earnings), but no one source of capital should be linked with a specific project. The reason is that the firm raises capital as an entity. It cannot continually finance by borrowing (debt) without building its equity base by retaining earnings or by the selling of additional equity shares. For most firms these specific sources of funds vary somewhat over time. Moreover, since the cost of capital is fundamentally concerned with valuing the firm as a whole, we must use an *overall* cost of capital as the base acceptance criterion for proposed projects, even though the firm may employ one type of financing for one project and another type for another project. It is the firm's overall cost of capital that establishes the lower boundary for the firm's opportunity costs.

To measure the firm's overall cost of capital, we must consider the costs of each method of obtaining capital. These are *explicit* costs rather than opportunity costs. For example, the explicit cost of debt capital is interest, and the explicit cost of equity capital is dividends. The explicit cost of any financing source is a discount rate, which equates the present value of the funds *received* by the firm (net of underwriting and other costs at time $t = 0$) to the present value of the expected future funds outflows. Such outflows may be interest, repayment of debt principal, dividends, or premium redemptions of convertible securities. So, for any source of capital, the explicit financing cost can be determined by solving the following equation for k:

$$P_0 = C_0 + \frac{C_1}{1+k} + \frac{C_2}{(1+k)^2} + \cdots + \frac{C_t}{(1+k)^t} \tag{6.1}$$

where: P_0 = the net amount of funds received by the firm at time $t = 0$,
C_0 = underwriting and other flotation costs at time $t = 0$,
C_t = future cash outflows for $t = 0, 1, \ldots, N$,
k = the explicit cost of that source of capital.

In what follows we describe measuring the explicit costs of each source of capital and combining these costs into an effective cost for the firm. Note, however, that our constant concern is the present and future costs of capital and not the firm's historic costs. While historic costs can and do affect the firm's ability to raise new capital, and therefore its cost, our focus is in the incremental cost of new capital. Except as they may affect the cost of any retained earnings, past costs of financing have no bearing.

In calculating the explicit costs of each source of capital, all money amounts are stated on an after-tax basis so that the firm's overall cost of capital is after-tax as well. Once the explicit source costs are determined, they are combined into the firm's weighted average, or overall, cost of capital. The weighted average cost of capital (WACC) then embodies the concept of a minimum attractive rate of return, or a threshold value for judging the acceptability of projects. Adjustments are often made based on the riskiness of the projects relative to the typical risks of the firm's projects.

6.3 COST OF DEBT CAPITAL

Borrowed funds, or debt capital, have many sources. Short-term loans may be obtained from banks and investment companies by borrowing on promissory notes or open lines of credit. Long-term loans are obtained from financial underwriting firms and the public by offering bonds or other forms of debt for sale. In either case the methodology is the same. The major difference between the cost of *debt* capital (short or long term) and the cost of other forms of funding is that interest payments on debts *are deductible from ordinary income* in many countries, and so the net cost of debt capital differs on before– and after–tax bases. To calculate the cost of debt, we simply use Equation (6.1) with *after-tax* cash flows.

6.3.1 Short-term capital costs. The cost of short-term debt financing is the effective annual (periodic) interest rate, after income taxes. Here, there are usually no flotation costs, so the problem is to convert a nominal interest rate into an effective cost of capital, which considers both the number of compounding periods and the tax rate. As in Section 2.6, we let

r = the nominal annual interest rate,
M = the number of compounding periods per year,
k_t = the effective after-tax annual interest rate,
T_e = effective income tax rate.

Then:

$$k_t = \left[\left(1 + \frac{r}{M}\right)^M - 1\right](1 - T_e) \qquad (6.2)$$

gives the loan's effective annual after-tax cost of capital.

Example 6.1 Effective interest rate after tax
$20,000 is borrowed at an annual interest rate of 6% with quarterly interest. If the effective income tax rate is 40%, what is the loan's effective interest rate or effective cost of capital?

$$k_t = \left[\left(1 + \frac{0.06}{4}\right)^4 - 1\right](1 - 0.40) = 0.0368$$

6.3.2 Capital costs for bonds. A bond is a written promise to pay, which is sold by the firm into the financial market. The firm receives immediate cash proceeds and promises to pay the face value of the bond at its maturity plus regular interest payments at a nominal rate of interest based on the bond's face value. Interest payments are usually made semiannually, but other periods can be used. When interest payments are made more often than annually, then the effective interest rate must be used, as for short-term securities in Section 6.3.1 and Example 6.1.

The price (P) actually received from the sale of a bond usually differs from its face value. If the price received is less than the bond's face value, the bond is being sold at a discount; and, if the price received exceeds its face value, the bond is being sold at a premium. When the price paid for a bond differs from its face value, the premium or discount must be amortized by the issuer for federal income taxes in the United States.

Using straight line amortization over the number of payment periods, one MNth of the total discount or premium is added to or subtracted from taxable income for each period. To put this more concretely, let F be the bond's face value at maturity; then the per period amortization amount (either discount or premium) is $(F - P)/MN$. If the bond is sold at a discount ($F > P_0$), then amortization is an expense to be subtracted from taxable income; and if it is sold at a premium ($P_0 > F$), then it is added to taxable income.

Calculating the after-tax cost of capital for a bond requires solving a present value function that relates the bond's present and future cash flows. To formulate this equation, let

P_0 = the present value of net cash inflow to the firm
F = the bond's face value to be paid in N years upon the bond's maturity
N = the number of years to maturity
k = the unknown after-tax effective cost of capital
r = the coupon or nominal rate of interest on the face value, F
M = number of interest payments per year (for annual payments, $M = 1$; for semiannual payments, $M = 2$)
$k_M = \sqrt[M]{1 + k} - 1$, the effective interest rate per compounding period. While this is theoretically correct and slightly more accurate than approximations, other sources will often use the approximate rate of k/M
S = selling expenses of the bond at $t = 0$, such as advertising and legal fees
T_e = effective tax rate of the firm

Now, upon sale of the bond(s) to the underwriter (and then to the public), the issuing firm receives P units of currency and a bill for S units for selling expenses, so the net after-tax cash inflow to the firm is $P_0 = P - S$. The selling expense, like the discount or premium, is not a deductible expense in the year of payment; rather, it must be amortized over the bond's life. So it is a prepaid expense that results in tax savings in future years. The periodic interest payments paid by the firm to the bondholder are also deductible expenses against ordinary income.

So we may write the present value function in terms of the unknown cost of capital factors as follows:

$$P_0 = F(P/F, k_M, MN) + [F(r/M)](1 - T_e)(P/A, k_M, MN) \\ - (F - P + S)(T_e)(P/A, k_M, MN) / (MN) \quad (6.3)$$

This equation is then solved for the unknown after-tax cost of capital, k. Note that Equation (6.3) consists of four cash flows (or their equivalents in terms of tax savings):

1. P_0, the present net cash inflow from the bond's sale,
2. F, the future after-tax cash outflow (the bond's face value at maturity),
3. $[F(r/M)](1 - T_e)$, the net after-tax outflows due to the periodic payment of interest at an annual rate r every M periods per year. $[F(r/M)]$ is the required interest payment of the bond.
4. $(F - P + S)(T_e /MN)$, the after-tax cash equivalent of the tax consequences due to the amortization of the bond discount (or premium) and the selling expenses. Most textbooks omit the tax consequences of the discount or premium and the selling expenses, including the fact that it needs to be amortized over time.

The following example demonstrates the approach.

Example 6.2 After-tax cost of capital from a bond
Riley Manufacturing will issue a series of bonds whose face value is $1000 and whose interest rate is 8% per year, payable semiannually as a nominal rate of return. The bond series matures in 10 years. The bonds have been bid on by several underwriters, and the highest bid received has been $910, less selling expenses of $2.00 per bond. If the company's effective tax rate is 40%, what is the effective after-tax cost of this new borrowed capital to Riley Manufacturing?

Solution
P_0 = $910 − $2 = $908 per bond
S = $2 per bond
r = 0.08 per year
M = two periods per year (semiannual interest payments)
N = 10 years
F = $1000 per bond
T_e = 0.40; $(1 − T_e) = 0.60$

The interest payment is $[F(r/M)] = [1000(0.08/2)] = \$40$ per period. From Equation (6.3),

$$\$908 = \$1000[P/F, k_M, 2(10)] + \$40(0.60)[P/A, k_M, 2(10)]$$
$$- (\$1000 - \$910 + \$2)(0.40)[P/A, k_M, 2(10)] / (2\times 10).$$

This reduces to:

$$\$908 = \$1000(P/F, k_M, 20) + \$24(P/A, k_M, 20) - \$1.84(P/A, k_M, 20)$$
$$\$908 = \$1000(P/F, k_M, 20) + \$22.16(P/A, k_M, 20)$$

With a financial calculator, $k_M = 2.82\%$, or $k = (1.0282)^2 - 1 = 5.72\%$.
The value of amortizing the price discount and selling expenses is

$$1.84(P/A, k/2, 20) = \$27.84.$$

While comprising only 3% of the current value of the bond, it is a significant cost.
Note that this can be approximated by finding the pre-tax rate on the bond and multiplying by $(1 − T_e)$. Using the pre-tax semiannual interest payment of $40 and the $908 and $1000, the pre-tax rate is 4.72% semiannually or 9.66% annually. If we multiply this rate by 0.60, the approximate after-tax rate is 5.80%. This approximation is shown, because it is used in some sources and because it shows how the discount in the bond's initial price raises the interest rate and the tax deduction lowers it.

6.4 COST OF PREFERRED STOCK

Preferred stock is a security that is senior to common stock, at least in its claim on the firm's after-tax earnings. Unlike a bond, there is no guarantee of a return to the shareowner of preferred stock so there is no prior or preferential claim on the firm's assets as there is with a bond; however, if dividends are declared, then preferred shareholders generally have prior claim to the firm's after-tax income.

Because preferred stock represents ownership in the firm and there is no contractual requirement for the redemption of stock, its cost is:

$$k_P = \frac{D_P}{P_0} \qquad (6.4)$$

where D_P = the preferred stock dividend, paid from *after-tax* earnings, and P_0 = the present value (price) of the preferred stock issue.

So, if a firm sells a 9% preferred stock issue ($100 par value) with a current price of $96 per share, the cost of the preferred issue would be:

$$k_P = \frac{0.09(\$100)}{\$96} = 0.09375 \text{ or } 9.38\%$$

This cost is not adjusted for income taxes because dividends are paid from after-tax earnings. So the explicit cost of preferred stock is usually significantly greater than the cost of debt financing.

Note that Equation (6.4) does not amortize any issuance costs, such as printing and clerical costs. Since there is no finite life for preferred stock, the expenses of its sale simply reduce the price received by the firm.

Not all companies have preferred stock. It is often used where a company is in need of cash, and arranges special financing. For example, at one time the US Treasury owned over $2 billion in preferred stock in General Motors as part of the US government's Troubled Asset Relief Program (Mitchell and Bater, 2010).

6.5 COST OF EQUITY CAPITAL (COMMON STOCK)

While the cost of equity capital is difficult to measure exactly, several valuation models, based on simplified assumptions, provide relatively easy approximations. We present a *dividend valuation* model in which a constant stream of dividends are to be paid, and two growth models that require growth in the dividend stream. This is followed by the Capital Asset Pricing Model (CAPM), which does not require the use of dividends.

6.5.1 Dividend valuation model. The financial stock market establishes the present value of a common stock by daily trading activities, in which willing buyers and sellers exchange share ownership. If we assume for the moment that a stock price remains constant, equity shareholders would receive only dividends from the firm for their investment. The income stream is the series of cash dividends to be paid. From the firm's standpoint the stock's present traded price (value), P_0, is related to the infinite stream of future dividends, D_t, by the relationship:

$$P_0 = \sum_{t=0}^{\infty} \frac{D_t}{(1+k_e)^t} \qquad (6.5)$$

If uniform dividends are assumed, that is, constant D, then Equation (6.5) becomes:

$$P_0 = D \sum_{t=0}^{\infty} (1+k_e)^{-t} = \frac{D}{k_e} \qquad (6.6)$$

for $-1 < k_e < \infty$. In other words, from Equation (6.6), the cost of equity capital is the dividend-price ratio, or:

$$k_e = \frac{D}{P_0} \qquad (6.7)$$

This equation is based on the assumptions that future dividends will remain unchanged, the stock price is constant, and the shareholders' tax situation may be ignored.

6.5.2 The Gordon-Shapiro growth model. The Gordon-Shapiro (1956) model also estimates the equity cost of capital using a stream of dividends and the current market price. However, unlike the dividend valuation model in which dividends are assumed to remain constant over time, the Gordon-Shapiro model incorporates a growth factor for earnings. Prospective investors value potential growth in earnings and therefore place a higher value on shares with earning growth than on shares without growth.

The starting point for the Gordon-Shapiro growth model is Equation (6.5). We do not assume a constant D as before; rather, we assume that a portion of total annual earnings will be retained by the firm and the balance paid as dividends. So, subsequent earnings will be based in part on prior retained earnings, and subsequent dividends will also be partly based on prior retained earnings.

Let
E_0 = after-tax earnings per share at $t = 0$
b = fraction of earnings retained by the firm (not paid as dividends)
D_0 = dividends paid at $t = 0$

Then the fraction of earnings retention is simply:

$$b = \frac{E_0 - D_0}{E_0} \qquad (6.8)$$

We assume that b is constant over time, although it may be reevaluated each time the cost of capital is calculated. Initially, retained earnings are simply bE_0, and the dividends paid are $(1-b)E_0$.

We now let B_0 equal the initial book value of the stock per share, and r_b is the simple rate of earnings to book value at $t = 0$, or $r_b = E_0/B_0$. Again, r_b is assumed constant, but it may be reevaluated each time the cost of capital is estimated.

In each period we assume that the retained earnings will be added to the stock's book value and that the total book value, $B_t + bE_t$, will earn at the rate r_b for the firm in the next period. So the amount $r_b(bE_t)$ is the *earnings on retained earnings*. Consequently, at $t = 1$ we have:

$$E_1 = E_0 + r_b b E_0 = E_0(1+r_b b).$$

Similarly, at $t = 2, \ldots, N$ we have:

$$E_2 = E_1 + r_b b E_1 = E_1(1+r_b b)$$

$$\vdots$$

$$E_N = E_{N-1} + r_b b E_{N-1} = E_{N-1}(1+r_b b)$$

which leads to the relationship:

$$E_t = E_0(1+r_b b)^t. \tag{6.9}$$

In general, the dividends paid are:

$$D_t = E_t - bE_t = (1-b)E_t; \tag{6.10}$$

and if Equation (6.9) is multiplied by the factor $(1-b)$, we have:

$$(1-b)E_t = (1-b)E_0(1+r_b b)^t \tag{6.11}$$

Now, the left side of Equation (6.11) is simply the dividends paid in year t, and the right side is $D_0(1+rb)^t$, where D_0 is the dividend paid at $t = 0$. So, the relationship between the initial and later dividends is:

$$D_t = D_0(1+r_b b)^t \tag{6.12}$$

Substituting this value into Equation (6.5) gives the basic valuation equation of the Gordon-Shapiro model:

$$P_0 = \sum_{t=0}^{\infty} D_0(1+r_b b)^t (1+k_e)^{-t}. \tag{6.13}$$

If we convert Equation (6.13) to continuous compounding to simplify the summation, we have:

$$(1+r_b b)^t \sim e^{r_b b t} \quad \text{and} \quad (1+k_e)^{-t} \sim e^{-k_e t}$$

so that:

$$P_0 = D_0 \int_{t=0}^{\infty} e^{r_b b t} e^{-k_e t} dt = D_0 \int_{t=0}^{\infty} e^{-t(k_e - r_b b)} dt$$

or

$$P_0 = D_0 \left(\frac{1}{k_e - r_b b} \right) \tag{6.14}$$

Solving Equation (6.14) for k_e, we have:

$$k_e = \frac{D_0}{P_0} + r_b b \qquad (6.15)$$

and if we substitute the equivalents $r_b = E_0 / B_0$ and $b = (E_0 - D_0) / E_0$, then:

$$k_e = \frac{D_0}{P_0} + \frac{E_0 - D_0}{B_0} \qquad (6.16)$$

So the Gordon-Shapiro growth model for equity cost of capital consists of two terms all stated in initial values per share. The first is simply the dividend divided by the market price, which corresponds to the basic model in Section 6.5.1; the second is the growth term, which is the ratio of retained earnings to book value. In other words, if growth is provided for by retention of earnings, then the basic cost of equity capital is increased by the ratio of retained earnings to book value.

Equation (6.16) is now usually expressed as:

$$k_e = \frac{D_1}{P_0} + g \qquad (6.17)$$

where D_1 is the dividend expected to be paid at the end of year 1 and g is the expected growth rate of dividends.

Equation (6.17) assumes that there are no flotation costs—expenses that must be paid to sell additional stock. Unfortunately this is generally not true; flotation costs average about 7% of the cost of the stock (Lee et al., 1996). Where there are flotation costs (f_c), Equation (6.17) may be rewritten as:

$$k_e = \frac{D_1}{P_0 + f_c} + g \qquad (6.18)$$

6.5.3 The Solomon growth model. A model similar to the Gordon-Shapiro growth model was suggested by Solomon (1963, pp. 55–62), which in our present notation is:

$$k_e = \frac{D_0}{P_0} + \frac{E_0 - D_0}{P_0} \qquad (6.19)$$

The principal difference is that the denominator of the growth term in Solomon's model is based on *market value* (P_0) rather than on *book value* (B_0) as in the Gordon-Shapiro model, Equation (6.16). If the firm's stock is valued on the market at a price higher than the stock's book value, then Solomon's estimate of the cost of capital is lower. On the other hand, if the firm's shares are depressed in value on the market, Solomon's estimate of the cost of capital exceeds that of Gordon-Shapiro. In either case, a conservative estimate of the cost of capital is the higher value of k_e.

Example 6.3 Cost of equity
The present after-tax earnings of Royal Machine are $3.40 per share of common stock, which sells for $36.00 per share. Current dividends are $1.67 per share. The ratio of retained earnings to total earnings is expected to remain constant in the future. The stock's present book value is $19.50 per share. Calculate the firm's cost of equity capital (a) by the Gordon-Shapiro model and (b) by the Solomon model.

Solution
a. Gordon-Shapiro model
$D_0 = \$1.67$ per share
$P_0 = \$36.00$ per share
$B_0 = \$19.50$ per share
$E_0 = \$3.40$ per share
From Equation (6.16),
$k_e = D_0/P_0 + (E_0 - D_0)/B_0 = \$1.67/\$36.00 + (\$3.40 - \$1.67)/\$19.50$
$k_e = 0.0464 + 0.0887 = 0.1351$, or 13.5%.

b. Solomon model
From Equation (6.18),
$k_e = D_0/P_0 + (E_0 - D_0)/P_0 = \$1.67/\$36.00 + (\$3.40 - \$1.67)/\$36.00$
$k_e = 0.0464 + 0.0481 = 0.0945$, or 9.45%.

In both cases the equity cost of capital, k_e, is an *after-tax* rate since the firm's earnings and dividends are both after-tax amounts.

There are two major problems with the dividend growth models in today's market. First, many growth companies intentionally reinvest their earnings and pay no dividends. When there are no dividends, as is true with many companies, the dividend growth models simply cannot be used. Second, dividend paying companies manage those dividends very closely. Dividends are paid by firms with "permanent" operating cash flows (Jagannathan et al., 2000), and act as a signal to the market that those cash flows are continuing. Dividends are managed to be fairly stable with slow and steady increases, and are often much less volatile than actual cash flows or stock prices. Dividend growth models should be used only with caution, because they *may* reflect the firm's ability to manage the dividend more than the actual success of the firm.

6.5.4 Note on book value of stock. In the absence of treasury stock (see Section 6.5.9), the balance sheet net worth of the firm's *common* stock (which in this case includes retained earnings) may be defined as

Net worth (common) = total assets − total liabilities − preferred shares.

The *book value*, B_0, per share of the common stock is then simply:

$$B_0 = \frac{\text{net worth of common stock}}{\text{number of shares of common stock}}$$

This definition of B_0 is sufficiently precise to calculate the firm's cost of equity capital.

6.5.5 Capital Asset Pricing Model (CAPM). The Capital Asset Pricing Model deals with the risks of investments as stand-alone purchases and as items in a portfolio of investments. Some basic features of the CAPM are discussed so that another model for defining k_e can be defined. The theoretical derivation of this model will receive more attention in Chapter 16.

A few investments are considered risk-free. The *risk-free rate*, r_f, is the minimum interest rate for a time period; the interest rate required for an investment that has no risk. Any investment that contains risk will have a required interest rate that is higher than the risk-free rate. The risk-free rate will vary depending on the length of the investment; the risk-free rate for a 10-year investment should be higher than the risk-free rate of a 6-month investment because of the longer time horizon. One definition of the risk-free rate is the rate paid by US Treasury bonds; the 10-year bond rate is usually used for comparing most stocks (Titman and Martin, 2007, p. 138). A few other countries around the world whose debt is highly rated meet this same definition. This is the basis to which all other interest rates are compared.

The theoretically perfect portfolio of stocks is defined as the *market portfolio*; a mixture of all of the stocks that are currently available in the stock market. The risk of this market portfolio is known as the *market risk*, and this market risk requires a certain return. Because the market risk is higher than zero, the market carries a required *market rate*, r_M. Most portfolios do not have as many stocks as the market portfolio, so each stock that is present impacts the risk of the portfolio. A stock portfolio does not require a huge number of stocks; much of the diversifiable risk can be minimized with as few as 10 to 11 stocks, as long as they are well chosen. Just as the market is widely diversified, a well-balanced investment portfolio is also well diversified. A primary conclusion of the Capital Asset Pricing Model is that the risk of an individual stock depends on its contribution to the risk of a portfolio of stocks.

When the stock market goes up (increases in value), many stocks move up also. These stocks are positively correlated. Some stocks actually tend to move in the opposite direction of the market averages, and these are considered to have a negative correlation with the market. Some stocks tend to move up (or down) more than the market average, and some stocks tend to change less than the market average. The *beta coefficient*, β, is an indicator of an individual stock's risk relative to the market, and is defined as:

$$\beta_i = \left(\frac{\sigma_i}{\sigma_M} \rho_{iM}\right) \qquad (6.20)$$

where β_i = the beta coefficient
σ_i = standard deviation of stock i's return
σ_M = standard deviation of the market's return
ρ_{iM} = the correlation between stock i's return and the market return

The beta coefficients for a vast number of stocks are readily available in financial publications and their websites. It is important to note that even though a stock has a beta of 1.0, this does not necessarily mean that the volatility of the stock is the same as the volatility of the market. CAPM is largely a model that was created as a way to build portfolios such that risks from individual stocks mostly cancel out. While some stocks are well correlated with the market, many stocks have different risk characteristics than the market as a whole, and basic model assumptions such as Brownian motion do not fit well for individual stocks (Derman, 2011).

The market risk premium is the added interest required by investors to invest in the market instead of in risk-free investments. It is simply the difference between the market rate, r_M and the risk-free rate, r_f. So the market risk premium is defined as:

$$RP_M = r_M - r_f. \tag{6.21}$$

We can measure the risk premium of an individual stock by multiplying the market risk premium by its beta coefficient, or:

$$RP_i = \beta(RP_M) = \beta(r_M - r_f) \tag{6.22}$$

The required return for a given stock is then the risk-free rate plus the individual stock's risk premium, or:

$$k_e = r_i = r_f + \beta(r_M - r_f) \tag{6.23}$$

Equation (6.23) is known as the Security Market Line, and is an alternative way of identifying the required return for a stock. This is also a method of determining the cost of equity, because $k_e = r_i$.

The Capital Asset Pricing Model is widely taught, and appears in most introductory financial textbooks. In spite of this, the model is not accurate in many cases. Fama and French (1992) demonstrated that firm size (as measured by market capitalization) and book-to-market ratios provide more accurate predictions of a stock's future returns. Moreover, beta is calculated different ways, and is dependent on granularity and the time frame reviewed. Beta can be calculated using daily, monthly, or quarterly data, and this difference in granularity creates different results. Also, various research firms use stock information over different time periods. While the usual market comparison is the S&P 500 index, this is not universally used. Due to these differences, publications may list a significantly different value of beta for the same firm. For instance, three different research reports gave three different beta results for General Electric on the same day ranging from 1.1 to 1.88.

Beta is a measure of systematic risk; it measures the risk of the firm (or a portfolio), relative to the risk of the market as a whole. The resulting cost of equity combines the sources of equity, including retained earnings, and common stock.

Pension obligations may or may not be fully reported in the balance sheet, (when they are not they are known as off-balance sheet pension liabilities). Jin, Merton, and Bodie (2006) report that when pension obligations are funded and reported, the firm's stock price accurately reflects the funded pensions. However, in many companies this funding is quite large, and may even approach the market capitalization of the firm. This is not properly reflected in the firm's cost of capital, which typically fails to recognize the assets that are in place. When funded pension liabilities are quite large, the cost of capital is overestimated, and the discount rate required for new projects is too high.

If pension obligations are not fully reported, then the off-balance sheet financing can affect the financial analysis (including the financial ratios) of a firm. When underfunded pension liabilities are not reported, total debt is understated, affecting all debt-related financial ratios. The presence of off-balance sheet obligations improperly lowers the cost of debt. Failure to report liabilities on the balance sheet is usually done for only one reason: to hide bad news from potential creditors and investors.

Example 6.4 Cost of equity using CAPM
Royal Machine has a beta of 1.2. The long-term return on US Treasury bonds is 4.5%, and the market return is 11.0%. What is the cost of capital for Royal Machine using the CAPM method?

$$k_e = r_i = r_f + \beta(r_M - r_f)$$
$$k_e = 0.045 + (1.2)(0.110 - 0.045) = 0.123 \text{ or } 12.3\%$$

6.5.6 Cost of retained earnings. For many firms, much of their financing is derived from retained earnings. These retained earnings might appear to be free to the firm, since the firm generated the retained earnings in the first place. This is not true for two reasons. First, the firm does not own the retained earnings—the *shareholders do*. Second, there is an opportunity cost, that is, the dividend foregone by the shareholders *out of their own earnings*.

The cost of retained earnings is obtained by solving Equation (6.18) or Equation (6.23) for k_e, using the current market price for P_0. k_e represents the minimum return that investors *expect* the firm to receive on its investments.

If the firm cannot generate projects whose return is at least k_e, then the firm should distribute its current earnings to its shareholders so they can invest in other firms earning at least k_e. Alternatively, the firm could repurchase some of its stock to earn k_e. If the firm were to invest in projects with expected returns *lower* than k_e, then the expected dividends to shareholders would decline, and the shareholders would suffer a loss in expected future wealth. So k_e is the minimum opportunity cost of continuing dividends that will provide at least the current level of expected future shareholder's wealth.

With taxation, however, the shareholder may lose some of the dividend to taxation. Tax law regarding personal income taxes on shareholders varies from year to year. Currently, dividends (on stocks held at least 60 days) are taxed at a 15% rate for people in the 25% marginal income tax bracket or higher. For those paying less than a 25% marginal tax rate, dividends are not taxed. Short-term capital gains (for stocks held for one year or less) are taxed at the same rate as income. Long-term capital gains (for stocks held for more than one year) are currently taxed at the same rate as dividends. Federal tax law typically sets taxes on long-term capital gains lower than dividends.

Income taxes are of little or no consequence to some shareholders such as pension funds, tax-exempt foundations, and universities. These shareholders may have investment opportunities that approach or equal the before-tax return on equity capital of the firms in which they own stock. Thus, there is no minimum yield of the firm at which all shareholders are better off if the firm reinvests its earnings rather than paying greater dividends.

A second approach to evaluating the cost of retained earnings is Solomon's (1963, pp. 53–55) "external yield criterion." This implies that the firm should evaluate *external* investment opportunities and that it should use as an opportunity cost the yield presented by the best foregone external investment opportunity. Solomon also suggests that, in any reasonable financial market, the firm can invest after-tax earnings in another firm.

A word of caution should remove any possible argument about connecting a particular project to an explicit cost of capital, such as retained earnings. None of the preceding analysis implies that paying dividends *never* represents the best alternative to internal investment. Financial theorists agree that a stable dividend policy enhances the market value of a firm's stock and hence reduces the cost of equity capital in the long run. Thus, retention is not required even if there are investment opportunities that provide returns better than k_e. Here we have established a correct criterion for investment in projects that must be exceeded for approval of an internal project. Once the project is approved, then financing it is a separate issue. For example, a project's return may exceed the screening standard, k_e, but priority on retained funds may be given to the payment of dividends, and the project may actually be financed from external funds. As far as the investment decision is concerned, the appropriate standard for measuring the cost of retained earnings remains k_e.

6.5.7 Treasury stock. Many firms routinely purchase their own stock at market prices. This is done for a variety of reasons, including employee profit sharing, executive stock options, or as a strategic investment by the firm as noted above. Stock that has been purchased without being retired is known as *treasury stock*. Accounting rules do not consider treasury stock as an asset, because it is stock in its own firm. It is listed as negative equity, because its purchase decreases the available cash in the firm. Treasury stock is valued at its purchase price, not at its issue price (like common stock), so the presence of treasury stock complicates the value of a firm's common stock and retained earnings.

Treasury stock complicates the determination of k_e. Lamdin (2001) points out that if a firm repurchases stock and if k_e is based on the dividend valuation model [Equation (6.18)], the cost of equity will be estimated too low. However, the Security Market Line [Equation (6.23)] will normally provide a higher and more accurate result. Research supports this conclusion; CAPM tends to provide a higher and more stable estimate of the cost of capital than does the dividend model, in spite of being an imperfect model (Prescott, Whittaker, and Eschenbach, 2002).

6.6 WEIGHTED AVERAGE COST OF CAPITAL (WACC)

The explicit costs of capital for each individual source (short-term debt, bonds, equity, and retained earnings), may be combined to find the firm's average or effective cost of capital. The method weighs each cost and then calculates the firm's weighted average cost of capital, expressed as a rate of return. The weighting depends on how much of total capital is represented by each source of capital.

Example 6.5 Weighted average cost of capital
To illustrate the mechanics of calculating the weighted average cost of capital, suppose that Royal Machine's capital structure is:

	Amount	Proportion
Short-term debt	$500,000	0.05
Bonds	1,000,000	0.10
Preferred stock	1,500,000	0.15
Common stock	6,000,000	0.60
Retained earnings	1,000,000	0.10
	$10,000,000	1.00

6.6 Weighted Average Cost of Capital

Suppose the firm's after-tax explicit costs of capital are, respectively:

	After-Tax Cost
Short-term debt	3.08%
Bonds	5.56
Preferred stock	10.00
Common stock	12.30
Retained earnings	12.30

The explicit source costs are weighted by their relative amounts, and the firm's weighted average cost of capital is calculated as follows:

(1) Source of Financing	(2) Proportion	(3) Explicit Cost	(4) Weighted Cost = (2)(3)
Short-term debt	0.05	0.0308	0.0015
Bonds	0.10	0.0556	0.0056
Preferred stock	0.15	0.1000	0.0150
Common stock	0.60	0.1230	0.0738
Retained earnings	0.10	0.1230	0.0123
		Weighted average cost of capital:	0.1082

Because common stock and retained earnings carry the same cost of capital, they can be combined into one item, equity. The above procedure for determining the weighted average cost of capital can be summarized in Equation (6.24):

$$\text{WACC} = w_d k_d (1 - T) + w_{ps} k_{ps} + w_e k_e \qquad (6.24)$$

where WACC is the weighted average cost of capital
w_d = proportion of debt
w_{ps} = proportion of preferred stock
w_e = proportion of common equity
k_d = cost of debt (before tax)
k_d = cost of preferred stock
k_d = cost of common equity
T = tax rate

So one can approximate the firm's *historic* cost of capital by using a weighted average in which the weights are the relative proportions of capital from each source. This definition of WACC assumes a constant cost of equity, a constant cost of debt, a constant tax rate, and a constant ratio between the amount of equity and debt. This use of both debt and equity to finance the company is known as the capital structure of the firm. It is measured using the leverage ratio, L, also known as the debt-equity ratio:

$$L = \text{Total debt} / \text{Total equity} \qquad (6.25)$$

Research shows that this approach is valid as long as the company maintains all of these as constant (Miles and Ezzell, 1980). The problem is maintaining the leverage ratio, because the firm would need to issue more debt any time the value of equity increased, and reduce debt any time the value of equity decreased. Similarly, any time the firm borrows money (increases debt) then equity would need to raise capital to maintain a constant ratio. In reality, firms keep their debt/equity ratio fairly stable by using retained earnings as a major source of equity.

The value of debt can be easily measured. It includes all forms of debt, including any off-balance sheet accounts. The cost of debt is simply the weighted average of all debt instruments, averaging their rate to maturity. Under US Generally Accepted Accounting Principles, bonds may be valued at market price if they are available for sale, or valued at book value if they are being held to maturity. There are also two ways of looking at the value of equity: book value and market value. The book value of an asset is its purchase price minus depreciation; a fully depreciated asset has a book value of zero. The book value of a firm's equity is the firm's total assets minus the firm's liabilities, based on the balance sheet equation:

$$\text{Assets} = \text{Liability} + \text{Equity} \qquad (6.26)$$

The market value of an asset is the price for which it can be sold. The market value of a firm is the total number of common shares outstanding times the price of the stock. The shareholders' equity on the firm's balance sheet may be completely different than the market value of its shares. How different, and does it matter? A recent (2012) survey of the 30 firms that make up the Dow Jones Industrial Average showed an average market/book (also known as price/book) ratio of 3.64. In other words, the average company had a market value 3.64 times the company's book value. The highest of the companies was 15.96 (Boeing) and the lowest was 0.46 (Bank of America). Market values tend to give a much higher weighting to equity than do book values because the market value of equity varies more from book than does the market value of debt. The WACC can change considerably depending on which value is used.

There are many proponents for using market value as the primary method. The real value of an asset is its market value, and many view the value of a firm as its market value (Ross, Westerfield, and Jaffe, 2010). This also fits with the meaning of the WACC, which is market driven. The expected rate of return of equity is that rate which the market requires in order for investors to make an investment.

However, when making capital structure decisions, such as whether to increase the firm's amount of debt, managers take credit ratings into account (Kisgen, 2006). Rating agencies such as Standard and Poor's tend to use book values to assign value to a firm's assets (Kisgen, 2006; S&P, 2008). Market values of firms tend to be driven by the stock market, tend to have a short time horizon, and can be extremely volatile. Because rating agencies focus on book value, many firms do the same.

6.7 MARGINAL COST OF CAPITAL

6.7.1 Market values imply a marginal cost approach. Once the historic cost of capital has been calculated, the critical question is, does this figure represent the firm's *real* cost of capital for evaluating new projects? Several authors have argued in the past that the weighted historical average cost is the correct value to judge the acceptability of new projects. At one point in time, the clear consensus was that a marginal cost of capital should be used for judging new project acceptability.

Use of a rate lower than the marginal cost of capital would permit acceptance of projects that would lower the firm's future value (or, equivalently, the value of its common stock). Past costs of financing, used to calculate the weighted average cost of capital are sunk costs and have no bearing on the actual cost of financing new projects. It is the expected rise in future earnings that increases the firm's value. For future projects to raise earnings and dividends of the firm, the *last-added* future project—which has the lowest rate of return of accepted projects—must *have a rate of return at least equal to the cost of the last-added increment of capital used to finance it.* Otherwise, the firm's future value would decrease, since a project would have been accepted that would provide a return on the *incremental* investment *smaller* than the cost of the *incremental* capital used to finance it.

If book values of debt and equity are used, this concern is very real. However, market values reflect the marginal cost of capital. Market value of debt is based on the market price of the debt instruments and the interest due to maturity. This is marginal cost of debt. The market value of equity involves the market capitalization of the firm, which changes continuously based on the buying and selling of stock. The stock price and dividend determines the stock's return (the cost of equity), and so the market constantly updates both the cost of equity and the proportion of equity. *This is an advantage to the use of market values: they reflect the marginal cost of capital.*

The remainder of Section 6.7 explains the marginal cost of capital concept—relying on an article by Arditti and Tysseland (1973).

6.7.2 Marginal cost-marginal revenue approach. The firm is considering investment projects (A, B, ... , H), all of which are *independent;* that is, the net cash flow stream from each project does not depend on that of any other project. We also assume that a unique rate of return exists for each project (e.g., $i_A, i_B, i_C, ..., i_H$) and that $i_A > i_B > i_C > ... > i_H$. Each project requires an initial expenditure, which is to be financed by the capital to be raised.

Figure 6.1 is a typical marginal cost – marginal revenue approach to determining the *cutoff* point for cost of capital and internal rate of return. We will explore what happens when the budget is constrained (not all projects can be funded) in Chapter 8. The costs of investment in the projects are given on the horizontal axis (cumulated), and the marginal revenue (each project's internal rate of return in descending order) is given on the vertical axis. Also plotted are both average and marginal costs of capital. Note that through project C the marginal cost of capital (MCC) and the average cost of capital (ACC) are equal; thereafter, the marginal cost of capital (MCC) increases at discrete intervals. The average cost of capital (ACC) also increases, but more slowly.

In an *ideal* world where only expected return and not risk need be considered, the cutoff point for project selection should be the total funds expended at either Q_0 or Q_1, since one would be indifferent toward project E because returns exactly equal the marginal cost of capital. At either Q_0 or Q_1, the *marginal* rate of return equals the *marginal* cost of capital. For convenience, we assume that E is accepted, but project F definitely would *not* be accepted since its marginal cost of funds exceeds its expected rate of return.

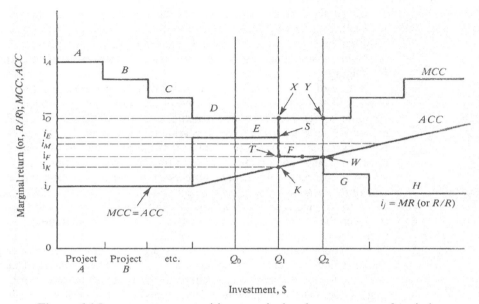

Figure 6.1 Investment opportunities, marginal and average cost of capital

6.7.3 A discounted cash flow approach. Referring again to Figure 6.1, if we stop at Q_1, where the project investment opportunity curve (i_A, i_B, ... , i_H) intersects the marginal cost of capital curve (MCC), project E is accepted, project F is rejected, and the marginal cost of capital is i_E.

Looking at the effect on the firm's value of having accepted projects A–E, the firm and its shareholders are better off by the sum of the NPVs from projects A–E when discounted at the average cost of capital ACC_K. So the effect on the firm's value is computed using the average cost of capital since this is the average cost of the bundle of funds that financed this bundle of projects. *We do this even though the higher i_E is used as the marginal cutoff rate.*

Another, more intuitive approach is to view the firm's total prospective return in dollars as the area $ORSQ_1$ (which is the sum of the incremental interest return rates times incremental funds used, the cost of the funds used as the area $OJSQ_1$, (which is the dollar area under the marginal cost curve), and the *net* dollar return as the area $JRSJ$ (which is simply the total dollar return for projects A–E less the dollar cost of the funds used). Hence, an approximate average total internal rate of return for the project bundle equals the area $ORSQ$, divided by the funds used, OQ_1; likewise, an approximate average cost of capital equals the area $OJSQ$, divided by the funds used, OQ_1; and finally, an *average* net rate of return equals the area $JRSJ$ divided by the funds used, OQ_1. In every instance, the three average interest rates apply to the entire bundle of projects A through E, but the cutoff rate is the marginal cost of capital, not the average. This approach is intuitive and approximate rather than exact because the projects may have different horizons and/or cash flow patterns.

To repeat, the marginal cost of capital curve is used as the cutoff rate by comparing it with the project's rate of return on the marginal investment opportunity schedule. To evaluate the effect of accepting the entire project *bundle* upon the firm's value, the *average* cost of capital is used, that is, the weighted average cost of all the funds that finance the *bundle* of accepted projects.

This may be clarified by posing the opposite problem of what would happen if we were to cut off at the average cost rate. For example, suppose we were to cut off at an investment of $Q2$, using the average cost of capital rate, and accept project F. Then the total contribution of project F in dollars would be the area Q_1TWQ_2, or in terms of the rate of return and amount of capital used, $i_F(Q_2 - Q_1)$. The corresponding total cost of the funds used is the area Q_1XYQ_2, or in terms of the cost of capital and amount of capital used, $\bar{O}(Q_2 - Q_1)$. Obviously, the net benefit in dollars from adding project F is $(i_F - \bar{O})(Q_2 - Q_1)$, which is negative since the cost of capital $(\bar{O})$ exceeds the rate of return (i_F). Hence, project F reduces the earlier net benefit from selecting projects A–E. Since project F reduces the net present value of the previously accepted projects, it should be rejected, even though it appears marginally acceptable at the firm's average cost of capital. This conclusion is more precisely demonstrated in the next section.

6.7.4 Mathematical approach to marginal cost of capital. The firm's market value at any time t_0 may be represented by

$$V_0 = \frac{\bar{Y}}{1 + k_0} \tag{6.27}$$

where $\bar{Y}$ = the firm's future expected net cash flow stream from projects already accepted
k_0 = the firm's average cost of capital used to capitalize the income stream Y

Now, the firm's value could also be represented by

$$P_0 = S_0 + D_0 \tag{6.28}$$

where S_0 = the market value of the firm's common stock at t_0,
D_0 = the market value of its debt at t_0.

We assume that only common stock and debt are in the firm's capital structure. For simplicity, the model has only one time period. The more general N-period case is straightforward, but cumbersome algebraically and not useful for teaching. Our formulation is a variation of Modigliani and Miller's (1958), except that their income stream is in perpetuity and they assume a constant k.

The firm invests C in a project whose expected rate of return is r. The firm's new value is:

$$V_1 = \frac{\bar{Y} + C(1+r)}{1 + k_1} \tag{6.29}$$

where k_1 is the *new* average cost of capital that capitalizes the modified expected income stream.

The firm's new value could also be represented by:

$$V_1 = S_1 + D_1 + C \tag{6.30}$$

where S_1 = the new market value of the old common stock
  D_1 = the new market value of the old bonds
  C = either new stock or new debt, or both.

Under what conditions will project C be accepted? *Only if it will not diminish the value of the firm's common stock,* namely, if, and only if,

$$S_1 \geq S_0.$$

So there is a minimal rate of return, r, for the new investment C that satisfies the condition. To find this rate of return, we first subtract Equation (6.26) from Equation (6.28) to obtain:

$$V_1 - V_0 = S_1 + D_1 + C - S_0 - D_0.$$

We then assume that $D_1 = D_0$, since the new project's risk is unlikely to differ sufficiently from that of the firm's present projects to change the market value of the old bonds. So we drop the offsetting debts and transfer C to the left side, to get:

$$V_1 - V_0 - C = S_1 - S_0.$$

Earlier, we assumed a shareholder benefit condition, $S_1 \geq S_0$, so:

$$V_1 - V_0 - C \geq 0$$

or

$$V_1 - V_0 \geq C \tag{6.31}$$

is the *equivalent criterion* for an acceptable prospective project. Then, substituting Equation (6.29) for v_1 and Equation (6.27) for V_0, we have:

$$\frac{\bar{Y} + C(1+r)}{1+k_1} - \frac{\bar{Y}}{1+k_0} \geq C$$

which may be written as:

$$\left(\frac{\bar{Y}}{1+k_0}\right)\left(\frac{1+k_0}{1+k_1} - 1\right) + \frac{C(1+r)}{1+k_1} \geq C.$$

On multiplying through by $(1 + k_1)$, we have:

$$\left(\frac{\bar{Y}}{1+k_0}\right)[(1+k_0) - (1+k_1)] + C(1+r) \geq C(1+k_1).$$

Recalling that $\bar{Y}/(1+k_0) = V_0$ by definition, then:

$$V_0[(1+k_0) - (1+k_1)] + C(1+r) \geq C(1+k_1).$$

Dividing through by C and simplifying, we obtain:

$$\frac{V_0}{C}(k_0 - k_1)(1 + r) \geq (1 + k_1).$$

Finally, solving for r, we have:

$$r \geq k_1 + \frac{V_0}{C}(k_1 - k_0) \tag{6.32}$$

Since k_1 is the new average cost of capital and r was defined [Equation (6.29)] as the minimum rate of return required on investment, then from Equation (6.32) *r is obviously greater than* k_1, if $k_1 > k_0$.

The easier of the two cases is when the new investment C does not change the firm's risk characteristics, and the market evaluates k_1 as equal to k_0; then $r = k_1 = k_0$. In the second case, the new investment, C, *increases the firm's risk* and the market evaluates the firm's stock such that $k_1 > k_0$. Then, from Equation (6.29) the rate of return from C must exceed k_1. In other words, the new project must generate not only additional income, but it must generate sufficient *additional* income to raise the expected return on *total* assets from k_0 to k_1.

Example 6.6. Marginal cost of capital
Suppose that the following values apply to a particular firm:
$$k_0 = 0.145$$
$$k_1 = 0.150$$
$$V_0 = \$1{,}000{,}000$$
$$C = \$100{,}000.$$
What would be the minimum rate of return for C to be acceptable? The answer may be calculated from Equation (6.32):

$$r \geq 0.150 + (\$1{,}000{,}000/\$100{,}000)(0.150 - 0.145)$$

or

$$r \geq 0.20.$$

The firm must earn more than 20% on this new investment because it must earn enough to raise the average rate of return on *total* investment from 14.5 to 15.0%. For each dollar invested in C, there are $10 of old investments returning 14.5%. Therefore, each dollar of C must earn 5% above 15%.

This example also illustrates that accepting C would diminish the firm's value if its expected rate of return were 19%. To avoid diluting the common stock, i.e., to prevent its price from falling, V_1 must exceed $V_0 + C$. Since $V_0 + C = \$1{,}000{,}000 + \$100{,}000$, we can write $V_1 \geq \$1{,}100{,}000$. But setting $r = 0.15$ in Equation (6.26) and letting $\bar{Y} = (1 + k_0)V_0 = (1.145)(\$1{,}000{,}000)$, we have:

$$V_1 = \frac{1.145(1{,}000{,}000) + 1.19(100{,}000)}{1.15} = \$1{,}099{,}130,$$

which diminishes the firm's value.

When weighted with the old k_0, this required $r = 0.20$ gives the new k_1, and therefore $r = 0.20$ must be the marginal cost of capital:

Capital Amount	Relative Amount	Cost of Capital	Weighted Average
$1,000,000	0.909	0.145	0.1318
100,000	0.091	0.200	0.0182
$1,100,000	1.000		0.1500 = k_1

So the new project, if it returns only the new average cost of capital (15%), should be rejected; whereas if it returns the marginal cost of capital ($r = 0.20$), it should be accepted.

6.8 NUMERICAL EXAMPLE OF THE MARGINAL WEIGHTED AVERAGE COST OF CAPITAL

6.8.1. Calculation of the present weighted average cost of capital. This example demonstrates in two parts (1) how a firm's weighted average present cost of capital may be determined from balance sheet and market data and (2) how its marginal cost of capital can be estimated from the amounts and explicit costs of new capital.

Example 6.7 Calculation of the present weighted average cost of capital

The balance sheet of the ABC Company is shown in Figure 6.2. The current market (traded) price for the common stock is $29 per share, and ABC has paid dividends on it of $1.00 per share for last year. Slightly lower dividends were paid in prior years, and this growth in dividends is expected to continue over the next few years. The company's ordinary income tax rate for federal and state taxes is 40%. Calculate the company's weighted-average cost of capital.

Solution

The first step is to calculate the explicit costs of each capital source:

a. *Bonds:* For simplicity, the face value rather than the current yield to maturity (see Equation (6.3)) has been used. From the balance sheet, the face value, F is $10,000 per bond, the coupon interest rate is $r = 8\%$, interest is paid semiannually, and maturity is $N = 10$ years. Since no selling costs or amortizable discounts are mentioned, the bond's explicit after-tax cost is calculated from Equation (6.2):

$$k_i = [(1 + 0.08/2)^2 - 1](1 - 0.40) = 0.0490 = \underline{4.90\%.}$$

b. *Preferred Stock.* From the balance sheet, the preferred dividend rate is 5% on the par value of $100 per share, which through Equation (6.5) is also the after-tax cost of this source.

c. *Common Stock.* The common stock's cost is based on its book value, which is obtained from the balance sheet:

$$BV_0 = \text{total assets} - \text{total liabilities} - \text{preferred stock}$$
$$= \$1,600,000 - \$665,000 - \$100,000 = \$835,000$$

The book value *per share* is then

$BV_0/\text{number of shares} = \$835,000/30,000 = \$27.83$ per share.

Next, *retained earnings per share* is obtained from the balance sheet, recognizing that retained earnings equals total earnings (E_0) minus common dividends (D_0), or

$$E_0 - D_0 = \text{retained earnings/number of shares}$$
$$= \$40,000/30,000 \text{ shares} = \$1.333 \text{ per share.}$$

To calculate *total earnings per share* (E_0) we add back the dividends paid:

$$E_0 = (E_0 - D_0) + D_0$$
$$= \$1.333 + \$1.00 = \underline{\$2.333.}$$

Summarizing, we have
$D_0 = \$1.00$ (given in the problem)
$P_0 = \$29.00$ (given in the problem)
$E_0 = \$2.3333$ (calculated above)
$BV_0 = \$27.83$ (calculated above)

Using the Gordon-Shapiro growth model Equation (6.16), the cost of equity capital is

$$k_e = \frac{D_0}{P_0} + \frac{E_0 - D_0}{BV_0'} = \frac{\$1.00}{\$29.00} + \frac{\$1.3333}{\$27.83} = 0.0824.$$

The last step is to calculate the weighted-average cost of capital, 6.85%, as presented in Section 6.6.

Source of Financing	Amount	Proportion	Explicit After-Tax Cost of Source	Weighted Cost
Bonds	$500,000	0.348	0.0490	0.01705
Preferred stock	100,000	0.069	0.0500	0.00345
Common stock	835,000	0.583	0.0824	0.04804
	$1,435,000	1.000		0.06854

6.8.2 The future weighted average cost of capital after provision for new capital. Financial managers recognize that excess common stock, relative to outstanding debt, can *lever* additional debt without necessarily increasing the firm's overall cost of capital. In this case debt capital can be obtained without adding equity capital, and the firm's average cost of capital will drop. However, as the debt-to-equity ratio increases, at some point added equity capital is required by lending institutions that refuse more debt capital due to the firm's equity (ownership) being too "thin." So a growing firm that continually needs new capital will both borrow and seek new equity.

All values in thousands (000)

ABC Company
BALANCE SHEET
Year Ended June 30, 20XX

ASSETS

Current Assets:

Cash on hand and in banks		$102,000	
Accounts receivable		348,000	
Raw material, in-process and finished goods inventories (at cost)		300,000	
Total, current assets			$ 750,000

Depreciable Assets:

Plant and equipment (at cost)		1,200,000	
Less: Allowance for depreciation		350,000	850,000

Total Assets: $1,600,000

LIABILITIES AND NET WORTH

Current Liabilities:

Accounts payable		$150,000	
Employees' withholding taxes payable		5,000	
Federal income tax payable		10,000	$165,000

Long-Term Liabilities:

ABC Co. corporate bonds, face value $10,000 each; bearing 8% interest payable semiannually; maturity = 10 years			500,000

Total Liabilities: $665,000

Net Worth:

1,000 shares (issued), 5% preferred stock, par value $100 per share		100,000	
30,000 shares (issued) common stock, par value $20 per share		600,000	

Surplus:

Retained earnings, year just ended	$ 40,000		
Retained earnings, prior years	195,000	235,000	
Total, net worth:			935,000
Total, Liabilities + Net Worth:			$1,600,000

Figure 6.2 Balance sheet of the ABC Company

How new capital should be raised depends, first, on the prevailing market rates for bonds, preferred stock, and common stock; second, on the firm's financial history; third, on the firm's financial structure (current ratio of borrowed to equity capital); and fourth, on the market's subjective evaluation of the firm's riskiness.

6.8 Numerical Example of the Marginal Weighted Average Cost of Capital

Various theories have been proposed about the existence of an *optimal* capital structure—that is, whether a firm, by balancing its debt-to-equity ratio, can minimize its cost of capital. The traditionalist theory maintains that an optimal capital structure can be built that will minimize the firm's cost of capital [see Solomon (1963), Durand (1959), Schwartz (1959), and Leland (1994)]. In opposition, the *Modigliani-Miller hypothesis* says that the firm's cost of capital is independent of its capital structure and, therefore, no minimum cost of capital exists. Modigliani and Miller (1958, 1959, 1961) have attempted empirical substantiation and have found evidence supporting it.

When the Modigliani-Miller hypothesis was first introduced, it caused a furor among financial theorists, who attacked not only the theory, but also the supporting empirical work. Thereafter, Modigliani and Miller added to their theoretical developments and performed additional empirical investigations, which again tended to support their theories. More recent empirical tests, which remedy the statistical defects in the earlier studies, support the traditional (minimum cost of capital) hypotheses.

Example 6.8 Future weighted average cost of capital
Returning now to the problem in Section 6.8.1, suppose that the ABC Company proposes to obtain an additional $550,000 as follows:

a. $150,000 through a new series of bonds that will bear interest at 9% payable quarterly. Maturity dates average 20 years, and the bonds are expected to sell at par (face value).
b. $300,000 through a new issue of common stock, which according to the company's financial advisors can be sold at or above the present market price of the company's existing stock.
c. $100,000 from retaining earnings from the company's existing operations.

What will ABC's weighted average cost of capital be after adding this capital?

Solution
a. The *new* bonds have an after-tax cost of capital from Equation (6.2) that is

$$k_t = \{[1+(0.09)/4]^4\}(1 - 0.40) = \underline{0.05585.}$$

b. The after-tax cost of capital for the new stock matches that of the old stock, assuming that (i) the price will not change and (ii) the dividends will continue to grow at the same rate. So,

$$k_e = 0.0824.$$

c. The cost of retained earnings is assumed to be the same as the cost of equity capital, for the reasons stated in Section 6.5.6. So,

$$k_r = 0.0824.$$

d. The *new weighted-average cost of capital* is calculated to be 7.03%, as shown:

Source of Capital	Present Amount	Projected Increment	Future Amount	Future Ratio	Explicit After-Tax Cost	Weighted Cost
Bonds (1)	$500,000	—	$ 500,000	0.2519	0.0490	0.0123
Bonds (2)	—	$150,000	150,000	0.0756	0.0558	0.0042
Preferred	100,000	—	100,000	0.0504	0.0500	0.0025
Common	835,000	300,000	1,135,000	0.5718	0.0824	0.0471
Retained earnings		100,000	100,000	0.0504	0.0824	0.0042
Total	1,435,000	$550,000	1,935,000	1.0001		0.0703

6.8.3 The marginal cost of capital. We now have the firm's old and new costs of capital, $k_0 = 6.85\%$ and $k_1 = 7.03\%$. Both are weighted averages, and neither is the firm's marginal cost of capital, which is estimated as follows.

First, we assume that the new capital budget of $500,000 will fund a bundle of projects, which may be viewed as *one single project* whose effective rate of return exceeds the average cost of the incremental capital. To calculate the bundle's marginal cost of capital, we use Equation (6.30) to obtain:

$$r \geq k_1 + (P_0/C)(k_1 - k_0).$$

Recalling that $V_0 = S_0 + D_0 = (\$29/\text{common share})(30,000 \text{ shares}) + (\$100/\text{preferred share})(1,000 \text{ shares}) + \$500,000 \text{ debt} = \$1,470,000$.

$$r \geq 0.0703 + \frac{1,470,000}{550,000}(0.0703 - 0.0685)$$

or

$$r \geq 0.0751.$$

Thus, the marginal cost of capital, about 7.51%, is higher than the average cost of capital, 7.08%. The marginal cost generally exceeds the average cost of capital, because additional increments of capital usually come at incrementally higher costs. And it is the higher rate that should be used in calculating the present value of a bundle of projects, at least to define a minimum discount rate.

6.9 MARR AND RISK

The weighted average cost of capital represents the minimum acceptable interest rate (MARR) for an average project. Multidivision firms often have risks, returns, and costs of capital that vary from one division to another. Using a common corporate hurdle rate for all divisions is not appropriate, and can lead to incorrect decisions and failure to maximize shareholder wealth (Block, 2003). Many firms try to isolate their costs of capital for each business unit or division by estimating divisional WACCs (Titman and Martin, 2007). By comparing individual divisions to other companies that operate in the same business, some firms identify divisional costs of capital.

Project managers are known to be optimistic on potential revenues, and have a tendency to underestimate costs. This is turn yields project returns that are less than originally forecast. To compensate, many companies require a minimum acceptable rate of return (MARR), or hurdle rate, which is higher than the WACC. Higher risk projects are given a proportionally higher hurdle rate. The assignment of hurdle rates is usually subjective. No matter which mathematical technique is used, assigning a hurdle rate to an individual project is usually less than objective, and subject to some form of criticism.

6.10 WACC AND THE PECKING ORDER MODEL

In theory, every firm manages its weighted average cost of capital, maintaining an optimum mix of debt and equity. The use of debt lowers the WACC because debt is tax deductible. At the same time, large amounts of debt are considered risky and will raise the WACC. In theory, once the optimal debt ratio and WACC are determined, a firm will use both debt and equity in the proper proportions in order to maintain its cost of capital. However, this may not be the situation in most firms.

Myers and Majluf (1984) developed a model regarding the firm's potential sources of capital and reached several conclusions. They recommended that firms will and should avoid issuing stock; the best way to raise equity is through retained earnings. If outside funding is needed, it is preferable to use debt. This idea later became known as the pecking order model (Fama and French, 2002). The pecking order develops because the cost of issuing new stock overwhelms the costs related to dividends and debt. Because of these costs, firms requiring capital should first use retained earnings, then use low interest debt, then higher cost debt, and finally as a last option, issue new equity.

Block (2011) performed a survey of Fortune 1000 companies to determine what approach, minimizing weighted average cost of capital or pecking order, firms actually use. Of those responding, 71.2% cited the pecking order model as the primary method used. The primary reason for avoiding the issuance of new stock was that firms viewed this as a negative signal to the market (investors would view the stock as overvalued). The fact that additional stock dilutes the value of existing shares was a distant second place reason.

Myers and Majluf (1984) recommended that firms employ *financial slack*, in order to implement the pecking order approach. This is done by maximizing the firm's liquid assets such as cash, marketable securities, and unused debt capacity. Block (2011) found that 62.6% of firms surveyed used financial slack on a regular basis, and another 21.0% used it part of the time. Another way of building financial slack is to reduce the payment of dividends to shareholder, maximizing retained earnings.

Note that even if the pecking order model better matches the goals of the firm, the weighted average cost of capital is still the relevant cost of capital—it just may not have been strictly minimized.

6.11 SUMMARY

In this chapter, we have considered the firm's financing function: how it raises capital and how to identify the *costs* of doing so. Through opportunity costing, we identified after-tax costs of capital for the firm's various sources of capital: short-term borrowing, long-term bonds, preferred stock, and equity (common) stock. For equity capital the Gordon-Shapiro and Solomon growth models were presented, as well as the capital asset pricing model (CAPM). The cost of retained earnings was shown to match that for equity funding, because retained earnings rightfully belong to the shareholders and retention is equivalent to paying the dividends out and then reacquiring them at the current market price.

Thereafter, we showed how to obtain the firm's weighted-average cost of capital, both before and after raising additional capital. Finally, we demonstrated how to find the firm's *marginal rate of return* for new projects to be financed by the new capital.

REFERENCES

ARDITTI, FRED D., and MILFORD S. TYSSELAND, "Three Ways to Present the Marginal Cost of Capital," *Financial Management*, **2**(2) (Summer 1973), pp. 63–67.

BARGES, ALEXANDER, *The Effect of Capital Structure on the Cost of Capital* (Prentice-Hall, 1963).

BIERMAN, HAROLD JR., "We Cannot Measure the Cost of Equity Capital Exactly," *Public Utilities Fortnightly* (August 16, 1984), pp. 31–35.

BLOCK, STANLEY, "Divisional Cost of Capital: A Study of its Use by Major U.S. Firms," *The Engineering Economist*, **48**(4) (2003), pp. 345–362.

BLOCK, STANLEY, "Does the Weighted Average Cost of Capital Describe the Real-World Approach to the Discount Rate?" *The Engineering Economist*, **56**(2) (2011), pp. 170–180.

CALLEN, JEFFREY L., "Estimating the Cost of Equity Capital Using Tobin's q," *The Engineering Economist,* **33**(4) (Summer 1988), pp. 349–358.

DERMAN, EMANUEL, *Models. Behaving. Badly.* (Free Press, 2011).

DURAND, DAVID, "Costs of Debt and Equity Funds for Business: Trends and Problems of Measurement," chapter in *Conference on Research in Business Finance*, published by National Bureau of Economic Research, 1952.

FAMA, EURGEN F., and KENNETH R. FRENCH, "The Cross-Section of Expected Stock Returns," *The Journal of Finance*, **47**(2) (June 1992), pp. 427–465.

GORDON, MYRON J., and ELI SHAPIRO, "Capital Equipment Analysis: The Required Rate of Profit," *Management Science,* **3**(1) (October 1956), pp. 102–110.

JAGANNATHAN, MURALI, CLIFFORD P. STEPHENS, and MICHAEL S. WEISBACH, "Financial Flexibility and the Choice Between Dividends and Stock Repurchases," *Journal of Financial Economics*, **57**, (2000), pp. 355–384.

JIN, LI, ROBERT C. MERTON, and ZVI BODIE, "Do a Firm's Equity Returns Reflect the Risk of its Pension Plan?" *Journal of Financial Economics*, **81,** (2006), pp. 1–26.

KISGEN, DARREN J., "Credit Ratings and Capital Structure," *The Journal of Finance*, **61**(3) (June 2006), pp. 1035–1072.

Lamdin, Douglas J., "Estimating the Cost of Equity for Corporations that Repurchase: Theory and Application," *The Engineering Economist*, **46**(1), (2001), pp. 53–63.

LEE, INMOO, SCOTT LOCHHEAD, JAY RITTER, and QUANSHU ZHAO, "The Costs of Raising Capital," *Journal of Financial Research*, **19**(1) (Spring 1996), pp. 59–74.

LELAND, HAYNE E., "Corporate Debt Value, Bond Covenants, and Optimal Capital

Structure," *The Journal of Finance*, **49**(4) (September 1994), pp. 1213–1252.

MILES, JAMES A., and JOHN R. EZZELL, "The Weighted Average Cost of Capital, Perfect Capital Markets, and Project Life: A Clarification," *The Journal of Financial and Quantitative Analysis*, **15**(3) (September 1980), pp. 719–730.

MILLER, MERTON H., AND FRANCO MODIGLIANI, "Dividend Policy, Growth, and the Valuation of Shares," *Journal of Business*, **34**(4) (October 1961), pp. 411–433.

Mitchell, Josh, and Jeff Bater, "GM to Buy Back $2.1 Billion of Preferred Stock from U.S." Wall Street Journal, (October 28, 2010).

Modigliani, Franco, and Merton H. Miller, "The Cost of Capital, Corporation Finance, and the Theory of Investment," *American Economic Review*, **48**(3) (June 1958), pp. 261–297.

Modigliani, Franco, and Merton H. Miller, "The Cost of Capital, Corporation Finance, and the Theory of Investment: Reply," *American Economic Review*, **49**(4) (September 1959), pp. 655–669.

MYERS, STEWART C., and NICHOLAS S. MAJLUF, "Corporate Financing and Investment Decisions When Firms Have Information That Investors Do Not Have," *Journal of Financial Economics*, **13** (1984), pp. 187–221.

PRESCOTT, LISA, JOHN WHITTAKER, AND TED ESCHENBACH, "Shifting *i*'s Are Not a Firm Foundation," 2002 Annual Conference proceedings, American Society for Engineering Management, (October 2002), pp. 434–438.

ROSS, STEPHEN A., RANDOLPH W. WESTERFIELD, and JEFFREY JAFFE, *Corporate Finance*, 9th ed. (McGraw-Hill Irwin, 2010).

SCHWARTZ, ELI, "Theory of the Capital Structure of the Firm," *Journal of Finance,* **14**(1) (March 1959), pp. 18–39.

SOLOMON, EZRA, "Measuring a Company's Cost of Capital," *Journal of Business,* **28** (October 1955), pp. 240–252.

Solomon, Ezra, *The Theory of Financial Management* (Columbia University Press, 1963).

STANDARD & POOR'S, *Corporate Ratings Criteria 2008*, (McGraw-Hill, 2008).

TITMAN, SHERIDAN, and JOHN D. MARTIN, *Valuation* (Pearson Addison Wesley, 2007).

PROBLEMS

6-1. The Shortcash Manufacturing Company will borrow $100,000 from its bank at an interest rate of 8% per year. The company's effective ordinary income tax rate is 40%. Calculate the effective after-tax annual interest rate of the loan if interest payments are made:
(a) Quarterly (b) Monthly (c) Semiannually

6-2. If Shortcash Manufacturing (problem 6-1) must keep at least $20,000 on deposit in the bank at all times as a liquidity requirement, what is the effective annual after-tax interest rate being paid on the *available* portion of the loan, if the interest payments are made quarterly, monthly, or semiannually?

6-3. What effective annual interest rate is being paid by savings and loan associations who pay their depositors "5¼% interest compounded daily"? *(Note:* A savings and loan association normally pays income taxes at corporate tax rates. Assume that the effective tax rate is 40%.)

6-4. Assume you borrowed $20,000 from a savings and loan association to remodel your kitchen under these terms: (a) The $20,000 is immediately discounted by 1% (that is, you are charged *one point*) and you actually receive $19,800 but owe $20,000. (b) The $20,000 is to be repaid at 9½% annual interest through 240 equal monthly installments. Assuming that you can deduct the interest portion of your payments and that your effective tax rate is 25%, what is the effective annual rate of interest you will pay?

6-5. Calculate the before-tax interest rate for problem 6-4, and compare this to the after-tax cost of money for a savings and loan association (problem 6-3). Estimate the annual gross profit of a savings and loan company that averages $25,000,000 in outstanding loans each year.

6-6. A new series of bonds to finance equipment is to be issued by Delay, Linger, and Wait Airlines. Each bond will have a face value of $10,000 and will bear a coupon rate of 8% per year paid quarterly, and the bond series matures in 15 years. The bid price by the successful underwriter is $9000 per bond, less marketing expenses of $5.00 per bond. If the airline's effective income tax rate is 42%, what is the effective after-tax cost of this new borrowed capital?

6-7. If the bonds in problem 6-6 are purchased by you for $9000, and your effective tax rate is 34%, what effective rate of return do you receive?

6-8. Iron Enterprises (a US owned firm) will be issuing bonds to finance the rebuilding of a European manufacturing facility. The bonds will pay interest semiannually at a nominal rate of 12%. The bonds have a face value of $20,000, and they mature in 10 years. Iron's effective tax rate is 46%. The issuing house expects the bonds to sell at a premium price, $20,500, but their fee of $750 per bond is subtracted from this. What is the effective, annual after-tax cost (interest rate) of the bonds to Iron?

6-9. If you buy the bonds in problem 6-8, (a) what is your before-tax rate of return? (b) If your effective tax rate is 28%, what is your after-tax rate of return?

6-10. Corporations often obtain new money by selling *convertible debentures,* which permit the owner to convert the debenture (bond) into a stated number of common stock shares at some specified future date. Because of this conversion right, the offering price of convertible debentures is frequently higher than conventional bonds of similar face value and interest rate. The *conversion premium* is the difference between debenture's initial offering price and the present value of an equivalent conventional bond if held to maturity.

Consider a convertible debenture of the BCD Company, which has a face value of $1000 at maturity in year 10. It will pay 6% annual interest, payable each 6 months. The initial offering price of the debenture (the market price) is now $1000. If your time value of money is 10% (before taxes) and your tax rate is 30%, what is your conversion premium if you bought the debenture?

6-11. New Terrain Construction is planning on "financing" its new equipment for this year with a 6-year lease. The equipment would cost $750,000 if purchased. The lease fees will be 20% of the first cost each year. Operations and maintenance costs are the same for leased and owned, except that New Terrain must spend an extra $7500 per year on insurance for the leased equipment. The machine's salvage value at 6 years would normally be 15% of its first cost. Assume that lease and insurance payments are made at the start of each year. (a) What is the before-tax cost of this source of funds? (b) If the firm's effective tax rate is 46%, what is the after-tax rate of this source of funds?

6-12. In problem 6-11, New Terrain could pay 22% of the first cost each year for a lease purchase. Thus, they would receive title to the machine after making the payment at the beginning of year 6. What is the (a) before-tax and (b) after-tax cost of these funds?

6-13. Some industrial firms offer their customers terms similar to "2% for 10 days or net 30 days." Assume that a firm's customers would pay all bills within 30 days, but they pay in a week to receive the 2% discount. (a) What before-tax interest rate is the firm paying to receive the funds earlier? (b) If the firm's effective tax rate is 39%, what is the effective after-tax annual interest rate the firm is paying for the funds?

6-14. One way a firm can "borrow" funds is by changing when it pays its accounts payables. Rather than receiving the 2% discount for prompt payment (for paying in a week), it can pay 1½% per month for 3 months. Interest is added to the amount due on the 10th if payment has not been received by that date. If a firm with an effective tax rate of 39% pays at the end of month 3, what effective annual interest rate has it paid for the funds?

6-15. If a firm's sources of funds have been as listed in the following table, what is the firm's historical weighted average cost of capital?

Source of Financing	Amount ($ million)	After-Tax Cost
Short-term debt	$1	7.8%
Bonds	1.5	6.2%
Preferred stock	2	11.0%
Common stock	12	14.1%
Retained earnings	3	14.1%

6-16. The common stock of General Spaceship Enterprises is presently traded on the New York Stock Exchange at a mean price of $65.00 per share. Last year, the company paid out $720,000 in dividends on 360,000 shares of common stock, and the present ratio of dividends to total earnings is expected to remain constant. The company's operating statement for the past year discloses a profit of $1,440,000 after income taxes. The book value of the company's equity capital (common stock + retained earnings) is $24 per share. The company owes $5,000,000 in corporate bonds, which pay interest at 6% per annum paid quarterly. The bonds mature in 8 more years. The company's effective ordinary income tax rate is 55%.
  (a) What is the company's present cost of equity capital, using the Gordon-Shapiro model?
  (b) What is the company's weighted-average cost of capital?
  (c) What is the common stock dividend, in dollars per share, expected to be 3 years from now (that is, payable at the end of the third year)?

6-17. The balance sheet of the B & B Company at the end of its current fiscal year follows. Cash dividends have averaged $0.25 annually per share of common stock for the last 4 years. Earnings for the last 12 months were $825,000 after income taxes. The company's effective tax rate is 40%. On the American Exchange (Chicago) the price for the company's common stock has ranged between $5.50 and $7.25 per share in the last 6 months, while the preferred stock price has averaged $9.05 per share.
  What is the company's present weighted-average cost of capital, using the Gordon-

Shapiro model to estimate the cost of equity capital?

Assets		Liabilities and Net Worth	
Current Assets:		Current Liabilities:	
Cash	$250,000	Current installments on long-term debt payment	$ 50,000
Receivables	1,600,000	Accounts payable	100,000
Inventories	1,950,000	Short-term bank loan	250,000
		Federal income tax payable	200,000
Fixed Assets:		Other Liabilities:	
Plant and equipment, net after depreciation	1,700,000	6.0% bonds due in 20 years[a]	600,000
		Net Worth:	
		Preferred stock (100,000 shares)[b]	800,000
		Common stock (1,000,000 shares outstanding)	3,000,000
		Surplus and retained earnings	500,000
Total Assets	$5,500,000	Total liabilities and net worth	$5,500,000

[a] Bonds are each $1000 face value, mature in 20 more years, and bear interest at the rate of 6% per year; interest is paid quarterly.
[b] Preferred stock pays a dividend of $0.47 per share annually.

6-18. The B & B Company (problem 6-17) is contemplating a series of new projects whose $1,000,000 cost will be financed as follows:
- $500,000 (nominal amount) by a new series of bonds (series B), with face value of $1000 each, bearing an interest rate of 8% per year and maturing in 15 years. Interest will be paid semiannually. In addition, the successful bond underwriter bid $960 per bond, less $10 selling expense, for this series B issue.
- $300,000 (actual amount) by the sale of a new preferred stock issue, to pay 6% dividends per year.
- The balance of the $1,000,000 by retained earnings.

(a) After the proposed financing is accepted and marketed, what will the company's weighted-average cost of capital be?
(b) What is the company's marginal cost of capital for the $1,000,000 worth of new projects?
(c) How does your answer for (b) compare with that given by Equation (6.32)? How do you explain the difference?

6-19. If the risk-free rate is 4% and the market rate is estimated at 11%, what is the market risk premium?

6-20. A stock has an estimated beta value of 1.05. Treasury bonds with a maturity of 10 years are paying 3% interest. The market rate is estimated at 11%. What is the required return for this stock?

6-21. A stock has an expected return of 12.10 percent and a beta of 1.12. The expected return on the market is 10.90 percent. What must the risk-free rate be?

PART TWO

Deterministic Investment Analysis

7

Economic Measures

7.1 INTRODUCTION

In earlier chapters *present value* (present worth) and *rate of return* have been used as measures of wealth to analyze projects. Our concern was not the criteria's appropriateness, but rather simply defining the project's *worth*. Now, we are concerned with the methods for evaluating and comparing projects.

In economic analysis of the firm's potential projects, the objective is to determine a project's acceptability from the economic data available and from a managerial standpoint. Should a project be executed by the firm to become a part of the firm's productive activities, or should it be rejected? This question is answered in two general situations. The project selection problem either is *constrained* by certain stated (or implied) assumptions concerning the selection process or is *unconstrained*. For example, if the firm is selecting executable projects from a larger set of candidate projects, but a limit on total expenditures will not permit execution of all projects, then the selection problem is constrained. Similarly, 10-, 12-, 14-, or 16-story buildings on a particular site are *mutually exclusive* alternatives. The selection problem again is constrained, this time by the non-independence of the four alternatives, since only one can be executed. On the other hand, if there are no constraining factors, then the project selection problem is unconstrained.

In the unconstrained selection problem, we determine the economic desirability of a single project without comparison to competing or alternative projects. However, the merit of the project is judged by comparison with the firm's opportunity cost. The principle of opportunity costing is always present, even in judging single projects under unconstrained conditions. The firm must compare two alternatives: *accept the project* and *do nothing*. The latter choice always involves other uses of the funds that could be invested in the project. Single, unconstrained projects must be measured against the best alternative use of the firm's funds: either the firm's cost of capital (as a minimum basis) or marketplace opportunities for external investments with similar risks. So while a decision to accept or reject a single, isolated project may appear to be made without a formal stated comparison, there is an implicit comparison with the firm's opportunity cost base, which is its marginal investment rate.

In the constrained selection problem often two or more projects (or alternatives) are compared, with the objective being the selection of the most desirable alternative. Typically, textbook engineering economy problems emphasize this comparative selection technique. A common problem posed to the student, for example, is to find the minimum cost alternative from several possible solutions, only one of which can be executed. This analysis is always a constrained selection problem, since execution of one method or process automatically precludes the execution of any other alternative.

This chapter investigates selection criteria that have theoretical merit, particularly their underlying assumptions and limitations of application. Some criteria—payback, net present value, and internal rate of return—are widely used, but often they are misapplied or misunderstood. The reasons appear to be that (1) the *assumptions* that underlie these criteria and limit their applicability are not fully understood and (2) the economic conditions *assumed to apply* in the problem are inconsistent with the measure's underlying assumptions. This chapter investigates underlying assumptions and clarifies the conditions of application. Again, we consider only the unconstrained case.

7.2 ASSUMPTIONS FOR UNCONSTRAINED SELECTION

In all the methodologies presented in this chapter, certain basic assumptions are made about the investment projects and the firm. These assumptions are often called *perfect market conditions* and the *essential assumptions of certainty*. These assumptions are generally traced to the pioneer works of Irving Fisher, *The Theory of Interest* (1954), Friedrich and Vera Lutz, *The Theory of Investment of the Firm* (1951), and Jack Hirshleifer, "On the Theory of Optimal Investment Decision" (1958). The essential assumptions are:

1. There is a perfect capital market, and the supply of funds is unrestricted.

The detailed conditions of *a perfect capital market* were discussed in Section 2.1.1 as follows: In a perfect capital market, each buyer or seller of securities trades in such small amounts (relative to the total market) that no one buyer or seller has any appreciable effect on prices. This means that by selling securities, a firm can raise as much cash as it wants at the going interest rate. Moreover, the firm, as a buyer of securities, can invest as much surplus cash as it wants at the market interest rate. Since the firm has (or can obtain) the funds required for all *profitable* investments, there is no need to rank investment projects by their profitability. Under these conditions, the firm makes all profitable investments, and the market rate of interest accurately measures the firm's marginal investment opportunities. This perfect capital market also requires that there be no costs associated with market transactions.

2. There is complete certainty about investment outcomes.

The firm has certainty about investment outcomes. This includes all relevant *present and future* knowledge about (a) the projects under present consideration and (b) all possible future projects and the investment opportunities. Moreover, the relevant knowledge is *exact,* or deterministic. There is no risk, no uncertainty, and no probability. This certainty applies to all variables: cash flows, discount rates, timing, economic lives, etc. Because of this certainty, firms need not allow for uncertainty, nor do security buyers distinguish between bonds and stocks, for example.

3. Investment projects are indivisible.

Unlike stock or bond purchases of multiple shares or bonds at a constant unit price, *a project* is generally indivisible—it must be executed completely or not at all. This distinguishes the *portfolio* or *investment* analysis of financial practitioners and the project analysis of engineers and industrial managers. In project analysis, each project is executable only in its entirety. Fractional parts of a large project can sometimes be considered as discrete, separately executable projects. The relationship between the separate parts creates a constrained selection situation, but it preserves the *executable entity principle*. Financially, this implies that firms must commit funds by discrete amounts, each representing the investment for a particular project (or combination of projects).

4. *Investment projects are independent.*
The profitability of an independent project does not affect and is not affected by the profitability of any other project. There is no economic, technical, or other dependence in any elements of their cash flow streams. This assumption precludes consideration of mutually exclusive projects, which by definition are *dependent*.

These four assumptions describe the ideal situation. They are relaxed in later Chapters, but in Chapter 7 they are the assumptions underlying project analysis. As a result, the firm decides "go" or "no go" on each discrete project, without considering any other project.

7.3 SOME MEASURES OF INVESTMENT WORTH (ACCEPTANCE CRITERIA)

Four different measures of investment worth, or acceptance criteria, are analyzed because they are currently used in business practice or because good theoretical arguments can be advanced in their favor. These are the *payback method, net present value (net present worth), internal rate of return,* and *profitability index (benefit-cost method).*

Obviously, other acceptance criteria could be examined. Many practitioners will recognize the omission of the following five criteria:
1. Proceeds per dollar of outlay (total proceeds divided by investment)
2. Average annual proceeds per dollar outlay (like 1, except proceeds and investments are averaged)
3. Average income on book value (an accounting measure based on the declining value of a depreciating investment)
4. Average income on cost (like 3, except that cost is substituted for book value)
5. Return on investment (ROI) procedures—typically a return on investment for each year is determined by dividing the year's annual income by its average annualized investment and then averaging the annual returns on investment to obtain the project's overall ROI

These methods are not investigated because they are incorrect. They either do not consider the time value of money or they account for it incorrectly (for example, the ROI methods). While this chapter focuses on unconstrained selection, most of these five criteria have been advocated for use in selecting among competing projects as well. These five criteria are not valid for either the constrained or unconstrained cases.

7.4 THE PAYBACK PERIOD

The *payback* period is a relatively simple and frequently used criterion for judging an investment's economic worth. Formally, the payback period is the length of time required for the stream of cash inflows to the firm from a project to equal the cash outlay(s) required by the investment. The payback period, θ, is defined by:

$$\sum_{t=0}^{\theta} Y_t = 0 \qquad (7.1)$$

where Y_t = net cash flows to or from the project at times $t = 0, 1, 2, \ldots, \theta$. (Normally, $Y_0 < 0$ and the other $Y_t > 0$, so that θ is then determinate. The periods 0–1, 1–2, etc., can be subdivided so that θ can be estimated for fractions of a year.)

Ordinarily payback is used as *a limit* rather than as a direct criterion. Thus, a firm might establish a maximum payback period for a given class of projects and then reject all projects whose payback periods exceed that maximum. For example, the firm might adopt maximum payback periods of 5 years for new construction, 3 years on new labor saving machinery, and 2 years on new product tooling.

Payback period can be used as a ranking device among projects. When used this way, projects with the shortest payback periods have the highest rankings. For example, the three hypothetical projects shown in Table 7.1 would be ranked as given in the last column.

Table 7.1 Cash flows and payback periods of three hypothetical projects

Project	Initial Investment	Net Cash Proceeds		Total Proceeds	θ = Payback Period (years)	Ranking
		End of Year 1	End of Year 2			
A	$10,000	$10,000	–	$10,000	1	1
B	10,000	10,000	$1,100	11,100	1	1
C	10,000	5,762	5,762	11,524	1.74[a]	3

[a] The fractional year assumes that the end-of-period cash flow for year 2 can be prorated.

The payback period, as an isolated selection criterion, is clearly not very reliable. For example, it fails to distinguish between projects A and B, since it ignores B's longer cash flow stream. Obviously, ignoring cash flows that occur after payback is a poor idea. Payback's second defect is ignoring the time value of money. For example, A returns nothing but its initial cash flow, yet it is highly ranked. A third defect is often displayed by the payback criterion: It may incorrectly rank projects. For example, it ranks projects A and B above C, whereas it is entirely possible that C would be preferred because of the higher *total* cash inflows for C. So it would appear that payback is not very reliable.

Nevertheless, payback continues to be widely applied to investment decisions. While some academic writers have dismissed payback as misleading and worthless for the reasons indicated, at the same time businessmen and other academicians have advocated payback at least as *a secondary* criterion. There are cases where payback period is acceptable or even worthwhile.

For example, a plant manager might use an 18-month payback period for cost reduction projects. This may leave some good projects undone, but a forklift with a 10-year life that can be justified using an 18-month payback is a *good* investment. But a reconfiguration of the assembly line for a product with a 2-year life-cycle could only be properly analyzed with measures that properly consider the time value of money.

Firms going through an economic crisis may be willing to invest in cost savings projects, but only if the payback period is very short. An organization may simply not have a year or two to recover an investment; crisis situations require that an investment be recovered and provide cash savings *fast*.

When the payback period is exceptionally short, payback may be the best way to communicate to management. A pharmaceutical project returned $15 million on a three-stage investment totaling $90,000. In this case, net present value can be confusing, and the internal rate of return of over 10,000% is nearly meaningless. However, the initial investment of $27,000 was paid for in less than a day—the payback period was clear to all. In this particular project, validation and formal approval required three weeks. By the time the project was validated and approved, it had already returned about fifteen times the initial investment. Payback period was the most meaningful way to communicate the savings to management. In the case of very short payback periods, there is power in demonstrating that an investment has paid for itself, and that the future cash flows are pure profit.

Finally, there are inherent differences between financial and capital project investments. Financial investments commonly have a ready and open trading market, where the investor's decision at any time is to leave the funds in an investment or to withdraw them. The situation is different with physical assets and projects. Not only is the market value for used physical assets substantially below their *going concern* value *in the project,* but the range of operational alternatives is quite different. Basically, the firm is limited to (1) committing additional resources (expanding the project's scope), (2) withdrawing resources (contracting its scope), (3) leaving the resources at a static level, and (4) *always* operating the project within the limits of the project's resources, regardless of their levels. The randomness and the timing of the economic outcomes focus the manager's attention on the resource levels. Thus, payout information is extremely important to the manager—it is the rate at which uncertainty is resolved, with a consequent lifting of the resource constraints. Work by Narayanan (1985) and Statman (1982) supports the proposition that fast resolution of uncertainty makes the manager look good. Thus, principal agent theory suggests that the popularity of the theoretically incorrect payback theory is due to differences in the goals of managers and owners.

7.4.1 Payback rate of return was developed by Gordon (1955), and later generalized by V. L. Smith (1961).

Evaluate an investment proposal (project) with the following attributes:

P = the initial investment at time $t = 0$
F = the salvage value at time $t = N$
N = the project's life years
i = the effective interest rate per year
t = time in periods *(t = 0, 1, 2, . . . , N)*
Y = uniform end-of-period *cash flow* generated by the project.

This simple project has a single investment of P dollars at $t = 0$, and it returns a uniform stream of Y dollars per year, over the life of N years beginning with $t = 1$. In Chapter 2 the depreciable portion of an investment $P - F$, was related to a uniform series of end-of-period amounts, A, by Equation (2.30), which is repeated here:

$$A = (P - F)\left[\frac{i(1+i)^N}{(1+i)^N - 1}\right] + Fi \qquad (2.30)$$

If the salvage value, F, is small with respect to the investment, P, then we can consider $F = 0$, and the entire investment P is recovered, and Equation (2.30) reduces to:

$$A = P\left[\frac{i(1+i)^N}{(1+i)^N - 1}\right] \qquad (7.2)$$

Solving Equation (7.2) in terms of P/A and setting our project cash inflow $Y = A$ essentially inverts Equation (7.2), so that our payback, θ, is defined as:

$$\theta = \frac{P}{Y} = \frac{(1+i)^N - 1}{i(1+i)^N} \qquad (7.3)$$

(The units, years, derive from the units of i ($/$-year) in the denominator.) The ratio of investment to net cash inflow, or payback period, establishes the length of time for investment recovery and *defines the value of an effective interest rate, i,* via Equation (7.3).

What does this interest rate mean? As we saw in Chapter 2, the effective interest rate assures the firm not only of the *return of its original capital,* but also of receiving *interest on the unrecovered investment balances* at that interest rate.

Smith (1961, pp. 220–230) arrives at substantially the same economic interpretation of θ by using a continuous compounding approach. A capital constraint and Kuhn-Tucker conditions indicate an optimal marginal solution, in which the payback period's *reciprocal* defines an equilibrium intersection point between the cost of funds invested and the rate of return on funds inflow. The equilibrium point defined by the payback reciprocal is the interest rate:

$$\frac{1}{\theta^*} = \frac{k + r}{1 - e^{-rN}} \qquad (7.4)$$

where
- $k =$ the marginal profitability of the last dollar's worth of available capital, which is the firm's marginal cost of capital.
- $r/(1 - e^{-rN}) =$ incremental capital recovery rate on the last dollar's worth of investment in the project, with annual interest rate r on the project investment.

In summary, the payback ratio (or period), $\theta = P/Y$, provides (1) the recovery over the project's life of the invested capital and (2) its reciprocal, $1/\theta = Y/P$ defines an *effective interest rate, i,* on the periodic declining investment balances during the project life. If this interest rate exceeds the firm's marginal attractive rate of return (or, under conditions of certainty, the firm's cost of capital), then the project should be accepted. This concept will be demonstrated by example.

7.4 The Payback Period

Example 7.1 Payback period and rate of return
A project requires an initial investment of $10,000 and returns $2500 each year for 10 years. The firm's cost of capital is 15%. (a) What is the payback period? (b) What is the payback rate of return? (c) What is the effective return rate? (d) Should the project be accepted?

Solution

a. Payback period $= \theta = P/Y = \$10,000/(\$2500/yr) = 4$ years.

b. Payback rate of return $= 1/\theta = 0.25 = 25\%$ per year.

c. $\theta = \frac{P}{Y} = \frac{(1+i)^{10}-1}{i(1+i)^{10}} = 4.0 = (P/A, i\%, 10)$

where i is unknown. This can easily be solved with a financial calculator,

$i(N,A,P,F) = i(10, 2500, -10000, 0) = 0.2141$ or 21.41%

or spreadsheet,

=RATE(NPER, PMT, PV, FV)
=RATE(10, 2500, -10000, 0) = 21.41%

d. Since $(i = 21.41\%) > (k = $ cost of capital $= 15\%)$, the project should be accepted. Note that the decision is made on $i > k$, not on $1/\theta > k$.

7.4.2 Discounted Payback. The payback method can be modified to incorporate the time value of money. Each cash flow is discounted to a present worth to determine the discounted payback period—which is always longer than the payback period where $i > 0$. The discounted payback period is an improvement, but cash flows that occur after the payback continue to be ignored, making this method less useful than NPV or IRR in most cases.

The discounted payback period is used much less frequently than the payback period. Note also that an alternate approach to calculating the discounted payback period is to find the time when the unrecovered investment (Section 7.8) or the project balance (Section 7.10) returns to zero after the initial investment.

Example 7.2 Discounted payback period
Using the cash flows from Example 7.1, determine the discounted payback period using an interest rate of 15%.

Period	Cash Flow	PW of CF_t	Cumulative PW
0	−10,000	−10,000.00	−10,000.00
1	2,500	2,173.91	−7,826.09
2	2,500	1,890.36	−5,935.73
3	2,500	1,643.79	−4,291.94
4	2,500	1,429.38	−2,862.56
5	2,500	1,242.94	−1,619.62
6	2,500	1,080.82	−538.80
7	2,500	939.84	+401.04
8	2,500	817.25	
9	2,500	710.66	
10	2,500	617.96	

The discounted payback is greater than 6 years, and less than 7 years. Specifically, the payback period is 6 + (538.80/939.84) = 6.57 years. Note that this is significantly longer than the 4 year payback period from Example 7.1.

7.5 CRITERIA USING DISCOUNTED CASH FLOWS

Many project selection criteria are eliminated from serious consideration because the *magnitudes* and *timing* of all cash flows are not correctly taken into account. This is essentially true of all but three—net present value, internal rate of return, and profitability index (*excess present value index* or *benefit-cost ratio*). These three criteria can be properly applied to the selection problem. These are the *rational criteria* because they account for (1) the project's entire cash flow stream and (2) the time value of money.

There has been much advocacy, argument, and disagreement in the research literature concerning these three criteria (and others). The root of much, if not all, of the disagreement lies in whether selection is under constrained conditions or not. At first, the importance of a constraint in the selection problem was not realized, and until Weingartner (1963a) published his Ford Foundation prize-winning dissertation, "Mathematical Programming and the Analysis of Capital Budgeting Problems," no substantial progress was really made. Weingartner pointed out that a linear-programming formulation of the capital budgeting constrained problem resulted in an optimal selection of projects (under certainty conditions), when the objective function *maximized the net present value criterion.*

Since then, several outstanding articles have carefully and critically examined the assumptions and limitations of these three rational criteria and their relatives. Early examples include Lorie and Savage (1955), Solomon (1956), Renshaw (1957), Swalm (1958), Bernhard (1962), and Beranek (1964). These compared two or more criteria but did not comprehensively evaluate and compare their assumptions and limitations. Bernhard (1971) published a comprehensive comparison and critique of eight criteria[1]:

    1a. Present worth [net present value]
    1b. Equivalent "annual" worth

---

1. Bernhard also discussed the MAPI urgency rating, which is omitted here.

2a. Benefit-cost ratio[2]
2b. Net benefit-cost ratio
2c. Solomon's average rate of return
3. Internal rate of return [or, "yield," "investor's method," "marginal efficiency of capital," etc.]
4. Eckstein's benefit-cost ratio[3]

Bernhard showed that criterion 1b is simply a variation of 1a, and that 2b and 2c are equivalent to 2a. So, we need consider only the three basic criteria: net present value (present worth), benefit-cost ratio, and internal rate of return. However, 1b has some advantages when the lives differ between alternatives. It is worth noting that most defects in the discounted cash flow ratios and internal rates of return are remedied if an incremental approach is used for analysis of mutually exclusive projects.

In addition to Bernhard's and Weingartner's work, three other articles are particularly noteworthy. The first is the classic by Jack Hirshleifer, "On the Theory of Optimal Investment Decision" (1958), which justifies net present value as a selection criterion. The other two, by Teichroew, Robichek, and Montalbano (1965, 1965), rigorously analyze the internal rate of return (IRR) criterion.

The analysis of constrained and unconstrained project selection has also been complicated by disagreements over the reinvestment assumption used with IRR. Lohmann (1988) showed that IRR made the same reinvestment assumption as NPV, but it did so through comparing rates rather than explicit calculations at that reinvestment rate. This has been summarized clearly by Park and Sharp-Bette (1990):

> Use of PV implies reinvestment at the rate used for computing PV. Since the other criteria, when used properly, give the same project selection, it can be argued that their use also implies reinvestment at the same rate.

These three criteria of NPV, IRR, and benefit-cost ratios can all be applied to the unconstrained project selection problem. However, the latter two require incremental analysis for the constrained mutually exclusive problem. Also, projects may have multiple sign changes that can cause severe problems for all three criteria.

7.6 THE NET PRESENT VALUE CRITERION

The general form for net present value was developed in Chapter 2, as Equation (2.2):

$$P_0 = \sum_{t=0}^{N} \frac{Y_t}{\prod_{j=0}^{t}(1+i_j)} \tag{2.2}$$

---

2. Alternative benefit-cost ratios include the net benefit-cost ratio (Bernhard's 2b) and Eckstein's benefit-cost ratio (Bernhard's #4). They are all derivable from net present value and are discussed more fully in Section 7.7 and footnote 3.

3. Eckstein's benefit-cost ratio (1958) is (in our notation) as follows: Let
$b = \sum_{t=1}^{N} b_t(1+i)^{-t}$ = present value of cash *in*flows from t = 1 to N
$c = \sum_{t=1}^{N} c_t(1+i)^{-t}$ = present value of cash *out*flows from t = 1 to N
C = initial cost of project at $t = 0$
$B = b - c$ = difference in present values of cash inflows and outflows.
Then, Eckstein's benefit-cost ratio is defined as EBC = $\frac{B+c}{C+c}$

where P_0 = net present value
Y_t = net cash flow at the end of period t
i_j = the interest (discount rate) for period j
N = the project's life
j = points in time prior to t, i.e., $j = 0, 1, 2, \ldots, t$
t = the point in time under consideration; i.e., $t = 0, 1, 2, \ldots, N$.

In general, the interest rates, i_j do not need to be equal. Usually, however, one assumes $i_0 = 0$ and $i_1 = i_2 = i_3 = \ldots = i_N = i$, so that Equation (2.2) reduces directly to Equation (2.3):

$$P_0 = \sum_{t=0}^{N} \frac{Y_t}{(1+i)^t} \quad (2.3)$$

Applying Equation (2.3) to find the net present value is straightforward. The result is *a point* or single estimate of P_0 at a particular interest rate i. This point estimate is informative, because the positive, negative, zero, or indeterminate value can be used for decision making. For the unconstrained case, the general rule is to accept the project if net present value is positive; otherwise, reject it. The detailed development (due mainly to Hirshleifer, 1958) that leads to this decision criterion follows.

7.6.1 Production-consumption opportunities of the firm. The expression for net present value was derived in Sections 2.1.2 and 2.1.3, under perfect market conditions. We saw that a firm can borrow and lend funds in the financial market. This shifts consumption opportunities between periods, but because the borrowing and lending rates are identical (perfect capital market), this does not increase the firm's present value (see Section 2.1.2 and Figure 2.1). The reason is that given an initial *endowment position* of incomes at times $t = 0$ and $t = 1$, the *market interest line* with constant slope $[= -(1 + i)]$, which passes through the endowment position, defines a single present value (W, in Figure 2.1). Thus, the firm can borrow and lend to shift its consumption capability in time, but it cannot increase its present worth above its endowment position.

A second activity will increase the productive firm's net worth. By acquiring some intermediate factors of production, such as buildings, equipment, or tools, the firm shifts income between periods and *also increases the worth of the firm.*

The economist knows that roundabout production methods that depend on intermediate "tools" will produce more product, in the long run, than "hand" methods. Also, investments often bring large initial returns, but eventually a sort of diminishing return effect sets in. Then, additional increments of investment bring ever-decreasing returns.

The following is not strictly analogous, but illustrates the diminishing returns effect. Suppose a firm has available several projects in which it can invest funds and thereafter produce consumption products with the "tools" it purchased. Between now ($t = 0$) and tomorrow ($t = 1$), each productive project, j ($j = 1, 2, 3, \ldots$), provides a rate of return Equation (7.5) to the firm of:

$$i_j^* = \left(\frac{\Delta Y_1 - \Delta Y_0}{\Delta Y_0}\right)_j = \left(\frac{\Delta Y_1}{\Delta Y_0}\right)_j - 1 \quad (7.5)$$

where $(\Delta Y_0)_j$ is the investment in project j at time $t = 0$, $(\Delta Y_1)_j$ is the return cash flow at time $t = 1$, and i_j^* is the simple one-period rate of return. The firm's potential projects are arranged by decreasing rate of return and then plotted. The *cumulative* cash flows, $\sum(\Delta Y_0)_j$ at time $t = 0$ and $\sum(\Delta Y_1)_j$ at time $t = 1$, are represented on exchange axes, which is the *exchange map* of the firm's investment opportunities, such as the piecewise linear function $0Z$ in Figure 7.1.

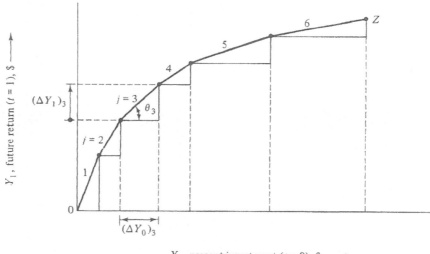

Figure 7.1 Discrete productive opportunity curve

The piecewise-linear curve $0Z$ is the firm's *productive opportunity curve*. Note that its slope, for any project, is defined by the project's ratio $(\Delta Y_1/\Delta Y_0)_j$. This slope is related to the project's rate of return by Equation (7.5). That is, for any project, the slope $\theta_j = \tan^{-1}(\Delta Y_1/\Delta Y_0)_j = \tan^{-1}(1 + i_j^*)$. Hence, projects can be arranged in order of decreasing i_j^* and the cumulative effect becomes one of diminishing returns with increasing total investment (the individual rates of return, proportional to θ_j, diminish as more projects are selected).

Assume that the number of productive opportunities available becomes very large and the cumulative cash flow required for investment, $\sum(\Delta Y_0)$ at $t = 0$, is derived solely from the firm's *savings*. Under these assumptions, the productive opportunity curve, $0Z$ in Figure 7.1, becomes continuous. And, when it is placed on the firm's consumption exchange map, Figure 7.2, the productive opportunity curve assumes a reverse or a left-handed shape. The source of investment funds is foregone or negative consumption. So in Figure 7.2 the *productive* opportunities for exchange form the curve WZ. Assuming that the firm starts with only present-time funds (i.e., its endowment is present wealth, W, on the $t = 0$ axis), it can invest in *productive* activities any portion of its current funds *saved*, say $Y_w - Y_0 = \sum(\Delta Y_0)$, and then receive at $t = 1$ the increased future amount, $\sum(\Delta Y_1)$. If the firm lends the same saved amount, $Y_w - Y_0$, it can receive at $t = 1$ only the amount Y_1'. (The WZ curve assumes that at least *some* productive opportunities better than the market rate of interest exist.)

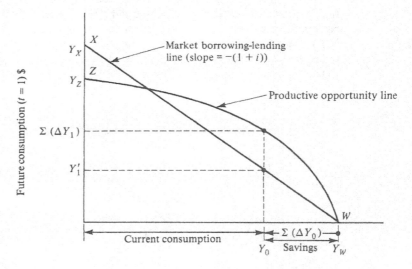

Figure 7.2 Productive and borrowing-lending opportunities

How much "savings" (e.g., $Y_w - Y_0$) should the firm spend on intermediate productive tools? As long as the slope of the productive opportunity line exceeds that of the market borrowing-lending line, it can choose any spending level. In Figure 7.3, the firm could choose any investment-production alternative between W and R^*, which is the point of tangency of the capital market line whose slope is $-(1 + i)$. At R^*, if the firm had further savings to invest, it would switch from investing in more capital goods to lending at a higher rate in the capital market.

The choice of a particular investment consumption point is fixed by the firm's *indifference curves*.[4] These express the firm's indifference between combinations of "today's" and "tomorrow's" consumption (income) levels. Suppose that the firm's

---

4. An indifference curve expresses a decision maker's indifference of choice among combinations of two or more goods. All choices on the curve have constant value or utility. For example, below, an individual would be indifferent between the commodity sacks of apples and oranges shown on indifference curve A: A1, A2, A3 have the same value. However, any of the sacks represented by curve B is better than any represented by curve A. Indifference curves representing intertemporal choice are typically convex to the origin.

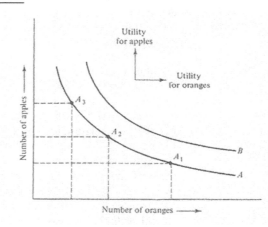

indifference curves were represented by the family $A_1, A_2, A_3, \ldots$, in Figure 7.3. The firm desires as high an indifference curve as possible, i.e., A_3 or beyond. Along the productive opportunity line WR^*Z, the highest-valued indifference curve touched by the former is A_2, at point A'. Since the firm maximizes its utility for exchange at A', it would invest its savings, $Y_W - Y_0$, in intermediate productive goods at $t = 0$, and at $t = 1$, it would receive the amount Y'_1; as income from the sale of the product. If the firm followed this policy, its net present value would increase from Y_W (the endowment point) to Y'_W.

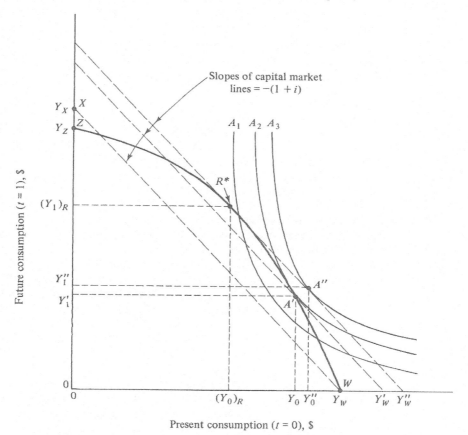

Figure 7.3 Choice of productive exchange alternatives

But the solution at A' can be bettered if the firm can also borrow in a perfect capital market. By investing more in capital goods, the firm can move from A' along the productive opportunity line to the point R^*, the tangency point. The firm now moves *in a reverse direction* (borrowing) along the financial market line to point A'', which lies on a higher-level indifference curve (A_3) with an even higher net present value, Y''_W.

Thus, the firm invests in two steps. First is the productive solution. The firm invests until it reaches point R^*, where the slope of the productive opportunity line equals that of the market line. That is, R^* is the point at which:

$$\frac{dy_1}{dy_0} = -(1 + i^*) = -(1 + i) \tag{7.6}$$

where dy_1/dy_0 is the slope of the productive opportunity line, i^* is the incremental rate of return, and i is the market interest rate. From Equation (7.6), $i^* = i$, which implies that further investments in intermediate goods should cease when $i^* = i$ (i.e., at point R^*).

The second step is to "finance" the investment, or part of it, by borrowing (i.e., moving along the financial market line to point A''). This point maximizes the firm's utility. The investment in intermediate goods is accomplished by first "saving" (at $t = 0$) the amount $Y_W - (Y_0)_R$ and then by borrowing the amount $Y_0'' - (Y_0)_R$ in the financial market. The net "saving," or equity funds, used in purchasing the intermediate goods is $Y_W - Y_0''$. This derivation demonstrates that the firm maximizes utility by *maximizing net present value*, Y_W''.

One essential point of this optimization is that the productive optimum, R^*, is fixed solely by (1) the productive opportunity curve WZ and (2) the market rate of interest. The fact that the *production optimum* can be established independently of the firm's borrowing or lending in the financial market is the *separation theorem*. It follows from the initial assumption of perfect and costless financial markets. Because maximum present value (wealth) is reached at R^*, *which is independent of how much money is borrowed or lent* in the financial market, present value can be used for project selection.

7.6.2 The present value criterion for project selection. A project is simply an opportunity for the firm to invest funds. In the simple model assumed above, a project is a sequence of dated cash flows, Y_0 and Y_1, at times $t = 0$ and $t = 1$. If Y_0 is negative and Y_1 is positive, the firm is investing by sacrificing present income for future income. The firm can also divest an operating project. In this case, Y_1 would be negative and Y_0 would be positive; i.e., a future cash inflow entitlement is sacrificed for a present cash inflow.

In either investment or divestment, the project's *present value, P_0*, is defined as:

$$P_0 = Y_0 + \frac{Y_1}{1 + i} \tag{7.7}$$

Present value, P_0, is the wealth increment associated with a project. This, in conjunction with perfect capital markets and the separation theorem, implies that the objective of investment in (or divestment from) intermediate productive goods is wealth maximization. The following decision rule results:

Present Value Rule 1: Accept any project (investment or divestment) for which present value, P_0, is positive; reject any project for which P_0 is negative.

Our concern here is only single projects, but the present value rule can be easily applied to multiple projects, some of which may not be independent. If the cash flows of projects are interdependent, the solution is to form *sets* of projects and then choose the set that maximizes present value. For example, A, B, and C may not be independent; however, sets A, B, C, AB, AC, BC, and ABC can be compared with the present value rule. The decision rule can be stated as follows:

Present Value Rule 2: If two projects (or combinations of projects) are mutually exclusive, accept that which has the greater present value, P_0.

7.6.3 Multiperiod analysis. The principles developed for the $t = 0$, 1 case also apply to decisions involving more periods. For example, to add another period, indifference curves become indifference *shells,* and the financial market *line* becomes a financial market *plane* (as in Section 2.1.3). The *optimal productive opportunity is* again determined at the point of tangency of the financial market plane with the productive opportunity *surface,* independent of the firm's choice to borrow or lend. Hence, the separation theorem again applies, and the present value rules stand. Similarly, the method can be extended by analogy so that finally Equations (2.2) and (2.3) become the bases for the above present value decision rules.

7.6.4 Characteristics of net present value. If we drop the subscript on P_0, then P now stands for net present value. Equation (2.3) is a function of the interest rate, i, project life, N, and the series of cash flows, $Y_0, Y_1, Y_2, \ldots, Y_N$. Thus, P as a function of i is

$$P(i) = \sum_{t=0}^{N} \frac{Y_1}{(1+i)^t} \tag{7.8}$$

If $Y_0 < 0$ and all other $Y_t > 0$, with N finite and known, then $P(i)$ can be expanded to

$$P(i) = -Y_0 + Y_1(1+i)^{-1} + Y_2(1+i)^{-2} + \cdots + Y_N(1+i)^{-N} \tag{7.9}$$

If i is a continuous variable (as assumed), then $P(i)$ is continuous in i and is differentiable. The first derivative of $P(i)$ is

$$\frac{dP(i)}{d(1+i)} = -\frac{Y_1}{(1+i)^2} - \frac{2Y_2}{(1+i)^3} - \frac{3Y_3}{(1+i)^4} - \cdots - \frac{NY_N}{(1+i)^{N+1}} \tag{7.10}$$

which is negative for all values of $-1 < i < \infty$. Furthermore, the second derivative is

$$\frac{d^2P(i)}{d(1+i)^2} = +\frac{2Y_1}{(1+i)^3} + \frac{6Y_2}{(1+i)^3} + \frac{12Y_3}{(1+i)^5} + \cdots + \frac{N(N+1)Y_N}{(1+i)^{N+2}} \tag{7.11}$$

which is always positive for all values of $-1 < i < \infty$. Thus, in the range of i from -1 to infinity, this project's present value function is *monotonic decreasing.* The negative first derivative indicates the function decreases as i increases, but at a decreasing rate since the second derivative is positive (the function is convex). Figure 7.4 is a typical trace of $P(i)$ for a simple project.

The interest rate, i, was restricted to the range $-1 < i < \infty$ for Equations (7.10) and (7.11) to be valid. If $i \to \infty$, then the limit in Equation (7.9) is:

$$\lim_{i \to \infty} P(i) = -Y_0$$

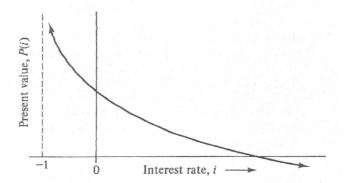

Figure 7.4 Trace of *P(i)* for a simple investment

so that *P(i)* is asymptotic to this initial investment value and the derivatives disappear. On the other hand, the future value, *F(i)*, can be found by multiplying Equation (7.9) by $(1 + i)^N$ so that:

$$F(i) = -Y_0(1 + i)^N + Y_1(1 + i)^{N-1} + Y_2(1 + i)^{N-2} + \cdots + Y_N. \qquad (7.12)$$

However, we have that:

$$P(i)|_{i=-1} = \frac{F(i)}{(1 + i)^N} = \frac{Y_N}{\underline{(1 + i)^N}} \longrightarrow 0$$

so that *P(i)* is indeterminate at $i = -1$ since the division of Y_N by zero is not defined. The usual interpretation of this phenomenon is that an interest rate of $i = -1$ corresponds to a complete loss of capital (loss of Y_0).

A simple investment project should be accepted whenever $P(i) > 0$. The present value of the cash inflows exceeds the present value of the cash outflows, and rule 1 applies. Since *P(i)* is a function of *i*, then how much can *i* increase? Obviously, *i* can increase until $P(i) = 0$, when the marginal acceptance point in reached; beyond this point $P(i) < 0$ and the project should be rejected by rule 1. The *i* at which $P(i) = 0$ is the project's internal rate of return. Designated by the symbol *i**, it will be more fully discussed later in this chapter.

Usually net present value for an independent project is calculated for some known (or assumed) interest rate. Opportunity cost principles require selecting only those projects that will not make the firm worse off financially. This principle dictates the minimum interest rate as being the marginal cost of capital, *k*, under our assumed conditions of certainty.

The net present value function, *P(i)*, is not always a monotonic decreasing function in *i*. Depending on the magnitudes and signs of the cash flows and on the project life, the present value function may be an Nth degree polynomial in *i*. For example, oil fields frequently generate cash flow streams in which *net* cash outflows (from the firm to the project) occur during the early years' initial developmental expenditures, then again at some intermediate period when secondary recovery with forced injection of water or gas to raise the producing pressure is added, and again at the project's end because of abandonment expenses. The result is a series of interspersed negative and positive cash flows.

An example is the following:

End of Year, t	Net Cash Flow, Y_t
0	$-1,000
1	+ 800
2	+ 800
3	− 200
4	+ 350
5	− 100

Figure 7.5 displays the $P(i)$ function for the above data over the range of $-0.80 \leq i \leq 10.00$. Note that $P(i)$ tends asymptotically toward $P_0 = -1,000$ as i becomes large and that a second root occurs at approximately $i_2 = -0.727$ (the first root, $i_1 = 0.380$). This is a fifth-degree polynomial, and the remaining three roots (not shown here) are $i_3 = -1.915$, and $i_4 = i_5$ which are imaginary, occurring at the vertical asymptote $i = -0.9695$. This cash flow series has four shifts in the sign of the cash flows, and the two roots greater than -1 satisfy Descartes' Rule (with 4 sign changes there are 4, 2, or zero roots for $i > -1$).

Since the imaginary roots are not usable, and $i_3 < -1$, the region of interest for $P(i)$ is $-0.9695 < i < \infty$. Accordingly, the project would be accepted if the firm's cost of capital were in the range $0 < k \leq 0.380$ and rejected otherwise.

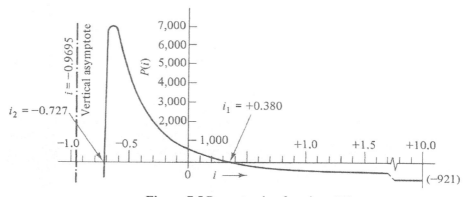

Figure 7.5 Present value function, $P(i)$

Obviously, the cost of capital would not be less than zero, and beyond $k = 0.380$ the project's net present value is negative implying nonrecovery of the investment and interest at the cost of capital rate.

7.7 THE BENEFIT-COST RATIO CRITERIA

There are two common formulations for the *benefit-cost ratio* (*B/C*), defined by Bernhard (1971) as the *benefit-cost ratio* and *net benefit-cost ratio*. Another form, Eckstein's benefit-cost ratio, was discussed in footnote 3. Other engineering economy authors typically use the same terminology, but *profitability index* and *excess present value index* are also used. At times the different names are used to distinguish between identical criteria applied in the public (benefit/cost) and private (profitability index) sectors.

The difference between the benefit-cost and net benefit-cost ratios can be illustrated simply. Let P_j be the net present value of the cash flows of project j at time $t = 0$ and C_j be its present cost at $t = 0$. Then, total benefits (at $t = 0$) are $P_j + C_j$, and the following definitions may be stated:

1. *Benefit-cost ratio* (*B/C*) is the ratio of *total benefits* to present cost:

$$\frac{B}{C} = \frac{P_j + C_j}{C_j} = \frac{P_j}{C_j} + 1 \tag{7.13}$$

2. *Net benefit-cost ratio* (*NB/C*) is the ratio of *net present value* to present cost:

$$\frac{NB}{C} = \frac{P_j}{C_j} \tag{7.14}$$

The difference between Equations (7.13) and (7.14) is simply unity and the following decision rules result:

If *B/C*	or	if *NB/C*,	the rule is
> 1		> 0	Accept
= 1		= 0	Indifferent
< 1		< 0	Reject

These criteria were advanced by some early writers for constrained project selection, who argued that net present value did not discriminate among large and small investments; another criterion was needed to rank projects based on net present value *per dollar outlay*. Then, the most *efficient* projects could be chosen first.

As it turns out, neither ratio is acceptable for ranking projects under constrained conditions, which will be examined in Chapter 9. In the absence of constraints, the ratios are acceptable but no more informative than net present value itself. Hence, there is no particular advantage in using benefit-cost over the simpler NPV.

Weingartner (1963b) derives the *NB/C* criterion from the linear programming dual of the project selection problem (see Chapter 9.) Weingartner indicates that *NB/C*, under unconstrained conditions, will accept and reject the same projects as NPV, with no apparent advantage for the longer *NB/C* method. When capital expenditures are limited in the current period, the *NB/C* can rank projects correctly on their efficient use of capital. However, it cannot aggregate several small projects to displace one large project, thereby perhaps using the total available capital more effectively (the "lumpiness" effect). When cash outflows occur after $t = 0$, then *NB/C* is completely ineffective and must be redefined. Finally, Weingartner (1963a and 1963b) shows that *NB/C* is inaccurate when used with interdependent or mutually exclusive projects.

The benefit-cost ratios really add nothing to the NPV criterion for the unconstrained problem. The guidance of the present value per dollar outlay is the same as a positive net present value that signifies accept and a negative one that signifies reject. We shall reexamine these ratios for the constrained case in Chapter 9.

7.8 INTERNAL RATE OF RETURN

One difference between the application of present worth and internal rate of return (IRR) criteria is the necessity in the latter case to specifically address whether a cash flow series represents a loan or an investment. Firms prefer investments that have high IRRs and loans that have low IRRs. Since each series can be converted to the other through a multiplication by -1, we simplify the language by assuming that only investment proposals need be considered.

7.8.1 Defining the internal rate of return. The net present value and all benefit-cost methods depend on an external interest rate, such as the firm's cost of capital. The internal rate of return method also discounts cash flows, but it seeks to avoid choosing an arbitrary interest rate. Instead, the focus in finding an unknown interest rate that is *internal* to the project. An interest rate that makes the flow's present value equal to zero is found. Thus, IRR is the value i^*, such that:

$$P(i) = \sum_{t=0}^{N} Y_t (1+i^*)^{-t} = 0 \qquad (7.15)$$

Except for very simple cash flows, the IRR must be found by a search technique (automatic in Excel or financial calculators) since it is an unknown root (or roots) of a polynomial in i. By letting $R^t = (1+i^*)^{-t}$, Equation (7.15) can be expanded in terms of R:

$$Y_0 R^0 + Y_1 R^1 + Y_2 R^2 + \cdots + Y_N R^N = 0$$

Since this is a polynomial of degree N in terms of R, there may be one or more values of i^* that will make Equation (7.15) true.[5] If real, these values are known as the

---

5. Numerous papers address whether a unique internal rate of return can be predicted by the cash flow stream. Descartes' rule of signs and the number of sign changes in the cash flow stream define an *upper limit* on the number of positive roots of Equation (7.15). This is discussed even in some elementary engineering economy texts (see Newnan et al., 2012, pp. 256-257). Soper (1959) develops a sufficient condition for a unique root to exist in the interval of interest ($-1 < i^* < \infty$), but determining uniqueness requires a trial-and-error search for the single root. Kaplan (1965) demonstrates how Sturm's theorem (1829) can be applied to find the *exact* number of *distinct* roots in the interval ($-1, \infty$). The Sturm-Kaplan method can also be used (most effectively with a computer) to find the root's numerical values (e.g., see Kaplan, 1967). Note that if $P(i^*)$ has repeated roots, they are counted as one in the Sturm-Kaplan method (see Kaplan, 1965 and Turnbull, 1952). Norstrøm (1972) develops a simpler condition for uniqueness of the root, *but in the interval* ($0 < i^* < \infty$). Norstrøm shows that if the *cumulative* cash flow stream (undiscounted) changes sign only once and if the final cash flow is positive, this is a sufficient (but not necessary) condition for a unique root, i^*, to exist. Bernhard (1977) shows that even if Norstrøm's sufficiency condition holds in the (0, ∞) interval, Soper's condition may be violated in the (-1, 0) interval, with a consequence that the root is not unique and, hence, $P(i) > 0$ is not guaranteed. Bernhard (1979) provides a more general method for detecting uniqueness in the ($-0, \infty$) interval. Eschenbach (1984) demonstrates that comparison of specific alternatives (subscriptions or memberships with only a first cost) with different length lives typically involves a unique, meaningful IRR in spite of multiple sign changes and failure to satisfy any of the sufficiency conditions. Hajdasinski (1983, 1988, 1989) provides yet another root finding approach.

project's *internal* rates of return, or *yield*. The criterion is select the project if i^* (the IRR) exceeds the firm's marginal investment rate; otherwise, reject. Because some authors assert otherwise, it is important to reiterate the point that this comparison of the IRR with the firm's marginal investment rate ensures that the same reinvestment assumption is being made with present worth and internal rate of return analyses.

This project acceptance criterion is popular because the IRR is established by the project's expected cash outlays and inflows, and it is internal or entirely dependent on the project's parameters. The IRR or rate of return is a single number that can be used to describe a project's attractiveness. In contrast, to properly judge a project's present value two additional values are needed—the interest rate and the initial investment.

7.8.2 The fundamental meaning of internal rate of return. To develop the economic interpretation, consider first a numerical example. Suppose a single project has estimated cash flows as follows:

End of Period, t	Cash Outflow (−) or Cash Inflow (+), Y_t
0	−10,000
1	+ 2,000
2	+ 4,000
3	+ 7,000
4	+ 5,000
5	+ 3,000

For the undiscounted case ($i = 0$), Figure 7.6 graphically represents the cash flow stream. Note that any resultant vector (ordinate) from the zero cash level defines either an unrecovered or an over-recovered investment at time t. For example, at time 2 the unrecovered investment is $−4000, and at time 3 the over-recovery is $+3000.

Economically meaningful interest rates must satisfy ($-1 < i < \infty$) and, for practical purposes, ($0 < i < \infty$), since a negative interest rate implies either partial or no recovery of capital. A positive interest rate, $i^* > 0$, modifies Figure 7.6 as each unrecovered investment balance, F_b, earns compounded interest at the rate i^* during the following period. This increases the magnitude of the unrecovered investment balance, F_t, just before the cash inflow, Y_{t+1}, is applied to produce the new unrecovered investment balance, F_{t+1}. This sequence is illustrated in Figure 7.7.

The unrecovered investment (or project balance) at time $t + 1$ can be found from the recursive relationship

$$F_{t+1} = F_t(1 + i) + Y_{t+1} \qquad (7.16)$$

where F_t = unrecovered investment at time t
F_{t+1} = unrecovered investment at time $t + 1$
Y_{t+1} = cash flow at time $t + 1$
i = applicable interest rate.

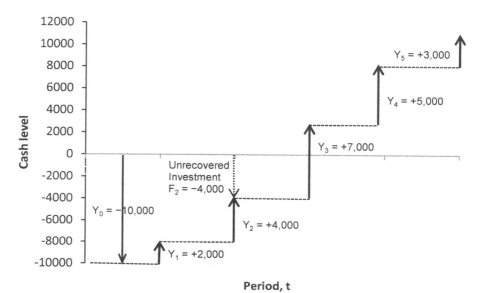

Figure 7.6 Undiscounted cash flow stream

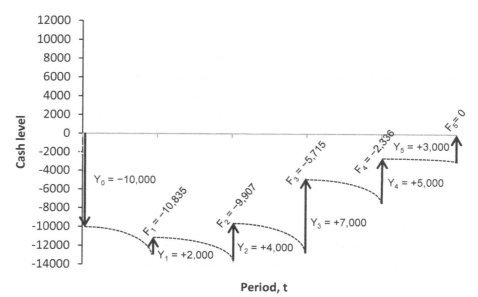

Figure 7.7 Cash flow stream discounted at IRR = $i^* = 28.35\%$

To motivate an economic interpretation of IRR we first note that $i^* = 28.35\%$ for the previous example. Now apply this same interest rate to the unrecovered investments $F_0, F_1, \ldots, F_4$ to calculate the interest earned by each during the succeeding periods. The model for these calculations is Equation (7.16). The calculations are given in Table 7.2, and Figure 7.7 illustrates the sequence to scale.

Table 7.2 Unrecovered investments, $i^* = 28.35\%$

End of Period	Unrecovered Investment at Beginning of Period t	Interest Earned from t to $t+1$	Sum	Cash Flow at End of Period t (Time $t+1$)	Unrecovered Investment at Beginning of Period $t+1$
t	F_t	$F_t i^*$		Y_{t+1}	$F_{t+1} = F_t(1.2835) + Y_{t+1}$
0	-	-		$-10,000	$-10,000
1	$-10,000	$-2,835	$-12,835	2,000	-10,835
2	-10,835	-3,072	-13,907	4,000	-9,907
3	-9,907	-2,809	-12,716	7,000	-5,716
4	-5,716	-1,621	-7,337	5,000	-2,337
5	-2,337	-663	-3,000	3,000	0

One should note particularly in Table 7.2 that the unrecovered investment balance at the end is zero. Thus, the IRR as calculated by Equation (7.15), $i^* = 28.35\%$, not only is the interest rate at which the net present value of the cash flow stream is zero, but is also the interest rate that exactly recovers the investment over the project's life *plus* a return on the unrecovered investment balances during that life. Hazen (2003) uses a similarly defined investment stream (opposite in sign) and treats each root for the IRR as a per-period rate of return on that stream, as in Table 7.2. With this approach multiple roots or rates of return (see section 7.11.1) can coexist because the overall cash flow can be interpreted as different rates of return on different investment streams.

The IRR is often misinterpreted as a rate of return on the *initial* investment. This is not so, and that fact can be demonstrated two ways. First, consider the cash flow stream in the previous example. If the IRR, $i^* = 28.35\%$, applies to the initial investment only, then the interest earned in each period is $-2,835. This changes the unrecovered balances, and, indeed, the positive cash flows fail to recover the initial investment plus interest at the assumed rate of 28.35%, as detailed in Table 7.3. A second example could be constructed for any project that requires investments spread over a several year startup (as many large projects do).

Table 7.3 Initial investment recovery ($10,000 × 28.35%)

End of Period	Unrecovered Investment at Beginning of Period t	Interest Earned from t to $t+1$	Cash Flow at End of Period t (Time $t+1$)	Unrecovered Investment at Beginning of Period $t+1$
t	F_t	$F_0 i^*$	Y_{t+1}	$F_{t+1} = F_0(1.2835) + Y_{t+1}$
0	-	-	$-10,000	$-10,000
1	$-10,000	$-2,835	2,000	-10,835
2	-10,835	-2,835	4,000	-9,670
3	-9,670	-2,835	7,000	-5,505
4	-5,505	-2,835	5,000	-3,340
5	-3,340	-2,835	3,000	-3,175

7.8 Internal Rate of Return

The fundamental economic meaning of IRR is the rate of interest earned on the *time-varying, unrecovered* balances of investment, such that the final investment balance is zero. This was shown numerically in Table 7.1 and Figure 7.7. This can also be shown algebraically. Let F_t = the unrecovered investment balance at time t, and N = the project's life. Since $F_0 = Y_0$ at $t = 0$, then from the relationship [Equation (7.16)] one obtains

$$
\begin{aligned}
F_0 &= Y_0 \\
F_1 &= F_0(1 + i^*) + Y_1 = Y_0(1 + i^*) + Y_1 \\
F_2 &= F_1(1 + i^*) + Y_2 = Y_0(1 + i^*)^2 + Y_1(1 + i^*) + Y_1 \\
&\vdots \\
F_N &= F_{N-1}(1 + i^*) + Y_N \\
&= Y_0(1 + i^*)^N + Y_1(1 + i^*)^{N-1} + Y_2(1 + i^*)^{N-2} + \cdots + Y_N
\end{aligned}
\tag{7.17}
$$

For exact capital recovery, $F_N = 0$. After setting Equation (7.17) to zero and dividing by $(1 + i^*)^N$, one obtains:

$$0 = Y_0 + Y_1(1+i^*)^{-1} + Y_2(1+i^*)^{-2} + \ldots + Y_{N-1}(1+i^*)^{1-N} + Y_N(1+i^*)^{-N}$$

or

$$P(i^*) = \sum_{t=0}^{N} Y_t(1 + i^*)^{-t} = 0$$

which defines i^* as the internal rate of return.

As will be detailed in Section 9.5, since IRR does not measure the return on initial or total investment, the internal rates of return from two or more mutually exclusive projects should not be compared to establish a preference ordering. A given project's IRR merely states the rate of interest earned on the time-varying unrecovered investment balances for that project. Moreover, IRR is independent of the absolute level of investment, so it has meaning only when the level of investment is separately considered. Thus, comparing IRRs for two or more projects ignores the absolute investment levels. Consider two projects, A and B, which have cash flow streams as follows:

Table 7.4 Cash flows for two mutually exclusive projects

End of Period, t	Project A Cash Flow	Project B Cash Flow
0	$-1,000	$-10,000
1	475	4,380
2	475	4,380
3	475	4,380

A's IRR equals 20%, and B's equals 15%. In the absence of budget or other constraints, project B should logically be preferred over A. In spite of its lower rate of return, its larger size increases the firm's wealth (ignoring the time value of money) by 3($4380) − $10,000 = $3140; whereas A provides only 3($475) − $1000 = $425.

To compare the two mutually exclusive alternatives, the incremental IRR must be calculated. Project B requires $9000 more in investment, and it produces $3905 in incremental income each year for three years. IRR = RATE(3,3905,−9000) = 14.44%. If this is an acceptable rate of return, then the larger Project B should be accepted.

7.8.3 Conventional and nonconventional investments (and loans). The intuitive definition of an investment has two parts. First, the cash flow series begins with a cash outflow or outflows (the investment). Second, the sum of the cash flows is positive so that the total of the returns exceeds the investment. For a loan, both parts of the intuitive definition are reversed.

However, there are projects which must be evaluated where this intuitive definition does not work. Consider for example, projects where the investment is made, but the returns never pay back the initial investment. Intuitively, and as shown in the top left quadrant of Figure 7.11 in Section 7.10, these projects have a negative rate of return. A firm might implement such a project despite its negative PW and IRR because of its strategic importance. For example, it might be the first project in a new market area—new product, geographical region, or type of client. Also in Chapter 12, when uncertainty is addressed it is common for negative PW or IRR values to correspond to the bad outcomes in an array of possibilities.

Note that there are subsidized loans where meeting conditions, such as building a project in an economically depressed area, can mean that the initial borrowing is larger than the total of the repayments. This is a loan with a negative rate of return, which can be a very attractive source of funds.

As discussed earlier, firms prefer investments that have high IRRs and loans that have low IRRs, and we are in general simplifying the presentation by assuming that only investment proposals need be considered. However, we will find that when we cannot clearly define a series of cash flows as an investment or a loan, then the roots for the PW equation may not yield a useful IRR, and that we *may not even be able to tell whether we would like a low or a high IRR. In some cases, even the usefulness of the PW value may be in question.*

Let us define types of investment. The cash flow stream is classed as *a conventional* investment if it begins with one or more negative cash outflows followed by a single sign change to a series of one or more positive cash inflows. A *conventional* loan begins with one or more positive cash flows, which are paid back by one or more negative cash flows (or cash outflows). Again, there is a single sign change, in this case from positive to negative cash flows. A nonconventional investment intersperses the positive and negative cash flows. Examples of these different *investment* possibilities include:

Table 7.5 Example cash flows for conventional and nonconventional investments.

Type of Investment	Sign of Cash Flow at Time =			
	0	1	2	3
Conventional	−	+	+	+
Conventional	−	−	+	+
Nonconventional	−	+	+	−
Nonconventional	−	+	−	+

For conventional investments the key part of the intuitive definition of an investment is that it begins with one or more negative cash flows. If the sum of all cash flows is positive, then the IRR is positive. If the sum is negative, then the IRR is negative.

For nonconventional investments, the existence and interpretation of a project's internal rate of return and its classification as a pure or mixed investment or mixed investment/loan (see next section) depend on the cash flow stream itself and the recovered investment balance stream. More particularly, it is the number of sign changes in the net cash flows over time and in the *unrecovered* project balances over time.

7.8.4 Conventional investments and internal rate of return. The present value of a conventional investment is stated in Equation 7.18, where $v = 1 + r$:

$$P(r) = Y_0 + Y_1 v^{-1} + Y_2 v^{-2} + \cdots + Y_N v^{-N} \tag{7.18}$$

Descartes' rule of signs says that since a conventional investment has only one sign change in its cash flows, Equation (7.18) can have at most one root, $v > 0$. The case for $Y_0 < 0$, $Y_t \geq 0$ ($t = 1, 2, \ldots, N$) will be proved. The more general case for $Y_r < 0$ ($r = 0, 1, \ldots, m$), $Y_\theta \geq 0$ ($\theta = m + 1, m + 2, \ldots, N$) can be proved by the same method but with more labor. From Equation (7.18):

$$\frac{dP(r)}{dv} = -Y_1 v^{-2} - 2Y_2 v^{-3} - \cdots - NY_N v^{-(N+1)} \tag{7.19}$$

which is always < 0 since $Y_t \geq 0$ ($t = 1, \ldots, N$) by assumption and $v > 0$. Moreover:

$$\frac{d^2 P(r)}{dv^2} = 2Y_1 v^{-3} + 6Y_2 v^{-4} + \cdots + N(N+1)Y_N v^{-(N+2)} \tag{7.20}$$

which is always greater than 0 for the same reasons. Thus, $P(r)$ is strictly decreasing (the first derivative is negative) and strictly convex (the second derivative is positive) for $Y_0 < 0$, $Y_t \geq 0$ ($t = 1, \ldots, N$), and $r > -1$. Furthermore, from Equation (7.18):

$$\lim_{r \to \infty} P(r) = P(\infty) = Y_0$$

which is negative by assumption, also from Equation (7.18):

$$\lim_{r \to -1} P(r) = Y_N / (1+r)^N$$

Which is positive, since Y_N is positive by assumption as is $(1 + r)$. Then the strictly monotonic decreasing function $P(r)$ will provide at most one root, r^*, greater than minus one.

We conclude from this reasoning that conventional investments have at most one real value of $i^* = r^*$ in the region $(-1 < r \leq \infty)$, then we are justified in considering i^* to be the internal rate of return for the project.

7.9 NONCONVENTIONAL INVESTMENT

7.9.1 Nonconventional investment defined. When a stream of cash flows have more than one sign change, it is necessary to enforce the stipulation that an investment project has a total sum of cash flows that is positive—that is, the first cash flow is negative and more is returned than is invested.

Then investment projects may be classified *as pure* or *mixed* depending on the *unrecovered investment balance* stream. This unrecovered investment balance or *project balance* was defined earlier in Section 7.8.2. The unrecovered investment balance at the end of period t ($t = 0, 1, \ldots, N$), with interest earned at rate i, is given by:

$$F_t(i) = Y_0(1+i)^t + Y_1(1+i)^{t-1} + \cdots + Y_t \tag{7.21}$$

As an investment we have $Y_0 < 0$, but all other Y_j ($j = 1, 2, \ldots, t$) are unrestricted as to sign. Thus, the unrecovered investment balance $F_t(i)$ can be positive, negative, or zero at any time $t \neq 0$. If it is negative, then the firm has committed (lent) $F_t(i)$ dollars to the project for the next period from time $t + 1$. In other words, the firm has money invested in the project. On the other hand, if the unrecovered investment balance is positive at time t, then the firm has over-recovered its investment and is actually borrowing from the project during the period from t to $t + 1$. Finally, $F_t(i) = 0$ means that the firm has exactly recovered its invested funds at time t together with interest at rate i on the unrecovered investment balances up to time t.

We can now define pure and mixed investments. *A pure* investment has project investment balances, calculated at the project's internal rate of return, i^*, that are either zero or negative throughout the project's life, N. In symbolic terms the investment is pure if, and only if,

$$F_t(i^*) = \sum_{j=0}^{t} Y_j(1+i^*)^{t-j} \leq 0 \quad (t = 0, 1, \ldots, N-1) \tag{7.22}$$

and

$$F_N(i^*) = \sum_{t=0}^{N} Y_t(1+i^*)^{N-t} = 0$$

Thus, for a pure investment the firm does not borrow from the project at any time during the project's life and exactly recovers its investment $[F_N(i^*) = 0]$ at the end of the project's life, earning interest at the IRR value, i^*, in the interim periods.

7.9 Nonconventional Investment

In contrast, *a mixed* investment may be defined as any investment that is not a pure one. Thus, a mixed investment is a project for which $F_t(i^*) > 0$ for some values of t and $F_t(i^*) \leq 0$ for all other values of t. A mixed investment contains both unrecovered and over-recovered investment balances. During the project's life the firm is both an investor in (or lender to) the project (there are unrecovered investment balances) and a borrower from the project (there are over-recovered investment balances).

In fact, if the positive values of F_t dominate the graph of the investment balances, then the nonconventional investment is better classified as a mixed loan/investment or even a mixed loan, or a pure loan. In these cases the loan portion of the cash flows appears to be dominating the project's economic value.

Using these definitions, any investment could be classed as conventional, or nonconventional and pure, or mixed. In fact all conventional investments are also pure investments. Thus, Figure 7.8 contains only four classifications. These are the basis for the study of investments and of their internal rates of return as investment criteria.

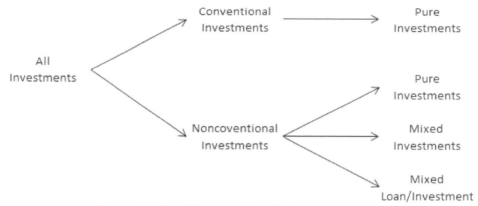

Figure 7.8 Classification of the firm's investments

One group of nonconventional investments that often are pure investments is those projects that include one or more shutdowns for maintenance or for project expansion. These maintenance or expansion costs can be large enough for the net cash flows to be negative (particularly if cash flows are evaluated on a monthly or quarterly basis), yet the project balances often satisfy the requirements for a pure investment.

Before presenting examples of these different possibilities, the next section demonstrates that all conventional investments are pure investments to clarify what is meant by a pure investment. The next section emphasizes graphical approaches that help develop a deeper understanding of the economic value of the example cash flows.

7.9.2 Conventional, pure investments. The demonstration that all conventional investments are pure investments goes as follows. A conventional investment is defined as one in which the initial $Y_t < 0$ ($t = 0, 1, \ldots, m$), successive $Y_t \geq 0$ ($t = m+1, m+2, \ldots, N-1$), and $Y_N > 0$. Let i^* be the internal rate of return, where $i^* > -1$ is assumed. Note: worthwhile investments will have an i^* above 0, but some conventional investments do *not* pay back and have $i^* < 0$. However, $1 + i^*$ is greater than zero.

Since the investment balance at any time t at interest i is given by Equation (7.21), the investment balance at any time, $t \leq m$ with interest i^* is negative. That is:

$$F_t(i^*) = \sum_{j=0}^{t} Y_j(1 + i^*)^{t-j} < 0$$

since $i^* > -1$ and all $Y_t < 0$ for $t = 0, 1, \ldots, m$ are assumed. Now, suppose that under the influence of some positive Y_t in succeeding periods the investment balance, $F_t(i^*)$, becomes positive in the region ($t = m + 1, m + 2, \ldots, N - 1$). It follows that if any intermediate balance $F_t(i^*) > 0$ at any time $t \neq N$, then $F_{t+1}(i^*) > 0$ also, since $i^* > -1$ and $Y_t \geq 0$ have been assumed. Similarly, if $F_{t+1}(i^*) > 0$, then $F_{t+2}(i^*) > 0$, and so forth, until ultimately the final end-of-project investment balance $F_N(i^*) > 0$ also. The implication that $F_N(i^*) > 0$ contradicts the requirement that $F_N(i^*) = 0$ in order for $i^* > 0$ to exist (see Section 7.8.2). Hence, we must conclude that no $F_t(i^*)$ in the intermediate region ($0 \leq t \leq N - 1$) can be positive. This conclusion is identical to the first requirement of a pure investment, and since $F_N(i^*) = 0$ is also required for $i^* > -1$ to exist, a conventional investment thus conforms to the definition of pure investment [Equation (7.22)]. We conclude, therefore, that all conventional investments are pure investments.

7.9.3 Analyzing nonconventional investments. Mathematical and algorithmic approaches can be used, and in the second edition of this text they were featured. In this edition, those approaches have been replaced by graphical techniques that rely on the computational power of spreadsheets. Simply calculating or graphing the PW for a range of interest rates can answer questions about the existence and number of solutions to the PW = 0 equation. This is demonstrated in the next section for four examples that correspond to the four investment classifications shown in Figure 7.8.

More importantly graphs also give us a sense of perspective. How large are the present worth values for realistic interest rates? Is the present worth negative (or positive) for all but a very small range of values for the discount rate? In Figure 7.12, the PW of project D is shown to be positive for a wide range of interest rates. However, the maximum PW is $900 on a project with over $20,000 in cash flows, and for realistic (but high) rates of 25% to 40% the PW is much smaller. This set of cash flows is the Lorie-Savage problem, which is analyzed in detail in Section 7.11.

Once a root to the PW equation is identified, the next step is to use the tool of investment or project balances over time. If the tabulated values or graph shows that the investment is pure, then the root is a useful IRR. See for example Project B in the following discussion. However, as detailed in the discussion of Project C, the graph of project balances may indicate whether a calculated root is useful even for mixed investments.

Analyzing both graphs for a potential project develops understanding of the project's economic value. In addition, the graphs are more understandable for presentations to managers, to peers, to subordinates, and to the public.

7.9.4 Numerical examples. Numerical examples of the four investment classifications highlight their differences. Project A is a conventional investment. Projects B and C are nonconventional investments with an expansion cost of $20,000 in year 3. Project B will be shown to be a pure investment, and project C a mixed investment. Project D is nonconventional, but it will not pass our tests to be classed as an investment. Projects A, B, and C have unique real roots to the internal rate of return equation $F(i^*) = 0$, while project D has two real roots in the region ($-1 < i^* < \infty$). The cash flows are given in Table 7.6.

7.9 Nonconventional Investment

Table 7.6 Cash flows for four projects

End of Period	Net Cash Flows (Y_t)			
	Project A	Project B	Project C	Project D
0	−10,000	−10,000	−10,000	−1,600
1	+2,000	+6,000	+8,000	+10,000
2	+4,000	+5,000	+8,000	−10,000
3	+7,000	−20,000	−20,000	–
4	+5,000	+15,000	+15,000	–
5	+3,000	+10,000	+10,000	–

For projects A and B there is only one root for the IRR equation, and all of the investment balances are ≤ 0, so the focus is on the calculated values that would be graphed if needed.

For the conventional investment (project A); there is one sign change and the IRR equation has one real root, IRR = i^* = 28.35%. As was shown in Figure 7.7 and Table 7.2, the investment balances are negative until the final investment balance which is 0. Since these investment balances conform to Equation (7.22), project A is a pure investment.

Project B has interspersed negative and positive cash flows, thus it is a nonconventional investment. On solving IRR equation $F(i^*) = 0$, we obtain two imaginary roots and three real roots: $i_1 = 0.12974$, $i_2 = -2.30$, and $i_3 = -1.42$. Since only the region ($-1 < i^* < \infty$) is valid for internal rates of return, we shall consider $i_1 = i^* = 0.1297$ to be the project's IRR. Table 7.7 is the spreadsheet that computes the F_t values in column D. This spreadsheet breaks the calculation down into the steps for columns C and D to support graphing of the investment balances, which is Figure 7.9.

Table 7.7 Project B's investment balances at $i = 12.97\%$

	A	B	C	D	E	F
1	Year	B	F_t start	F_t end		
2	0	−10,000	0	−10,000		
3	1	6,000	−11,297	−5,297	=C3+B3	
4	2	5,000	−5,985	−985		=D2*(1+$B$8)
5	3	−20,000	−1,112	−21,112		
6	4	15,000	−23,852	−8,852		
7	5	10,000	−10,000	0		
8	IRR	12.97%				

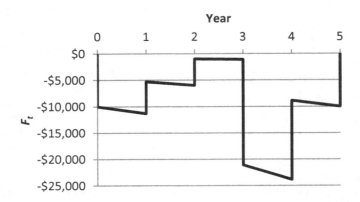

Figure 7.9 Investment balances for project B at 12.97%

Table 7.8 shows how a spreadsheet can be constructed in order to draw the graph. The left most column has the *x*-axis values, while the right most column is the *y*-axis values. Each time period must have two entries, since each has a vertical line corresponding to the cash flow at that time. Note that once the initial formulas have been entered, they can be copied down and the intervening *zero* rows do not affect the calculations.

Table 7.8 Project B's investment balances for spreadsheet graphing

	CF	F_t start	F_t end	F_t for B
0		0		0
0	−10,000		−10,000	−10,000
1		−11,297	0	−11,297
1	6,000	0	−5,297	−5,297
2		−5,985	0	−5,985
2	5,000	0	−985	−985
3		−1,112	0	−1,112
3	−20,000	0	−21,112	−21,112
4		−23,852	0	−23,852
4	15,000	0	−8,852	−8,852
5		−10,000	0	−10,000
5	10,000	0	0	0

Again, for project B the intermediate investment balances are all negative and the final balance is zero, so project B is a pure investment, even though it is a nonconventional investment. Moreover, more than one sign change in the cash flow stream does not necessarily indicate a mixed investment, nor does it indicate that the cash flow stream has more than one real value for IRR in the region ($-1 < i^* < \infty$).

For project C, the interspersing of negative and positive cash flows identifies it also as a nonconventional investment, but is not a pure investment. On solving $F(i^*) = 0$, we obtain five real roots, of which three are identical: $i_1 = i_2 = i_3 = 0.285$, $i_4 = -2.308$, and $i_5 = -1.419$. Thus, there is one real root of interest in the region ($-1 < i^* < \infty$), namely, $i_1 = i^* = 0.2850$. Using this value for i^*, we obtain from Equation (7.19) the investment balances:

$F_0 = -10,000$

$F_1 = -10,000(1.285) + 8000 = -4850$

$F_2 = -4850(1.285) + 8000 = +1767$

$F_3 = +1767(1.285) - 20,000 = -17,729$

$F_4 = -17,729(1.285) + 15,000 = -7782$

$F_5 = -7782(1.285) + 10,000 = 0$

As illustrated in Figure 7.10, the investment balance at time $t = 2$ is positive, whereas all other investment balances are zero or negative. Thus, the firm has committed (lent) funds to the project following $t = 0$, 1, 2, and 4 but has overdrawn (borrowed) excess funds from the project from $t = 2$ to $t = 3$. The project is a mixed investment since it includes both investing (lending to the project) and financing (borrowing from the project). The project both uses invested capital supplied by the firm and supplies excess funds for the firm to use elsewhere.

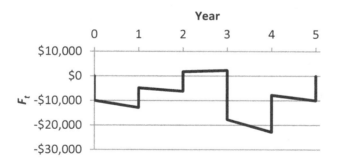

Figure 7.10 Project C unrecovered investment balances

A high rate of return is good for investments and a low rate is good for loans. But a mixed investment is neither, and as a result some have concluded that the calculated IRR, i^*, is not useful. However, if the positive project balances have relatively short periods and relatively small magnitudes, (as in Figure 7.10) then the IRR may be useful for decision-making.

In fact, if present worth values are graphed for interest rates greater than -1, the curves for projects A, B, and C are all monotonically decreasing with a single value for the IRR. If projects B and C are compared, we see that project C has $5000 more in total cash flows split between periods 1 and 2. Not surprisingly, this increases the IRR from

12.9% to 28.5%. Even though project C is a mixed investment when evaluated at the 28.5% rate, the value can be used as an IRR and exactly the same decisions will be made as with present worth.

Example D is the "Lorie-Savage pump problem," which will be discussed in more detail in Section 7.11. This is a nonconventional mixed loan/investment, because the initial cash flow is negative like an investment, but the sum of the cash flows is negative—like a loan. This cash flow series is *not* an investment. On solving $F(i^*) = 0$, we obtain two values for i^*: 0.25 and 4.00. The investment balances at both rates indicate mixed loan/investments where the largest investment balances shown in Table 7.9 are positive as in a loan.

Table 7.9 Project D's investment balances

	Investment Balances at	
	$i^* = 0.25$	$i^* = 4.00$
$F_0(i^*) =$	−1600	−1600
$F_1(i^*) =$	+8000	+2000
$F_2(i^*) =$	0	0

We now face a dilemma: When does the internal rate of return exist and when is it useful? Projects A, B, and C had a rate of return internal to the project itself, but for project D a useful rate of return internal to the project did not exist. From these examples we intuitively associate the usefulness of IRRs with pure investments and with mixed investments where there is a single value for the IRR. The next section will show that there are mixed investment/loans with multiple roots for the IRR equation, where the positive value is an IRR useful for decision-making.

7.10 ROOTS FOR THE PW EQUATION

7.10.1 Using the Root space for *P*, *A*, and *F*. In section 2.2 *P*, *A*, and *F* were defined as a single cash flow at time 0, a uniform end-of-period series from period 1 to period *N*, and a single cash flow at the end of period *N*. By restricting possible cash flows to a *P*, an *A*, and an *F*, which has the same *N* as the *A*; it is possible to describe the PW equation and roots that result for all possible combinations of values for the cash flows (Eschenbach, Baker, and Whittaker, 2007). In other words, for any *P*, *A*, *F*, and *N*, it is possible to define the number and sign for the roots of $PW(i) = 0$.

There are many industrial projects where the economic model, at least for initial analyses is at this level of detail. Other projects have different patterns, particularly as more detailed and accurate cash flow models are built. As simple examples, projects A, B, and C from the previous section do not fit this model. However, project D can be restated as having a *P* = −$1600 at time 0, an *A* = $10,000 for periods 1 and 2, and an *F* = −$20,000 in period 2. This is an *N* = 2 model.

To support two dimensional figures, the value of *P* is assumed to be −1. Then, the values of *A* and *F* are simply scaled to match. For example, consider project D above, where the three cash flows are divided by 1600. The roots are unchanged, but the values are now *P* = −$1 at time 0, an *A* = $10,000/1600 = $6.25 for periods 1 and 2, and an *F* =

−$20,000/1600 = −$12.5 in period 2.

Without loss of generality, we can assume that $P = -1$, since we are interested in investments. (For loans we simply have a negative multiplier for all of our cash flows.) Then Figure 7.11 (Eschenbach, Whittaker, and Baker, 2008) can be constructed. In this diagram the vertical axis is the value of the final cash flow, F. The horizontal axis is the value of the uniform cash flow every period, A.

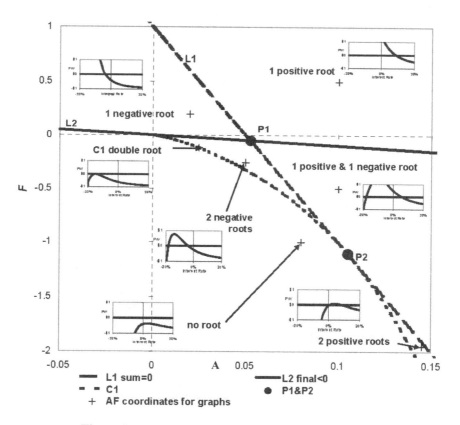

Figure 7.11 Root space for P, A, and F with PW curves

The next section will define the different lines, curves, and points. For this section let us focus on the 7 embedded figures for PW vs. i. These figures define the number and sign of the roots to PW(i) = 0.

At the top, there are 2 embedded figures for a single root, the IRR. There is no question that applying the criteria (1) is PW positive at i^* and (2) is the IRR > i^* lead to the same decision.

The bottom left embedded figure, shows a PW(i) that is negative for all i. While the IRR is indeterminate, there is no confusion about the recommended decision—reject the project. When using the IRR criteria, if the PW is negative (or positive) for all i, then the decision is clear even though the IRR *value* is indeterminate.

The three remaining embedded figures in the lower left describe multiple roots—a double root, two negative roots, and two positive roots. These will be examined further in Section 7.11, but for now we will note that they do not satisfy our definition of an investment, since the sum of the cash flows in all three cases is negative. They are to the left of L1, which is the line of those points where the sum is zero.

This leaves the embedded figure on the right side showing a positive and a negative root for PW(i) = 0. If the larger root is defined as the IRR, then again applying the criteria (1) is PW positive at $i^* > 0$ and (2) is the IRR $> i^*$ lead to the same decision. This region has $P = -1$, $A > 0$, and F sufficiently negative so that the total cash flow in period N is negative. Thus, there are two sign changes in the cash flows, two roots, and the project is a mixed investment. Since the final cash flow is negative and the final investment balance is zero, the investment balance in period $N-1$ must be positive; thus, the project is a mixed investment.

Thus for problems with only P, A, and F cash flows that have a single sign change, the IRR and PW criterion lead to the same decisions. If the set of cash flows has two sign changes, but satisfies the requirement to be an investment (total cash flows positive), then using the larger root for the IRR leads to the same decisions as the PW criterion. Note: this point is also made in Ben-Horin and Kroll (2012).

7.10.2 Defining the root space for P, A, and F. Figure 7.12 defines the lines and points, while Equation 7.23 will define curve C1. Eschenbach, Baker, and Whittaker (2007) presents the proofs of the different relationships. As noted in the previous section, L1 is defined by the condition where the sum of the total cash flows over the project's life equals zero. Since there are N cash flows with the value A, the slope of line L1 is $-N$. Points above L1 correspond to total cash flows being positive, so that at least one root is positive. However, for points below L1 where the sum of the cash flows is negative, there are more possibilities.

Line L2, which is close to the horizontal axis separates the Figure 7.12 into combinations with a single root (above the line) and ones with two or no roots (below the line). This line is defined as $F = -A$, and it determines the sign of the total cash flow in period N. The slope of L2 equals -1. P1 is the intersection of the two lines, and P2 is the tangency point of C1 and L1.

Curve C1 is well-behaved. The slope of C1 matches L2 as it approaches the origin and its slope matches L1 at the point P2. C1 is defined by two conditions which are that the PW = 0 and the PW is maximized at the double root. The second condition can also be stated as the slope of the present worth equation versus $i = 0$ at the double root. Thus, the relationship between F and A along curve C1 is defined by Equations 7.23a and b, solved in the following order.

$$A = \frac{N}{(N-1)(P/A, i, N) - (P/G, i, N)} \tag{7.23a}$$

The easiest way to use this equation is to take a value of i, solve for the corresponding value of A, then substitute that into Equation 7.23b to find the value of F, or use the Excel equation for the future value, which is =FV(i, N, A, -1).

7.10 Roots for the PW Equation

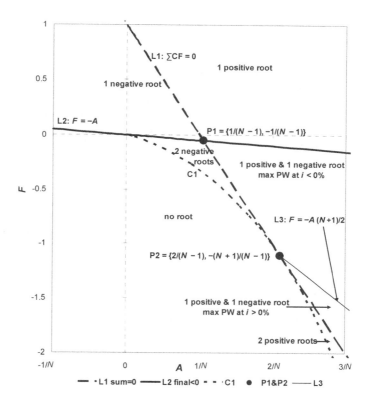

Figure 7.12 Root space equations

$$F = \frac{1 - A(P/A, i, N)}{(P/F, i, N)} \tag{7.23b}$$

The values are tabulated in Eschenbach, Baker, and Whittaker (2007) for $N = 20$, which is also the basis for Figure 7.11. F must take on very negative values for double roots above 10%. To move, for example, from a double root of 25% to one of 30%, A must increase by 15% to 0.38, but F must nearly double to –50.94. This is not a typo; the final cash flow is nearly 51 times the initial investment for a double root of 30%.

Figure 7.12 also includes a ray, L3 that subdivides the region with one negative and one positive root. This corresponds to whether the maximum value for the PW equation occurs for negative or positive interest rates. If the maximum present worth value occurs at a positive interest rate, then that peak is graphed when the project is analyzed over positive interest rates. This makes the presence of the negative root obvious. If the maximum present worth occurs at a negative interest rate, then the graph of an investment's PW over positive interest rates looks like the curve of a classic investment with one root.

7.10.3 Practical implications of the root space for *P*, *A*, and *F*. The original articles by Eschenbach et al. (2007 and 2008) emphasize the implications of the root space with respect to cash flow series with two positive double roots. However, the more important result shown in the two previous sections is that the IRR criteria and the PW criteria lead to the same result for evaluation of investments (where total cash flows are positive). The only requirement is that the IRR be defined as the larger positive root if there are two roots.

These results apply to a limited set of possible cash flows defined by an initial investment, a uniform series, and a final cash flow. However, it is likely that similar results apply to cash flows where the uniform series is actually a gradient or even a curve of values such as project A in Section 7.9.4. In fact Ben-Horin and Kroll (2012) argue that multiple IRR values are rarely relevant in the real world. Even though multiple roots exist, one can be identified as the IRR and used to get the same decisions as when using the PW criterion.

Fortunately the test is easy. Graph the PW vs. *i* to find the number of roots and to examine if the PW is positive for appropriate interest rates. If there is only one positive root for the IRR, then using it should lead to exactly the same decisions as using the PW criteria.

It must be emphasized that these conclusions differ from those in the previous version of this text. That discussion emphasized algorithmic approaches and concluded that the presence of a second root for the PW(i) = 0 equation was a fatal flaw for use of the IRR measure. It then described a more complex approach that included two interest rates: a cost of capital measure and an investment measure. While this type of approach has generated numerous academic papers, these more complex measures have little usage in practice.

Much of the discussion of these approaches has focused on sets of cash flows, such as project D, which may not satisfy the tests to be an investment, where there may be two or more positive roots for the PW equation. These are examined in the next section.

7.11 INTERNAL RATE OF RETURN AND THE LORIE-SAVAGE PROBLEM

7.11.1 Multiple positive roots for rate of return. Lorie and Savage (1955) pointed out that a unique solution does not exist for the IRR for some cash flow streams, and they posed the famous "pump problem" to illustrate this. Should a larger pump be installed in an existing oil well? The larger pump will pump a fixed amount of oil faster than the existing pump. Solomon (1956) provided the cash flow numbers shown in Table 7.10 for the now-famous Lorie and Savage problem. In words, investing $1600 in a larger pump causes the $10,000 to be realized one year earlier. Should the investment in the larger pump be made?

As shown in Section 7.9.4 and Figure 7.13, project D has two roots: 25% and 400%. Moreover, Table 7.9 showed that the pattern of investment balances was a mixed loan/investment at both rates. Detailed treatment of the Lorie-Savage problem was postponed until this section, as these cash flows do *not* satisfy one condition to be considered as an investment—the sum of the cash flows is not positive, it is a negative $1600.

7.11 Internal Rate of Return and the Lorie-Savage Problem

Table 7.10 Lorie and Savage oil pump cash flows

End of Year	Cash Flow with Existing Pump	Cash Flow with Larger Pump	Incremental Cash Flow
0	0	−1,600	−1,600
1	+10,000	+20,000	+10,000
2	+10,000	0	−10,000

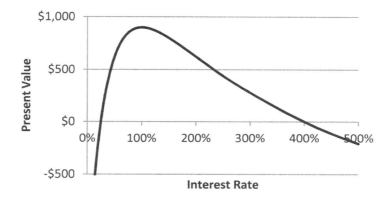

Figure 7.13 Present value function for project D

The most common examples of similar cash flow patterns are found in projects that are terminated by abandonment expenditures in the project's last year that exceed the year's positive revenues. Examples include the closure of an oil field, the shutdown of a nuclear power plant (disposal of radioactive wastes), and the restoration of strip-mined lands. As shown in Eschenbach, Baker, and Whittaker (2007) and summarized near the end of Section 7.10.2, there are two positive roots when the terminal expenditures exceed the initial investment by a substantial amount.

This example illustrates that we have not yet finished specifying the conditions under which the internal rate of return exists and can be used to judge a project's worth. The first general method of solution appeared in the later of two related publications by Teichroew, Robichek, and Montalbano (1965, 1965). In earlier editions of this text, this was presented as the necessary and sufficient conditions for validity of the internal rate of return; however the last two sections have shown that the condition is sufficient, but *not* necessary. A useful IRR that produces the same conclusions as PW can be defined in many cases where there are multiple roots and/or a mixed investment/loan.

7.11.2 Return on invested capital. The following figure of merit for mixed investments has been proposed for when i^* cannot be used. This procedure will involve the calculation of a *return on invested capital* (RIC), r, that is a function of a second interest, k. Let k be the interest rate the firm imputes as its cost of "borrowed" money—used when the project is a source of capital which is applied when investment balances are positive. (The rate k might be interpreted, for example, as the firm's cost of capital or the minimum acceptable discount rate. However, at this point, we define it only as an *imputed* cost for the firm's use of money.) Then r is the interest rate when the project is an investment for the firm with negative investment balances.

In the following development, we assume positive interest rates ($0 < i < \infty$) and an initial negative cash flow ($Y_0 < 0$). Other cash flows may be zero, negative, or positive, except that the last cash flow $Y_N \neq 0$ (if $Y_N = 0$, the project could be shortened to make $Y_N \neq 0$). The development with positive interest rates can use intuitive proofs, but the method can be extended to the entire region ($-1 < i < \infty$), as Teichroew, Robichek, and Montalbano (1965, 1965)] point out.

As we have seen, a mixed investment is both an investment and a source *of* funds for the firm. In this case the investment balance stream can be analyzed as a function of *two* interest rates. The "borrowing rate" k is applied when the investment balances are positive; the rate of return on invested capital, RIC or r, is applied when the investment balances are negative.

What we seek is a functional relationship between r and k for a given project cash flow stream. One approach is (1) to express the project's investment balances, period by period, in terms of r and k; (2) to apply the end condition that $F_N(r, k) = 0$ (the firm recovers its investment exactly at rate r at the end of N years, given an imputed cost k); and then (3) to solve $F_N(r, k) = 0$ for r in terms of k. Note that the end condition $F_N(r, k) = 0$ defines an implicit relationship between r and k, which is the function we seek.

The algorithm for determining the return on invested capital, r, as a function of k is as follows:

Step 1: Confirm the cash flows represent a mixed investment.
Step 2: Recursively calculate $F_t(r, k)(t = 0, 1, 2, \ldots, N)$ according to the rule
$$F_0(r, k) = Y_0$$

$$F_t(r,k) = \begin{cases} F_{t-1}(1+r) + Y_t & \text{if } F_{t-1} < 0 \\ F_{t-1}(1+k) + Y_t & \text{if } F_{t-1} > 0 \end{cases} \text{ for } t = 1 \text{ to } N-1 \quad (7.24)$$

Step 3: Solve the equation $F_N(r, k) = 0$ for r.
Applying this to project D (the Lorie-Savage oil-pump problem) we get.

Step 1: Table 7.9 demonstrated that project D is a mixed investment for both potential values of $i^* = 0.25$ and 4.00.

Step 2: Recursively calculate $F_t(r, k)$:
(a) $F_0(r, k) = Y_0 = -1600$.
(b) $F_1(r, k) = -1600(1 + r) + 10{,}000$ at r since $F_0 \leq 0$
            $= 8400 - 1600r$.
(As long as r is less than 5.25, that is 525%, then $1600r < 8400$ and $F_1(r, k) \geq 0$. Hence, $F_1(r, k)$ is compounded with $(1 + k)$ in the next step.)
(c) $F_2(r, k) = (8400 - 1600r)(1 + k) - 10{,}000$ at k since $F_1 \geq 0$.

Step 3: Solve $F_N(r, k) = 0$ for $r(k)$ and, since $F_N(r, k) = F_2(r, k)$,
$F_N(r, k) = F_2(r, k) = 0 = (8400 - 1600r)(1 + k) - 10{,}000$

$$\therefore r = 5.25 - \frac{6.25}{1+k} \quad (7.25)$$

Equation (7.25) makes explicit the desired implicit relationship between the return on invested capital, r, and the imputed cost of source funds, k—for the Lorie-Savage problem. Figure 7.14 illustrates the solution of Equation (7.25).

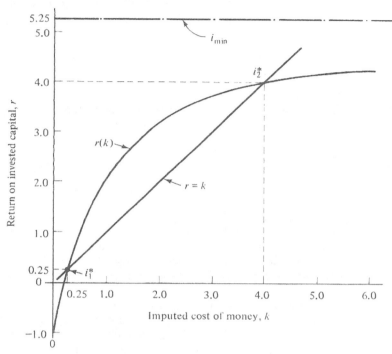

Figure 7.14 Return on invested capital (r) as a function of imputed cost of money (k), for project D. (*Source:* Teichroew, et al. 1965, p. 169.)

Four conclusions can be drawn from the functional relationship established between r and k. First, r is a monotonically increasing function of k. This is not unexpected, since a higher credit to the project for the use of its excess funds raises the return realized on the invested funds. Second, if we set $k = r$ in Equation (7.25), then $k = r = i^* = (0.25, 4.00)$. In Figure 7.14 the function $r(k)$ crosses the line $r = k$ at i^*_1 and i^*_2. Along the line $k = r = i$, the final value function in Equation (7.25) reduces to the form of:

$$F_N(i, i) = Y_0(1 + i)^N + Y_1(1 + i)^{N-1} + \cdots + Y_N,$$

which is exactly the form of Equation (7.21), from which i^* can be obtained when $F_N(i, i) = 0$. *For mixed investments, the roots, i^*, are the values of the return on invested capital, r, when the cost of money, k, is assumed to be equal to r.* Since this would be the implicit assumption if i^* were used for mixed investments, i^* cannot be used.

Third, the function $r(k)$ asymptotically approaches the limiting value for i^*, i_{min}, that is calculated in algorithmic approaches. For project D from Equation 7.25:

$$r(k) = 5.25 - \frac{6.25}{1+k}$$

Thus, as $k \to \infty$, $r(k)$ approaches 5.25.

Fourth, simply relying on the PW criteria does *not* mean what we expect. For project D, the acceptance region using PW is defined as $(0.25 \leq k$ and $r \leq 4.00$; and $k < r)$. For project D the net present value function

$$P(k) = -1{,}600 + 10{,}000(1+k)^{-1} - 10{,}000(1+k)^{-2} \qquad (7.26)$$

was illustrated in Figure 7.13. (1) The two roots of $P(k) = 0$, (0.25 and 4.00) are by definition the i^* values. (2) $P(k)$ reaches a maximum value of +900 at $k = 1.0$. However, using the relationship between r and k helps us understand the meaning of the $900 present worth at an interest rate of 100%. First Equation 7.25 can be restated at Equation 7.27.

$$\therefore k = \frac{1+r}{5.25 - r} \qquad (7.27)$$

Thus, if $r = 100\%$, it is easy to calculated the imputed cost of capital of 47%. Thus, if the $900 value for the PW at 100% is used for decision-making, we are assuming that the correct rate for investments is an unrealistically high 100% and that borrowing at the 47% imputed cost of capital is also acceptable. Neither is likely to be true.

7.11.3 Present worth and the Lorie-Savage problem.
The conclusion that the present worth criteria can be unreliable for decision-making is sufficiently surprising that it is worthwhile to look at the Lorie-Savage problem in other ways.

There is no question that if possible it would be desirable to pay for the pump at the end rather than the beginning of the year. This would shift the $1600 expense from time 0 to time 1. This more attractive project would have a net cash flow of $8400 in period 1 and a cash outflow of −$10,000 in period 2. This is a loan with an interest rate of 19.05%. [$0 = 8400 - 10{,}000/(1+i)$ → $i = 10{,}000/8400 - 1$] This is a bad loan at a high interest rate—even though no interest was charged to delay the payment of $1600 by a year.

Another approach is to say that loans should be at 10%, and then ask how much of the year 1 cash flow represents the borrowing that is *paid back* by the final cash flow of $10,000. At 10% the amount that could be borrowed for a year to be paid back is $9091 (= $10,000/1.1). Then we ask if $1600 should be invested to receive $909 (= $10,000 − $9091). The answer is: of course not.

The fundamental issue when we find the present worth of the Lorie-Savage problem at any single interest rate is that the cash flows are a mixed investment/loan and the loan portion is more significant. It is well recognized that acceptable rates for loans and investments are different. Thus, a mixed loan and investment must be evaluated at *two* interest rates not *one*. This principle is true whether IRR or PW is used as the decision criteria.

7.12 SUBSCRIPTION/MEMBERSHIP PROBLEM

The subscription/membership problem is briefly described here as a real-world problem, where one condition that is almost always satisfied is enough to ensure that a unique, useful IRR exists—in spite of many sign changes and an environment where present worth is of little value as a criteria.

This problem is included in this chapter even though (like the Lorie-Savage problem) it is based on the comparison of mutually exclusive alternatives. Our reason for doing so is to emphasize a key point, *when done properly present worth, internal rate of return, and benefit-cost criterion lead to the same recommendations.* However, in some circumstances one measure may be much easier to apply. When different recommendations are indicated by two measures—at least one of them is being applied incorrectly or with different assumptions.

Consider the problem of buying a magazine subscription or joining an organization. The cost is paid up-front before the service is received. In many cases, a lower cost per year is used to encourage payment for more than one year at a time. Table 7.11 summarizes a case where $50 buys a 1-year subscription, buying 3 years at a time saves $5 per year, and buying 5 years at a time saves another $5 per year.

Table 7.11 Example subscription problem

Life	1 year	3 years	5 years
Cost	$50	$135	$200
Cost/year	$50	$45	$40

As shown in Eschenbach (1984), a lower cost per year for the longer life is the only condition needed in order to ensure a single useful IRR for each incremental comparison. In this case the easiest way to compare the 5-year price of $200 with the 3-year price of $135 is to calculate the equivalent annual cost (EAC) of each at the same interest rate. If this is done in a spreadsheet, then GOAL SEEK can be used to calculate the interest rate at which these two choices are equivalent.

If an interest rate of 11.55% is used, then both the 1- and the 3-year subscriptions have an annual cost of $55.78. If an interest rate of 13.74% is used, then both the 3- and the 5-year subscriptions have an annual cost of $57.90. Note that the comparing the two EACs for 3 and 5 years implicitly assumes that subscriptions will be purchased for 15 years (the least common multiple of 3 and 5 years).

A present worth comparison would require an explicit choice of a 15-year life with repetitive purchases of new subscriptions until the lives match. As shown in Figure 7.15, the comparison of the 3- and 5-year choices has 5 changes in sign in the series of cash flows.

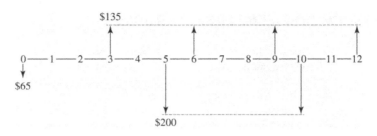

Figure 7.15 Cash flow diagram to compare $135 for 3-year life with $200 for a 5-year life.

7.13 SUMMARY

Under the *essential assumptions of certainty* and *perfect capital markets,* we have examined four criteria for evaluating the economic acceptability of an unconstrained single project. These criteria are (1) payback, (2) net present value, (3) benefit-cost ratio, and (4) internal rate of return.

As a special case of the payback method, the payback rate of return has a rational interpretation (an imputed interest rate) under certain special conditions. It is the interest rate that would be earned by a project whose cash inflows are level over time and whose salvage value is zero. Other than this highly limiting interpretation, early payback probably expresses only a subjective preference for early liquidity. While payback values are correlated with other criteria, when they indicate different decisions, they are wrong.

Net present value is shown, via the separation theorem, to be the logical criterion for project selection under perfect capital market conditions. Not only can the optimal production or investment decision be made, but the subsequent financing decision can also maximize the utility of the decision maker.

The benefit-cost ratio for a single project is shown to be only a special case of net present value that adds calculation effort without increasing information content.

Internal rate of return (IRR) is examined in detail. The IRR's fundamental meaning is developed, that is, that it is an interest rate earned on an investment's time-varying, unrecovered balances, such that complete and exact recovery occurs at the end of the project life. It is demonstrated that IRR is **not** a rate of interest on the initial investment.

For typical investments with a single sign change in the pattern of cash flows (*pure* investments), it was easy to show that all criteria that properly consider the time value of money lead to the same decisions. For some mixed investments, both PW and IRR will indicate the same decision (as long as the IRR is defined as the larger of two roots.

For problems such as the Lorie-Savage problem, it has been shown that cash flows that are a mix of investments and loans may need to be evaluated using distinct interest rates for investment and for loan parts—whether evaluated using the project's return on invested capital, which is a function of an external interest rate for loans or as a present worth with appropriate rates for investments and for loans.

REFERENCES

BEN–HORIN, MOSHE, and YORAM KROLL, "The Limited Relevance of the Multiple IRRs," *The Engineering Economist*, **57**(2) (April–June, 2012), pp. 101–118.

BERANEK, WILLIAM, "A Note on the Equivalence of Certain Capital Budgeting Criteria," *The Accounting Review*, **39**(4) (October 1964), pp. 914–916.

BERNHARD, RICHARD H., "Discount Methods for Expenditure Evaluation–A Clarification of Their Assumptions," *Journal of Industrial Engineering*, **18**(1) (January–February 1962), pp. 19–27.

BERNHARD, RICHARD H., "A Comprehensive Comparison and Critique of Discounting Indices Proposed for Capital Investment Evaluation," *The Engineering Economist*, **16**(3) (Spring 1971), pp. 157–186.

BERNHARD, RICHARD H., "Unrecovered Investment, Uniqueness of the Internal Rate, and the Question of Project Acceptability," *Journal of Financial and Quantitative Analysis*, **12**(1) (March 1977), pp. 33–38.

BERNHARD, RICHARD H., "A Simpler Internal Rate of Return Uniqueness Condition Which Dominates That of de Faro and Soares," *The Engineering Economist*, **24**(2) (Winter 1979b), pp. 71–74.

BIERMAN, H., JR., and S. SMIDT, *The Capital Budgeting Decision*, 2nd ed. (Macmillan, 1971).

DEGARMO, E., PAUL, WILLIAM G. SULLIVAN, and JOHN R. CANADA, *Engineering Economy*, 7th ed. (Macmillan, 1984), some material omitted from 8th ed.

DRYDEN, MYLES, "The MAPI Urgency Rating as an Investment Ranking Criterion," *Journal of Business*, **33**(4) (October 1960), pp. 327–341.

ECKSTEIN, OTTO, *Water Resource Development: The Economics of Project Evaluation* (Harvard University Press, 1958).

ESCHENBACH, TED G., "Multiple Roots and the Subscription/Membership Problem," *The Engineering Economist*, **29**(3) (Spring 1984), pp. 216–223.

ESCHENBACH, TED, ELISHA BAKER, IV, and JOHN WHITTAKER, "Characterizing the Real Roots for P, A, and F with Applications to Environmental Remediation and Home Buying Problems," *The Engineering Economist*, **52**(1) (Winter 2007), pp. 41–65.

ESCHENBACH, TED, JOHN WHITTAKER, and ELISHA BAKER IV, "Teaching Students about Two Positive IRR Roots for the PW Equation: Theory and Case Studies," *Proceedings of the 2008 IERC*, Vancouver, May 2008, CD.

FISHER, IRVING, *The Theory of Interest* (Macmillan, 1930; reprinted, Kelley and Millman, Inc., 1954).

GORDON, MYRON, "The Payoff Period and the Rate of Profit," *Journal of Business*, **28**(4) (October 1955), pp. 253–260.

HAJDASINSKI, MIROSLAW M., "A Complete Method for Separation of Internal Rates of Return," *The Engineering Economist*, **28**(3) (Spring 1983), pp. 207–250.

HAJDASINSKI, MIROSLAW M., "On Bounding the Internal Rates of Return of a Project," *The Engineering Economist*, **33**(3) (Spring 1988), pp. 235–271.

HAJDASINSKI, MIROSLAW M., "A Post Scriptum to Bounding the Internal Rates of Return of a Project," *The Engineering Economist*, **34**(4) (Summer 1989), pp. 339–346.

HARTMAN, JOSEPH C., and INGRID C. SCHAFRICK, "The Relevant Internal Rate of Return," *The Engineering Economist*, **49**(2) (Spring 2004), pp. 139–158.

HAZEN, GORDON B., "A New Perspective on Multiple Internal Rates of Return, *The Engineering Economist*, **48**(1) (Winter 2003), pp. 31–51.

HIRSHLEIFER, J., "On the Theory of Optimal Investment Decision," *Journal of Political Economy,* **66**(5) (August 1958), pp. 329–352; reprinted in Ezra Solomon, ed., *The Management of Corporate Capital* (The Free Press, 1959), pp. 205–228.

KAPLAN, S., "A Note of a Method for Precisely Determining the Uniqueness or Non-uniqueness of the Internal Rate of Return for a Proposed Investment," *Journal of Industrial Engineering,* **16**(1) (January–February 1965), pp. 70–71.

KAPLAN, S., "Computer Algorithms for Finding Exact Rates of Return," *Journal of Business,* **40**(4) (October 1967), pp. 389–392.

LOHMANN, JACK R., "The IRR, NPV and the Fallacy of the Reinvestment Rate Assumptions," *The Engineering Economist,* **33**(4) (Summer 1988), pp. 303–330.

LORIE, JAMES H., and LEONARD J. SAVAGE, "Three Problems in Capital Rationing," *Journal of Business,* **28**(4) (October 1955), pp. 229–239.

LUTZ, FRIEDRICH, and VERA LUTZ, *The Theory of Investment of the Firm* (Princeton University Press, 1951).

MAGNI, CARLO ALBERTO, "Average Internal Rate of Return and Investment Decisions: A New Perspective," The Engineering Economist, **55**(2) (Spring 2010), pp. 150–180.

MAO, J. C. T., *Quantitative Analysis of Financial Decisions* (Macmillan, 1969).

NARAYANAN, M. P., "Observability and the Payback Criterion," *Journal of Business,* **58**(3) (July 1985), pp. 309–323.

NEWNAN, DONALD G., Ted G. Eschenbach, and Jerome P. Lavelle, *Engineering Economic Analysis,* 11th ed. (Oxford University Press, 2012).

NORSTRØM, CARL, J., "A Sufficient Condition for a Unique Nonnegative Internal Rate of Return," *Journal of Financial and Quantitative Analysis,* **7**(13) (June 1972), pp. 1835–1839.

PARK, CHAN S., and GUNTER P. SHARP–BETTE, *Advanced Engineering Economics* (Wiley, 1990).

RENSHAW, ED, "A Note on the Arithmetic of Capital Budgeting Decisions," *Journal of Business* **30**(3) (July 1957), pp. 193–201.

SMITH, VERNON L., *Investment and Production* (Cambridge, MA: Harvard University Press, 1961), Chapter 9, pp. 219–241.

SOLOMON, EZRA, "The Arithmetic of Capital Budgeting Decisions," *Journal of Business,* **29**(2) (April 1956), pp. 124–129.

SOPER, C. S., "The Marginal Efficiency of Capital: A Further Note," *The Economic Journal,* **69**(273) (March 1959), pp. 174–177.

STATMAN, MEIR, "Persistence of the Payback Method: A Principal–Agent Perspective," *The Engineering Economist,* **28**(2) (Winter 1982), pp. 95–100.

SWALM, RALPH O., "On Calculating the Rate of Return on an Investment," *Journal of Industrial Engineering,* **9**(2) (March–April 1958), pp. 99–103.

TEICHROEW, D., A. A. ROBICHEK, and M. MONTALBANO, "Mathematical Analysis of Rates of Return Under Certainty," *Management Science,* **11**(3) (January 1965), pp. 395–403.

TEICHROEW, D., A. A. ROBICHEK, and M. MONTALBANO, "An Analysis of Criteria for Investment and Financing Decisions under Certainty," *Management Science,* **12**(3) (November 1965), pp. 151–179.

TURNBULL, H. W., *Theory of Equations,* 5th ed. (Oliver & Boyd, 1952).

WEINGARTNER, H. MARTIN, *Mathematical Programming and the Analysis of Capital Budgeting Problems* (Prentice-Hall, 1963a).

WEINGARTNER, H. MARTIN, "The Excess Present Value Index: A Theoretical Basis and Critique," *Journal of Accounting Research,* **1**(2) (Autumn 1963b), pp. 213–224.

PROBLEMS

7-1. An *investment-type* project has the following cash flows:

End of Period, t	Net Cash Flow, Y
0	$-3000.00
1	791.39
2	791.39
3	791.39
4	791.39
5	791.39

(a) By examining the unrecovered investment balances, determine the type of this investment.
(b) Determine the internal rate of return if it exists; otherwise, find R(k).

7-2. For each of the following cash flow streams, calculate the net present value, given that the minimum attractive rate of return is 15%.

	Project Cash Flows				
EOY	A	B	C	D	E
0	$-2000	$-2000	$-2000	$-2000	$-2000
1	+597	+425	+500	—	—
2	597	525	500	—	+1000
3	597	625	927	—	1424
4	597	725	527	—	1000
5	597	825	127	+3525	—

7-3. Determine the *payback period* for each project cash flow stream in problem 7-2. What projects would be desirable investments if the payback period is not to exceed 3 years?

7-4. (a) Calculate the *payback rate of return* for each project in problem 7-2.
(b) Calculate the *internal rate of return* for each project in problem 7-2.
(c) Calculate the *benefit-cost ratio* for each project in problem 7-2.

7-5. For the cash flow streams in problem 7-2, calculate the NPVs for each using nominal interest rates of 0, 5, 10, 12, 15, 20%, and then, using the IRRs determined previously, plot net present value versus interest rate for each cash flow stream.

7-6. Tests indicate that a proposed oil well, if completed, will probably produce 30 barrels of oil per day for the next 10 years. It will cost $1000 dollars a day to pump the well. The oil company must pay the landowner $30 for every barrel of oil pumped. Crude oil is expected to sell for $90 per barrel for the next 10 years. The well will cost $1,000,000 to drill. This is a first cost and will be amortized on a straight-line basis with a life of 10 years and no salvage value. The installed cost of the pumping equipment at time $t = 0$ is $750,000. This cost will be depreciated on a straight-line basis with a life of 10 years and no salvage value. What is the net present value, internal rate of return, and payback period if the company's tax rate is 40% and the minimum attractive rate of return is 8%? Assume that profits and expenses that occur during a year occur at the end of the year and that amortization of the

$1,000,000 cost of the well is in lieu of allowable depletion expense.

7-7. A company has an opportunity to increase its production of widgets. The price of widgets will remain at $10 apiece for the next 10 years. The additional operating expenses include a fixed amount of $40,000 for overhead expenses plus $5 per widget in production and selling costs. The project will last 10 years. To execute this project the company must purchase a new widget machine for $300,000. This machine will be depreciated using straight-line depreciation with a life of 10 years and no salvage value. How many widgets must be sold each year so that the project will have an internal rate of return of 10%? Assume an effective tax rate of 48%.

7-8. Given the following cash flow stream, (a) demonstrate that it is a pure investment, and (b) find the internal rate of return, i^*.

End of Period, t_j	Net Cash Flow, Y_j
0	$-50,000
1	-32,000
2	-1,500
3	25,000
4	35,000
5	45,000
6	40,000
7	35,000

7-9. An investment is assumed to produce the following cash flows:

End of Year, t	Y_t = Cash Flow
0	$-19,000
1	10,000
2	17,000
3	-7,000
4	-6,000
5	12,000
6	10,000

(a) Solve this cash flow system for its internal rates(s) of return.
(b) Is the investment conventional or nonconventional? Pure or mixed? Why?
(c) If the firm's minimum attractive rate of return (MARR) is 15%, should it select this investment based on a comparison of the project rate with the MARR?

7-10. The Stripper Petroleum Company has an opportunity to purchase an oil well from another company. The well is being sold because it doesn't produce enough oil to make it profitable for the present owner to operate it. Stripper can purchase the well for $50,000, and for an additional expenditure at the time of purchase of $150,000, the following net cash flows from the produced oil can be realized:

Time, t	Net Cash Flow, Y_t
1	$+ \quad 10,000$
2	$+ \ 1,000,000$
3	$+ \quad 10,000$
4	$+ \quad 20,000$
5	$+ \quad 10,000$

At the end of the fifth year, however, Stripper will be required to spend an additional $1,190,000 to plug the well and restore the premises.

(a) Can Stripper make a decision based on the internal rate of return criterion, assuming their minimum attractive rate of return is 14%? If so, what is their decision?
(b) If the answer to (a) is no, write the recursive end-of-period unrecovered balance equations and determine the return on invested capital.

7-11. The Rolling Wheels Harvester Company would like to decide whether it should accept a project with the following cash flows:

End of Year, t:	0	1	2	3	4	5
Cash Flow, Y_t:	$-5000	+1000	+4000	+5000	+2000	-6000

(a) Solve the problem by the internal rate of return (IRR) method if the IRR is procurable. Otherwise, solve for the return on invested capital (RIC) function.
(b) If the company's minimum attractive rate of return is 15%, should the project be accepted? Why?
(c) If the company's minimum attractive rate of return is 24%, should the project be accepted? Why?

7-12. What is the separation theorem? Under what conditions does it apply? What might happen if it were not applicable?

7-13. What are the present value rules? Under what conditions will the application of the present value rules always lead decision-makers to correct choices of projects? When might the present value rules fail?

7-14. Advocates of the internal rate of return method of project selection often state that IRR "does not require the use of an externally determined interest rate" for a decision-maker to select or reject a project. When and why may this be a false statement?

7-15. Prove, where a unique positive IRR exists, that it and the present value criteria lead to the same conclusion.

7-16. Construct a cash flow diagram for a mixed investment that has an initial cash flow of −$10,000 and three different positive roots for the IRR equation.

7-17. (a) For problem 7-10, how sensitive is the RIC to changes in the final cash flow? (b) In the initial cash flow? (c) How sensitive are the roots for the internal rate of return equation to change in the final cash flow? (d) In the initial cash flow?

7-18. (a) For problem 7-9, does a smaller negative cash flow in year 3 increase or decrease the positive internal rate of return? (b) How sensitive is the positive internal rate of return to changes in the cash flow in year 3? (c) How large a magnitude can the cash flow in year 3 have, before there is not a single, positive IRR?

7-19. Northwest Coal is considering the operation of a strip mine in a state with very strong reclamation laws. As a result, the costs when the mine is closed are very large and very uncertain. The mine will cost $80 million to open and the annual net return is expected to be $15 million per year for 10 years. The firm's minimum attractive rate of return is 10%. (a) How large can the disposal costs be, such that the project will have a positive present worth? (b) What is the internal rate of return if the disposal costs are half that breakeven value? (c) What roots exist if the disposal costs are one and a half times the breakeven v?

7-20. (a) In problem 7-20, how sensitive are the answers to the value of the firm's MARR (10%)? (b) What happens if that value is 6% or 15%? (c) Graph the breakeven values for the disposal cost for MARR values between 0 and 20%.

7-21. For the following project graph (a) the present worth, (b) the internal rate of return, and (c) the benefit-cost ratio as a function of the project's estimated life. The first cost is $800,000, the net annual return is $150,000 per year, and the firm's interest rate is 10%. The range for project life is 8–20 years. Place all three relationships on the same graph.

8 Replacement Analysis

 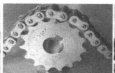  

8.1 INTRODUCTION

Replacing an asset entails two decisions: (1) Determining when to retire an asset currently in use, at which time it ceases to provide service, and (2) Selecting another asset to acquire in order to continue the required service. Problems which only consider the retirement of a current asset are referred to as *abandonment* problems while those which also consider the subsequent purchase of another asset are called *replacement* problems. In this chapter, we are concerned with tactical replacement problems, where assets and their replacements are assumed to perform the same type and level of service over some time horizon, which may be infinite.

Motivation for replacing an asset comes in many forms. A machine may fail beyond repair, forcing its replacement. The tolerances of a current machine may not meet the needs of a manufacturing system, leading to modifying or replacing the equipment. While these are valid reasons for replacement, they are not the focus of this chapter. Rather, we are concerned with decisions in which a viable alternative is to keep the defender in its current state. Clearly, the assets described above cannot be kept in their current states, as desired operations cannot continue.

The decisions studied here are a special case of the mutually exclusive investment options studied earlier in this text as the choice is to either keep the asset or replace it with another asset. This in itself defines a difference between replacement problems and typical investment problems in that a viable option is to keep an asset in its current state (which requires no additional capital investment). Replacement economy problems are further differentiated from typical investment problems in three ways. First, the solutions are generally in the form of a policy defined by a sequence of "keep" or "replace" decisions for each period over the horizon. This significantly differs from "one-time" investment decisions discussed earlier in this text. Second, the horizons often examined are infinite, or indefinite. Given these horizons of required service, a current asset (defender) and its possible replacements over time (challengers) must be considered when making replacement decisions. Third, the immediate defender is typically nearing the end of its economic life, and the potential replacements have much longer economic lives.

In this chapter, we are generally concerned with the following motivations for economic replacement:
1. **Deterioration:** Operating and maintenance (O&M) costs of assets generally increase and salvage values decline with age and wear. Further, as assets age, they are more prone to fail, leading to unplanned (and costly) replacements.
2. **Obsolescence:** New assets available on the market are, in general, technologically advanced and operate more efficiently than previously purchased assets. This

translates into lower O&M costs (when compared to current assets) and assets that may retain their value longer, possibly at the expense of higher purchase costs. The new assets may also have additional capabilities with economic value.

These categories capture the essence of the economic tradeoffs in replacement analysis.

Research in the area of asset replacement is vast. As such, various authors have tried to categorize the different problems studied according to problem characteristics, such as the horizon time, assumptions on technological change, number of assets, budgeting constraints, etc. (Luxhoj and Jones, 1986). In this chapter, we divide the problems studied according to the number of assets being analyzed and their interaction. Further we examine different solutions approaches based on the given horizon and cost assumptions.

Single asset or multiple independent asset problems are known as *serial replacement* problems. In these problems, the replacement of an asset does not influence the replacement (or economic analysis) of any other asset. For example, a single pump that services a dry dock or different power tools used in a machine shop.

Multiple asset problems in which the assets are not independent are not easily categorized. In the case where the assets operate in parallel but are economically interdependent, the problems are termed *parallel replacement* problems (Vander Veen, 1985). Economic interdependence means that the replacement decisions cannot be made independently, as the assets are related economically. Examples of these problems include traffic lights, as there is motivation to replace them simultaneously due to the fixed cost of assigning work crews to their removal and installation. Forklifts operating in a warehouse are economically interdependent if a capital budget constrains replacement decisions or fleet discounts are available for new purchases.

For cases in which the assets may or may not operate in parallel, the problems may be characterized as *series* or *series-parallel replacement* problems (Stinson and Khumawala, 1987). Examples include a production flowshop where many of the processes have multiple machines in parallel in order to increase throughput. Another is a telecommunications network with routers and signal boosters where component selection and configuration define system capacity and reliability. In these problems, the capacity of the system is generally defined by the configuration of the assets, which is problem specific. Thus, we focus on serial and parallel replacement problems, as they are more general in nature.

Problems are further categorized according to finite or infinite horizons, depending on the length of service required by the assets. The horizon length and assumption(s) of technological change strongly influence the solution method that should be chosen to evaluate replacement policies.

8.2 INFINITE HORIZON STATIONARY REPLACEMENT POLICIES

The easiest policy to implement is a stationary policy, as it does not change with time. A stationary replacement policy is defined when an asset and its subsequent replacements are kept for equivalent amounts of time, often referred to as their economic life. For the case of stationary costs (no technological change), a stationary policy is optimal. Many researchers have identified optimal stationary policies under conditions of technological change, but few of these are guaranteed to be the global optimum, as discussed later in this section.

8.2 Infinite Horizon Stationary Replacement Policies

8.2.1 Stationary costs (no technological change). We first consider the most basic replacement problem which assumes no technological change over an infinite horizon. That is, all challengers available over time are economically equivalent (and may or may not be equivalent to the defender).

Given the assumption of stationary costs, it should be clear that once the defender is replaced, the optimal policy is to replace the challenger at equal intervals, as the economics of this decision do not change with time. This was first shown by Preinreich (1940). The length of time to keep the asset is referred to as the *economic life* of the asset. Note that this does not presume that the defender should be kept for this amount of time, but merely the challenger.

We can use this definition of economic life to solve two problems:
1. Challenger identical to defender with repeating challenger
2. Challenger different than defender with repeating challenger.

The analysis is the same as it entails two decisions: (1) determining the economic life of the challenger and (2) determining how long to keep the defender. Once the defender is replaced, the challenger and all ensuing challengers are replaced at the stationary economic life. These questions are addressed in the following two sections.

Optimal life of challenger. Under the assumption of stationary costs, or repeatability, the economic life of an asset is found by determining the age which minimizes the equivalent annual (periodic) costs (EAC) of owning and operating an asset. This is because the EAC over N periods of service is equivalent to the EAC over an infinite horizon when the asset is replaced every N periods. In general, annualized ownership costs decline with each year of use as the initial purchase cost is spread over the period of use. However, annualized O&M costs generally rise with time. Thus, the economic life is the optimal tradeoff between capital costs (purchase and salvage value) and operating costs.

Example 8.1 Life of the challenger
Consider the costs of a challenger described in Table 8.1 with the respective EAC for each possible length of service life (the purchase price is $15,000 and the interest rate is 12%).

Table 8.1 EAC for challenger

Age	O&M	SV	EAC
1	$500	$11,000	$6,300
2	$1,000	$9,000	$5,366
3	$2,000	$7,000	**$5,281**
4	$4,000	$5,000	$5,607
5	$8,000	$2,500	$6,472
6	$16,000	$500	$7,930

For illustration, note that the EAC for keeping the asset for two years is:

$EAC_2 = \$15,000(A/P, .12, 2) + \$500 + \$500(A/G, .12, 2) - \$9000(A/F, .12, 2) = \$5366$.

As the data shows, the economic life of the challenger is 3 years, as it reaches a minimum EAC of $5281. Thus, under the assumption of stationary costs, the optimal policy is to replace the challenger every three periods.

The costs from Table 8.1 are shown in Figure 8.1 (the O&M costs and capital costs are annualized with the 12% interest rate). As can be seen in the figure, the economic life corresponds to the time period where the EAC is minimized. Note that it is usually assumed that replacements occur at some given time period, such as a quarter or year, such that the solution corresponds to an integral number of periods, and not necessarily the minimum of the curve (if the functional form of the curve is known with respect to continuous time).

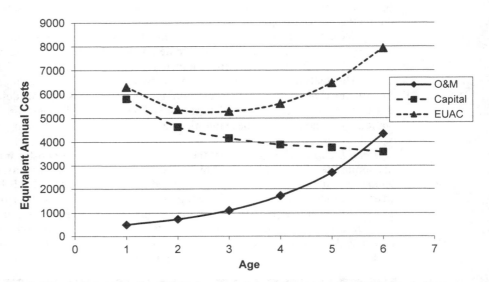

Figure 8.1 Annualized O&M and capital costs comprising EAC for challenger

As with Table 8.1, the economic life is generally computed iteratively, in that the EAC is calculated for all possible service lengths until the minimum is found. If the EAC cost function is convex such that it has a unique minimum as in Figure 8.1, the economic life N is identified by locating the first age $N + 1$ where the EAC is greater than that of N. For many assets, the annualized cost curves shown in Figure 8.1 are typical.

Optimal life of defender. The economics of the defender, identical to the challenger or not, are different since it is assumed that it has already been used for a number of periods. The question to be answered is when should the cycle determined in Section 8.2.1 begin.

If an asset has t years of its service life remaining, there are $t + 1$ possible decisions:

1. Salvage defender now and start the challenger cycle
2. Keep defender for one year and then start the challenger cycle.
...
$t+1$. Keep defender for the maximum t years and then start the challenger cycle.

A number of solution methods have been proposed in the literature, including calculating the marginal costs of retaining the defender for each additional year or the

EAC for the defender over its remaining life (Matsuo, 1988). We illustrate the possible decisions and illustrate how an optimal choice can be determined, regardless of any cost assumptions.

Example 8.2 Life of the defender

A currently owned asset (defender) was purchased 4 years ago for $12,000 and is currently worth $4000 on the open market. The asset can be kept in service for another 2 years, with O&M costs and salvage values over those 2 years given in Table 8.2. Note that the salvage values include any removal costs. The defender must be replaced on or before it reaches its maximum physical life of 6 years. It is assumed that the challenger described in Example 8.1 is the only asset available for replacement in each period for the foreseeable future.

Table 8.2 Costs for defender

Age	O&M	SV
4	—	$4000
5	$6,500	$1500
6	$10,000	$250

The question to be answered is when to replace the challenger and start the optimal economic life cycle of 3 years with the new challenger. The three possible choices are illustrated in Figure 8.2. Note that in each case, after the current asset is sold, the optimal replacement cycle of the challenger with an EAC of $5281 commences and continues indefinitely.

It is easily seen that the cash flow diagrams are equivalent beginning with year 3 (they are always equivalent after year t, where t is the maximum number of remaining years for the defender). Thus, the cash flows for each option beyond year two can be ignored and the optimal decision is the minimum net present value of costs between year 0 and year 2. The present value (or any other measure of worth) can be computed over the first 2 years of each of the cash flow diagrams, as given here for the example. Note that costs are considered positive here as with previous (Example 8.1) calculations:

$PV(12\%)_0 = -\$4000 + \$5281(P/F, .12, 1) + \$5281(P/F, .12, 2) = \$4925$
$PV(12\%)_1 = (-\$1500 + \$6500)(P/F, .12, 1) + \$5281(P/F, .12, 2) = \$8674$
$PV(12\%)_2 = \$6500(P/F, .12, 1) + (-\$250 + \$10,000)(P/F, .12, 2) = \$13,576$

The minimum cost decision is to salvage the defender immediately for $4000, purchase the new challenger and start the stationary replacement policy (3-year replacement cycle). This can be a cumbersome method, as $t+1$ calculations must be made. However, it guarantees an optimal solution.

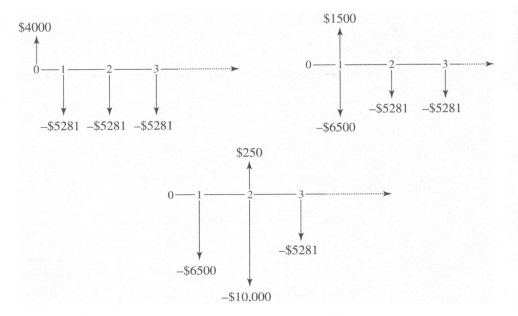

Figure 8.2. Cash flow diagrams for defender replacement options

If the cost structure for the defender is convex, then marginal costs of retaining the defender may be analyzed and compared to the EAC of the challenger. The marginal cost for retaining the defender an additional period is:

$$MC_t = S_{t-1}(1+i) - S_t + O\&M_t \tag{8.1}$$

Consider the marginal cost of keeping the defender for one year. The cost, at time period one is:

$$MC_1 = \$4000(1.12) - \$1500 + \$6500 = \$9480$$

This is clearly greater than the EAC of the challenger cycle, so the defender should be replaced immediately, assuming the marginal costs increase for the remaining periods. The marginal cost for keeping the defender from year one until year two (at time two) is:

$$MC_2 = \$1500(1.12) - \$250 + \$10,000 = \$11,430$$

This method arrives at the same conclusion. However, one must be cautioned when making decisions based on marginal costs. In this case, the marginal costs were increasing with the remaining life of the defender such that the decision is to replace the defender at the age where the marginal costs exceed the EAC of the challenger. One would expect increasing marginal costs under normal cost assumptions, but this assumption is not always warranted. Consider the following example, which assumes that an asset has a warranty for a number of periods.

Example 8.3 Defender with warranty

A defender has the remaining expected costs as given in Table 8.3 for another seven years of possible service. The asset has three years remaining on its warranty, which keep O&M costs down and salvage values relatively high. The EAC of the challenger is $2,685 and the interest rate is 10%.

Table 8.3 Costs for defender with three years remaining on warranty

Year	O&M	SV
Present	—	$6000
1	$500	$5500
2	$500	$5000
3	$500	$4500
4	$1500	$2000
5	$1700	$1000
6	$2000	$500
7	$2400	$0

Given the defender data, the marginal costs of retaining the defender between one and seven years are given in Table 8.4.

Table 8.4 Marginal costs for retaining defender in Example 8.3

Year	MC
1	$1600
2	$1550
3	$1500
4	$4450
5	$2900
6	$2600
7	$2950

It is clear from Table 8.4 that the marginal costs are not increasing, and in fact, are not monotonic. However, the marginal costs to extend the defender are less than the challenger's EAC for the first 3 years and larger for the last 4. The optimal point to replace is after 3 years, when the marginal cost to extend exceeds the challenger's EAC.

Note that to check the optimality of this choice, the 8 possible decisions should be evaluated as presented earlier. Calculations are left to the reader to show that replacing the defender after 3 years is optimal with a PW of −$10,257.

Note that the defender is not analyzed like the challenger. For the challenger, we computed the minimum EAC. However, this method cannot be used for the defender because it requires incorrect assumptions. It requires assuming that we can find another identical used asset and install it for the current salvage value of $6000. This salvage value has subtracted any removal and sale costs from the used asset's current market value. If we could find an identical used asset at that market value, we would have to add any purchase and installation costs to find the used asset's first cost. Assuming the salvage value of the existing defender equals the first cost of a hypothetical used replacement leads to erroneous calculations and conclusions. Thus, we omit the approach of calculating the EAC for the defender's economic life.

In most cases comparing the marginal cost to extend service and the challenger's minimum EAC is the easiest way to find the best time to replace. However, if the defender's marginal cost to extend is declining or erratic, then we recommend computing the PW over the problem's horizon for the $t + 1$ alternatives.

8.2.2 Technological change and stationary results. In industries where changes in technology are slow (or nonexistent), the results from the previous section provide optimal replacement decisions over an infinite horizon. However, challengers generally "improve" with time. This situation is generally referred to as technological change. In this situation, the economics of owning an asset changes with each replacement. Purchase prices, O&M costs, and/or salvage values may differ with each available challenger over time.

It should be clear that under the assumption of technological change, or more formally, nonstationary costs, the optimal replacement policy over time is not stationary. That is, it may not be optimal to repeatedly replace an asset at the same age over an infinite horizon. Rather, it may be advantageous to sell assets sooner in order to take advantage of a newly arriving technology, or to retain an asset longer while waiting for better technology to arrive. This is more readily obvious if innovation, and thus the availability of improved assets over time, is sporadic.

This problem involves finding the minimum cost replacement schedule over an infinite horizon. This involves minimizing the net present value of an infinite stream of cash flows that results from the replacing assets over time. Let us consider a general case. Assume that a challenger is available at time zero (current time period) for the purchase price $P(t) = P(0)$. The O&M costs over its lifetime are equal to cash flows $F(t,n)$, where n represents the age of the asset and t represents the period in which the asset was purchased. Similarly, define $S(t,N)$ as the salvage value of an N-period old asset purchased at time t. Thus, if the asset is purchased at time zero and retained for N periods, the net present value of the decision, assuming a discrete periodic interest rate i, would be:

$$P(0)+\sum_{n=1}^{N}\frac{F(0,n)}{(1+i)^n} - \frac{S(0,N)}{(1+i)^N} \qquad (8.2)$$

We now introduce technological change. Technology is generally assumed to progress either regularly (Bean et al., 1985) or through breakthroughs (Hopp and Nair, 1991). These are, respectively, often referred to as continuous and discontinuous technological change. For example, the designs of new automobiles may be "tweaked" every year, resulting in improved fuel efficiency each year (say an improvement of 0.5% per year). However, completely revamped designs only occur every five years or so. These breakthroughs lead to much bigger jumps in fuel efficiency (say on the order of 10%). For some industries, such as automobiles, these breakthroughs are predictable, because consumers are made aware of impending innovations. However, in other industries, breakthroughs are much more unpredictable, such as the development of alternative sources of power to drive automobiles. It is known that research is being conducted in this area, but it is unknown when the technology will be implemented.

We focus on continuous technological change and assume that improvements occur every period according to some known function. This assumption of continuous technological change is assumed for mathematical convenience and the fact that the ability to forecast breakthroughs is limited. Fortunately, discounting minimizes the influence of future cash flow uncertainty. Given equation (1), we assume that our three cash flow components ($P(t)$, $F(t,n)$ and $S(t,n)$) change with time, but as a function of the original (time zero) asset. More directly:

$$P(t) = f(P(0))$$
$$F(t,n) = g(F(0,n))$$
$$S(t,n) = h(S(0,n))$$

where $f(\cdot)$, $g(\cdot)$ and $h(\cdot)$ represent functions with known form. If we assume a stationary replacement policy such that each asset is replaced at age N over an infinite horizon, the best stationary policy minimizes the following cost function:

$$\sum_{t=0}^{\infty}\left[P(tN) + \sum_{n=1}^{N}\frac{F(tN,n)}{(1+i)^{tN+n}} - \frac{S(tN,N)}{(1+i)^{tN+N}}\right] \quad (8.3)$$

In the first cycle, when $t = 0$, the cash flows assume that an asset is purchased at time zero and retained for N periods. In the second cycle ($t = 1$), it is assumed that an asset is purchased at time N and retained for another N periods, with all costs discounted to time zero. Substituting a functional form of the cash flows based on the first cycle cash flows, different assumptions on technological change can be evaluated.

Note that we have restricted our search to the best stationary policy, which we cannot guarantee is the best of all replacement policies. However, it would be very difficult to find the best policy with this approach, as we would have to identify (and check) all possible combinations of asset lives over the horizon. By limiting our search to stationary policies over an infinite horizon, we only have to find the minimum of N infinite cash flow streams. It should be noted that many researchers in this area assume Equation (8.3) is defined by continuous cash flows and continuous compounding in order to take advantage of solution methods from differential calculus (see Yatsenko and Hritonenko (2008, 2010), among others).

Researchers have examined various functions for technological change. Terborgh (1949) (and subsequently Smith, 1966) assumed constant purchase prices and salvage values of subsequent challengers but a linear decrease in the initial operating costs of new assets. Grinyer (1973) substituted negative exponential functions for the linear functions to model technological change, illustrating that linear technological change generally leads to an increased economic life over time. Oakford (1970) studied a variety of technological change functions, including geometrically bounded increases in asset efficiency. Fraser and Posey (1989) closely examined Terborgh's model and Oakford's various models of technological change (linear, geometric, and geometric bounded functions). They showed that only the Oakford model with geometric data is guaranteed to have an optimal stationary policy (although instances exist where the others have optimal stationary policies). In general, the modeling of technological change leads to the need for other solution methods. Other technological change cost modeling is discussed in Jones and Tanchoco (1987), Regnier et al. (2004), Rogers and Hartman (2006), and Yatsenko and Hritonenko (2009).

8.3 NONSTATIONARY REPLACEMENT POLICIES

Stationary replacement policies are highly desirable in practice because they are easy to implement. Unfortunately, they are not optimal for most nonstationary cost problems. However, the challenge of finding optimal solutions to infinite horizon, nonstationary cost problems is especially problematic due to the problem's combinatorial nature. Thus, if technological change is slow and/or there is great uncertainty in forecasts, the assumptions of a repeating challenger in the future may be warranted and the analysis from section 8.2.1 is generally sufficient. However, for cases where technological change leads to highly nonstationary (yet predictable) costs or a solution is required over a finite horizon, one is advised to turn to other solution techniques.

In these situations, dynamic programming is the preferred solution method from the literature. Dynamic programming, developed by Bellman (1957), is used to analyze sequential decision processes, such as keep or replace decisions for an asset over some horizon. The model defines a system according to state variables, such as the asset's age (as age is often used to describe how much an asset will cost to operate in a given year). Decisions map transitions from one state to another state through stages. The stages represent time periods in which keep/replace decisions are made. All possible states comprise the model's state space.

Dynamic programming is preferred for replacement analysis because it can evaluate keep or replace decisions at each period over a finite horizon and does not require a stationary policy for solution. Thus, all solution combinations can be evaluated. Furthermore, no restrictions are required on costs or functions of technological change, as these assumptions do not change the computational complexity of the dynamic programming algorithm.

The disadvantages of dynamic programming include that many people are unfamiliar with it and there is no readily available software package for its implementation. While this may be true, the decisions evaluated with dynamic programming can generally be visualized with a graph similar to a decision tree and if desired, may be evaluated with different solution approaches, such as linear programming or complete enumeration. We illustrate two dynamic programming approaches in this section and also illustrate how other mathematical programming techniques, such as linear programming, may be used.

Specifically, dynamic programming can solve the following problems:
1. Keep or replace decisions in each period over a finite horizon.
2. Optimal time-zero decision for an infinite horizon problem, assuming the costs are bounded (to be explained later).

8.3.1 Age-based state space approach. Bellman (1955) presented the first dynamic programming formulation for the single asset replacement model. In this model, the state of the system is the asset's age and transitions from period to period depend on the keep or replace decision. A "keep" decision transitions from the state of an age n asset to the state an age $n + 1$ asset. A "replace" decision transitions from an age n asset to an age 1 asset, as the n-period old asset is salvaged and a new asset is purchased. (Purchasing new assets may be easily generalized to allow for purchasing used assets.)

The transitions are illustrated in Figure 8.3, with the nodes labeled with states (asset age) and arcs labeled with decisions (K for keep and R for replace). Given this figure and a cost for each arc (decision), the decision maker's goal is to find the minimum discounted cost path, representing a sequence of keep and replace decisions, from stage (time) zero to the end of the finite horizon. It is assumed that the asset is sold at the end of the horizon.

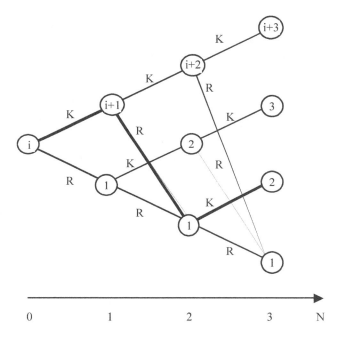

Figure 8.3 Decision network for Bellman's dynamic programming formulation for the single asset replacement problem

Assuming the finite horizon is $N = 3$, a viable solution to the replacement problem is to keep the n-period old asset for one period, replace it, and retain the newly acquired asset until the end of the horizon, at which time it is age 2. This sequence of decisions is highlighted as a path in Figure 8.3. The cash flow diagram for this path (KRK) is given in Figure 8.4.

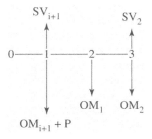

Figure 8.4 Cash flow diagram for KRK replacement solution

One could enumerate each path in Figure 8.3, draw the associated cash flow diagram as in Figure 8.4, and evaluate each alternative. This results in a maximum of 2^N cash flow diagrams (sequence of alternatives) to evaluate, assuming that there is one challenger per period. Each alternative has the same horizon of N periods, which simplifies analysis.

Due to the inefficiency of complete enumeration, dynamic programming (Bellman, 1957), any shortest path algorithm (Ahuja et al., 1993), or linear programming (Chvatal, 1983) may be used to find the minimum cost path from time period zero through time period N. Rather than detailing a solution procedure, we turn to an alternative formulation of the single asset replacement problem which is more efficient.

8.3.2 Length of service state space approach. Wagner (1975) proposed a different dynamic programming formulation for the replacement problem by defining the period as the state variable. This is similar to the approach used by Wagner and Whitin (1958) in their algorithm for lot-sizing. With this state space representation, there is only one possible state per period; however, the number of decisions for each state increases. The decisions for a given period are whether to keep an asset for one, two, ... , or M periods, where M is the asset's maximum service life (see Figure 8.5).

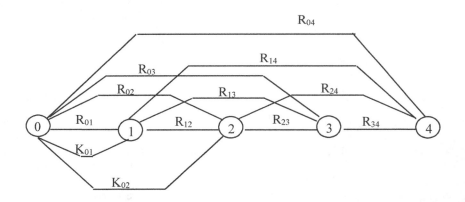

Figure 8.5 Decision network for Wagner's dynamic programming formulation for the single asset replacement problem

In the figure, the nodes represent states (time period) and the arcs represent decisions of how many periods to retain an asset. At time zero, the arcs represent the decisions of keeping the defender for one (K_{01}) or two (K_{02}) periods, assuming the defender has a maximum of two years of service life remaining, or replacing the defender with a challenger for a number of periods (R_{01}, R_{02}, R_{03}). Exclusive of the arcs representing the decisions to keep the time zero defender, the number of arcs emanating from a node is equal to the asset's maximum service life or the horizon's number of remaining periods. The arcs carry the cost of the respective decisions. As before, the optimal solution is the minimum cost path from node 0 to node 3. It can be found either through enumeration or linear programming, but we illustrate the dynamic programming recursion here.

8.3 Non-Stationary Replacement Policies

Example 8.4 Dynamic programming

In Examples 8.1 and 8.2 the defender was 4 years old and the available challengers over the next few years were identical (see Table 8.1). Further assume that the asset is only needed for four years of service. The decision network then matches Figure 8.5, with costs for each arc given in Table 8.5. Note that all costs are discounted to time zero with 12% annual interest.

Table 8.5 Present value of costs for each decision in Wagner's network

Decision (Arc)	Cost (Decision)
K_{01}	$4,464
K_{02}	$13,576 (see Figure 8.6)
R_{01}	$1,625
R_{02}	$5,069 (see Figure 8.7)
R_{03}	$8,685
R_{04}	$13,032
R_{12}	$5,022
R_{13}	$8,097
R_{14}	$11,326
R_{23}	$4,484
R_{24}	$7,229
R_{34}	$4,004

The costs in Table 8.5 are computed by finding the net present value of the life-cycle costs of the asset over the given time period. For example, the decision to keep the defender for two periods (K_{02}) is defined by the cash flow diagram in Figure 8.6. The net present value is $13,576.

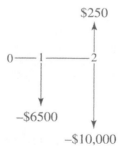

Figure 8.6 Keeping the defender for two periods (K_{02})

Replacing the defender now and keeping the challenger for two periods (R_{02}) has the cash flow diagram shown in Figure 8.7 which has the net present value of $5069 (see Table 8.5).

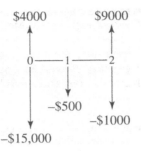

Figure 8.7 Replacing the defender now and keeping the challenger for two periods (R_{02})

The costs along arc R_{13} are the same as those of R_{02} (without the salvage value from the defender), but are discounted an additional period back to time zero. The sum of costs along any path from node 0 to node 4 (sequence of replacement decisions) is the net present value cost of that path.

As noted earlier, any shortest path algorithm may be used to find the optimal solution. Dynamic programming can either be solved forwards (called reaching, (Denardo, 1984)) or backwards (Bellman, 1957). We illustrate a forward dynamic programming recursion here as it will also be used in the subsequent section.

Define $C(t)$ as the minimum (net present value) cost of reaching node t from any other node. With this definition, our goal is to determine $C(N)$ and the optimal decisions to reach state N. Finding the optimal solution entails solving the following recursion:

$$C(t) = \min_{j=1\ldots\min\{t,M\}} \{C(t-j) + c_{t-j,t}\} \quad t > 0$$
$$C(0) = 0$$

Note that the value of $C(0)$ is merely a boundary condition that allows for the recursion's complete solution. The state $C(t - j)$ is used in the solution of state $C(t)$, where j represents the number of periods of the asset's service. The decision's cost is given by c. Given that $C(0) = 0$, the solution examines each node in succession.

Consider node 1 in Figure 8.5. The only node that can reach node 1 is node 0 via arcs K_{01} or R_{01}. Thus, the minimum cost path from node 0 to node 1 is the cheaper of these two paths, or:

$$C(1) = \min\begin{Bmatrix} C(0) + c(K_{01}) \\ C(0) + c(R_{01}) \end{Bmatrix} = \min\begin{Bmatrix} 0 + \$4464 \\ 0 + \$1625 \end{Bmatrix} = \$1625(R_{01})$$

The minimum cost path to state (node) 1 is via arc R_{01} at the present value cost of \$1625. Note that this only represents the optimal decision at time zero if the optimal path to node N is via node 1. This cannot be determined until node N is reached in the recursion.

Proceeding to the next stage, there are three arcs which lead to node 2. Two arcs (K_{02} and R_{02}) emanate from node 0 and one arc (R_{12}) emanates from node 1. The minimum cost of reaching node 2 is thus:

8.3 Non-Stationary Replacement Policies

$$C(2) = \min \begin{Bmatrix} C(0) + c(K_{02}) \\ C(0) + c(R_{02}) \\ C(1) + c(R_{12}) \end{Bmatrix} = \min \begin{Bmatrix} \$0 + \$13{,}576 \\ \$0 + \$5069 \\ \$1625 + \$5022 \end{Bmatrix} = \$5069\,(R_{02})$$

Note that the previously calculated values of $C(0)$ and $C(1)$ are required in the recursion. The minimum cost path to node 2 is via node 0 (replacing the defender with the challenger for two periods), at a total net present value cost of $5069.

Continuing the recursion at node 3:

$$C(3) = \min \begin{Bmatrix} C(0) + c(R_{03}) \\ C(1) + c(R_{13}) \\ C(2) + c(R_{23}) \end{Bmatrix} = \min \begin{Bmatrix} \$0 + \$8685 \\ \$1625 + \$8097 \\ \$5069 + \$4484 \end{Bmatrix} = \$8685\,(R_{03})$$

results in retaining the challenger for three periods. The final period's calculations follow:

$$C(4) = \min \begin{Bmatrix} C(0) + c(R_{04}) \\ C(1) + c(R_{14}) \\ C(2) + c(R_{24}) \\ C(3) + c(R_{34}) \end{Bmatrix} = \min \begin{Bmatrix} \$0 + \$13{,}032 \\ \$1625 + \$11{,}326 \\ \$5069 + \$7229 \\ \$8685 + \$4004 \end{Bmatrix} = \$12{,}298\,(R_{24})$$

The minimum cost path to node 4 is via node 2. Retracing our steps, the best path to node 2 was via node 0. Thus, the optimal decision is to sell the defender at time zero, keep the respective challenger for two periods (R_{02}) and then keep the final challenger for the remaining two periods (R_{24}). The total net present value cost is $12,298. Note that implementing the stationary solution of an economic life of three years for the challenger (as found in Example 8.1) leads to the path defined by arcs R_{03} and R_{34}, at a cost of $12,689, which is $391 greater than the optimal decision. Replacing the defender at time zero and keeping the challenger for four periods costs $13,032. Further note that the recursion evaluated all of these solutions. The finite horizon of four years has changed how long the challenger should be kept from the Example 8.1 solution for an infinite horizon.

Although the previous example was solved assuming stationary costs, there are no restrictions on what costs can be assigned to arcs. For example, if challengers are assumed to become more efficient each period, this can be represented with changes in the forecasted costs. For example, consider Example 8.3 where the cost of R_{01} (salvaging the defender and retaining the challenger for one period) was $1625. This value comes from the $5625 cost of using the challenger for one year less the defender's $4000 salvage value. If challengers "improve" 5% per year, then the costs for R_{12} become (.95)($5625) = $5344 and R_{23} becomes (.95)($5344) = $5077 before discounting. More complicated modeling of continuous or breakthrough technological change is allowed, as long as the representative cash flows can be forecast. If multiple challengers are modeled, they can be represented with multiple arcs between nodes. Preprocessing can eliminate inferior challengers, as only the minimum cost arc between any two nodes must be retained for the dynamic program.

For the case of one challenger in each period over N periods with a maximum service life of M periods, a maximum of M arcs exit a node (excluding arcs representing retaining the defender). Thus, a maximum of MN evaluations must be made in the recursion. For the case of K challengers, this results in KM arcs and thus a maximum of KMN calculations. Note that preprocessing can reduce the network to a maximum of M arcs per node, regardless of the number of challengers, as challengers with similar service lives are merely parallel arcs between similar nodes. This method is more efficient than Bellman's (1955) approach with the state space consisting of an asset's age. In Bellman's recursion, there is a maximum of M nodes in each time period and two decisions per node. This doubles the maximum calculations to $2MN$. For the case of K challengers, the number of nodes increases to KM per period, totaling $2KMN$ maximum calculations, which is also double the alternate formulation. Note that simple preprocessing cannot be implemented with Bellman's recursion due to the state space representation.

Due to ease of computation, Oakford, Lohmann, and Salazaar (1984) and Bean, Lohmann, and Smith (1985) have used Wagner's model for generalized models which incorporate technological change.

8.3.3 Applying dynamic programming to an infinite horizon problem.
Solving dynamic programs requires a finite horizon. In order to use dynamic programming in an infinite horizon setting, we must restrict our search to the *optimal time zero decision*. Generally, this suffices in practice as the time zero decision is implemented now and the problem is reexamined later with newer data.

To find the optimal time zero decision, we must identify a finite horizon such that the time zero decision does not change for any horizon greater than or equal to that horizon, including an infinite horizon. This horizon is often called the planning, forecast, or decision horizon (Bean and Smith, 1984; Chand and Sethi, 1982; Sethi and Chand, 1979), as costs only need to be estimated over this horizon.

Bean and Smith (1984) prove that if the discount rate is greater than the rate of technological change then present values of costs and revenues are bounded, and there exists a planning horizon. Furthermore, the horizon can be found efficiently, as illustrated in a replacement application in Bean et al. (1985). The algorithm follows:

> Step 0. Select an arbitrary horizon of length N (preferably small). The maximum service life of all assets (challengers) available over the entire horizon is denoted as M.
> Step 1. Solve the associated dynamic program (as in the previous section). Denote the optimal time zero decision (K_{01}, K_{02}, R_{01}, R_{02}, etc.) as Decision(N).
>
> For AGE = 1 to M
> Step 2. Increment $N = N + 1$.
> Step 3. Solve dynamic program and record time zero Decision(N).
>   IF Decision(N) = Decision($N - 1$), CONTINUE
>   ELSE, set horizon to N and GO TO STEP 1.
> NEXT AGE
> Optimal time zero solution found.

The algorithm terminates when M consecutive, identical time zero decisions are found. Bean et al. (1985) stated that in general, the horizon needs to be about twice the value of M to guarantee the time zero decision. Thus, it seems appropriate to initiate the algorithm with HORIZON = M and expect to perform roughly M iterations. Note that solving the dynamic program with reaching, as shown previously, allows for straightforward computations, as previous period solutions need not be recomputed. That is, solving the dynamic program of horizon N is nested in the solution of a horizon of length $N + 1$. Thus, each iteration of the algorithm will only require a maximum of M calculations (one for each new decision (arc) defined by the new state (node) $N + 1$).

8.3.4 Solving with linear programming. Dynamic programming is very appealing when solving a problem with an infinite horizon—especially if costs and benefits change over time. But it is not required for finite horizon solutions, as in Section 8.3.2. Rather, linear programming may be used to solve, with general costs, and is often preferred due to the availability of software packages such as Solver (in Microsoft EXCEL), LINDO, EXPRESS, and CPLEX.

To solve Example 8.4 as a linear program, we must define the appropriate variables, objective function, and constraints. As with Wagner's dynamic programming approach, we define the variables according to how long an asset is retained, as either $R_{t,t+n}$ to denote retaining a challenger for n periods or $K_{0,n}$ to denote retaining the defender at time zero for n periods. The cost of each decision (variable) equals the costs on the respective arcs in Figure 8.5.

Our objective is to minimize the path's total net present value cost over the horizon, or:

$$\min \$4464 K_{01} + \$13{,}576 K_{02} + \$1625 R_{01} + \$5069 R_{02} + \ldots + \$4004 R_{34}$$

To define our constraints, we note that a solution would allow one unit of "flow" from node 0 in Figure 8.5 to node 4. The defined path would determine how long each asset is retained, and each period's keep and replace decisions. To accomplish this mathematically, we require a single unit of flow to exit node 0, or:

$$K_{01} + K_{02} + R_{01} + R_{02} + R_{03} + R_{04} = 1$$

Similarly, we require a single unit of flow to enter node 4, or:

$$R_{04} + R_{14} + R_{24} + R_{34} = 1$$

Finally, we require the flow to be balanced at each intermediate node. In other words, at nodes 1, 2, and 3, the flow in must equal the flow out, or:

$$K_{01} + R_{01} - R_{12} - R_{13} - R_{14} = 0$$
$$K_{02} + R_{02} + R_{12} - R_{23} - R_{24} = 0$$
$$R_{03} + R_{13} + R_{23} - R_{34} = 0$$

We require that each of the K and R variables take on a value of 0 or 1. However, this is a network flow formulation, so we are guaranteed that the solutions will integer (0 or 1). Thus, we must only restrict the solutions to be non-negative:

$$K_{01}, K_{02}, R_{01}, R_{02}, R_{03}, R_{04}, R_{12}, R_{13}, R_{14}, R_{23}, R_{24}, R_{34} \geq 0$$

Solving this linear program with the costs from Example 8.4, we get the same decision as with the dynamic program such that $R_{02} = R_{24} = 1$.

8.4 AFTER-TAX REPLACEMENT ANALYSIS

Replacement analysis involves two additional cash flows that must be considered in after-tax analysis. These include the annual effects of depreciation, which is an expense that reduces tax liabilities as described in Chapter 5, and taxes or credits from the gain or loss on the asset's sale. We do not consider capital gains here, as it is rare that industrial equipment will appreciate with age and use. (A capital gain is taxed when an asset is sold for a value greater than its purchase price.) Rather, we are concerned with the gain or loss from the sale of an asset, which is defined as the difference between the book value and salvage value of the asset at the time of its disposal. According to current United States tax law, this gain is charged as ordinary income and taxed accordingly.

Under current US tax law, the tax on the sale of an asset is only charged when an asset is retired, not when it is replaced (this is termed a like-for-like exchange in the US Tax Code, Section 1031, 1954). In a replacement situation, any residual book value from the defender is transferred to the challenger and no taxes are paid on the transaction. Thus the challenger's initial book value and its depreciation schedule depend on the defender's replacement time.

As illustrated in Hartman and Hartman (2001), if an asset of age n with book value B_n is sold for a salvage value of S_n, then the acquired asset's purchase price (P) initial book value is:

$$B_0 = P + B_n - S_n$$

Thus if the asset is sold at a gain ($S_n > B_n$), then the new asset's initial book value decreases while the opposite is true if the defender is sold at a loss. It is not until an asset is retired (sold without replacement) that taxes are paid on the sale's gain or loss. In an infinite horizon problem, this would never occur. In a finite horizon problem, the tax is paid at the horizon.

Note that this greatly complicates analysis. Considering stationary costs, assume an asset is purchased and its salvage value is always less than its book value. In this situation, the residual value ($B_n - S_n$) is always positive, as it is always sold at a loss. This increases the challenger's initial book value (when compared to the defender) and under the assumption of time-invariant costs, the salvage values will also be smaller than all book values. It can be shown (Hartman and Hartman, 2001) that if an asset is continually sold at a loss, then the capital costs are nonincreasing with each replacement cycle. Similarly, assets sold for gains result in nondecreasing capital costs.

Ignoring the transferal of residual book value was examined in two extreme examples in Hartman and Hartman (2001). They compared the correct model to one in which a tax was levied at the time of each asset sale (replacement). If an asset is continuously sold at a gain (example given was aircraft engines which retain their value despite high operating and maintenance costs), the incorrect model will generally have a longer economic life in order to defer taxes on gains from asset sales. For assets continually sold at a loss, the incorrect model will generally have a shorter economic life that takes advantage of the tax credits while the correct model retains the asset longer. It is common that salvage values drop quickly, and steeply, in industries with high

8.4 After-Tax Replacement Analysis

technological change or fast deterioration. While correctly modeling the transferal of residual book values complicates traditional economic replacement models, it is required to properly capture current United States tax law and all of its cash flow implications.

Example 8.5 After-tax cash flows

We reexamine the optimal solution to Example 8.4, which was to replace the defender at time zero and replace the new challenger after two periods of service with a second challenger used for the final two periods. The before-tax cash flows are given in Figure 8.8.

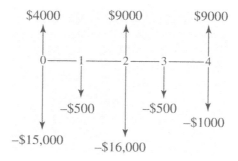

Figure 8.8 Before-tax cash flows for optimal decisions in Example 8.4

Assume MACRS depreciation for a 5-year asset, a tax rate of 35%, and that revenues are unaffected by asset choice. The defender, originally purchased for $12,000 has depreciated for four years when it is sold for $4000. Including the half-year convention, its depreciation charges over the four years are $2400, $3840, $2304 and $691. Subtracting these from the initial book value of $12,000, its final book value is $2765. Thus, the asset is sold at a gain of $1235, which decreases the challenger's initial book value at time zero from $15,000 to $13,765. Table 8.6 details the after-tax cash flows for retaining the challenger from period 0 through period 2.

Table 8.6 After-tax cash flows (positive costs) for retaining challenger for two periods

Age	O&M	Depreciation	Sale	ATCF
1	$500	$2753[1]	—	−$639[2]
2	$1000	$2202[3]	$9000	−$7919[4]

1. Depreciation is based on $13,765 initial book value ($15,000 − $1235) and half-year convention such that $(.20)(\$13{,}765) = \$2753$.
2. ATCF = $-(.35)(\text{Dep}) + (1 - .35)\text{O\&M} = -(.35)(\$2753) + (.65)(\$500) = -639$.
3. Depreciation is based on $13,765 initial book value and half-year convention due to premature asset sale, or $(0.32)(\$13{,}765)/2 = \$2202$.
4. ATCF = $-(.35)(\text{Dep}) + (1 - .35)\text{O\&M} - \text{Sale Revenue} = -(.35)(\$2202) + (.65)(\$1000) - \$9000 = -\$7919$.

The book value of the 2-year-old salvaged challenger is $8810, such that it is sold for a loss of $190, which is transferred to the initial book value of the newly purchased asset. The cash flows for the third and final asset are given in Table 8.7.

Table 8.7 After-tax cash flows (positive costs) for challenger over final two periods

Age	O&M	Depreciation	Sale	ATCF
1	$500	$3038[1]	--	−$738[2]
2	$1000	$2430[3]	$9000	−$9453[4]

1. Depreciation is based on $15,190 initial book value ($15,000 + $190).
2. ATCF = −(.35)(Dep) + (1 − .35)O&M
3. Depreciation is based on $15,190 initial book value and half-year convention due to premature asset sale.
4. ATCF = −(.35)(Dep) + (1 − .35)O&M − Sale Revenue − .35(Loss). The book value at the time of sale is $9722, resulting in a loss of $722.

These after-tax cash flows are summarized in Figure 8.9.

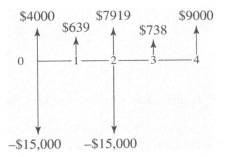

Figure 8.9 After-tax cash flow diagram for optimal decision from Example 8.4

This present value of the ATCF costs for the decisions in Figure 8.9, assuming an after-tax MARR of 10%, is $9260. This example illustrates the changes in cash flows from the incorporation of taxes. To determine the optimal after-tax decision, each possible path of replacements must include the transferal of gains or losses to the initial book value of the newly acquired asset.

The dynamic programming formulations presented earlier must be altered in order to solve the after-tax replacement problem, because an asset cannot be economically characterized solely by its age and/or the time period. Rather, the asset's book value must also be included for proper costing. Figure 8.10 illustrates the respective network in which the shortest path can be found as illustrated earlier. Comparing the networks shows that although there are still two decisions (arcs) exiting each state (node), the number of nodes in each period increases to keep track of different book values. This says that two assets of the same age in different time periods, even under stationary costs, are not economically equivalent. The representative network is somewhat harder to solve due to the increased size of the state space.

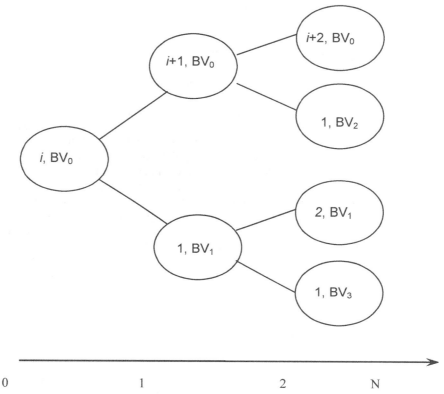

Figure 8.10 Decision network with an age and initial book value state space

8.5 PARALLEL REPLACEMENT ANALYSIS

Many applications use multiple assets. If these assets are economically independent, then serial replacement algorithms determine the optimal replacement policy. If the replacing an asset impacts other assets, then they are economically interdependent and replacement decisions must be determined simultaneously.

A distribution company that owns its delivery vans generally faces a parallel replacement problem. The vans essentially operate in parallel (independently) but they are economically interdependent if the company receives fleet discounts for bulk purchases (replacements). A waste management company would have similar issues with garbage trucks as would a railway with boxcars. Fleet discounts are generally referred to as economies of scale models in the literature (Jones et al., 1991, and Chen, 1998). They have also been examined under increasing demand (Rajagopalan, 1998, and Chand et al., 2000), budget constraints (Karabakal et al., 1994) and fluctuating demand (Hartman, 2000).

Fleet replacement with quantity discounts is merely one example of a parallel replacement problem. Capital budgets force assets to compete for funding. While quantity discounts lead to large group acquisitions, assets purchased together also fail together, leading to dis-economies of scale (Jones and Zydiak, 1993).

Note that this research does not encompass all multiple asset replacement problems. Here, it is assumed that the assets operate in parallel (i.e. a fleet) such that the system's capacity equals the sum of each asset's capacity. This does not include many manufacturing applications, where the system configuration defines its capacity. Unfortunately, many of these applications are problem specific, and thus not discussed here.

The most common problem studied in the literature is parallel replacement under economies of scale. A fixed charge is paid in each period in which a replacement occurs, regardless of the number of assets replaced. If the fixed charge is high, there is motivation to replace assets simultaneously. Note that other economies of scale models exist, as in Childress and Durango-Choen (2005).

The reason for less research in this area, when compared to serial replacement analysis, is due to the combinatorial nature, and thus difficulty, of these problems. For a single asset problem with one challenger, there are only two decisions for a given asset in each period. However, for a two-asset problem, the decisions are now whether to keep one asset, keep the other asset, keep both assets, or replace both assets.

Consider Figure 8.11 which depicts the possible replacement decisions for three homogeneous assets, two of age 1 and one of age 2, over a three-period horizon. They are depicted as the vector (2,1,0), as they may be retained until they are age 3, at which time they must be replaced.

Figure 8.11 illustrates the difficulty of solving parallel replacement problems, as the number of possible decisions grows exponentially with the number of assets. However, the work of Jones et al. (1991) drastically reduced this state space through the introduction of two rules.

The first rule, termed the No-Splitting Rule (NSR), designates that no "cluster" of assets "split" in any time period. That is, a group of same aged assets in the same time period are either kept or replaced together (under the assumption of constant demand and no capital budgeting constraints.) This is an intuitive rule, as the splitting of a cluster means that two similar assets will have different costs over their remaining lives (and the lives of their replacements). If these remaining lives are envisioned as paths, only one of these paths can be optimal and thus splitting cannot be optimal. Hartman (2000) and Chand et al. (2000) generalized this rule to the case of nondecreasing demand. (This rules does not hold in all economies of scale models, as in Childress and Durango-Cohen (2005), but is true for fixed cost charge models.)

The impact of the NSR is that the network of decisions can be reduced to those in Figure 8.12. The network can be further reduced with the Oldest-Cluster-Replacement-Rule, which states that a cluster cannot be replaced unless all older clusters are also replaced. The OCRR requires the following cost assumptions:

- Nondecreasing O&M costs over the asset's service life.
- Nonincreasing salvage values over the asset's service life.
- Nondecreasing sum of O&M costs and salvage values over the asset's service life.

8.5 Parallel Replacement Analysis 243

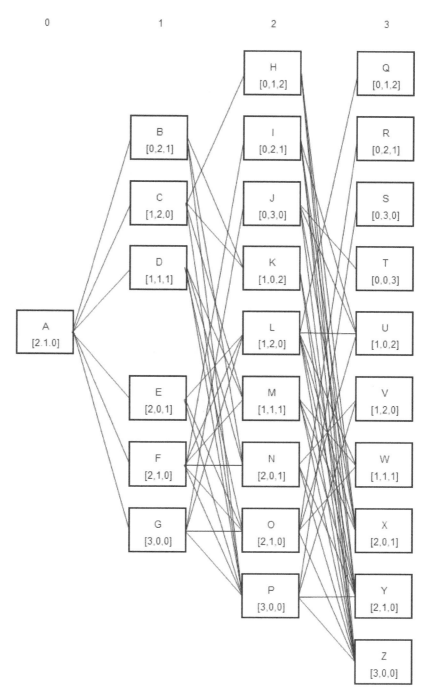

Figure 8.11 Decisions for three possible assets over a three-period horizon

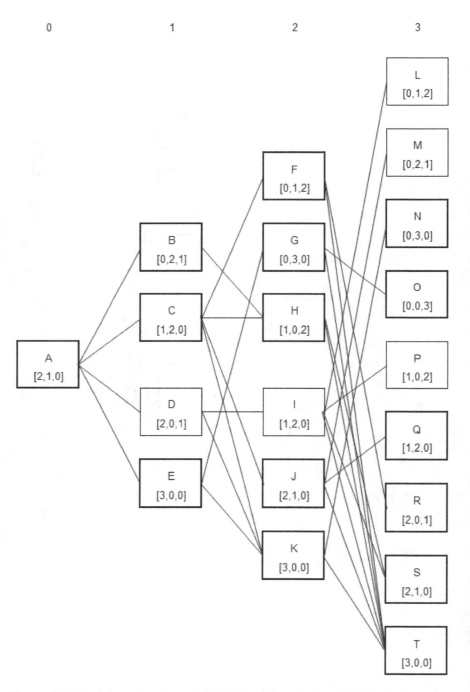

Figure 8.12 Decision network under NSR (regular nodes) and OCCR (bold nodes)

This further reduces the network of decisions to those bolded in Figure 8.12. It should be noted that Hopp et al. (1993) and Chand et al. (2000) slightly altered the last assumption in proving the same result (OCRR). Interestingly, Jones et al (1991) reported that randomly generated numerical examples produced solutions which followed the OCCR rule, regardless if the third assumption was assumed. However, they were not able to prove OCRR without the final assumption.

Tang and Tang (1993) further modified the final assumption to prove the All-Or-Nothing-Rule (AONR), which states that all assets are replaced in the same period. The AONR rule further reduces the possible states, as there are only two possible decisions at each node in the network, much like the single asset replacement problem. Once the first replacement has occurred, a single cluster of same aged assets exists.

The figures illustrate that parallel replacement problems under economies of scale can also be solved as shortest path problems. Thus, the networks can be analyzed with dynamic programming or integer programming, as the fixed charge requires the use of binary variables.

8.6 SUMMARY AND FURTHER TOPICS

A few authors have tried to categorize the literature through different taxonomies. The reader is directed to Luxhoj and Jones (1986) and their references. This chapter examined the serial, or single asset, replacement problem under infinite and finite horizons under stationary costs and technological change. Under stationary costs, solutions can be found incorporating the economic life of an asset, or the time period in which minimizes the EAC of an asset. For problems involving technological change, optimal replacement schedules can be found using dynamic programming. Taxes and multiple assets further complicate the analysis.

In addition to the literature referenced in this chapter, there has been significant work in the area of stochastic replacement analysis. Elmaghraby (1959) and Lohmann (1986) considered probabilistic data in replacement analysis. For dynamic programming applications with stochastic implications, see Brown (1993), Hartman (2001) or Tan and Hartman (2010). Deterioration is often modeled as a Markov process, as assets deteriorate from a state to a state, and not necessarily with each period. Derman (1963) was the first to consider this problem in a replacement context, while Hopp and Nair (1994) combined Markovian deterioration and technological change modeling.

REFERENCES

AHUJA, R. K., T. L. MAGNANTI, and J. B. ORLIN, *Network Flows* (Prentice-Hall, 1993).

BEAN, JAMES C., JACK R. LOHMANN, and ROBERT L. SMITH, "A Dynamic Infinite Horizon Replacement Economy Decision Model," *The Engineering Economist*, **30**(2) (1984), pp. 99–120.

BEAN, JAMES C., JACK R. LOHMANN, and ROBERT L. SMITH, "Equipment Replacement Under Technological Change," *Naval Research Logistics*, **41**(1) (February 1994), pp. 117–128.

BEAN, JAMES C. and ROBERT L. SMITH, "Conditions for the Existence of Planning Horizons," *Mathematics of Operations Research*, **9** (1984), pp. 391–401.

BELLMAN, R. E., "Equipment Replacement Policy," *Journal of the Society for Industrial and Applied Mathematics*, **3**(1) (September 1955), pp. 133–136.

BELLMAN, R. E., *Dynamic Programming* (Princeton University Press, 1957).

BROWN, M. J., "A mean–variance serial replacement decision model: the correlated case, *The Engineering Economist*, **38**(3) (1993), pp. 237–247.

CHAND, SURESH, TIM MCCLURG, and JIM WARD, "A Model for Parallel Machine Replacement with Capacity Expansion," *European Journal of Operational Research*, **121**(3) (March 2000), pp. 519–531.

CHAND, SURESH, and SURESH SETHI, "Planning Horizon Procedures for Machine Replacement Models with Several Replacement Alternatives," *Naval Research Logistics*, **29**(3) (1982), pp. 483–493.

CHEN, ZHI-LONG, "Solution Algorithms for the Parallel Replacement Problem under Economy of Scale," *Naval Research Logistics*, **45**(3) (April 1998), pp. 279–295.

CHILDRESS, S., and P. DURANGO–COHEN, "On Parallel Machine Replacement Problems with General Replacement Cost Functions and Stochastic Deterioration," *Naval Research Logistics,* **52**(5) (2005), pp. 409–419.

CHVATAL, VASEK, *Linear Programming* (W. H. Freeman, 1983).

DENARDO, ERIC V., *Dynamic Programming, Models and Applications*, (Prentice-Hall, 1984).

DERMAN, C., "Inspection-maintenance-replacement Schedules under Markovian Deterioration," in *Mathematical Optimization Techniques*, (University of California Press, 1963), pp. 201–210.

ELMAGHRABY, SALAH E., "Probabilistic Considerations in Equipment Replacement Studies," *The Engineering Economist* **4**(1) (Summer 1958), pp.1–31.

FRASER, JANE M. and JACK W. POSEY, "A Framework for Replacement Decisions," *European Journal of Operational Research*, **40**(1) (May 1989), pp. 43–57.

GRINYER, PETER H., "The Effects of Technological Change on the Economic Life of Capital Equipment*,"* *AIIE Transactions*, **5**(3) (1973), pp. 203–213.

HARTMAN, JOSEPH C., "The Parallel Replacement Problem with Demand and Capital Budgeting Constraints," *Naval Research Logistics*, **47**(1) (February 2000), pp. 40–56.

HARTMAN, JOSEPH C., "An Economic Replacement Model with Probabilistic Asset Utilization," *IIE Transactions*, **33**(9) (2001), pp. 717–729.

HARTMAN, JOSEPH C., and RAYMOND V. HARTMAN (2001), "After-Tax Replacement Analysis*,"* *The Engineering Economist*, **46**(3) (2001), pp. 181–204.

HOPP, WALLACE J., PHILIP C. JONES, and JAMES L. ZYDIAK, "A Further Note on Parallel Machine Replacement," *Naval Research Logistics*, **40** (1993), pp. 575–579.

HOPP, WALLACE J., and S. K. NAIR, "Markovian Deterioration and Technological Change," *IIE Transactions*, **26**(6) (1994), pp. 74–82.

HOPP, WALLACE J., and S. K. NAIR, "Timing Replacement Decisions under Discontinuous Technological Change," *Naval Research Logistics*, **38** (1991), pp. 203–220.

JONES, M. S., and J. M. A. TANCHOCO, "Replacement Policy: Impact of Technological Advances," *Engineering Costs and Production Economics*, **11**(1) (April 1987), pp. 79–86.

JONES, P. C., and JAMES L. ZYDIAK, "The Fleet Design Problem*,"* *The Engineering Economist*, **38**(2) (Winter 1993), pp. 83–98.

JONES, P. C., J. L. ZYDIAK, and W. J. HOPP, "Parallel Machine Replacement*,"* *Naval Research Logistics*, **38** (1991), pp. 351–365.

KARABAKAL, NEJAT, JACK R. LOHMANN, and JAMES C. BEAN, "Parallel Replacement under Capital Rationing Constraints," *Management Science*, **40** (1994), pp. 305–319.

LOHMANN, JACK R., "A Stochastic Replacement Economic Decision Model," *IIE Transactions*, **18** (1986), pp.182–194.

LUXHOJ, JAMES T., and MARILYN S. JONES, "A Framework for Replacement Modeling Assumptions," *The Engineering Economist*, **32**(1) (1986), pp. 39–49.

MATSUO, HIROSHI, "A Modified Approach to the Replacement of an Existing Asset," *The Engineering Economist*, **33**(2) (1988), pp.109–120.

OAKFORD, ROBERT V., *Capital Budgeting: A Quantitative Evaluation of Investment Alternatives*, (The Ronald Press Co., 1970).

OAKFORD, ROBERT V., LOHMANN, JACK R., and SALAZAR, A., "A Dynamic Replacement Economy Decision Model," *IIE Transactions*, **16** (1984), pp. 65–72.

PREINREICH, GABRIEL A.D., "The Economic Life of Industrial Equipment," *Econometrica*, **8**(1) (January 1940), pp. 12–44.

RAJAGOPALAN, SAMPATH, "Capacity Expansion and Equipment Replacement: A Unified Approach," *Operations Research*, **46**(6) (Nov.–Dec. 1998), pp. 846–857.

REGNIER, EVA, GUNTER SHARP, and CRAIG TOVEY, "Replacement under Ongoing Technological Progress," *IIE Transactions*, **36**(6) (2004), pp. 497–508.

ROGERS, JENNIFER, and JOSEPH C. HARTMAN, "Equipment Replacement under Continuous and Discontinuous Technological Change," *IMA Journal of Management Mathematics* **16**(1) (January 2005), pp. 23–36.

SETHI, SURESH, and SURESH CHAND, "Planning Horizon Procedures for Machine Replacement Models," *Management Science*, **25**(2) (1979), pp, 140–151.

SMITH, VERNON L., *Investment and Production: A Study in the Theory of the Capital–Using Enterprise*, (Harvard University Press, 1961).

STINSON, J. P., and B. M. KHUMAWALA, "The Replacement of Machines in a Serially Dependent Multi–Machine Production System," *International Journal of Production Research*, **25**(5) (1987), pp. 677–688.

TAN, CHIN HON, and JOSEPH C. HARTMAN, "Equipment Replacement under an Uncertain Finite Horizon," *IIE Transactions,* **42**(5) (2010), pp. 342–353.

TANG, JEN, and KWEI TANG, "A Note on Parallel Machine Replacement," *Naval Research Logistics*, **40**(4) (June 1993), pp. 569–573.

TERBORGH, GEORGE W., *Dynamic Equipment Policy*, (McGraw–Hill, 1949).

UNITED STATES GOVERNMENT, *Internal Revenue Code*, (1954) as updated and amended.

VANDER VEEN, DAVID J., *Parallel Replacement Under Nonstationary Deterministic Demand*, Ph.D. Dissertation (1985), The University of Michigan.

WAGNER, HARVEY M., *Principles of Operations Research*, (Prentice-Hall, Inc., 1975).

WAGNER, HARVEY M., and THOMSON M. WHITIN, "Dynamic Version of the Economic Lot Size Model," *Management Science*, **5**(12) (1958), pp. 89–96.

YATSENKO, Y., and N. HRITONENKO (2008), "Properties of Optimal Service life under Technological Change," *International Journal of Production Economics* **114**(1), pp. 230–238.

YATSENKO, Y., and N. HRITONENKO (2009), "Technological Breakthroughs and Asset Replacement," *The Engineering Economist* **54**(2), pp. 81–100.

YATSENKO, Y., and N. HRITONENKO (2010), "Discrete-continuous Analysis of Optimal Equipment Replacement," *International Transactions in Operational Research* **17**(5), pp. 577–593.

PROBLEMS

8-1. How are abandonment and replacement problems related?

8-2. What are the two most common reasons for preemptively replacing an asset before it fails?

8-3. When is it appropriate to use dynamic programming to solve single asset replacement problems?

8-4. As an asset ages, in general, what cost functions (increasing, decreasing, constant) do you expect the operating and maintenance and salvage values to follow?

8-5. When is it appropriate to replace an asset at its economic life, defined as its minimum EAC?

8-6. A company provides mid-size sedans to its middle management for company use. The cars cost $18,000 each and can be maintained for a maximum of eight years. Operating and maintenance costs are estimated at $600 the first year and are expected to increase $600 per year of use. The car's salvage value is expected to decline 15% per year of use.

(a) Assuming the company uses a 15% interest rate for discounting, what is the sedan's economic life?

(b) Given your solution to (a), how often should the company purchase new sedans? What assumption(s) did you make in answering this question?

(c) If the company purchases a large quantity of sedans at once, they can get a discount on the purchase price. How would this alter your analysis in (a)? Will the economic life increase or decrease? Explain.

8-7. A car dealer would like to win over the fleet business for problem 8-6. They are offering similar terms, but improving the warranty such that no O&M costs are to be paid in the first three years. What is the car's new economic life and how much can the company save on an annual basis?

8-8. Often in industrial settings, purchased equipment can be highly specialized with little chance to sell it after it has been used. Can you make any statements about an asset's economic life given that its salvage value is zero after each possible year of use? Consider cases where the annual O&M costs are constant and increasing.

8-9. Similar to Problem 8-8, consider the case where O&M costs are essentially zero on an annual basis but salvage values decrease with age. This is possible if you are an equipment lessor and your clients must pay for all O&M costs while the assets are in use. Can you make any statements about the economic life?

8-10. A new sedan has come on the market. Its purchase price has increased to $19,500, but its O&M costs are expected to start at only $500 in the first year, increasing by only $400 per year while maintaining a similar salvage value decline (85% of previous year).

(a) What is the new sedan's economic life?

(b) If the currently owned asset (Problem 8-6) is 2 years old, should a replacement be made? If not, when should the replacement be made?

(c) Write the ATCF to your solution in (b) assuming MACRS depreciation, a 3 year class asset and a 40% tax rate.

(d) Repeat (b) if the asset is 6 years old.

8-11. Reconsider Problem 8-10b under the assumption of a five period horizon.
(a) Draw the network of decisions associated with Bellman's dynamic programming formulation.
(b) Draw the network of decisions associated with Wagner's dynamic programming formulation.
(c) Write down the linear programming formulation for (b).
(d) Solve the dynamic programming recursion from (b) to determine the optimal replacement decisions. How does this solution compare to the first five periods of Problem 8-10b?

8-12. A firm is revising its procedure to replace its heat exchangers. They currently own a two-year old exchanger, purchased at the price of $25,000. Where a is the age of the asset, annual O&M costs equal

$$F(a) = \$1000(1.10)^a$$

and salvage values are expected to decline sharply the first year to 1/2 the purchase price and the drop an additional 10% each year. The manufacturer's models steadily improve with each year of release, which is reflected in slightly higher purchase prices:

$$P(t) = \$25,000(1.02)^t$$

The O&M costs also change each year:

$$F(t,a) = \$1000(1.10 - .01t)^a$$

The salvage values follow the same functional form as before, noting that the purchase price is changing with time. Note that a heat exchanger cannot be used for more than five years and it is currently $t = 0$ (such that the challenger available at time 0 is equivalent economically to the defender).

(a) Derive an expression to determine the optimal stationary replacement policy for the challenger (and its subsequent challengers).
(b) Consider an 8-year horizon and determine the optimal replacement schedule over that time, assuming a 12% interest rate.
(c) Consider an infinite horizon. What is the optimal time zero decision for the firm? Support your answer.

8-13. Assume the supplier of heat exchangers from Problem 8–12 has announced a breakthrough in its technology. In 2 years (t = 2), they will release a new model with the following costs:

$$P = \$35,000$$
$$F(a) = \$100(1.12)^a$$

Furthermore, the salvage value is only expected to decrease 40% after the first year of use and then 10% each year thereafter. Once the model is released, its challengers are expected to follow as before, such that:

$$P(t) = \$35,000(1.02)^t$$
$$F(t,a) = \$100(1.12 - .01t)^a$$

Note here that the time t is relative to the availability of the new asset at time 2.

(a) How does the dynamic programming network from 8–12(b) change for this problem?
(b) Alter the network and rework your solution. Do the decisions change?
(c) Would you expect the amount of iterations required to solve 8–12(c) to change with this new technology? Explain.

8-14. Assume that the company uses a number of heat exchangers in its systems. They currently have five 1-year-old, four 2-year-old and five 3-year-old exchangers. The supplier has told them that they will provide a 15% unit price discount if more than five exchangers are purchased at a time. All other costs remain as in Problem 8–12 and the company requires 14 exchangers each period over the next five periods. Draw the related network, placing costs on each arc for decisions, which reflect the discounts. What should the company do at time period one? Recall that an exchanger can only be used for five periods.

8-15. Assume only one exchanger is required over an infinite horizon. The one exchanger currently owned is three years old and follows the costs from Problem 8–12. A replacement exchanger following the costs of Problem 8–13 is available for the foreseeable future. When should the current exchanger be replaced?

9

Methods of Selection Among Multiple Projects

9.1 INTRODUCTION

As long as the firm is able to consider projects individually, that is, as long as a given project is accepted or rejected on its own merits, any of the rational criteria—net present value, internal rate of return, or benefit-cost ratio—will signal the correct decision if applied correctly. In practice, however, the prerequisites of project independence, project indivisibility, and lack of capital rationing are rarely met. One usually finds that many projects in the candidate set are dependent and that not enough capital is available to undertake all acceptable projects. In some cases a maximum of one project may be undertaken, as the set of candidate projects may be mutually exclusive alternatives—which may or may not include a *do nothing* alternative.

To introduce project selection under constraints, this chapter considers a simple constrained problem known as the *Lorie-Savage* problem which comes from comparing two mutually exclusive alternatives. This simple problem sometimes develops *a ranking inconsistency* when the same projects are ranked by two different criteria. For example, net present value and internal rate of return sometimes produce different answers. Because of this inconsistency, serious questions arise about the ranking methods, which are addressed in this chapter.

There is an ongoing debate about which method is best for ranking projects under constrained conditions (budget limits). Some authors prefer IRR, some prefer NPV, and others prefer using mathematical programming techniques that use NPV as the variable to be optimized. Literature exists to support any of these choices. Companies use a variety of tools, but small companies tend to use the payback method (Block, 1997) while larger companies are more likely to use discounted cash flow techniques (Graham and Harvey, 2001). Over time, the payback method has been increasingly replaced by DCF methods. The question of ranking project under constraints is examined theoretically in this chapter.

9.2 PROJECT DEPENDENCE

Two projects are *economically independent if* the acceptance or rejection of one does not alter the other's cash flow stream and it does not affect the other's acceptance or rejection. Two independent projects would be a manufacturing firm's prospective investment in a special-purpose milling machine and its prospective investment in some office equipment.

Two (or more) projects are *economically dependent* if the acceptance or rejection of one alters the cash flow stream of any other or if the acceptance or rejection of any other project is affected. The most obvious dependency is that among *mutually exclusive* projects, where acceptance of any one project automatically (technically or economically) excludes acceptance of all others. Examples of mutually exclusive projects include alternative uses for a particular land site, alternative designs for a building, competing plant expansions where monetary resources exist to support only one, and R&D projects where technical expertise and time can tackle only one. Obviously, two mutually exclusive projects cannot both be accepted.

Two (or more) projects may be economically dependent without being mutually exclusive. In general, this *partial dependence* occurs when accepting two projects together changes the cash flow streams of at least one project. An excellent example is given by Bierman and Smidt (1988, p. 64): It would be technically feasible to build a toll bridge *and* to operate a toll ferry between two points on the opposite sides of a river. Cities such as New York and San Francisco have done this, but the two projects are dependent because the revenues of each are affected by the other. For partial dependence the total beneficial returns or positive cash flows are usually decreased (not increased), since the two projects are partial substitutes for each other.

A third type of dependence, called economic *complementarity,* sometimes occurs between projects. If the benefits expected when both projects are accepted are synergistic (either by increasing revenues or decreasing investments or costs), then the projects are economic complements. If each project complements the other, the dependency is symmetrical. This might occur, for example, with a computer-aided design project that decreases the time required to design a new product and a project that reduces the cost of producing the product. Together they might produce far greater profits than their sum if undertaken in isolation. In practice, however, asymmetrical dependence is more common. For this the acceptance of project B benefits project A, but the acceptance of A in no way benefits B. For example, consider a proposed investment in the crane itself with a short boom (project A) and an additional long boom for a crane (project B). Without the long boom (B), the crane (A) may be profitable, but with the long boom, the earning of the crane might be enhanced. Generally, asymmetric economic dependence also involves *technical contingency* between projects. As the example implies, it would be technically meaningless to acquire a long boom without a crane. Thus, the acceptance of project B implies the acceptance of project A, but not vice versa.

9.3 CAPITAL RATIONING

Rationing of capital occurs whenever the funds available for investment are limited so that the firm cannot accept all otherwise acceptable projects. (The reader will recall that *acceptability* is determined by one of the rational criteria—positive NPV, IRR > marginal investment rate, RIC > marginal investment rate, or $B/C > 1.0$.) Restrictions on funds for investment may be imposed either by management (internal to the firm) or by the capital market (external to the firm).

Internal capital rationing occurs when (1) management limits the total funds available in a given period or (2) management sets a cutoff rate for investments that is higher than the firm's imputed cost of money; that is, a minimum attractive rate of return (MARR) that exceeds the cost of capital. In either case, internal capital rationing causes the rejection of some projects that would otherwise be acceptable from a profitability standpoint.

External capital rationing occurs when the firm cannot obtain funds from the capital market in sufficient amounts at an economical price. The unlimited availability of funds as postulated in *a free* capital market simply does not exist, and the firm must compete for available funds. Several authors[1] postulate an upward sloping supply curve for capital beyond a certain critical amount of capital, as in Figure 9.1.

With an increasing cost *of* capital curve, the firm cannot raise an unlimited amount of capital at one time at a constant cost. The increased interest demanded by the financial market, beyond B_{crit} in Figure 9.1, is due to an increase in perceived risk by the firm's investors. After a period during which the firm demonstrates to its creditors and investors its ability to generate satisfactory returns, however, the risk premium disappears and the firm can again raise capital at the lower constant rate up to some new critical level.

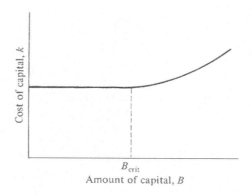

Figure 9.1 Capital supply curve

The rising capital supply curve implies that certain marginal projects with IRRs close to the increased cost of capital will no longer be acceptable. Thus, the firm must postpone or abandon the marginal projects that are graphed at the extreme right of Figure 9.2. Here, projects using incremental amounts of capital are cumulated according to decreasing IRRs (or RICs) on the same axes as the capital supply curve. The capital demand curve gives the incremental (marginal) rate of return for each increment of investment added. The intersection of the incremental rate of return curve with the marginal cost-of-capital curve determines the amount of capital to be raised and the amount of funds to be invested for the period. Thus, the external capital market can limit the capital available for investment, and the firm consequently must accept some and reject other investment opportunities.

Whether capital is rationed externally or internally, the firm must select some investment projects as better than others that it rejects, which on their own merits would be perfectly acceptable. However, as we will see in Section 9.12 internal rationing of capital leads to ranking on internal rate of return, while external rationing will still be approached on the basis of a weighted average cost of capital and net present value.

---

[1]. For example, see Duensenberry (1958), Chapter 5; Hyytinen and Toivanen (2005).

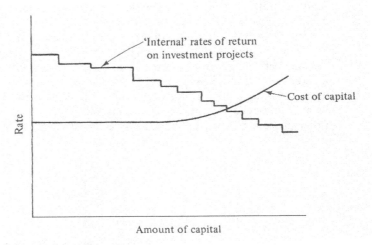

Figure 9.2 Marginal rates of return for investment projects and the marginal cost of capital

9.4 COMPARISON METHODOLOGIES

To compare two or more investment projects and to accept some and reject others—for whatever reason: mutual exclusivity, economic or technical dependence, capital rationing, or project indivisibility—the firm must select a rational methodology. There are two basic methods for selecting an optimal investment portfolio: (1) ranking and (2) mathematical programming. In the first methodology, candidate projects are ranked in decreasing order by their NPVs, IRRs, RICs, or *B/C* ratios. Projects are then accepted until the capital budget is exhausted. In the second methodology, an optimal set of projects is selected by a mathematical or integer programming procedure, so as to maximize some criterion subject to feasibility constraints (budget, manpower, liquidity, etc.).

When using NPV, projects must be compared using a common study period. If the project lives are not the same, then the projects must be assumed to repeat to reach a common period. So comparing two pumps where one has an expected life of 2 years and the other of 3 years must be compared with a study period of 6 years, and either three pumps or two pumps are purchased, one after another, until the 6-year study is complete. While this may or may not be realistic, it is often not possible to make better quality assumptions about future costs. The usual alternative is to use annual worth (AW) analysis for alternatives with different length lives. When the costs per year are compared, the implicit assumption is made that costs and lives repeat until the least common multiple of the individual project lives. This matches the explicit assumption for NPV.

The ranking methodology, originally proposed by Joel Dean (1951), is beset by a number of difficulties. First, the preference ranking obtained by using the IRR criterion may differ from the ranking obtained when NPV is used. Accepting projects in descending order of merit, therefore, results in a different set of projects being accepted with the IRR than with the NPV criterion. Rankings with the benefit-cost ratio can differ from both. This has been called the *ranking-error problem* (see, for example, Barney and

9.4 Comparison Methodologies

Danielson, 2004) and can be illustrated by a simple example.

Table 9.1 Characteristics of five projects

Project	Investment Required at $t = 0$	Annual Cash Flow $Y,(t = 1, ..., N)$	Project Life, N
F	$-12,000	$4,281	5
G	-17,000	5,802	10
H	-5,000	1,866	15
I	-14,000	2,745	20
J	-19,000	5,544	20

Example 9.1 Ranking using various methods

Consider five projects (F, G, H, I, and J) with required investments, expected cash flows, and project lives given in Table 9.1.[2] Salvage values of all investments are considered to be zero. These investments are ranked by three different criteria: internal rate of return, net present value, and benefit-cost ratio.

The ranking by internal rate of return is accomplished by calculating the IRR for each project; thus,

Project	IRR	Rank
F	$0 = -12,000 + 4281(P/A, i, 5) \Rightarrow (P/A, i, 5) = 2.80 \Rightarrow i^* = 0.23$	4
G	$0 = -17,000 + 5802(P/A, i, 10) \Rightarrow (P/A, i, 10) = 2.93 \Rightarrow i^* = 0.32$	2
H	$0 = -5000 + 1866(P/A, i, 15) \Rightarrow (P/A, i, 15) = 2.68 \Rightarrow i^* = 0.37$	1
I	$0 = -14,000 + 2745(P/A, i, 20) \Rightarrow (P/A, i, 20) = 5.10 \Rightarrow i^* = 0.19$	5
J	$0 = -19,000 + 5544(P/A, i, 20) \Rightarrow (P/A, i, 20) = 3.43 \Rightarrow i^* = 0.29$	3

Because project lives differ, we cannot use NPV to compare the projects, but must use AW (annual worth) instead. When the firm's marginal investment rate is 15%, ranking by annual worth leads to the following:

Project	Annual Worth	Rank
F	$AW_F = -12,000 (A/P, i, 5) + 4281 = \quad + \quad 701$	4
G	$AW_G = -17,000(A/P, i, 10) + 5802 = +2415$	2
H	$AW_H = -5000(A/P, i, 15) + 1866 = +1011$	3
I	$AW_I = -14,000(A/P, i, 20) + 2745 = \quad + \quad 508$	5
J	$AW_J = -19,000(A/P, i, 20) + 5544 = +2509$	1

The ranking by benefit-cost ratio is established similarly:

---

2. In this chapter all investments are assumed to be pure investments unless otherwise stated, so that the IRR criterion is applicable. This avoids having to consider the cost of capital as a variable, which would unduly complicate the presentation.

Project	Benefit-Cost	Rank
F	$B/C = 4281(P/A, i, 5)/12{,}000 = 1195$	5
G	$B/C = 5802(P/A, i, 10)/17{,}000 = 1713$	3
H	$B/C = 1866(P/A, i, 15)/5000 = 5910$	1
I	$B/C = 2745(P/A, i, 20)/14{,}000 = 1227$	4
J	$B/C = 5544(P/A, i, 20)/19{,}000 = 1826$	2

In this case the three criteria (IRR, AW, and B/C) produce three different rankings. If capital is not rationed, then all projects could be accepted. However, if capital is rationed, then a different decision is obtained from each criterion. For example, suppose the capital budget is $40,000. The IRR criterion would accept projects H and G, leaving $18,000 unspent; the AW criterion would accept projects G and J, leaving $4000 unspent; and the benefit-cost criterion would accept projects H and J leaving $16,000 unspent. The question, then, is, which criterion gives the "correct" decision? Other collateral questions are, why do the apparent inconsistencies in ranking exist, and what is the explanation for them? This ranking inconsistency was originally pointed out by Lorie and Savage (1955), and the basis is differing *implicit* assumptions about reinvestment of cash flows.

The second problem associated with these rankings is that of project indivisibility. For the example above, each technique could only accept two projects leaving from $4000 to $18,000 unspent. However, as Lorie and Savage point out (1955, p. 231), if the majority of investment projects are small relative to the total capital budget, then project indivisibility should have only minor consequences. If several indivisible investments are large compared to the total capital budget, then indivisibility becomes a serious problem and some form of 0–1 integer programming may be required.

Other problems are linked to various imperfections in fitting the perfect market model to real situations. For example, Woods and Randall (1989) have examined theoretically whether financial markets can properly calculate the net present value of future investment opportunities. They conclude that in most cases the NPV will exceed shareholder wealth at the time of project announcement.

9.5 THE REINVESTMENT RATE PROBLEM

Whenever investment projects must be compared and only some can be accepted, the method of comparison becomes relevant.[3] The reason for comparison is immaterial. If certain of the otherwise acceptable candidates are to be rejected, then comparison is inevitable.

Net present value and internal rate of return are two methods with different objective functions. One could argue that they are both correct, with different goals. When ranking on IRR, one maximizes the efficiency of capital spent; getting the most return for money invested. Using NPV guarantees the maximum wealth generated given a specified rate of return and a budget.

Internal rate of return is favored by many because of its intuitive appeal. Some

---

3. See Bierman and Smidt (1957).

may think of it as a return on initial investment, although this is an incorrect economic interpretation (see Chapter 7). However, its true basis includes selection of the marginal investment rate. On the other hand, net present value requires that a marginal investment rate be established in advance. The same is true of the benefit-cost method. As shown in Chapter 7, net present value maximizes shareholder wealth under perfect capital market conditions. Benefit-cost ratio possesses a further advantage in certain cases; namely, it measures the *relative* efficiency of providing future wealth per dollar of present investment.[4] Proponents of each method cite methodological advantages, such as those given above. Yet the ranking inconsistency problem remains, as shown in Example 9.1. This inconsistency of ranking is called the *reinvestment rate problem* since the inconsistencies are based on differing assumptions made about the *rate at which cash flows from projects are reinvested in alternative uses.*[5] If a *common* reinvestment rate is assumed, the ranking inconsistencies disappear.

Example 9.2 The reinvestment rate problem
An example by Solomon (1959) illustrates the reinvestment rate problem. The candidate set is the simplest nontrivial one—two projects, one to be accepted and the other rejected.[6] The firm must choose between two projects, each costing $100 (with all cash flows in $ thousands) at t = 0. Project K returns $120 at t = 1, and project L returns $201.14 at t = 5.

Applying the IRR method gives the following internal rates of return:

$$K: \quad 0 = -100 + 120(P/F, i^*, 1) \Rightarrow i_K^* = 0.20$$
$$L: \quad 0 = -100 + 201.14(P/F, i^*, 5) \Rightarrow i_L^* = 0.15$$

Since $i_K^* = 0.20$ is greater than $i_L^* = 0.15$, it is argued that K should be accepted and L rejected.

Suppose also that the firm's capital availability is limited and that management has imposed a marginal investment rate of 10%. At a discounted rate of 10%, the projects' net present values are:

$$K: \quad P_K = -100 + 120(P/F, i^*, 1) = +9.09$$
$$L: \quad P_L = -100 + 201.14(P/F, i^*, 5) = +24.89$$

Thus, if the firm maximizes net present value, it should accept L and reject K. This decision is the direct opposite of that signaled by the IRR criterion; and hence, it is argued, the IRR method does not guarantee maximization of net present value or the

---

4. See Schwab and Lusztig (1969).
5. Some writers strongly insist that the reinvestment rate has no place in economic studies of alternatives. See Lohman (1988), for example. The weight of the theoretical evidence, however, appears to support the reinvestment rate explanation of the ranking inconsistency problem. See, for example, Solomon (1956), Bernhard (1962), and Beidleman (1984).
6. Solomon (1956, 1959) uses this example to illustrate mutual exclusivity. As Bierman and Smidt (1957) point out, however, it is not necessary that this be due to economic or technical mutual exclusivity. All that is necessary is that a requirement be imposed to accept one and reject the other. This can result from the imposition of capital budget and project indivisibility constraints. The same reasoning can be extended to a candidate set of m (m = 2, 3, . . .) projects in which some are to be accepted and others rejected, since all can be compared in pairs.

future wealth of shareholders. However, as we will show in Section 9.8, finding the incremental internal rate of return between K and L indicates that at 10% L should be done. Thus, any difference in recommendations would be caused by using different reinvestment assumptions for different methods.

For large capital rationing problems (many projects and a lot of money), these two methods will most likely give similar answers. This is because, in optimization language, capital rationing is a knapsack problem: fill the knapsack with items that maximize value without spilling out of the pack. This now becomes an optimization problem. Caveats include whether you have to put full items in or if partial items are OK. If full items are required (a more realistic assumption), then the problem is hard to solve computationally due to the many possible combinations that may result. However, the problem has been studied in detail, and all knapsack problems, especially large ones, have solution "cores"—items that are always in the solution given a minimum knapsack size. These are the items that maximize the ratio of value to size, similar to maximizing IRR. That is why solutions tend to be similar.

9.6 THE REINVESTMENT ASSUMPTION UNDERLYING NET PRESENT VALUE

The goal at the end of the day is to have the most money; the most money beyond the cost of money if you use cost of capital for the interest rate, or the most money beyond the minimally accepted amount if you like to use MARR. That is the goal of net present value. It is not the goal of ranking on IRR, which has a different approach to establishing the MARR.

While IRR is good in terms of picking projects that are efficient, it does not efficiently spend capital. Only methods that examine combinations of investment options (itemized portfolios) or linear programming/integer programming approaches will spend all (or nearly all) of the capital allotted. This is because it will look at combinations of projects (these are the ones outside the core) that may have lower NPV returns, but also have lower spending, in order to get them in the knapsack.

The projects in Example 9.2 cannot be compared directly without an external reinvestment rate, because the cash inflows of K and L occur at different times. Assume units of $thousand or $millions. Both require a $100 investment, but is it better from the firm's standpoint to receive $120 at the end of 1 year (project K) or $201.14 at the end of 5 years (project L)? This question requires placing a value on the earlier receipt of $120. This value depends on what the firm intends to do in the interim with the earlier $120, in comparison to the later $201.14.

The principle of opportunity costing requires that the firm use the $120 in its best alternative use. The firm has at least four alternative uses: (1) It can pay $120 in dividends to its shareholders immediately; (2) it can retain the $120 in cash in a noninterest-bearing bank account; (3) it can invest the $120 in interest-bearing securities at an interest rate less than the firm's marginal investment rate,[7] say 8%; or (4) it can reinvest the $120 in another new candidate project at the firm's marginal investment rate of 10%. Obviously, alternatives 2 and 3 are inferior to alternative 4 under conditions of certainty, since alternative 2 provides no return and alternative 3 provides a lower rate of return. Furthermore, the imputed cost of alternative 1—paying out dividends—is the value of the

---

7. The risk of the interest-bearing securities is assumed to be less than or equal to that of projects accepted at the firm's marginal investment rate.

best foregone opportunity, that is, retaining the $120 and using it at 10% under alternative 4. The best alternative use of the $120 is reinvestment in another new project whose rate of return is at least the marginal investment rate. This assumes that such a new project is available at time t = 1.

The firm's cost of capital provides another way of looking at the cost of paying out dividends. For example, if the $120 were actually paid out to the shareholder and then simultaneously reobtained by the firm (through borrowing or new equity capital), the firm would incur a cost of obtaining the new $120. One could logically assume that the new $120 would be obtained at a cost equal to the firm's cost of capital. In any event, the payment of dividends has an imputed cost, and if the firm's marginal cost of capital is assumed to be its marginal investment rate, then alternatives 1 and 4 are equivalent uses of the $120. This reinvestment act places value on the receipt of the earlier cash inflow and makes the comparison of K and L possible. When the act of investing $100 in project K is combined with the act of reinvesting the $120 inflow at a reinvestment rate of 10%, the effective return on invested capital for K is

$$0 = -100 + 120(F/P, 10\%, 4)(P/F, i_K, 5)$$
$$= -100 + 120(1.464)(P/F, i_K, 5);$$
$$\therefore (P/F, i_K, 5) = 0.569 \Rightarrow i_K = 0.1193.$$

Comparing $i_K = 0.1193$ to $i_L^* = 0.15$, we see that now L is ranked ahead of K, which is the same ranking given by the NPV method. This approach mirrors that of Section 7.11 regarding return on invested capital in that it requires the explicit use of an external rate of return. In Section 9.8 we will use an incremental analysis to achieve the same result.

So what does the reinvestment rate assumption mean? Money we make on a project gets plowed back into the business and earns the MARR. This is actually a safe assumption as you would hope it would get plowed back and earn at least the MARR. If we thought the future would have lower returns, then we would assume a lower reinvestment rate which would increase the value of our current investment whereas if we thought the future would be better (higher rate), it would lower the value of a current project.

Now we will demonstrate that an external reinvestment assumption is implicit in the NPV method when projects with *unequal lives* are compared. Consider two competing projects, M and N, only one of which is to be accepted. Let M cost a_0 dollars at $t = 0$, and let N cost a_0'. Let M return Y_m dollars at the end of m years and N return Y_n dollars at the end of n years, assuming $n > m$. Also, let k equal the firm's marginal investment rate and j equal the firm's available reinvestment rate, where now it is assumed initially that $j \neq k$.

For the net present value method, with reinvestment for the shorter life, the NPVs of M and N are:

$$\text{M: } P_M = \frac{Y_m(1+j)^{n-m}}{(1+k)^n} - a_0$$

or

$$P_M = \frac{Y_m}{(1+k)^m}\left(\frac{1+j}{1+k}\right)^{n-m} - a_0 \tag{9.1}$$

and

$$\text{N: } P_N = \frac{Y_n}{(1+k)^n} - a_0' \tag{9.2}$$

Now, if the NPVs of M and N are stated without reinvestment, we obtain the conventional net present values:

$$\text{M: } P_M = \frac{Y_m}{(1+k)^m} - a_0 \qquad (9.3)$$

$$\text{N: } P_N = \frac{Y_n}{(1+k)^n} - a_0' \qquad (9.4)$$

Equation (9.4) matches Equation (9.2), so that reinvestment is not a factor for the longer life. However, the two equations for project M, Equations (9.1) and (9.3), are not the same. They differ by the factor $[(1+j)/(1+k)]^{n-m}$. The only way in which Equation (9.1) can be made equivalent to Equation (9.3), with $(n, m > 0)$, is for

$$\left(\frac{1+j}{1+k}\right)^{n-m} = 1,$$

which defines the implicit equality of j and k. In other words, *when we compare the NPVs of two unequal-lived projects by the conventional definition of NPV* [Equations (9.3) and (9.4)], *we implicitly assume that the reinvestment rate, j, is equal to the firm's marginal investment rate, k.*

If this common reinvestment rate is assumed for the IRR and NPV methods, then the ranking inconsistency is removed. For projects L and K in Example 9.2, the cash flows of project K are reinvested at the firm's *marginal investment rate* of 10% at the ends of years 1, 2, 3, and 4. Then, from Equations (9.1) and (9.2) the net present values are

$$P_K = -100 + \frac{120}{(1.10)^1}\left(\frac{1.10}{1.10}\right)^4 = +9.09$$

$$P_L = -100 + \frac{201.14}{(1.10)^5} = +24.89$$

which ranks L above K. Also, from Equations (9.1) and (9.2) the effective rates of return with reinvestment at the rate $j = k = 0.10$ are calculated by setting NPV = 0, as follows:

$$\text{K: } \frac{120}{(1+i_K)}\left(\frac{1.10}{1+i_K}\right)^4 - 100 = 0$$

or

$$(1+i_K)^5 = 1.2(1.10)^4 = 1.758$$

from which

$$i_K = (1.758)^{0.2} - 1 = 0.119$$

and for L:

$$\frac{201.14}{(1+i_L^*)^5} - 100 = 0$$

$$(1+i_L^*)^5 = 2.0114$$

9.6 The Reinvestment Assumption Underlying Net Present Value

from which

$$i_L^* = (2.0114)^{0.2} - 1 = 0.150$$

which also ranks L above K. Thus, assuming a common reinvestment rate implies that competing projects have identical rankings by both methods.

The conclusion that needs to be drawn is that two *unequal-lived* projects are strictly noncomparable, unless specific reinvestment assumptions are made for the *entire* period from the project's inception to the end of the longer life. The principle of opportunity costing requires reinvestment at a rate at least equal to the firm's marginal investment rate.

The reinvestment problem is not limited to competing projects with unequal lives. As the following example demonstrates, reinvestment must be considered when two *equal-lived* projects are compared, *if their cash flow streams differ*. Thus, reinvestment assumptions are implicit in comparing any two projects that are not identical.

Example 9.3 Different methods yield different ranking
Consider two 5-year investments, X and Y.

EOY	NCR(X)	NCF(Y)
0	$-37,900	$-37,900
1	+ 10,000	+ 2,000
2	10,000	6,000
3	10,000	6,000
4	10,000	10,000
5	10,000	28,000

The internal rates of return are $i_X^* = 10\%$ and $i_Y^* = 8\%$. Thus, X is more desirable according to the IRR criterion. If the marginal investment rate is above 3.7%, NPV provides the same ranking. If the firm's marginal investment rate is 3%, however, then the net present values are $P_X = \$7900$ and $P_Y = \$8230$, and the NPV criterion ranks Y as more desirable. Although the projects have identical investments and lives, *they do not generate cash inflows at the same rate*. It is the necessity for placing a value on the earlier receipt of money versus later receipt that introduces the reinvestment problem.

In conclusion, the reinvestment problem is present whenever two or more projects must be compared, with the objective being to select some and reject others. The question is not whether reinvestment must be considered, but rather what specific reinvestment assumptions are implicit in the several comparison criteria.

9.7 THE REINVESTMENT ASSUMPTION UNDERLYING THE INTERNAL RATE OF RETURN: FISHER'S INTERSECTION

Section 9.6 demonstrated that applying a common reinvestment rate to both the IRR and NPV methods results in consistent project rankings. The reinvestment rate that is used is the marginal investment rate since it is implicit in the NPV criterion. We may now ask if there is some higher reinvestment rate (or rates) that would cause the IRR and NPV methods to rank projects consistently. The answer is yes—provided that certain conditions exist.

We begin by graphically comparing the NPVs of the two projects, K and L in Example 9.2, as functions of the marginal investment rate, k (see Figure 9.3), using Equations (9.3) and (9.4). As the figure illustrates, the NPV method ranks K as more desirable at marginal investment rates greater than 0.138. This second ranking matches that yielded by IRR. So the IRR and NPV methods are consistent in their rankings for reinvestment rates higher than this *indifference* point, if it exists and if it occurs at a positive NPV. The indifference point, labeled A in Figure 9.3, is the reinvestment rate at which the two net present values are equal.

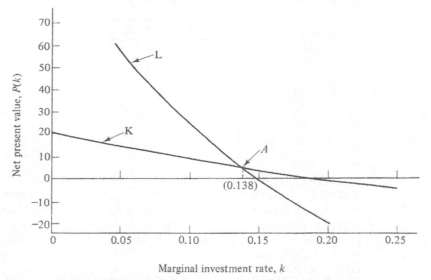

Figure 9.3 Fisher's intersection for two projects

The indifference point is called *Fisher's intersection* because its concept and method of determination are equivalent to Irving Fisher's (1930) *rate of return over cost*. Specifically, suppose project P generates net cash flows of $Y_0, Y_1, \ldots\ldots Y_n$ and project Q generates net cash flows of $Y'_0, Y'_1, \ldots\ldots, Y'_n$. Then the Fisher rate of return over cost, r_f is defined by:

$$\sum_{t=0}^{n} \frac{Y_t - Y'_t}{(1+r_f)^t} = 0 \qquad (9.5)$$

(see Fisher, 1930, pp.151–155).

9.7 The Reinvestment Assumption Underlying IRR

Applying Equation (9.5) to projects K and L, we determine r_f as follows:

End of Year	Cash Flow for L(Y_t)	Cash Flow for K(Y_t')	$Y_t - Y_t'$
0	−100	−100	0
1	—	+120	-120
2	—	—	—
3	—	—	—
4	—	—	—
5	+201.14	—	+201.14

from which we set:

$$\frac{-120}{(1+r_f)} + \frac{201.14}{(1+r_f)^5} = 0$$

resulting in $r_f = 0.138$. This is exactly the reinvestment rate at which K and L had the same net present value. This should be apparent from the Fisher formulation also, since r_f makes the incremental outflow ($Y_1 - Y_1' = -120$) exactly equivalent to the incremental inflow ($Y_5 - Y_5' = +201.14$).

The conclusion is, therefore, that if the firm's marginal investment rate or reinvestment rate is greater than the Fisher intersection, the nonincremental IRR and the conventional NPV methods give consistent rankings. In Section 9.8 we will see that using an incremental analysis of IRR will lead to results consistent with NPV at reinvestment rates below the Fisher intersection.

To summarize Sections 9.6 and 9.7, we may now say that

1. Comparing two projects by the NPV implicitly assumes reinvestment of positive cash flows, from their receipt until the end of the longer-lived project, *at the firm's marginal investment rate.*
2. Comparing two such projects by the non-incremental IRR implicitly assumes reinvestment *at a rate equal to or greater than the Fisher rate of return over cost.*
3. If the firm's reinvestment rate is less than the Fisher rate of return over cost, then the NPV and nonincremental IRR methods will give inconsistent rankings, unless the firm's marginal investment rate is used as a common reinvestment rate.
4. The ranking inconsistency problem is traceable to the differing assumptions concerning reinvestment that are made by the two methods.

The Fisher rate of return over cost may or may not exist. That is, the net present values of two competing projects, as functions of the marginal investment rate, k, may not intersect (and Fisher's intersection would not exist). If they do not intersect, then the IRR and NPV criteria signal identical rankings for all reinvestment rates. Another possible complication is that multiple intersections may exist; then, the firm is indifferent between the projects at more than one reinvestment rate, and interpretation is difficult. Mao (1966) has investigated the problem of the existence of Fisher's intersection extensively.

9.8 INCREMENTAL RATES OF RETURN

If certainty is assumed, then making decisions that are consistent with the economic principle of equating marginal benefits to marginal costs will usually result in an optional choice of investment alternatives. This principle was invoked in Section 9.3 when capital rationing was discussed, and it was illustrated in Figure 9.2. As applied to project selection, the marginal principle requires that projects be accepted *at the investment margin* until the last project accepted provides a marginal (incremental) rate of return just equal to the firm's marginal investment rate. Under perfect capital market conditions, the firm's marginal investment rate is also its marginal cost of capital. The words accepted *at the investment margin* mean *with the next increment of added investment.* This method is often called the *incremental rate-of-return method.* The incremental method is most clearly applicable when *mutually exclusive alternatives* are compared, as in Example 9.4.

Example 9.4 Incremental rate of return
Five mutually exclusive building designs are to be compared incrementally. All of the designs have a 15-year life and no salvage value, but their required initial investments and annual revenues differ. The marginal cost of capital (i_{CC}) is 15%.

Design	A	B	C	D	E
Total Investment (t=0)	335,000	500,000	725,000	885,000	940,000
Annual after-tax revenue	50,000	110,000	149,000	170,000	184,000

The stepwise solution for the incremental rates of return proceeds as follows.

Step 1: Calculate the IRR for each alternative and eliminate any alternative having an IRR less than the marginal cost of capital. So,
   A: $(P/A, i_A^*, 15) = 335,000/50,000 = 6.700 \Rightarrow i_A^* = 12.31\%$
   (Since $i_A^* = 12.3\%$ is less than $i_{CC} = 15\%$, drop alternative A.)
   B: $(P/A, i_B^*, 15) = 500,000/110,000 = 4.455 \Rightarrow i_B^* = 20.69\%$
   (Conditionally accept B into the candidate set.)
   C: $(P/A, i_C^*, 15) = 725,000/140,000 = 4.866 \Rightarrow i_C^* = 17.62\%$
   (Conditionally accept C into the candidate set.)
   D: $(P/A, i_{DA}^*, 15) = 885,000/170,000 = 5.206 \Rightarrow i_D^* = 17.50\%$
   (Conditionally accept D into the candidate set.)
   E: $(P/A, i_E^*, 15) = 940,000/184,000 = 5.109 \Rightarrow i_E^* = 17.92\%$
   (Conditionally accept E into the candidate set.)

Step 2: Arrange the conditionally accepted projects in order of ascending investment and, beginning with the lowest investment, calculate the incremental rate of return on each added increment of investment. If the incremental rate of return for a competing project exceeds the firm's marginal cost of capital, conditionally retain the competing project as the current most desirable one (defender); otherwise, reject it, and proceed stepwise to the next larger investment increment.

Arranging the conditionally accepted projects in order of ascending investment, we have

Design	Total Investment	Cash Inflow per Year
B	$500,000	$110,000
C	725,000	149,000
D	885,000	170,000
E	940,000	184,000

Since $i_B^* = 20.69\%$, which exceeds the marginal investment rate, the smallest increment (over doing nothing) is acceptable. Then C is compared with B by comparing the incremental investment, $725,000 - 500,000 = \$225,000$, with the incremental annual return, $149,000 - 110,000 = \$39,000$. The incremental rate of return, i_{C-B}^*, is found as follows:

$$(P/A, i_{C-B}^*, 15) = \frac{725,000 - 500,000}{149,000 - 110,000} = 5.769$$

$$\therefore i_{C-B}^* = 15.28\%$$

Since the incremental rate of return, $i_{C-B}^* = 15.28\%$, exceeds the firm's marginal cost of capital rate $i_{CC} = 15\%$, incremental investment is justified, and project C is the new defender.

Step 3: Calculate incremental rate of return for each successive project, basing the increments of investment and cash inflows on the current defender.

Calculate the incremental rate of return of project D (the challenger) over project C (the defender):

$$(P/A, i_{D-C}^*, 15) = \frac{885,000 - 725,000}{170,000 - 149,000} = 7.619$$

$$\therefore i_{D-C}^* = 9.97\%$$

Since i_{D-C}^* is less than $i_{CC} = 15\%$, the increment of investment between C and D is not justified, Project D is eliminated, and C remains the defender.

We now compare projects E and C. Thus:

$$(P/A, i_{E-C}^*, 15) = \frac{940,000 - 725,000}{184,000 - 149,000} = 6.143$$

$$\therefore i_{E-C}^* = 14.00\%$$

Since i_{E-C}^* is less than $i_{CC} = 15\%$, the incremental investment between C and E is not justified. The final decision, therefore, is to choose C as the most desirable of the mutually exclusive alternatives. C supplies the opportunity to invest the last increment at a marginal rate of return greater than the firm's marginal cost of capital.

It should be noted that the same decision results from choosing the alternative with the greatest net *present value*. To demonstrate, we calculate the net present values for the design alternatives:

$$P_j = Y_{0j} + A_j(P/A, i_{CC}, 15)(j = A, B, C, D, E).$$

Design, j	Y_{0j}	A_j	A_j (5.847)	P_j	
A	−335,000	50,000	292,000	−42,600	
B	−500,000	110,000	642,000	143,200	
C	−725,000	149,000	870,000	146,300	← Maximum
D	−885,000	170,000	994,000	109,000	
E	−940,000	184,000	1,075,000	135,900	

C maximizes the net present value, and in this instance, ranking by NPV and by the incremental rate of return method agree. Calculating incremental benefit-cost ratios would also reach the same results here. The incremental method, while being theoretically correct, fails if the incremental cash flow stream has *no answer* or a *multiple answer* to the incremental rate of return problem. So, as Bernhard (1971) points out, to avoid no answer and multiple answers, as well as to simplify the computations, it is probably more desirable to use the NPV procedure directly, so long as the predetermined marginal investment rate is known.

9.8.1 Incremental rate of return applied to the constrained project selection problem. The principle of marginality can be extended to certain constrained project selection problems, even when some of the projects are not themselves mutually exclusive. A method of handling project dependencies by integer programming was proposed by Weingartner (1963, pp. 11–12, 32–33). A manual version, devised by Fleischer (1966), will be described here. Dependent projects are grouped into financially (economically) mutually exclusive bundles and then incremental rates of return are calculated. The bundle providing the last incremental rate of return greater than the firm's cost of capital will contain the optimal subset of projects.

Example 9.5 Incremental rate of return
Consider the small business projects F, G, and H.

Project	Investment ($t = 0$)	Net Cash Inflows per Year	Life, Years	NPV
F	$−12,000	$+4,281	5	$2,351
G	−10,000	+4,184	5	4,025
H	−17,000	+5,802	10	12,119

The firm's marginal investment rate is assumed to be 15%, and initially we shall consider the three projects to be economically independent. What projects should be selected if successively smaller budget ceilings are imposed? The solution proceeds stepwise, as outlined in Example 9.5.

9.8 Incremental Rates of Return

Step 1: Identify the investments and cash inflows for all feasible project combinations. Each combination is an economically mutually exclusive bundle.

We consider all possible combinations[8] of projects F, G, and H by taking the projects one at a time, two at a time, and so forth:

Bundle	Component Projects	Investment Projects	Cash Inflow of Bundle
1	F	−12,000	+4281 for years 1–5
2	G	−10,000	+4184 for years 1–5
3	H	−17,000	+5802 for years 1–10
4	F, G	−22,000	+8465 for years 1–5
5	F, H	−29,000	+10,083 for years 1–5 + 5802 for years 6–10
6	G, H	−27,000	+ 9986 for years 1–5 + 5802 for years 6–10
7	F, G, H	−39,000	+14,267 for years 1–5 + 5802 for years 6–10

Step 2: Arrange the mutually exclusive bundles in order of ascending investment:

Bundle	Component Projects	Investment Projects	Cash Inflow of Bundle
2	G	−10,000	+4184 for years 1–5
1	F	−12,000	+4281 for years 1–5
3	H	−17,000	+5802 for years 1–10
4	F, G	−22,000	+8465 for years 1–5
6	G, H	−27,000	+9986 for years 1–5 +5802 for years 6–10
5	F, H	−29,000	+10,083 for years 1–5 + 5802 for years 6–10
7	F, G, H	−39,000	+14,267 for years 1–5 + 5802 for years 6–10

Step 3: Calculate the incremental rates of return from bundle to bundle and eliminate any bundle whose incremental rate of return is less than the firm's marginal

---

8. For a candidate set of m projects, the total number of combinations, M, taken r at a time where r = 0,1,2,..., m, is

$$M = \sum_{r=0}^{m} \binom{m}{r} = \frac{m!}{0!m!} + \frac{m!}{1!(m-1)!} + \cdots + \frac{m!}{m!0!} = 2^m$$

Note that for even "small" m, say m = 30, the number of possible combinations (bundles) becomes very large (i.e., $M = 2^{30} = 1,073,741,824$)!

investment rate. Make comparisons among bundles by comparing successive challenges with the best defender, the bundle that was last conditionally justified.

First, bundle 2 must be compared with a do-nothing alternative. Since $(P/A, i_2^*, 5) = 10{,}000/4{,}184 = 2.390$, $i_2^* = 31\%$, which exceeds $i_{MIR} = 15\%$. Bundle 2 should be accepted. Then, comparing 1 to 2 results in;

$$(P/A, i_{1-2}^*, 5) = \frac{12{,}000 - 10{,}000}{4281 - 4184} = 20.6$$

$$\therefore i_{1-2}^* = -34.06\% \quad \text{(The investment increment of \$2000 is never recovered by five cash flows of \$97.)}$$

Since i_{1-2}^* is less than $i_{MIR} = 15\%$, bundle 1 is eliminated. The same decision result can be reached by calculating the *incremental* net present value for the comparison:

$$\Delta P_{1-2} = -2000 + 97(P/A, 15\%, 5) = -1{,}675.$$

We need not actually calculate i_Δ^* since we can infer its magnitude from the sign of ΔP. This decision method will be used in the remainder of the example.

Continuing with step 3, we next compare bundle 3 with bundle 2:

$$P_3 = -17{,}000 + 5802(P/A, 15\%, 10) = +12{,}119.$$

$$P_2 = -10{,}000 + 4184(P/A, 15\%, 5) = 4015.$$

$$\Delta P_{3-2} = +8095 \Rightarrow i_{3-2}^* > i_{MIR}$$

Since $i_{3-2}^* > i_{MIR}$, the added investment is justified, and we accept bundle 3 as being more desirable than bundle 2. Continuing in the same manner, we obtain the following results:

Bundle	Component Projects	Investment in Bundle	Incremental Δ(NPV)	Incremental i_Δ^*	Action
0	None	$-0-$	—	—	—
2	G	$-10{,}000$	$4{,}025	$i_{2-0}^* = 31.0\%$	Accept
1	F	$-12{,}000$	$-1{,}675$	$i_{1-2}^* < 15\%$	Reject
3	H	$-17{,}000$	$+8{,}095$	$i_{3-2}^* > 15\%$	Accept
4	F, G	$-22{,}000$	$-5{,}745$	$i_{4-3}^* < 15\%$	Reject
6	G, H	$-27{,}000$	$+4{,}025$	$i_{6-3}^* > 15\%$	Accept
5	F, H	$-29{,}000$	-675	$i_{5-6}^* < 15\%$	Reject
7	F, G, H	$-39{,}000$	$+2{,}350$	$i_{7-6}^* > 15\%$	Accept

We may now determine which project bundles to accept at differing investment budget levels. Bundles 1, 4, and 5 are dominated by less expensive bundles. (They are rejected.) Thus, the decision rules are

For Investment Budget (B) in the Range	Select Bundle	Component Projects
$B < \$10,000$	0	(None)
$\$10,000 \leq B < \$17,000$	2	G
$\$17,000 \leq B < \$27,000$	3	H
$\$27,000 \leq B < \$39,000$	6	G, H
$B \geq \$39,000$	7	F, G, H

Again, the same decisions can be obtained by maximizing net present value among competing bundles within the available budget. Rank the projects again by bundles in order of increasing total investment, along with their net present values ($i = 15\%$):

Bundle	Component Projects	Total Investment	Net Present Value of Bundle
2	G	−10,000	+4,025
1	F	−12,000	2,351
3	H	−17,000	12,119
4	F, G	−22,000	6,376
6	G, H	−27,000	16,144
5	F, H	−29,000	14,470
7	F, G, H	−39,000	18,495

Then the following decision table is obtained by maximizing NPV within the budget ranges.

Budget Range	Decision Method	Decision
$B < \$10,000$	—	Reject all
$\$10,000 \leq B < \$17,000$	max(P_2, P_1)	Accept 2 (G)
$\$17,000 \leq B < \$27,000$	max(P_2, P_1, P_3, P_4)	Accept 3 (H)
$\$27,000 \leq B < \$39,000$	max($P_2, P_1, P_3, P_4, P_6, P_5$)	Accept 6 (G, H)
$B \geq \$39,000$	max[all P_j]	Accept 7 (F, G, H)

Thus, maximizing the net present value of economically mutually exclusive sets (bundles) of projects signals the same decisions as does the theoretically correct marginal rate of return method. However, Fleischer's method of defining mutually exclusive bundles and determining incremental rates of return between bundles, breaks down in general for two reasons (Bernhard, 1971). First, the incremental cash flow stream between two projects may default to the no answer or multiple answer cases when solving for the incremental rate of return. Second, if some projects have expenditures later than t = 0, it may not be possible to order the projects by increasing investment.

However, under certain conditions (i.e., pure investments and single expenditures at t = 0 for all projects), the net present value criterion recommends the same projects as does the marginal rate of return method; thus, both assume that marginal rate of return equals marginal cost. This is an important economic consequence.

9.8.2 Inclusion of constraints. Constraints, whether budgetary or physical, reduce the size of the feasible set of projects. After the feasibility constraints are applied, the optimization criterion is applied to the remaining feasible set, as illustrated by Example 9.6.

Example 9.6 Constraints

a. Because real firms do not have unlimited funds available for investment, a capital expenditures limit or *capital budget is* applied to the overall expenditures for new projects. Often the capital budget is a fixed sum of money. For example, in selecting project bundles, the capital budget might be $28,000. The question then is, which economically mutually exclusive bundle that costs less than $28,000 should be selected? The solution for our example begins by eliminating bundles 5 and 7 from the feasible set. Maximizing net present value results in selecting bundle 6, consisting of projects G and H and having $P_6 = \$16{,}415$.

b. Mutual exclusivity constraints can also be easily handled. For example, suppose that projects G and H are mutually exclusive alternatives; then bundles 6 and 7 are eliminated from the feasible set. If the same $28,000 budget constraint is applied, the remaining feasible set contains only bundles 2, 1, 3, and 4, and the optimal solution is bundle 3.

c. Conditional relationships between projects can also be handled by constraints that reduce the feasible set. For example, suppose that project G requires the execution of F, but not vice versa. That is, F may be done alone. This condition removes from the feasible set any and all bundles that contain G without F, that is, bundles 2 and 6. With this constraint and the budget constraint, the feasible set includes only bundles 1, 3, and 4, and the optimal choice is then bundle 3.

While other types of constraints can be designed to match other situations, these three illustrate how all such constraints reduce the size of the feasible set of projects. They also illustrate the basic nature of the project selection problem—that of selecting an optimal subset of projects from a larger, constrained set of candidate projects according to any objective of primary interest to the firm.

This rather roundabout route illustrates some of the problems involved with the ranking method of project selection. These problems are inevitably linked to the central reinvestment problem, to the problem of dependencies and constraints, and to the problem of project indivisibility. Under capital market assumptions that lead to an exogenously determined marginal investment rate, only Fleischer's bundling method of creating economically mutually exclusive alternatives, of all methods involving direct ranking of projects, leads to the selection of an optimal set of projects. Weingartner showed this in 1963 in his pioneer work on the application of mathematical programming to the capital budgeting problem. Thus, with this exogenous rate the more promising approach is to use mathematical programming. As we shall see, 0–1 integer programming successfully handles simultaneously the problems of capital rationing, project dependency, and project indivisibility, for the simple deterministic (certainty) model under perfect capital market assumptions.

We should note that the Lorie-Savage project selection problem, if it incorporates only monetary budget constraints, is trivial under perfect capital market assumptions, since the firm can borrow the funds to execute all projects having positive net present values. As Weingartner points out, however, other "budget" limits may be imposed based on space, manpower, throughput, machine capacity, and managerial availability. These constraints are formulated just as monetary constraints are, but not in money terms, so the nature of the Lorie-Savage formulation may not be changed, but its interpretation is. From this point of view, the Lorie-Savage formulation is of considerable interest.

One of Weingartner's major contributions was to demonstrate that integer programming correctly solves the Lorie-Savage problem, whereas the solution method provided by Lorie and Savage (1995) themselves fails under certain conditions. Lorie and Savage call for a trial-and-error choice of positive-valued parameters, $p_1, p_2, \ldots, p_t$, such that:

$$P_j - \sum_{t=1}^{T} P_t C_{tj} \tag{9.6}$$

is positive or zero for chosen projects and negative for rejected projects, where P_j is the net present value of project j, and C_{tj} is the present value of the outlays for project j in period t. The cause for failure of the criterion, Equation (9.9), is that no solution to it may be obtainable, as demonstrated by Weingartner.

9.9 THE WEINGARTNER FORMULATION

9.9.1 Objective function. The basic Weingartner (1963) formulation of the budget allocation problem maximizes a linear function of the present values of the cash flows of accepted projects. In our terminology, the objective is:

$$\underset{\forall j}{\text{Max}} \sum_{j=1}^{m} \sum_{t=0}^{N_j} Y_{tj} (1+i)^{-t}(x_j) \tag{9.7}$$

where Y_{tj} = cash inflow (outflow) from (to) the jth project at the end of the tth period; cash flows are + if inward from and − if outward from the project
m = the number of projects in the candidate set
N_j = the life of the jth project
i = the firm's marginal investment rate
x_j = the decision variable (which takes on values of 0 or 1 only)

In words, a subset of projects is to be selected by the decision variables, x_j, from a set of m candidate projects to maximize the net present value of the future economic contributions to the firm.

The choice of net present value as a selection criterion is deliberate. It provides, under certain conditions, the same answer as does the incremental method. Moreover, under assumptions of certainty and a perfect capital market (such as we have here), maximizing the firm's net present value also maximizes the firm's wealth (shown by the Fisher-Hirshleifer derivation of Chapter 7). Neither of the other nonincremental criteria (internal rate of return and benefit-cost ratio) will ensure wealth maximization.

One disadvantage is inherent in the net present value solution to the resource allocation problem. The discount rate, called the *marginal investment rate* in Equation (9.7), is *external to the model*. Usually the firm's marginal cost of capital, as calculated in Chapter 6, is used as a base for estimating the marginal investment rate. When this is done, then a conservative *reinvestment policy* is also established, since Section 9.6 the comparison of projects by NPV implicitly assumes reinvestment of released funds at the firm's marginal cost of capital. Practically, the implicit assumption is that the firm can continue to originate and adopt other new projects in the future that will absorb the released funds and in turn provide positive present values. This is not a stringent assumption, but it requires constant monitoring of ongoing activities to make certain that their rates of return exceed the firm's cost of capital or that the activities are canceled or divested.

If the assumption of a perfect capital market is modified, so that borrowing and lending interest rates are no longer equal, then the net present value formulation of the objective becomes meaningless. For example, there are formulations, such as Sharp and Guzman-Garza (1981), which explicitly consider the firm's capital structure and its impact on the cost of capital (see Chapter 6). Chapter 10 will address how some departures from the perfect capital market can be handled.

9.9.2 Constraints

a. The first set of constraints due to resource availabilities, by a series of inequalities:

$$\sum_{j=1}^{m} C_{tj}(x_j) \leq B_t \tag{9.8}$$

Where C_{tj} = the consumption of resource B by project j in the *t*th period.

B_t = the resource availability of the specified budget in the *t*th period

This constraint is often called the budget constraint since it is used to express the limited availability of investment capital for executing investment projects. For example, if only \$100,000 and \$75,000 in investment capital were available in periods 1 and 2, respectively, then there would be two such budget constraints:

$$\sum_{j=1}^{m} C_{1j}(x_j) \leq 100{,}000 \quad \sum_{j=1}^{m} C_{2j}(x_j) \leq 75{,}000.$$

Budget constraints need not be limited to the availability of money, since they can also be used to express the limited availability of manpower, construction crews, materials, and equipment (see Weingartner, 1963, pp. 125–127).

b. If several alternatives for a project or projects exist, then one or more mutual exclusivity constraints are required. The inequality of Equation (9.9) specifies that the projects included within it are mutually exclusive and that only one may be selected:

$$x_a + x_b + \cdots + x_k \leq 1. \tag{9.9}$$

Thus, if $a, b, \ldots, k$ are all alternative methods of doing a project, then each is made a separate project and the inequality in Equation (9.9) permits at most one to be selected. Thus, if any decision variable, say x_a, equals 1 in the selection process, then all others in this mutually exclusive set, $(x_b + \cdots + x_k)$ must equal 0.

As an example, suppose that in a candidate set of 25 projects, projects 9, 16, and 22 are alternative developments of a given tract of land. Only one can be done. Hence, the mutual exclusivity constraint would be $x_9 + x_{16} + x_{22} \leq 1$, which permits none or one of the three projects to be selected.

c. Two projects are contingent, if there is a conditional dependency between them. Project a is conditionally dependent on b when a's execution, although optional in itself, is operationally, functionally, or economically dependent on the execution of b. Project a is optional, but if selected then b must also have been selected.

Such dependencies are formulated as $x_a \leq x_b$ or, upon subtracting x_b from both sides,

$$x_a - x_b \leq 0 \tag{9.10}$$

where a is the dependent project and b is the one depended on. If $x_b = 1$ (acceptance of b), then either $x_a = 0$ or $x_a = 1$ is permitted, thus satisfying the constraint. With $x_b = 0$ (rejection of b), however, $x_a = 0$ is also required, since $x_a = 1$ is prohibited by the constraint. So a is accorded conditional acceptance, the condition being that b is accepted simultaneously. Neither, however, is required to be accepted (both $x_b = 0$ and $x_a = 0$ is a solution).

d. Sometimes two projects are strict complements of one another. In other words the execution of one requires the execution of the other. Thus, if projects c and d were strict complements, then the constraint is

$$x_c - x_d = 0$$

which requires that x_c and x_d both be zero-valued (neither accepted) or both unit-valued (both accepted). Instead of including this kind of constraint, the combined project, cd, can replace the component projects c and d. Thus, the constraint written above can be eliminated and the problem size is reduced. However, problem clarity and flexibility for considering other scenarios is also reduced. Thus, in the age of cheap computers the two-variable formulation is probably better.

e. Sometimes projects undertaken together have cash flows that are greater or smaller than the sum of the cash flows of the projects individually. For example, if projects e and f can be undertaken together as project *ef* with a net economic benefit, then e and f are complementary. With this condition, projects *e*, *f*, and *ef* are mutually exclusive—that is, we do not simultaneously undertake e and *ef*, f and *ef* or even e and *f*. The appropriate constraints are

$$x_e + x_{ef} \leq 1$$
$$x_f + x_{ef} \leq 1 \quad (9.11)$$
$$x_e + x_f \leq 1$$

or more simply $x_e + x_f + x_{ef} \leq 1$. In essence, *ef* is created with a new cash flow stream that expresses the economic cost or savings to be realized from undertaking the combined project, and then this combined project is made mutually exclusive with the parent projects.

f. The last constraint is the project indivisibility constraint

$$x_j = 0,1 \quad j = 1,2,\ldots\ldots,m \quad (9.12)$$

which requires that all of a project be selected ($x_j = 1$) or that none of a project be selected ($x_j = 0$). Fractions or partial acceptance of projects are not permissible.

The indivisibility constraint has technical and economic effects. First, this constraint converts what would otherwise be a linear programming problem into an integer programming one. This has implications for the solution methodology since integer programming is much more difficult than linear programming. Second, from an economic standpoint, this constraint permits the substitution of, say, two smaller projects for one larger project. This substitution effect is the basic economic advantage enjoyed by mathematical programming over ranking as a method of project selection.

9.9.3 The completed Weingartner model. The mathematical programming problem of the budget allocation type can be formally stated by combining Equations (9.7–9.12); thus,

$$\underset{\forall j}{\text{Max}} \sum_{j=1}^{m} \sum_{t=0}^{n} Y_{tj}(1+i)^{-t}(x_j) \quad (9.13)$$

subject to

$\sum_{j=1}^{m} c_{tj}(x_j) \leq B_t$ \quad (budget constraints; $c_{tj} \geq 0$)

b) \quad $x_a + x_b + \cdots + x_k \leq 1$ \quad (mutual exclusivity constraints)
$x_a - x_b \leq 0$ \quad (contingent projects)

$x_e + x_f + x_{ef} \leq 1$ \quad (complementary or overlapping projects)

$x_j = 0,1$

In addition there may be multiple constraints of each type. In fact there must be *m* indivisibility constraints. This is a 0–1 integer programming problem in which all relationships are linear, and which can be solved by any of several solution methods. Because the 0–1 or binary variables are a special case of integer programming, the most efficient solution methods are some type of branch-and-bound procedure. The necessary assumptions underlying this formulation of the simple capital budgeting problem are

1. All cash flows, Y_{tj}, and resource usages, c_{tj}, are deterministic (expected values measured with zero variance).
2. The discount rate, i, is the assumed known marginal investment rate of the firm, as estimated by the firm's marginal cost of capital.
3. Risk is zero for all projects in the candidate set.
4. There is no alteration of a project in the optimal set once the optimal set is determined (selected).
5. Reinvestment of all cash flows occurs at the firm's marginal investment rate.

9.9.4 Constrained project selection using Solver. The Weingartner formulation can also be performed with the aid of a personal computer using a variety of available software. A widely available program is Solver, available as an add-in for Microsoft Excel. The same information that is required to perform the Weingartner formulation may be put into a spreadsheet and quickly solved by the computer.

Example 9.7 Solver
Using the information from Example 9.6, a spreadsheet is constructed, as shown in Figure 9.4.

	B	C	D	E
2				
3	Project:	F	G	H
4	Investment	−12,000	−10,000	−17,000
5	Annual inflows	4281	4184	5802
6	Life, yr.	5	5	10
7	rate	0.15	0.15	0.15
8	Budget limit	30000		
9	NPV	2351	4025	12119
10	x	0	1	1
11	Portfolio NPV	16,144		
12				
13	Maximize C11			
14	Constraints			
15	Budget <=C8	27000		
16	x binary			

Figure 9.4 Spreadsheet for budget allocation problem

The input data is shown in the data block at the top of the sheet. The NPV is calculated for each of the three projects, F, G, and H. Each project is also given a factor x; either a 1

276   Chapter 9, Methods of Selection Among Multiple Projects

if it is chosen, or a 0 if it will not be chosen. The Portfolio NPV is simple the sum of each project multiplied by its x factor:

Portfolio NPV = {C8}*{C9} + {D8}*{D9} + {E8}*{E9}

If a project is not chosen as part of the portfolio, its x factor will be 0, and the project NPV will not be included in the portfolio total. The objective is to maximize the portfolio NPV (as before) with any required constraint. For the present example, we will simply constrain the total budget. Also, each value of x is restricted as a binary number, either 0 or 1. The Solver window is shown in Figure 9.5.

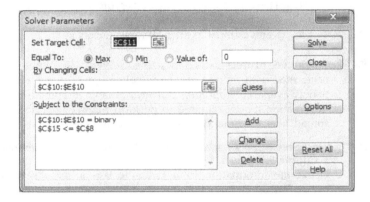

Figure 9.5 Solver window

In this way, simple constrained problems may be quickly solved. The budget constraint may be easily changed within the spreadsheet, and Solver may be repeated, to obtain different responses for different budget limits. In addition, any of the constraints outlined in Section 9.9.3 may be applied to the problem.

9.10 CONSTRAINED PROJECT SELECTION BY RANKING ON IRR

9.10.1 The opportunity cost of foregone investments. In previous sections we have focused on the development of models where the minimum attractive rate of return is determined externally to the project selection problem. In essence these models have been based on perfect market assumptions, so that the cost of capital can be used as the reinvestment assumption. In Chapter 10 we will expand on this model, relying on the tools of mathematical programming, and later in Chapter 16 we will account for the risk level of different projects through the capital asset pricing model.

In this section we discuss an alternative approach to the single-constraint budget allocation problem. This approach assumes that the minimum attractive rate of return is based on investment opportunities that must be foregone because of the capital constraint. This capital constraint is imposed by the firm's management and it intersects with the investment opportunity schedule to determine which projects should be undertaken and the firm's reinvestment rate for this period (which is assumed to apply in future periods as well). For simplicity in presentation, we have assumed that all projects are pure investment and not loan proposals.

The opportunity cost of foregone investments is cited by many authors as one basis for the minimum attractive rate of return. However, this development is considerably more complete. In Section 9.10.4 the conflicting assumptions of the cost of capital approaches and this IRR approach are clarified.

Most beginning finance texts also point out that IRR can sometimes not be determined but that NPV always can. Examples are often a variation on the Lorie-Savage oil pump problem. While it is technically true that an NPV answer can be found, the result is often not meaningful when there are multiple results for IRR.

9.10.2 Perfect market assumptions. Our first step is to review the necessary assumptions that underlie Figure 9.6, which depicts the perfect market model:

1. Investments are available in divisible amounts, so that a continuous function rather than a step function can be developed.
2. Investments are independent so that project interactions can be ignored.
3. Investment opportunities and financing sources are stable over time so that lows and highs do not have to be considered.
4. Risk is nonexistent; that is, the cash flows are deterministic for all projects.
5. Individual projects are *pure* investments so that complications introduced by multiple sign changes in the cash flow diagram are avoided.
6. Projects are available for simultaneous evaluation, at least for the determination of the appropriate discount rate.
7. There are no transaction costs for investing or financing.

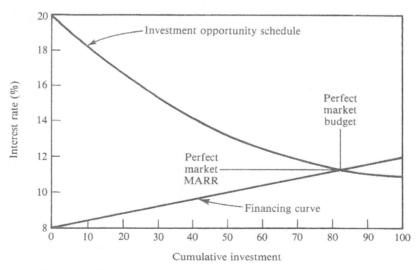

Figure 9.6 Perfect market model for constrained project selection

The financing curve in Figure 9.6 shows that as a firm borrows additional funds, the cost of financing increases. This is because lenders view high debt loads as risky. The investment opportunity schedule shows that as the firm invests money, higher interest investments are used first, and as they become unavailable lower interest investments are made.

Under the assumptions listed above, the intersection of the financing and investing curves in Figure 9.6 is clearly the point of optimal operation. At any total capital investment to the left of the intersection, the firm clearly could borrow funds, invest at a higher rate, and increase total income. To the right of the intersection decreases in the total investment would also increase total income. This figure is the basic justification for the Fisher's separation theorem, which says that investment and financing decisions rely on different criteria, and so can be separated (Fisher, 1930).

Neither the investment nor the financing curve will be truly smooth. However, theoretically, the divisibility and continuity assumptions are required to guarantee that the intersection point exists. If it exists, it defines an interest rate that can be used in evaluating the firm's investment opportunities. For whatever project (or incremental comparison) that is evaluated, capital is available at that rate to finance it.

9.10.3 Internally imposed budget constraint. However, consider the situation where the firm applies a capital constraint to the left of the perfect market intersection, as in Figure 9.7. The firm might choose to do this for any of the following reasons:

1. To compensate for the firm's higher risk, since the lender often has a secured claim on the firm's assets
2. To allow a margin to protect against overly optimistic estimates
3. To concentrate the firm's attention on the best projects
4. To allow more flexibility to pursue new projects

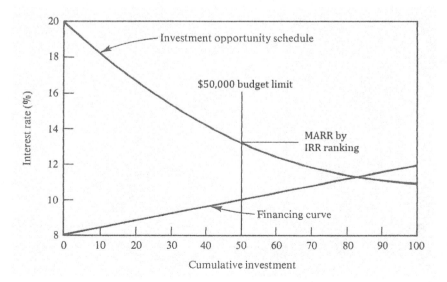

Figure 9.7 Ranking by IRR and constrained project selection

We note that Figure 9.7 is based on the same assumptions listed in Section 9.10.2

for the perfect market model, except for the following:

4. Only equivalent risk levels among the projects need by assumed, rather than completely known outcomes for each.
7. It is no longer necessary to assume no transaction costs, so long as the budget limit is far enough away from the intersection point of Figure 9.7.

With these assumptions, the capital constraint defines the best project that cannot be undertaken. And, due to assumption 3 this will also apply in future periods. All funds released by an accepted project will be reinvested in projects that return at least this rate of return. As before, the minimum attractive rate of return defines (or is defined by) the reinvestment assumption. Now, however, this is not based on the weighted average cost of capital. Instead, ranking by IRR clearly specifies that the firm can choose

1. the MARR and let that determine the capital budget,
2. the capital budget and let that determine the MARR, or
3. an interactive or iterative version of (1 and 2).

Because projects are not divisible and the investment opportunity schedule is not continuous, we must consider the possibility of defining the minimum attractive rate of return based on either the last *fundable* project or the first *foregone* project. Also, for firms whose investment decisions are made piecemeal over a year, we suggest the following. With reasonable stability in investment levels and the quality of proposed projects, data from the previous year can be accumulated and ranking on IRR used to define the minimum attractive rate of return.

9.10.4 Contrasting IRR and WACC assumptions. Figure 9.8 contrasts four different bases for the minimum attractive rate of return. The opportunity cost of foregone investment is above the various cost of capital measures, and if the capital constraint is even moderately severe, the difference is substantial.

Not only is the marginal cost of capital at this level of investment below the opportunity cost of foregone investments, it is even below the limiting interest rate defined by the perfect market model. The weighted cost of capital will be even lower than the marginal cost of capital.

Thus, ranking on IRR and the opportunity cost of foregone investments imply that discount rates that are based on the cost of acquiring funds are often far too low. Many PW analyses are flawed because they implicitly assume that future projects will return only the cost of capital rather than the much higher rate that is the opportunity cost of foregone investments. The chief requirement here is that the firm's investment opportunities possess a minimal degree of stability.

Since the cost of capital models are based on empirical models about returns to stockholders and lenders, a reasonable question would be why the higher rate implied by the opportunity cost of foregone investments is not reflected in those returns. One possible explanation is that the difference is absorbed by projects that do not provide the returns that are estimated when they are considered. A second explanation (Grant et al., 1990, p. 334) is that some returns of accepted projects are distributed to a firm's employees and customers in the form of higher wages, bonuses, and improved products.

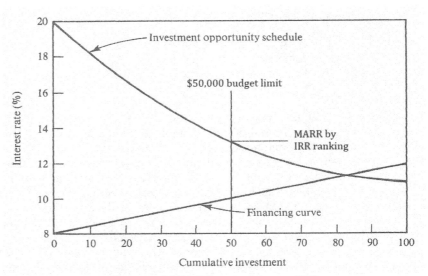

Figure 9.8 WACC vs. ranking on IRR

If ranking by IRR is required, why should NPV be considered at all? The answer is, of course, that NPV has some advantages in practice. For example, NPV facilitates dealing with project interactions ranging from simple choices between mutually exclusive alternatives to complex mathematical programming formulations. NPV is numerically simpler to calculate, although spreadsheets and company specific evaluation packages make NPV and IRR equivalent in this regard. NPV seems conceptually easier for many to understand. NPV supports evaluation of projects with multiple sign changes. And finally, NPV is better suited to the many situations where continuous review of proposals, rather than periodic rankings, takes place.

Thus, the key is identifying an approximate value for the minimum attractive rate of return. If the rate is too high, then money is left over at the end of the budget cycle. Much more commonly, the rate is set much too low. Then far too many projects pass the minimal standards, and the wrong selection criteria are applied. We suggest that once the proper rate is identified, it is unlikely to matter much if projects that differ by small percentages are switched in and out of the accepted set. Thus, at this point the second-order refinements available through mathematical programming may make sense.

If the discrepancy between the IRR and the NPV approaches is still bothersome to you, let us suggest the following. Net present value is best thought of as measuring a project's value against an assumed standard. A project's intrinsic worth is measured by its cash flows and its IRR. Its present value is a measure against a certain expectation of the future. Evaluating a project at 10% or 25% does not change its intrinsic value; it only represents different expectations about reinvestment opportunities in the future.

9.10.5 Summary of ranking on IRR. These results have relied, as most theory does, on a simplified model of the world. This model does not include fluctuating investment climates, multiple criteria, risk, and multiple roots. However, the first author believes that the principal conclusion would be strengthened, not impaired, by a larger dose of reality. It is easier to consider an IRR than an NPV in a multiple-criterion evaluation procedure, the IRR is easier to incorporate into calculations of risk/return tradeoffs, and ranking approaches are superior to the absolute go/no-go standards of NPV analysis in fuzzy decision situations.

It is worth noting that when alternative decision criteria are analyzed by simulation (White and Smith, 1986), two incremental criteria based on IRRs and annual equivalent excess of revenues minus costs divided by first cost are evenly matched and both are superior to other techniques.

With this model, ranking on IRR is a necessary precursor for calculations of present worth, since it establishes the minimum attractive rate of return. Furthermore, this model suggests that the cost of capital should be relegated to a subordinate status that defines a minimum condition that is rarely relevant.

9.11 SUMMARY

We have considered some of the fundamental problems involved in selecting an optimal set of projects from a larger set of candidate projects. Dependencies among projects, budget and other resource limitations, and project indivisibility all introduce complications. We have demonstrated that three rational criteria (net present value, nonincremental internal rate of return, and benefit-cost ratio) all lead to potential ranking inconsistencies when two or more projects are compared due to differing assumptions concerning the reinvestment of idle funds. With the proper choice of a common reinvestment rate, for example, through incremental analysis, the ranking inconsistency problem is resolved.

This latter finding leads to the choice of the net present value method as the preferred method of project selection when the minimum investment rate is exogenous. This led to the Weingartner formulation of the simple, constrained Lorie-Savage problem as an integer programming problem. Under assumptions of certainty and perfect capital markets, this formulation leads to an optimal choice of projects, if the firm's discount rate for future funds is its assumed known marginal cost of capital (determined by the methods developed in Chapter 6).

Finally, we analyzed the basis for using ranking on IRR as the preferred method for constrained project selection. This method is theoretically justified only when the perfect capital market assumptions are not justified, and when the future is enough like the present to warrant the assumption that the reinvestment rate defined by the "budget" constraint is stable over time.

REFERENCES

BARNEY, L. DWAYNE, JR., and MORRIS G. DANIELSON, "Ranking Mutually Exclusive Projects: The Role of Duration," *The Engineering Economist*, **49**(1) (2004), pp. 43–61.

BEIDLEMAN, CARL R., "Discounted Cash Flow Reinvestment Rate Assumptions," *The Engineering Economist*, **29**(2) (1984), pp. 127–139.

BERNHARD, RICHARD H., "Discount Methods for Expenditure Evaluation: A Clarification of Their Assumptions," *Journal of Industrial Engineering,* **13**(1) (January–February 1962), pp. 19–27.

BERNHARD, RICHARD H., "A Comprehensive Comparison and Critique of Discounting Indices Proposed for Capital Investment Evaluation," *The Engineering Economist,* **16**(3) (Spring 1971).

BIERMAN, HAROLD, and SEYMOUR SMIDT, *The Capital Budgeting Decision,* 7th ed. (Macmillan, 1988).

BIERMAN, HAROLD, and SEYMOUR SMIDT,"Capital Budgeting and the Problem of Reinvesting Cash Proceeds," *Journal of Business* **30**(4) (October 1957), pp. 276–279.

BLOCK, STANLEY, "Capital Budgeting Techniques Used by Small Business firms in the 1990s," *The Engineering Economist,* **42**(4) (Summer 1997), pp. 289–302.

DEAN, JOEL, *Capital Budgeting* (Columbia University Press, 1951).

DUESENBERRY, JAMES S.*, Business Cycles and Economic Growth* (McGraw-Hill, 1958).

FISHER, IRVING, *The Theory of Interest* (Macmillan, 1930; reprinted, Kelley & Millman, 1954), pp. 151–155.

FLEISCHER, GERALD A., "Two Major Issues Associated with the Rate of Return Method for Capital Allocation: The 'Ranking Error' and 'Preliminary Selection,' " *Journal of Industrial Engineering,* **17**(4) (April 1966), pp. 202–208.

GRAHAM, JOHN R., and CAMPBELL R. HARVEY, "The Theory and Practice of Corporate Finance: Evidence from the Field," *Journal of Financial Economics,* **60** (2001) pp. 187–243.

HYYTINEN, ARI, and OTTO TOIVANEN, "Do Financial Constraints Hold Back Innovation and Growth? Evidence on the Role of Public Policy," *Research Policy,* **34** (2005), pp. 1385–1403.

LOHMANN, JACK R., "The IRR, NPV and the Fallacy of the Reinvestment Rate Assumptions," *The Engineering Economist,* **33**(4) (Summer 1988), pp. 303–330.

LORIE, JAMES H., and LEONARD J. SAVAGE, "Three Problems in Rationing Capital," *Journal of Business,* **28**(4) (October 1955).

MAO, JAMES C. T., "The Internal Rate of Return as a Ranking Criterion," *The Engineering Economist,* **11**(4) (Summer 1966), pp. 1–13.

SCHWAB, BERNHARD, and PETER LUSZTIG, "A Comparative Analysis of the Net Present Value and the Benefit–Cost Ratio as Measures of the Economic Desirability of Investments." *Journal of Finance,* **24**(3) (June 1969), pp. 507–516.

SHARP, GUNTER P., and ARTURO GUZMAN-GARZA, "Borrowing Interest Rate as a Function of the Debt–Equity Ratio in Capital Budgeting Models," *The Engineering Economist,* **26(4)** (Summer 1981), pp. 293–315.

SOLOMON, EZRA "The Arithmetic of Capital–Budgeting Decision," *Journal of Business,* **29**(2) (April 1956).

SOLOMON, EZRA, ed., *The Management of Corporate Capital* (The University of Chicago Press, 1959).

WEINGARTNER, H. MARTIN, *Mathematical Programming and the Analysis of Capital Budgeting Problems* (Ford Foundation Award-Winning Dissertation) (Prentice–Hall, 1963; also Markham, 1967).

WHITE, B. E., and G. W. SMITH, "Comparing the Effectiveness of Ten Capital Investment Ranking Criteria," *The Engineering Economist,* **31**(2) (Winter 1986), pp. 151–163.

WOODS, JOHN C., and MAURY R. RANDALL, "The Net Present Value of Future Investment Opportunities: Its Impact on Shareholder Wealth and Implications for Capital Budgeting Theory," *Financial Management* **18**(2) (Summer 1989), pp. 85–92.

PROBLEMS

9-1. Three projects have the following net cash flows shown in the table.
(a) Using incremental cash flows, calculate the Fisher intersection rates for Projects A − C, A − B, and B − C.
(b) Calculate the internal rates of return (IRR) for A, B, and C.
(c) Plot $P(i)$ versus i in the interval $0 \leq i \leq 0.40$ and show the Fisher rates and IRR's obtained in (a) and (b) above.
(d) On the graph, designate the dominant (accepted) project, for each interval of i.

Time	Net Cash Flow		
t=	A	B	C
0	$-12,000	$-6,000	$-6,000
1	+3,180	+2,200	+2,070
2	+3,180	+2,228	+2,100
3	+3,180	+2,257	+2,128
4	+3,180	+2,286	+2,157
5	+3,180	+2,315	+2,185
6	+3,180	+2,343	+2,214
7	+3,180	+2,372	+2,241
8	+3,180	+2,401	+4,270

9-2. Four proposed projects are estimated to develop initial investments, net cash inflows, and project lives as follows:

Project	Initial Investment at Time $t = 0$	Net Cash Flow, $ per Year for $t = 1, 2, \ldots, N$)	N = Project Life, Years
A	$20,000	$+10,000	3
B	25,000	+16,000	3
C	40,000	+30,000	5
D	50,000	+20,000	10

(a) Assume that the firm's marginal investment rate is 15%. Formulate the capital allocation decision as an integer programming model in which net present value is maximized, subject to the following constraints: (i) the available budget at $t = 0$ is $80,000; (ii) each project is indivisible, and multiples of projects cannot be executed.
(b) Using the Fleischer bundling technique, solve the problem defined in (a) to obtain the optimal solution

9-3. For the data and constraints defined in problem 9-2, do the following:
(a) Assume that projects A, B, and C are mutually exclusive alternatives. Write the necessary constraint(s) and determine which project(s) should be chosen to maximize net present value?
(b) Suppose that the budget constraint and the 0, 1 constraint in problem 9-2 are the only operative constraints. Also, suppose that the execution of project C is conditioned upon the execution of project B but the acceptance of C is not mandatory. Write the necessary constraint to show this conditionality,
(c) and then construct a decision tree showing which project bundles, if any, should be eliminated from the feasible set of project combinations. With this constraint operative, which project combination should be selected? Is this constraint tight or loose, and why?

9-4. Five projects are available to a company for an investment opportunity (see table below). The company has a budget limit of $75,000 at time $t = 0$ for this investment and a marginal investment rate of 15%. The five projects are assumed to be economically independent of each other. There is a mutual exclusivity constraint between B and C; however, the execution of project A requires execution of project D although accepting D does not require acceptance of A.

DATA FOR CANDIDATE PROJECTS

Project:	A	B	C	D	E
Total investment at $t = 0$	$5,000	$15,000	$20,000	$25,000	$50,000
Annual net cash flow, $/yr	$1,000	2,500	6,500	5,000	15,500
Life, years	10	10	5	10	5
Salvage value	0	0	0	0	0

(a) Enumerate the 2^5 possible project combinations, and then define the feasible set of project combinations. For each feasible combination, calculate the total initial investment.
(b) Using the incremental rate-of-return method, rank the feasible combinations and determine the optimal project combination.
(c) Repeat (b), using the net present value criterion.

9-5. Two projects are being considered by a firm as investment opportunities, and only one can be selected. The expected cash flow streams are given in the table below. Using the net benefit-cost ratio and net present value as the selection methods, select the best project. If there is an inconsistency between the two decision methods, resolve it. The firm's marginal investment rate is 15%.

End of Period, t =	Project A	Project B
0	$-35,000	$-10,000
1	-10,000	-5,000
2	+20,000	+10,000
3	+25,000	+15,000
4	+45,000	+20,000
5	+50,000	+15,000

9-6. A company must choose between two investment proposals for making repairs to a building. Proposal A requires an initial investment of $10,000 with a return of $3,200 for 5 years. Proposal B requires an initial investment of $30,000 with a return of $9,300 for 5 years. The marginal investment rate is 15%.
(a) Evaluate the alternatives using the NPV method.
(b) Evaluate the alternatives using the benefit-cost criterion.
(c) Do the two methods yield conflicting results? If so, why?

9-7. A firm is faced with four projects competing for the available investment funds. Project A is a coil steel cut-to-length production line. This line will feed steel from a coil, straighten it, shear it, and stack it. Project B is an electric furnace for the foundry. Project C is a gas furnace for the foundry. Project D consists of an electric trolley car, to be used to load coil steel onto the proposed cut-to-length line, rather than using an existing crane. Management has restricted capital investments to $850,000 in the current year. Next year's limit on capital investments is expected to be $750,000. The marginal investment rate is 15%. The expected cash flows related to the projects are as follows:

	CASH FLOWS (10^3 DOLLARS)			
Time t =	Project A	Project B	Project C	Project D
0	$-500	$-250	$-400	$-100
1	-600	-100	50	-40
2	200	150	200	20
3	500	350	200	180
4	700	0	100	100
5	300	0	100	0

Formulate the problem as an integer program, and solve it using maximization of net present value as a criterion.

9-8. The Wichita Worms are a professional sports team. They are currently in a rundown arena. Attendance is dropping and the franchise appears headed for financial difficulty. Several alternatives are available to the owner: (i) leave the arena as is; (ii) remodel the present arena; (iii) build a new arena; (iv) make a trade for Joe Jock, superstar; (v) sell the franchise to a franchise-hungry Texas oilman. The owner has stated that if he leaves the arena as is, he will make a trade for Joe Jock. The following are the cost data (at times $t = 0$ and $t = 1$):

CASH EXPENDITURES (10^3 DOLLARS)		
Project	$t = 0$	$t = 1$
i	$ 10	$ 20
ii	60	40
iii	200	150
iv	125	130
v	200	200

A financial consultant has estimated that the net present values of the alternatives are, respectively, (i) $100,000, (ii) $175,000, (iii) $275,000, (iv) $150,000, and (v) $300,000. The capital budget available in the next 2 years is expected to be $200,000 for each year. (a) Write an equation to maximize a net present value, subject to the applicable constraints. (b) Generate the feasible solutions and find the optimal solution by maximizing net present value.

9-9. The Midwest Manufacturing Company is preparing its capital budget for the next 2 years. It has the following projects available:

Project	Investment at Time $t = 0$	Investment at Time $t = 1$	Expected Cash Flows
A	$-350,000	$ –	$ 90,000 for t = 1 to 9
B	−500,000	−300,000	$170,000 for t = 2 to 16
C	–	−300,000	$ 80,000 for t = 2 to 8
D	−300,000	−300,000	$140,000 for t = 2 to 10
E	−100,000	–	$ 45,000 for t = 1 to 6

Company financial analysts have determined that Midwest can provide $900,000 (at $t = 0$) and $700,000 (at $t = 1$) in the next 2 years for capital investment. The marginal investment rate is estimated to be 15%. (The cash flow generated from any accepted projects will not be available for investment in other projects of this set.) The design engineers have specified that project C is an extension of project B and hence cannot be undertaken separately. Decide which project(s) Midwest should accept by using the maximum net present value criterion.

9-10. Economists generally advocate the use of marginal rate of return on marginal investment as an effective and theoretically correct method for ranking investment opportunities (projects) in decreasing order of merit, to determine project acceptability. The ranking of acceptable projects is generally cut off with the project whose marginal rate of return is equal to the firm's marginal investment rate. Explain why it may not always be possible to use this method and, if used, why it may not result in an optimal set of projects.

9-11. Under what assumptions will the IRR, NPV, and B/C criteria rank all projects in the same order of priority?

9-12. If the projects in problems 9-2 and 9-3 are all independent and there is no budget constraint, how would you qualify your answers in problem 9-2(a) and (b)?

9-13. A manufacturer is considering the following projects to improve its production process. Because of a very rapid pace of development, a two year horizon is used in evaluation. The capital budget to fund the projects' first costs is $450,000.

(a) Rank the following projects from best to worst using the present worth index, payback period, and internal rate of return.
(b) Based on the opportunity cost of foregone investments, what is the MARR? Which projects should be done?
(c) If a reinvestment rate of 15% is assumed, which projects should be selected based on a present worth index?
(d) If payback is used, which projects should be selected?

Project	First Cost	Annual Return	PW at 15%
1	$195,000	$140,000	$32,599
2	297,000	205,000	36,270
3	85,000	60,000	12,543
4	33,000	24,000	6,017
5	225,000	150,000	18,856
6	135,000	100,000	27,571
7	390,000	255,000	24,556

9-14. In problem 8-13 using Fleischer's bundling approach, which projects should be selected?

9-15. Cold Iron Foundry is considering the following projects to improve their production process. Corporate headquarters requires a minimum 10% rate of return. Thus, all projects that come to the plant's capital screening committee meet that rate. The foundry can afford to do several projects, but not all, as the capital budget is $550,000.

(a) Rank the following projects from best to worst using the present worth index, payback period, and internal rate of return.

(b) Based on the opportunity cost of foregone investments, what is the MARR? Which projects should be done?

(c) If a reinvestment rate of 10% is assumed, which projects should be selected based on a present worth index?

(d) If payback is used, which projects should be selected?

Project	First Cost	Annual Return	Life (years)	PW at 10%
1	$230,000	$ 50,000	15	$150,304
2	320,000	70,000	10	110,120
3	110,000	40,000	5	41,631
4	50,000	12,500	10	26,807
5	240,000	75,000	5	44,309
6	160,000	33,000	20	120,948
7	420,000	140,000	5	110,710

9-16. In problem 9-15 using Fleischer's bundling approach, which projects should be selected?

9-17. How would your answers to problem 9-15 change if the budget were $500,000? Which projects should be done? What is the approximate MARR based on the opportunity cost of foregone investments?

9-18. A division of a firm is considering the following projects, under an executive policy that mandates a discount rate of 15%. Thus, all projects have had their present worth indexes computed at that rate. However, the available funding is likely to be a third of the $430M requested.

(a) Rank the following projects from best to worst.
(b) Choose about $225M worth of projects to recommend.
(c) What reinvestment assumption have you made and why?

Project	First Cost ($M)	Annual Return ($1000)	Life (years)	Present Worth Index at 15%	PW at 15%
1	$90	$20,000	20	1.39	$35,187
2	20	5,000	10	1.25	5,094
3	50	10,000	15	1.17	8,474
4	20	8,000	5	1.34	6,817
5	50	12,500	10	1.25	12,735
6	20	5,000	15	1.46	9,237
7	70	12,500	20	1.12	8,242
8	50	15,000	10	1.51	25,282
9	60	20,000	5	1.12	7,043

9-19. In problem 9-18, using Fleischer's bundling approach, which projects should be selected?

9-20. Redo problem 9-15 after-taxes. Assume the firm's tax rate is 34%, and that a 7% after-tax interest rate is used. Assume that all of the first cost must be depreciated. (a) Use straight-line depreciation. (b) Use SOYD depreciation. (c) Use MACRS depreciation, assuming that the project's life is the recovery period.

PART THREE

Investment Analysis Under Risk and Uncertainty

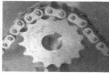

10
Optimization in Project Selection
(Extended Deterministic Formulations)

10.1 INTRODUCTION

Chapter 9 presented Weingartner's mathematical programming formulation of the constrained project selection problem. Under perfect capital market conditions, where it is assumed that funds can be borrowed to accept all projects with positive NPVs, monetary budget restrictions can be relaxed. Thus, the budget allocation problem was shown to be a nontrivial one only when it contains nonmonetary budget constraints (for example, limitations on manpower, space, and similar items).

However, the perfect capital market assumption is a fairly restrictive one. Thus, this chapter discusses modeling the capital allocation problem under the imperfect capital market assumption. Factors in the economy that may invalidate the perfect capital market assumption include trading costs, personal income taxes, and borrowing and lending interest rates that differ. Under these conditions the net present value criterion is invalidated, and maximizing NPV no longer selects an optimal set of projects.

It was principally this problem that led Weingartner to develop alternative formulations of the constrained project selection problem (1963), as well as the mathematical programming solution to the budget allocation problem. Since Weingartner's original work was published, project selection modeling has developed along two lines: deterministic and stochastic models. Stochastic models have been based on the Bernoulli principle or the expected utility hypothesis (which we shall develop in Chapter 11 and apply in Chapters 12 and 16), as well as assuming that input data from Wieingartner's models take on probability distributions (Sarper, 1993).

Deterministic modeling was discussed in Chapter 9 and will be considered further here. Beginning with Weingartner's basic horizon model (1963), additional early contributions include Weingartner, 1966; Baumol and Quandt, 1965; Bernhard, 1969; Byrne, Charnes, Cooper, and Kortanek, 1967; Manne, 1968; and Näslund, 1966. Most of this later research represents special cases of a generalized model developed by Bernhard (1969), which we shall examine. Since this time, there have been many articles written, analyzing the budget allocation problem using a variety of mathematical programming techniques. These include linear programming (such as Kumar and Lu, 1991; Chui and Park, 1998; and Kira et al., 2000), goal programming (such as Taylor et al., 1982), integer programming (such as Glover, 1975; and Meier et al., 2001), dynamic programming (such as Nemhauser and Ullmann, 1969; and Carraway and Schmidt, 1991), chance-constrained programming (such as Sarper, 1993; Kira et al., 2000; and Gurgur and

Luxhoj, 2003), and robust optimization (Kachani and Langella, 2005). There have also been literature reviews on the subject, such as White (1990) and Heidenberger and Stummer (1999).

This introduction suggests the chapter's organization. First, as a representative case of market imperfection, borrowing and lending interest rates are assumed to differ. This leads to Hirshleifer's demonstration that net present value and the separation theorem break down. Second, we shall consider Bernhard's generalized alternative formulation of the deterministic capital allocation problem. Third, goal programming is introduced to solve the capital allocation problem in the face of conflicting goals.

10.2 INVALIDATION OF THE SEPARATION THEOREM

Capital market imperfections, which we can expect in real markets, invalidate the separation theorem developed in Chapter 7. Choosing an optimal set of projects is no longer independent of the financing method. To determine net present value, we need to know the market rate of interest, or at least a MARR, but this may depend (in imperfect markets) on the financing method used, which in turn depends on the set of projects chosen. Therefore, the NPV cannot be defined, and the mathematical programming formulation of the budget allocation problem becomes useless. This is a direct consequence of relaxing the perfect capital market assumption.

The invalidation of the separation theorem is discussed at length in Ravid (1988). He focuses on how the firm's production and financial decisions interact, including the tax advantages of debt financing, bankruptcy risk, agency theory's relevance to the debt/equity mix, and the impact of information availability to management and shareholders.

Here we focus on the impact of capital constraints on the separation theorem. Hirshleifer (1958) extended the Fisher investment analysis to include different market borrowing and lending rates. As in Sections 7.6.1 and 7.6.2, where the separation theorem was developed, we rely on a two-time, one-period model (t = 0, 1) in which present income, Y_0, can be traded off for future income, Y_1. Also, we assume a set of intertemporal indifference curves (for example, an expected utility of consumption function) for the firm and a generalized type of production opportunity set (that is, with decreasing marginal productivity, but otherwise not of specific mathematical shape). The firm operates in a financial market where the lending rate of interest is i_l, which is less than the borrowing rate of interest, i_b.

Figure 10.1 describes three cases, but in only two (Figure 10.1a and b) do specified types of financial trading support an optimal solution. In the case of Figure 10.1c, the interest rate is found only after defining the optimal set, and it is this case that deals NPV its death blow.

First, consider Figure 10.1a, where the firm has a mediocre potential productive opportunity set, curve $W'X'$. The firm will move upward from point W (the assumed initial endowment) to R, the point of tangency between the lending rate of interest and the productive opportunity curve. In doing this, the firm will invest $(Y_W - Y_R)$ dollars. But it can increase its expected utility by lending $(Y_R - Y_A)$ dollars externally, arriving at point A', which is the point of maximum utility.

Suppose now that the firm has a superior productive opportunity set, such as curve $W'X'$ in Figure 10.1b. The firm will now invest $(Y_W - Y_{R'})$ dollars of its present endowment to reach point R', which is the point of tangency of the market borrowing line with the productive opportunity curve. This maximizes the exchange of present for future funds by investing in productive assets. Thereafter, the firm will borrow, at the market borrowing rate of interest, to reach point A', the point of maximum utility, without changing its NPV, $Y_{P'}$.

The point of Figure 10.1a is that with weak productive opportunity sets the firm should invest some funds in the potential productive assets and lend funds at i_l, whereas in Figure 10.1b, with strong productive opportunity sets, the firm should invest more in the productive assets and then recover its liquidity by borrowing at i_b.

There is a third case, where it is not clear what interest rate should be used or how the projects should be financed. Consider Figure 10.1c. Here the firm can choose from a set of productive opportunities, curve $W''X''$, which is tangent to the indifference curves, at an angle $\theta_i - \tan^{-1}(1 + i^*)$, which is between the tangents defined by θ_b and θ_1:

$$[\theta_1 = \tan^{-1}(1 + i_1)] < \theta_1 < [\theta_b = \tan^{-1}(1 + i_b)]$$

In this case, the rate for discounting Y_1 to a present value is unclear. In fact, this rate (i^*) cannot be discovered until the optimal solution is attained and so is of no assistance in reaching the solution. Because the discounting rate, i^*, depends on the optimal set of projects, NPV cannot be used to discover the optimal set, since NPV depends on a prior knowledge of the discount rate. So, abandoning the perfect capital market assumption leads to an inability to use NPV to select an optimal set of projects.

Finally, the separation theorem is invalidated by abandoning the perfect capital market assumptions. Because the discounting rate is not a previously known value but rather is an outcome of the optimal solution, one cannot say in advance how the optimal projects should be financed. In fact, determining the optimal set of projects and its method of financing is a joint problem in imperfect capital markets. In an imperfect capital market, the firm's present value [$Y(i^*)$ in Figure 10.1c] results from jointly determining the optimal set of projects and its method of financing. In short, it is the firm's time-preference expectancies that fix both the optimal set and the financing method. These expectancies are expressed in the expected utility functions (indifference curves). Operationally, this is not much help since one cannot easily determine the firm's indifference curves for the intertemporal exchanges[1].

10.3 ALTERNATIVE MODELS OF THE SELECTION PROBLEM

10.3.1 **Weingartner's horizon models**. The demonstrated breakdown of the NPV criterion and the associated invalidation of the separation theorem led Weingartner (1963) to develop the horizon models, which also use mathematical programming. These formulations follow the earlier lead of Charnes, Cooper, and Miller (1959) that used the firm's cost and revenue relations as data inputs, leaving some of the stream's elements and the internal discounting factors to be determined by the model. Weingartner's formulations concentrated on defining the firm's cash flow stream relationships by periods up to a

---

1. Bernhard (1971) investigates using a utility function instead of a discount rate. He demonstrates by linear programming that an optimal solution can be obtained, but the optimal use of resources (via the dual) depends on the original specification of the utility function.

finite time horizon, and the models were built using current (not present) values of the cash flows and resource constraints.

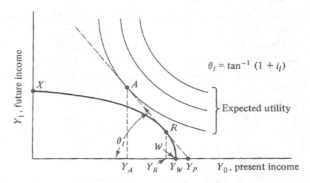

(a) Productive-exchange alternatives requiring lending

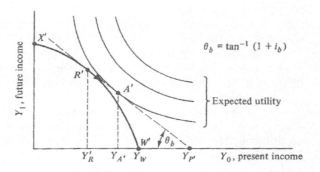

(b) Productive-exchange alternatives requiring borrowing

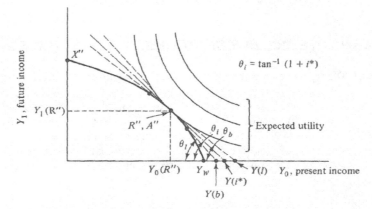

(c) Productive-exchange alternatives internally financed

Figure 10.1 Production-financing relationships in imperfect capital markets

Using this approach, Weingartner developed two forms of the capital allocation problem under the assumptions of certainty. First, the basic horizon model maximized the firm's terminal wealth, subject to restricted intertemporal transfer of funds at the assumed constant, but unknown, interest rate. Using only independent projects and by examining the linear program's dual he demonstrated that this formulation was equivalent to maximizing NPV at the assumed constant but unknown borrowing and lending interest rates. A surprising result of the basic horizon model is that, if mutually exclusive and contingent projects are considered, the optimal solution may include some project(s) with negative NPVs while rejecting other projects with positive NPVs (see Example 8.2, pp. 287–289 in Park and Sharp-Bette, 1990). The reason is that the time availability of internally generated funds is an important consideration.

Further forms of the horizon model developed by Weingartner extended the basic model. In these latter models, Weingartner relaxed the assumptions of a perfect capital market, by formulating terminal wealth models in which he (1) incorporated absolute limits on borrowing, (2) demonstrated that the firm's financial structure depends on its dividend policy when capital is restricted, (3) provided for a borrowing rate of interest that is a rising function of the amount borrowed, and (4) applied constraints on short-term borrowing to force (and determine) an optimal equity financing policy.

These are all important extensions to the theory of optimal resource allocation, especially since Weingartner provided numerical examples and solution methods for all his models, both with and without interdependencies between projects. Even though each model is worth reviewing from a historical standpoint, we do not because these models are special cases of the Bernhard generalized horizon model (examined next).

As summarized by Hayes (1989) the pure capital rationing problem can be reduced to the problem of maximizing terminal wealth. This development has not been without controversy, however, as evidenced by an earlier article by Hayes (1984) and related comments (Baum, 1985; Freeland and Rosenblatt, 1985). These papers, as well as others (including Freeland and Rosenblatt, 1978 and Rosenblatt, 1985), focus on the relationship between the discount rates and the dual variables.

10.3.2 The Bernhard generalized horizon model. Bernhard (1969) first formulates a generalized deterministic and mathematical model for capital budgeting, which we shall summarize. Second, he demonstrates that the deterministic models of Weingartner and others are either special cases of or closely related to the generalized model. We shall repeat some of Bernhard's special-case findings to demonstrate the model's properties. Third, Bernhard introduces uncertainty into the general model through chance-constrained programming. This topic is beyond this chapter's scope, but has shaped research in the area.

Bernhard's generalized mathematical programming model has an objective function, (0), and a constraint, (5), which may not be linear. If either is nonlinear it can sometimes be made linear or at least made monotonic nondecreasing and then approximated by piecewise linearization. In these cases, a solution is easy to find. However, because any optimal solution complies with the model's Kuhn-Tucker conditions[2], characteristics of the optimal solutions can be inferred, even without

---

2. The Kuhn-Tucker conditions are a series of partial differential relationships, derived by differentiating the objective function. They are somewhat analogous to the dual constraints of linear programming, and they have a comparable economic interpretation. The existence of the K-T conditions does

294   Chapter 10, Optimization in Project Selection

procuring the solution. We discuss some of Bernhard's statements about characteristics of the optimal solutions to certain special-case models.

10.3.3 Notation. We use the following notation, which differs slightly from Bernhard's:

j = index for the different projects $(1, 2, \ldots, m)$
t = index for the different time periods $(1, 2, \ldots, T)$
T = terminal time period
Y_{tj} = net cash flow from a unit of project j at t_j
$\hat{Y}_j$ = the time T present worth of post-T cash flows from a unit of project j
M'_t = amount of cash made available from projects and other sources outside this analysis at t
M' = time T present worth of post-T cash flows from these outside projects and sources
l_t = $1 + r_{lt}$, where r_{lt} = lending rate of interest from t to $t+1$
b_t = $1 + r_{bt}$ where r_{bt} = borrowing rate of interest from t to $t+1$
b_t $\geq$ l_t, for $t \leq T-1$; $b_t = l_t = \lambda_t = 1 + r_t$ for $t \geq T$
B_t = maximum value of w_t
d = total amount of scarce resource available
d_j = amount of scarce resource required by a unit of project j
C_t = cash requirement for period t, excluding liquidity requirement
c_t = fraction of outstanding debt that must be held as a liquid asset (liquidity required).

The variables in the general model are

x_j = number of units of project j to be undertaken
W_t = cash dividend paid to owners at t
w_t = cash to be borrowed from t to $t+1$; assume $w_0 = 0$
v_t = cash to be lent from t to $t+1$, above and beyond what must be lent to satisfy the liquidity requirement (1); assume $v_0 = 0$
G = the firm's *terminal wealth* at T after payment of W_T.

Optimal levels of these variables are indicated by an asterisk on the respective symbols; for example, x_j^*. Other notation will be introduced as needed.

10.3.4 Objective function. Most models developed prior to the publication of Bernhard's assumed that the firm's objective was to maximize the firm's terminal wealth, G, at T. Bernhard noted that widespread agreement existed that an appropriate objective of the firm was to maximize some function (usually a discounted sum) of all anticipated dividend payments to the firm's present shares. Combining the two concepts, Bernhard reasoned that if dividends were to be truncated at some finite horizon, T, then the terminal wealth, G, could be included in the objective function as a proxy for the post-T dividends. Thus, his objective was to maximize a function of the finite dividend payments stream, $W_1, W_2, \ldots, W_T$, plus the terminal wealth, G. Formally, the objective is to:

---

not necessarily guarantee the existence of an optimal solution, but an optimal solution will always comply with the K-T conditions. The K-T conditions are detailed in many operations research/management science texts.

(O) $$\text{maximize } f(W_1, W_2, \ldots, W_T, G). \tag{10.1}$$

It is assumed that $\partial F / \partial W \geq 0$ and $\partial F / \partial G \geq 0$; that is, $f(\ )$ is a nondecreasing function.

10.3.5 Constraints. The model has five sets of constraints. For each one, the associated Kuhn-Tucker (K-T) variable is indicated in brackets on its right. The general meaning and interpretation of the K-T variables are described in a later subsection.

1. Cash balance constraints including liquidity requirements. The firm must carry, in cash, at least $(C_t + c_t w_t)$, where $C_t \geq 0$ and $0 \leq c_t \leq 1$. At $t = 0$, it is assumed that $C_t = c_t = 0$. Thus, the cash requirement at any t consists of a constant, C, (which may vary between periods), plus a fraction of the outstanding debt, w_t which is the liquidity portion of the requirement. (This liquidity requirement is commonly made by lending institutions.) The entire amount, $(C_t + c_t w_t)$ earns interest from t to $t + 1$ at the lending rate of interest.

Thus, the general form of the cash balance restrictions is:

$$-\sum_{j=1}^{m} Y_{tj} x_j - l_{t-1}(v_{t-1} + c_{t-1} w_{t-1} + C_{t-1}) + (v_t + c_t w_t + C_t) + b_{t-1} w_{t-1} \\ - w_t + W_t \leq M'_t \qquad \text{for } t = 1, 2, \ldots, T. \tag{10.2a}$$

Beginning at the left, this relationship says that at t, the net cash outflow to projects$(-\sum_{j=1}^{m} Y_{tj} x_j)$; minus the cash inflow from $t-1$ loans, $l_{t-1}(v_{t-1} + c_{t-1} w_{t-1} + C_{t-1})$; plus the cash outflow for t loans, $(v_t + c_t w_t + C_t)$; plus the cash outflow for repayment of $t-1$ borrowing, $(b_{t-1} w_{t-1})$; minus the cash inflow from t borrowing, w_t; plus the cash outflow for t dividend payments, W_t; must as a sum be less than or equal to the cash available from outside sources, M'_t, at t. For convenience, the constants are transposed to the right so that Equation (10.2a) may be rewritten as:

(1) $$-\sum_{j=1}^{m} Y_{tj} x_j - l_{t-1} v_{t-1} + v_t + (b_{t-1} - l_{t-1} c_{t-1}) w_{t-1} \\ - (1 - c_t) w_t + W_t \leq M_t \quad \text{for } t = 1, 2, \ldots, T. \qquad [\rho_t] \tag{10.2b}$$

where

$$M_t = M'_t + l_{t-1} C_{t-1} - C_t.$$

2. Group payback restriction. To allow projects to start later than $t = 0$, Byrne et al. (1967) imposed a group payback restriction. Bernhard follows the same rule and requires that the net outlays to date (time = t') on the group of executed projects be zero or negative; that is:

(2) $$-\sum_{j=1}^{m} \sum_{t=1}^{t'} Y_{tj} x_j \leq 0, \quad \text{for some } t', \text{ such that } 1 \leq t' \leq T \qquad [\Psi] \tag{10.3}$$

In addition, Bernhard notes, one could add restrictions for several values of t', or a payback requirement on individual executed projects, such as:

$$-\sum_{t=1}^{t'} Y_{tj}x_j \leq 0 \quad \text{for } j = 1, 2, \ldots, m.$$

3. Scarce resource restriction. Weingartner (1963) imposed a "manpower" restriction, to illustrate how a scarce resource restraint is included. Bernhard uses the same general form:

(3) $$\sum_{j=1}^{m} d_j x_j \leq d \qquad [\nu] \quad (10.4)$$

Many such constraints could be written as needed, for example, for any given resource, or different restrictions in different periods.

4. Terminal wealth restriction. The firm's terminal wealth at T, following payment of dividend, W_T, is

$$G \equiv M' + \sum_{j=1}^{m} \widehat{Y}_j x_j + v_T + c_T w_T + C_T - w_T.$$

M' is the present worth of post-T cash flows from outside projects and sources. $\sum_{j=1}^{m} \widehat{Y}_j x_j$ is the analogous present worth of post-T cash flows from projects selected for execution, The sum, $v_T + c_T w_T + C_T$, is the amount lent out at T. From all these, w_T, the amount borrowed at T, must be deleted.

The terminal wealth constraint can be rewritten as:

(4) $$-\sum_{j=1}^{m} \widehat{Y}_j x_j - v_T + (1 - C_T)w_T + G = M \qquad [\phi] \quad (10.5)$$

where $M \equiv M' + C_T$

5. Terminal wealth horizon posture restriction. The terminal wealth, G, is constrained such that $G \geq K + g(W_1, W_2, \ldots, W_T)$ where K is a constant ≥ 0 and g is a function ≥ 0. That is, the firm cannot have a negative terminal wealth that is linked to past dividends. This allows the firm to pay post-T dividends in some way comparable to those paid at T and earlier.

As before, the horizon posture restriction is rewritten as:

(5) $$-G + g(W_1, W_2, \ldots, W_T) \leq -K \qquad [\theta] \quad (10.6)$$

10.3.6 Problems in the measurement of terminal wealth. There are some problems in establishing the firm's terminal wealth in restrictions 4 and 5 above. One problem concerns the firm's borrowing status at T (that is, whether it is a borrower, lender, or neither) and the applicable interest rate. The other problem is whether or not the amount borrowed at T is constrained. Taking these problems in order, we paraphrase Bernhard's analysis (1969):

Consider the present worth at T of post-T cash flows, $M' + \sum_{j=1}^{m} \widehat{Y}_j x_j$. If the periods beginning at T have only one market rate each for borrowing and lending, $\lambda_t = 1 + r_t$ then finding the present value of post-T cash flows is possible. In general, if all post-T cash flows are included, then:

$$M' = \sum_{t=T+1}^{\infty} \left(M'_t \prod_{\tau=T+1}^{t} \lambda_{\tau-1}^{-1} \right) \text{ and } \widehat{Y}_j = \sum_{t=T+1}^{\infty} \left(Y_{tj} \prod_{\tau=T+1}^{t} \lambda_{\tau-1}^{-1} \right)$$

and hence M' and $\widehat{Y}_j$ are defined beyond T.

As is typically true, if the borrowing rate exceeds the lending rate (that is, $b_t > l_t$) at T, then problems are encountered. The rate depends on whether the firm is a borrower or a lender at T. If the firm is a borrower at T, it may borrow hb_T^{-1} (the then present equivalent of h) and pay off the debt at $T + 1$ with h. But if the firm is a lender at T, the time T equivalent of h is hl_T^{-1}, which will be recovered by h at time $T + 1$. If h $\neq$ 0, then h_T is either hb_T^{-1} or hl_T^{-1} and that $hl_T^{-1} > hbT^{-1}$. Therefore, h cannot be known until one establishes the post-T lending and borrowing states and amounts. To avoid this, Bernhard simply assumes that there is a single market interest rate for both borrowing and lending; that is, $\lambda_t = b_t = l_t = (1 + r_t)$, for all post-$T$ periods.

This assumption does not remove the other obstacle, which is encountered if borrowing at T, w_T, is restricted. Weingartner (1963) suggests such a restriction, in which $w_T < B_T$, where $B_T \geq 0$. Suppose, for example, that $w_T = B_T = 0$, and suppose also that the firm is a "would be" borrower at $T + 1$—that is, even though no funds are borrowed at T, it anticipates borrowing h at $T + 1$. What is the worth of h at time T? It could be interpreted as hl_T^{-1} if it saves some lending at T, but otherwise it is zero. Thus, the value of hr again depends on the post-T financing pattern, which Bernhard notes is a product of the mathematical program rather than a previous assumption. Thus, if M' and $\widehat{Y}_j$, are to be defined in advance, then $w_T \leq B_T$ must be avoided.

10.3.7 Additional restrictions. Besides these five constraints of the Bernhard model, three others are added. Two are standard ones for capital budgeting models, and the third is an upper limit on borrowing. These restrictions are as follows.

6. Upper bounds on borrowing. Lending institutions will almost always limit the total borrowing of the firm to a fractional part (say 60%) of the firm's equity. Assuming that the firm's equity capital remains constant over time, as Bernhard does, this upper bound on borrowing is:

(6) $\qquad w_T \leq B_T \quad \text{for } t = 1, 2, \ldots, T - 1 \qquad [\beta_t] \quad (10.7)$

which matches Weingartner (1963).

7. Prohibition of multiple projects. In earlier models, we have required that mutually exclusive projects (or bundles of projects) be either rejected or accepted (i.e., $x_j = 0, 1$).

This requirement imposes great difficulties on solving, or even interpreting, Bernhard's model (see Weingartner, 1963).

To remove this integer requirement, Bernhard simply prohibits the execution of multiple projects by requiring that:

(7) $\qquad 0 \leq x_j \leq 1 \quad \text{for } j = 1, 2, \ldots, m \qquad [\mu_j]$ (10.8)

If some x's become fractional in the optimal solution, we assume that the binding constraint can be loosened to make the fractional project complete. The resulting solution is neither optimal nor feasible, but it may be near-optimal and probably quite realistic since reality is not rigidly fixed at deterministic values.

8. Nonnegativity restrictions. The final constraint is usually included in mathematical programs, and it is that all the decision variables be nonnegative. This restriction permits the use of commonly available algorithms. While some variables are already required to be nonnegative, others should be added; so

(8) $\qquad v_t, w_t, W_t \quad \text{for } t = 1, 2, \ldots, T$ (10.9)

In summary, the Bernhard generalized model maximizes the objective function (0), subject to constraints (1)–(8). In general, the model is difficult to solve unless the objective (0) and the terminal wealth constraint (5) are linear, which is a special case. Some pertinent economic features of the optimal solution are revealed by the model's Kuhn-Tucker conditions, even if the model cannot be solved. We consider some of Bernhard's findings below.

10.3.8 The Kuhn-Tucker conditions. The Kuhn-Tucker conditions are a set of constraints, one for each variable in the model, plus certain other restrictions on the K-T variables. The K-T variables, one per original constraint, are a generalized form of Lagrange multiplier and have the same economic interpretation. In general (for maximization problems) if a K-T variable is nonzero in the optimal solution, then the corresponding constraint in the original problem is tight, and it expresses the amount the objective could be increased if the constraint's right-hand-side value were loosened by one unit.

We present the steps for obtaining the K-T conditions for Bernhard's model, but not the complete procedure because it is quite lengthy. The first step is to convert all constraints into equalities (by adding or subtracting slack variables), and then to add converted constraints into the objective by using the appropriate K-T variables as Lagrange multipliers. To form a new Lagrange function, L(·), we write the original objective (0), reduced[3] by each constraint (1)–(8), multiplied by its own K-T variable:

$$L(v_t, w_t, x_j, W_t, G, \rho_y, \Psi, v, \phi, \theta, \beta_t, \mu_t) = \underbrace{f(W_t, G)}_{\substack{Original \\ objective(0)}} - \qquad (10.10)$$

---

3. If the sign of the multiplier (in general, λ), is taken as negative in formulating the Lagrange for maximization problems and $\lambda \geq 0$, then the optimal solution will contain λ's that are either zero or of the correct sign (+), so as to indicate an increase of the objective (that is, $\lambda = \partial f(\cdot)/\partial b$) if the constraint is tight.

$$\underbrace{\rho_t}_{\substack{K-T\\No.1}} \Bigg[\underbrace{-\sum_{j=1}^{m} Y_{tj}x_j - l_{t-1}v_{t-1} + v_t + (b_{t-1} - l_{t-1}c_{t-1})w_{t-1} - (1-c_t)w_t + W_t - M_t + \mu_{1t}^2}_{\text{First constraint,(1)}}\Bigg]$$

$$-\underbrace{\psi}_{\substack{K-T\\No.2}}\underbrace{\Big(-\sum_{j=1}^{m}\sum_{t=1}^{t'} Y_{tj}x_j + \mu_{2t}^2\Big)}_{\text{Second constraint (2)}}$$

$-v[\text{times constraint}(3)] - \phi[\text{times constraint}(4)]$
$-\theta[\text{times constraint}(5)] - \beta_t[\text{times constraint}(6)]$
$-\mu_j[\text{times constraint}(7)]$.

The second step is to differentiate the Lagrange with respect to each of its argument variables, thereby obtaining a set of partial derivatives. Each partial derivative equation is then set to zero, which is the necessary condition for an optimum to exist.

The third step is to solve this system of simultaneous equations. Bernhard remarks that unless the objective (0) and the terminal wealth posture restriction (5) are linear, the model is difficult to solve.

If the objective (0) is assumed to be concave and differentiable and if the left side of (5) is convex, then the system of simultaneous equations of the second step leads to a set of K-T conditions. Finding these K-T conditions is itself a tedious task that we shall not repeat here; instead, we rely on the results of Bernhard's paper. The K-T conditions for the model follow. In each case, the associated original variable is indicated in brackets:

(9) $\qquad -\rho_t + l_t\rho_{t+1} \leq 0 \quad \text{for } t = 1, 2, \ldots, T-1 \qquad [v_t] \quad (10.11)$

(10) $\qquad (1-c_t)\rho_t - (b_t - l_tc_t)\rho_{t+1} - \beta_t \leq 0$
$\qquad \qquad \qquad \qquad \text{for } t = 1, 2, \ldots, T-1 \qquad [w_t] \quad (10.12)$

(11) $\qquad -\rho_T + \phi \leq 0 \qquad [v_t] \quad (10.13)$

(12) $\qquad (1-c_T)\rho_T - (1-c_T)\phi \leq 0 \qquad [w_t] \quad (10.14)$

(13) $\qquad \sum_{t=1}^{T} Y_{tj}\rho_t + \widehat{Y}_j\phi - d_jv + \sum_{t=1}^{t'} Y_{tj}\Psi - \mu_j \leq 0 \qquad [x_j] \quad (10.15)$
$\qquad \qquad \qquad \text{for } j = 1, 2, \ldots, m$

(14) $\qquad \dfrac{\partial f}{\partial W_t}\bigg|_{W_t} - \rho_t - \theta\dfrac{\partial g}{\partial W_t}\bigg|_{W_t} \leq 0 \quad \text{for } t = 1, 2, \ldots, T \qquad [W_t] \quad (10.16)$

(15) $\qquad \dfrac{\partial f}{\partial G}\bigg|_G - \phi + \theta \leq 0. \qquad [G] \quad (10.17)$

For a discussion of these K-T conditions and the optimal solution, we quote Bernhard:

In the optimal solution, both for these K-T inequalities, (9)–(15), and for the original inequalities, (1)–(3) and (5)–(7), if the level of the associated variable, shown in brackets at the right, is greater than zero, the corresponding constraint must be satisfied as a strict equality. On the other hand, if the level of the associated variable is exactly zero, the corresponding constraint may or may not be satisfied as a strict equality. These relationships between variables and associated constraints are referred to by the term *complementary slackness*.

For a given constraint in the original model, the optimal level of the associated K-T variable represents on the margin, the rate at which the objective function (0) could be increased if there were a small increase in the right-hand-side constant.

As in the case of the original variables, all of the K-T variables are constrained to be ≥ 0, except for 4). Due to its association with a strict equality, (4) is, by that criterion, unconstrained in sign. But, from (11), $\phi \leq \rho_T$, and from (12), $\phi \geq \rho_T$. So,

(16) $$\phi = \rho_T.$$ (10.18)

From above, we know that $\rho_T \geq 0$. Therefore, $\phi \geq 0$.

Again, as in the case of the original variables, we will also use asterisks to indicate optimal levels of the K-T variables. So inserting (16) into (13), we find that:

(17) $$\mu_j^* \geq A_j \quad \text{for } j = 1, 2, \ldots, n,$$ (10.19)

where

(18) $$A_j^* = \sum_{t=1}^{t'} Y_{tj}(\rho_t^* + \Psi^*) \sum_{t=t'+1}^{T-1} Y_{Tj}\rho_t^* + (Y_{Tj} + \hat{Y}_j)\rho_T^* - d_j v^*$$ (10.20)

Note, in passing, that if there is no payback restriction or if it is not binding, $\Psi^* = 0$, and similarly, if there is no scarce resource restriction or if it is not binding, $v^* = 0$. So for the simpler case where these two restrictions are nonexistent or at least nonbinding,

(19) $$A_j^* = \sum_{t=1}^{T-1} Y_{tj}\rho_t^* + (Y_{Tj} + \hat{Y}_j)\rho_T^*$$ (10.21)

Returning to the general case, from conditions (7) and (13) and the complementary slackness relationships discussed above, we find that:

$$\begin{array}{ll} \text{if } x_j^* = 1 & \text{then } \mu_j^* = A_j^* \geq 0; \\ \text{if } 0 < x_j^* < 1 & \text{then } \mu_j^* = A_j^* = 0; \\ \text{and if } x_j^* = 0 & \text{then } \mu_j^* = 0 \geq A_j^*. \end{array}$$ (10.22)

That is, if, for a given project, j, $A_j^* > 0$, then that project should be accepted in full; if $A_j^* = 0$, then it does not make any difference in the objective function whether the project is accepted or rejected.

10.3 Alternative Models of the Selection Problem

So given the values of Y_{tj}, $\hat{Y}_j$, and d_j, for project j, if we can also learn the optimal values of the K-T variables, i.e., the ρ_t^*s, Ψ^*, and v^*, we may determine whether project j should be undertaken. Unfortunately, knowing the magnitudes of these K-T variables will, in general, require a complete programming solution. However, as illustrated below, some of their properties may, with less effort, be determined using the K-T conditions. Indeed, in some simple cases, a complete solution is possible using this latter approach.

10.3.9 Properties of ρ_t^*. The optimal Lagrange multiplier, ρ_t^*, is related to the cash balance and liquidity restriction, Equation (9.2b), via the K-T restrictions, Equations (9.11–9.17). If $\rho_t^* > 0$, then the objective will be increased at t at the margin if M_t', the cash made available from outside sources, is increased by \$1; if the liquidity requirement at $t-1$, C_{t-1}, is increased by \$1; if the liquidity requirement at t, C_t, is decreased by \$1; or if any combination of the above occurs. In essence, the dual variable ρ_t^* is related to the right side of Equation (9.2b) via

$$\rho_t^* \simeq \frac{\partial(\text{objective})}{\partial(M_t'; C_{t-1}; C_t)}.$$

Ordinarily, $\rho_t^* > 0$ will be true for all time periods, rendering Equation (9.2b) an equality. In this case, Bernhard shows that an extra dollar available in an earlier period can always be lent through to a later period when it has utility (for distribution to the shareholders), so it has utility in the earlier periods as well.

Conversely, for the pathologic case when some $\rho_t^* = 0$, then $\rho_{t+1}^* = 0$, and increases in terminal wealth, G, or in the dividend, W_t, will not increase the objective function value. The final consequence is that since Equation (9.12) reduces to $\rho_t^* \leq \beta$, then $\beta_t^* > 0$ also, which requires that the upper bound on borrowing be binding [Equation (9.7)]. The consequence of this pathologic case is that extra dollars beyond t have no utility in any period.

10.3.10 Special cases. In the second section of his article, Bernhard examines three special cases, which are summarized here. In case 1 the upper bound on borrowing, Equation (9.7), is either absent or nonbinding. In case 2a the borrowing and lending rates of interest in the same time period are equal; that is $b_t = l_t = \lambda_t$; and in case 2b the borrowing and lending rates in all periods are equal (that is, $b_t = l_t = \lambda_t = \lambda$). In both subcases, borrowing constraints are imposed. In case 3a the borrowing and lending rates are equal in a given period (that is, $b_t = l_t = \lambda_t$), but borrowing constraints are absent or nonbinding. Finally, in case 3b the interest rates are assumed constant; that is, $b_t = l_t = \lambda_t = \lambda$.

For case 1, which has inactive or absent borrowing constraints, the optimal solution will be characterized as follows: (1) All the cash balance constraints [Equation (9.2b)] will be equalities. (2) Excess funds from an earlier period can be freely lent into a later period. (3) Funds can be freely borrowed from a later period into an earlier period. (4) Due to the assumption that the borrowing rate exceeds the lending rate ($b_t > l_t$), the firm can engage in borrowing to maintain liquidity, but the value of a time t dollar is increased relative to a time $t + 1$ dollar.

In case 2a, when the borrowing and lending rates are equal for any given period and when there is a restriction on borrowing in a given period and, in case 2b, when the borrowing and lending rates are all assumed set to a constant value, λ, it can be shown that omitting payback and resource restrictions can lead to an optimal solution that contains projects that have a negative present value (or horizon value). That is, a project need not be rejected merely because it has a negative NPV based on the market rate of interest (Weingartner, 1963). The intuitive reason is that the project may generate cash inflows at times when they are most needed. The converse is also true: A project with positive NPV may be rejected. Note that case 2b is a perfect capital market model (borrowing and lending rates are equal and invariant among periods).

In both cases 3a and 3b the borrowing constraints are absent or nonbinding. When borrowing and lending rates are equal in a given period (case 3a), the firm may borrow to satisfy the liquidity constraint even though it has already lent funds elsewhere at the same rate. Thus, the firm can accept any project whose net present value calculated at the market rate, A, is positive, and it rejects any project whose NPV < 0. In case 3b, when $\lambda_t = \lambda$ for all periods, the existence of a capital restriction is meaningless since the firm can borrow as much money as necessary at the market rate, λ, to accept all projects whose NPV > 0. This confirms the NPV criterion for single projects developed in Chapter 7, together with its assumptions.

Bernhard also demonstrates how several of Weingartner's models are simply special cases of the generalized model and then shows that the rate of discount is not internally procurable from the model itself. Thus, he confirms and extends the earlier work of Baumol and Quandt (1965), who made this assertion after an abortive attempt to establish the appropriate discount within their model by using the dual.

Bernhard's generalized model is the basis of another paper (1971). Exact knowledge of the discount rate exists only in perfect capital markets; if net present value or incremental internal rate of return is used as a selection criterion, then the discount rate must be estimated from external sources. Thus, Bernhard's model confirms Hirshleifer's finding that in imperfect capital markets, the separation theorem collapses and optimal project selections cannot be made without considering the future utility of intertemporal funds exchange.

10.4 PROJECT SELECTION BY GOAL PROGRAMMING METHODS

Goal programming (GP) is mathematical programming that focuses on multiple conflicting objectives or goals. The principal restriction in applying GP methodology is that a target value or desired level of attainment is required for each goal. Then there is a second level of specification. Either relative weights that permit combining the deviations by which the various goals are under- or overachieved or an absolute priority or ordinal preference ordering is required. The GP rationale then provides a solution to the programming problem that minimizes the weighted absolute deviations or the absolute deviations in the specified priority order of goal attainment. Otherwise, GP is a "first cousin" to linear programming. For example, GP problems can be linearly constrained, they can be formulated and solved as integer [even binary (0, 1)] problems, their variables require the same nonnegativity constraints, and they can use a modified form of the simplex procedure to obtain solutions.

The GP model was first introduced by Charnes and Cooper (1961) and later considerably extended by Ijiri (1965) and Lee (1972). Ignizio (1976) combined the GP methodology with a Dakin algorithm for solving the 0, 1 case, which applies to project selection problems. Another presentation of integer-based goal programming is Lee (1985).

Because the weighted GP format is a simple extension of the ordinal format, we present the ordinal formats here. In fact, the weighted objective function is a simple linear combination that fits classic linear programming approaches. First, we give a brief description of the GP concept and format; second, we illustrate how a GP problem is solved graphically; and third, we formulate a simple project selection problem and summarize its solution.

10.4.1 Goal programming format. Unlike linear programs, the ordinal GP has a multidimensional objective function that seeks to minimize selected absolute deviations from a set of stated goals, usually within additional constraints. Each selected deviation in the GP objective carries an ordinal priority weight so that goals are attained (or approached as nearly as possible) in strict order of priority. In general, the format of the GP problem is

Find $X = (x_1, x_2, \ldots, x_j, \ldots, x_n)$ so as to obtain:

$$\text{Min: } Z = f(d^+, d^-) \tag{10.23}$$

subject to:

$$(1) \quad AX = B + d^+ - d^- \tag{10.24}$$
$$(2) \quad CX \leq D \tag{10.25}$$
$$(3) \quad X, d^+, d^- \geq 0 \tag{10.26}$$

where

(a) $X = (x_1, x_2, \ldots, x_j, \ldots, x_n)$ is the solution vector
(b) Equation (10.23) is the goal programming objective, which minimizes a function of the deviations from the stated goals
(c) Equation (10.24) states the original objectives, converted into goals by the inclusion of intentionally permissible deviations (d^+, d^-) from the right-hand-side targets (B_i); $i = 1, 2, \ldots m$
(d) Equation (10.25) shows the absolute linear programming constraints; $k = m + 1$, $m+2,\ldots,M$
(e) $f(d^+, d^-)$ is a linear, prioritized function of the permissible deviation variables
(f) d^+ is a vector of nonnegative variables that are the positive deviations from the objectives, Equation (9.24)
(g) d^- is a vector of nonnegative variables that are the negative deviations from the objectives, Equation (9.24)
(h) B is a vector of right-hand-side (RHS) target values, or aspiration
(i) C is a matrix of resource consumption coefficients
(j) A is a matrix of activity coefficients
(k) D is the vector of RHS bounds on the absolute constraints.

Quite often, the objective, Equation (9.23), takes the form:

$$\text{Min } Z = \{P_1[g_1(d_1^+, d_1^-)], P_2[g_2(d_2^+, d_2^-)], \ldots, P_i[g_i(d_i^+, d_i^-)]\}$$

where $g_i(d_i^+, d_i^-)$ is a linear function of the deviation variables, P_t is the ordinal priority level associated with $g_i(d_i^+, d_i^-)$, and $i \leq m$.

Note: In general the number of distinct priorities can be less than m; but this only complicates the subscript notation since some goals will have the same priority. Note that such goals should be measured in the same terms.

In ordinal goal programming, the priority levels, P_i, have a strict interpretation as the preemptive priority associated with the ith objective. Thus, P_1 is the top priority. Furthermore,

$$P_i >>> P_{i+1} >>> P_{i+2} \ldots \ldots$$

That is, satisfying priority level i objectives is immeasurably preferred to satisfying any other lower priority objectives. The solution procedure does not find a global optimum satisfying all constraints, as in linear programming. Rather, GP solutions find a feasible set of the best solutions to the priority 1 level subproblem; then within these solutions it finds that subset of solutions that best solves the priority 2 level subproblem, and so forth. GP permits intentional positive-, negative-, or zero-valued deviations away from the numerically valued goals. However, both d_i^+ and d_i^- cannot simultaneously take on positive values in the simplex procedure. So only d_i^+ or d_i^- can be in solution (that is, > 0) in the same problem tableau.

The preemptive prioritization within the objective function and the existence of permissible deviations, permit the GP procedure to "work" even when some (or all) of the goals are incompatible. Some goals will be attained (or even over attained), while others will be under attained (usually the lower priority goals). Over attained goals have the corresponding d_i^+ variable in solution (that is, $d_i^+ > 0$ and $d_i^- = 0$), whereas underattained goals have $d_i^+ = 0$ and $d_i^- > 0$. A simple example will illustrate the method of formulating and solving a goal programming problem.

10.4.2 An example of formulating and solving a goal programming problem. Consider the following simple mix problem, in which the firm wants to optimally allocate its resources to two products while prioritized but conflicting goals are met as nearly as possible.

The Chop-N-Block Company[4] produces a cheese knife and a wooden cutting board. Each product requires inputs and has a fixed sales price:

	Knife	Board
Direct Material Cost	$0.50	$1.00
Direct Labor Cost	0.50	1.00
Machine Time Required (hours)	0.50	0.25
Assembly Time Required (man-hours)	1.00	1.00
Sales Price per Unit	$3.00	$5.00

For simplicity we shall assume the following:

1. Consumer demand is strong and all production can be sold.
2. One production period is involved; that is, $t = 0, 1$.
3. Production is limited by two absolute restrictions: (a) Machine capacity is 8 hr per period, and (b) assembly capacity is 20 man-hours per period.

---

4. This problem is adapted from Baumgarten (1973).

4. Initial values of the firm's cash balance in the bank and its accounts receivable are $20 and $30, respectively. At $t = 0$, a bank loan to the firm of $10 is outstanding, and long-term bonds in the amount of $30 are owed.
5. Fixed expenses of $5 per period will be incurred in operating the firm, which must be paid in cash.
6. Cash dividends of $2.50 are to be paid in the first period, and $2.50 is to be spent for equipment replacement.
7. The firm sells its products on a one-period credit basis; that is, receivables for a prior period are collected in the next period. Labor costs, materials, and other expenses are paid in cash in the period used.
8. Taxes are nonexistent, and inventory balances are zero since all products can be sold as produced.
9. The firm's Board of Directors has approved a budget for plant expansion in period 2. These funds will be obtained by the sale of new bonds and new equity stock in period 2. To command a good market price for its proposed stock sale and to negotiate a favorable bonds sale, the firm's management judges that the firm must (a) continue to pay its current dividends of $2.50 and (b) make a satisfactory profit in period 1. The Board feels that investors would be satisfied with a net profit of $45 in period 1, and it feels the firm can attain that. Since the total of fixed expenses, dividends, and expenditures for equipment is $5 + $2.50 + 2.50 = $10 in period 1, the corresponding contribution to profit and overhead from production is the net profit plus $10, or $55 in period 1, from the production and sale of cheese knives and cutting boards.
10. The firm's president has established a policy of maintaining an end-of-period cash balance that exceeds the short-term liabilities by at least $2. In this case, the short-term liability is the bank loan of $10, so the policy goal for the cash balance is $10 + $2 = $12.
11. The contracts for the long-term bonds require that the firm's net working capital at the end of each period be at least twice the face value of the bond debt. Net working capital is defined as

Cash + Accounts receivable + Inventory − Short-term liabilities at the end of the period.

We now define the structural variables:

$$X_K = \text{number of knives to be produced}$$
$$X_C = \text{number of cutting boards to be produced.}$$

We also formulate the constraints and the goals of the problem. First, the technical constraints due to machine and assembly capacity restrictions are:

$$\left.\begin{array}{l} 0.5X_K + 0.25X_C \le 8 \quad \text{(machine capacity)} \\ 1.0X_K + 1.0X_C \le 20 \quad \text{(assembly capacity)} \end{array}\right\} \quad (10.27)$$

The contribution to profit and overhead (P&O) goal is the sales price minus the direct material and labor:
Contribution to P&O (knives) = $3.00 − $0.50 − $0.50

$$= \$2.00 \text{ per knife manufactured}$$

$$\text{Contribution to P\&O (boards)} = \$5.00 - \$1.00 - \$1.00$$
$$= \$3.00 \text{ per board manufactured}$$

We want the contribution to profit and overhead to equal or exceed the P&O goal of $55, so:

$$2.00 \, X_K + 3.00 \, X_C \geq 55.$$

To put this requirement into canonical form, we include the deviation variables on the right side:

$$2.00 X_K + 3.00 X_C = 55 + d_1^+ - d_1^- \tag{10.28}$$

and upon transposing the deviations we have:

$$2.00 X_K + 3.00 X_C - d_1^+ + d_1^- = 55. \tag{10.29}$$

Note that if $(d_1^+ > 0, d_1^- = 0)$ in the optimal solution, the P&O goal of $55 is *over* attained, and if $(d_1^+ = 0, d_1^- > 0)$, then it is *under* attained.

The end-of-period cash balance goal is determined as follows:

$20	+ $30	−$10	−$($1.0X_K + 2.0X_C$)	≥ $12 (9.30)
Beginning cash balance	Beginning accounts receivable collected	Cash fixed expenses + equipment expenditures + dividends paid	Cash production costs for direct labor and material	Desired end-of-period cash balance

Combining terms and adding the required deviation variables, we have:

$$-1.00 X_K - 2.00 X_C = -28 + d_2^+ - d_2^- \tag{10.30}$$

and upon multiplying by −1 and transposing the deviation variables to the left side, the end-of-period cash balance goal becomes:

$$1.00 X_K + 2.00 X_C + d_2^+ - d_2^- = 28 \tag{10.31}$$

It is necessary to be consistent with the signs attached to the structural variables when goals are formed. If cash flows are always signed (+) to represent inflows and (−) for outflows and if the desired relationship is always placed in the form:

$$AX \geq B$$

then the canonical form of Equation (10.24) can be used to convert desired relationships into goals. Table 10.1 gives the applicable procedures for forming the GP objective.

The third goal is the net working capital required by assumption 11. From Equation (10.31), the ending cash balance is $1.00X_K + 2.00X_C$. So the end-of-period net working capital is calculated as follows:

20	+ 30	$+(2X_K + 3X_C)$	−10	−10	≥ 2(30)
Beginning cash balance	Beginning accounts receivable	Net contribution to profit and overhead	Bank Loan	Cash fixed expenses equipment expenditures and dividends payment	Twice the bond debt

Combining terms and. adding the deviation variables to the RHS, we have

$$2.00X_1 + 3.00X_2 = 30 + d_3^+ - d_3^-$$

Table 10.1 Procedures for achieving objectives

If the Objective is to	Add on LHS	Goal Objective
(a) Equal or exceed B_i	$-d_i^+ + d_i^-$	Min d_i^-
(b) Equal or be less than B_i	(same)	Min d_i^+
(a) Equal B_i	(same)	Min $(d_i^+ + d_i^-)$

and upon transposing the deviations, the net working capital goal becomes:

$$2.00X_1 + 3.00X_2 - d_3^+ + d_3^- = 30. \qquad (10.32)$$

Now consider the GP objective function. Suppose the management of Chop-N-Block desires first and foremost (priority 1) to attain a net profit of $45 [see assumption 9 and Equation (10.25)]. Suppose also that the working capital goal [Equation (10.32)] is at a priority 2 level, and the end-of-period cash balance goal [Equation (10.31)] is at a priority 3 level. In addition, management considers that over attaining the end-of-period working capital goal is as good as meeting this goal exactly. Also, management wishes to have neither an excess nor a deficiency in the $12 end-of-period cash goal.

To formulate these preferences as a goal programming objective, we let P_1, P_2, and P_3 be ordinal priorities (that is, $P_1 >>> P_2 >>> P_3$, or there is no $k > 0$, such that $kP_2 \geq P_1$ or $kP_3 \geq P_2$). Then the GP objective can be written as:

$$\text{Min } Z = P_1 d_1^- + P_2 d_3^- + P_3(d_2^+ + d_2^-) \qquad (10.33)$$

where d_1^- = possible underattainment of profit goal

d_2^+ = possible overattainment of cash goal

d_2^- = possible underattainment of cash goal

d_3^- = possible underattainment of working capital goal

The objective in Equation (10.33) can be more fully interpreted as follows:

1. Management has set attaining its profit goal as its primary goal. Exceeding the profit goal is acceptable, but under attainment is unacceptable with priority 1—so $P_1 d_1^-$ is in the objective.
2. Without compromising the profit goal, management wants to equal or exceed the minimum working capital balance. This is expressed by minimizing $P_2 d_3^-$ in the objective.
3. Finally, after both of the goals above have been approached or satisfied, management wants cash to be used as fully as possible, with no excess or deficiency in the required end-of-period cash balance. This is assigned priority 3, and the term $P_3(d_2^+ + d_2^-)$ is included in the objective.

The multiple-goal problem can be stated formally as:

$$\text{Min } Z = P_1 d_1^- + P_2 d_3^- + P_3(d_2^+ + d_2^-)$$

subject to:

(1) $2.0 X_K + 3.0 X_C - d_1^+ + d_1^- = 55$ (profit and overhead goal)
(2) $1.0 X_K + 2.0 X_C - d_2^+ + d_2^- = 20$ (cash balance goal)
(3) $2.0 X_K + 3.0 X_C - d_3^+ + d_3^- = 30$ (working capital goal)
(4) $0.5 X_K + 0.25 X_C \leq 8$ (machine constraint)
(5) $1.0 X_K + 1.0 X_C \leq 20$ (assembly constraint)
(6) $X_i, d_i^+, d_i^- \geq 0$ (nonnegativity constraints)

The graphical solution is illustrated in Figure 10.2. Some features of this solution should be noted. First, the feasible area is enclosed by the absolute machine and assembly capacity constraints [Equation (10.27)] and the nonnegativity conditions.

Second, the optimum occurs at ($X_K = 0$, $X_C = 20$), which is a result of the priority 1 minimization of the negative deviation from the profit and overhead goal-in fact, $d_1^- = 0$ and $d_1^+ = \$5.00$ in the optimal solution. Thus, d_1^- was not only driven to its minimum (zero) as P_1 required, but $d_1^+ = 5$ (>0) came into solution, indicating that the profit and overhead goal was over attained.

As a result of this superordinate requirement, another goal was achieved while the third was not. The priority 2 working capital goal was overachieved by $30, as seen by solving Equation (9.32):

$$2.00(0) + 3.00(20) - d_3^+ + 0 = 30$$

or

$$d_3^- = 30 \quad d_3^+ = 0$$

The priority 3 cash balance goal was underachieved by $12, however, as seen by solving Equation (10.31):

$$1.0(0) + 2.0(20) + 0 - d_2^+ = 28$$

or

$$d_2^- = 12 \quad d_2^+ = 0$$

10.4 Project Selection by Goal Programming Methods 309

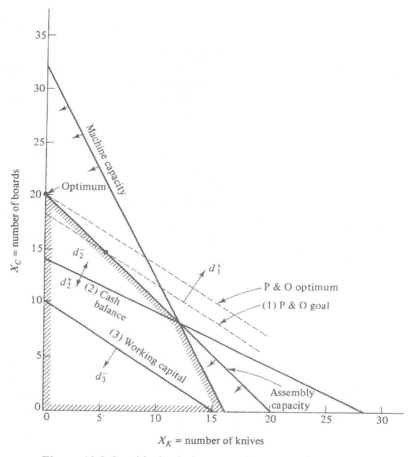

Figure 10.2 Graphical solution to goal programming problem

This negative deviation indicates that the ending cash balance is zero (the goal was $12, and it was underachieved by $12). Finally, the optimal solution indicates that there is excess machine capacity. This can be seen by defining S_4 as the slack. The machine constraint is rewritten as an equality, and then solved for S_4, in terms of $X_K = 0$, $X_C = 20$:

$$0.5 X_K + 0.25 X_C \leq 8$$
$$0.5(0) + 0.25(20) + S_4 = 8$$

or

$$S_4 = 3.$$

For more than two variables, the goal programming problem is solved as a modified linear program. The multiple objectives appear in the objective row, ordered by priorities. The objective variables with the highest priority are brought into solution first and optimized, and then the first objective is fixed. Note that the priority 2 objective is then optimized, subject to the deviation for priority 1. In turn, the lower priority objectives are treated in the same manner until all objective rows are optimized.

10.4.3 Project selection by goal programming. Our example for the application of goal programming to project selection in an imperfect capital market has a candidate set of 15 projects (some are interrelated), which are evaluated relative to nine prioritized goals (some conflicting).[5] Although cash flows and project lives are assumed to be known with certainty, the market discount rates for borrowing and lending are unknown. Since a discounting rate is unavailable in an imperfect capital market, all we can establish is a preference ordering—funds flowing at $t = 1$ are ordinally more valuable than similar amounts at $t = 2$, and so forth. [Baumol and Quandt (1965) and Bernhard (1971) suggest a utility function of temporal funds flows in lieu of a discount rate. Here, we have essentially done this.] From a multicriteria perspective, an ordinal preference ordering is equivalent to a lexicographic ordering, while the utility function approaches an additive weighted model. A discount rate would define a fixed relationship between the weights of adjacent periods.

Table 10.2 Estimated annual costs of projects

Project, j	Description	Cost in Year t ($\$ = \$10^4$)			
		$t = 1$	$t = 2$	$t = 3$	$t = 4$
1	Internal project	$50	$10	$5	$0
2*	Build foreign factory	100	0	0	8
3	Noise pollution control project	20	20	0	0
4	Noise pollution control project	75	2	2	1
5	Internal project	200	0	55	0
6*	Foreign distribution project	0	150	0	0
7	Internal project	0	90	30	0
8	Internal project	0	100	0	10
9 ‡	Clean air project	0	60	5	0
10	Internal project	0	10	86	4
11	Internal project	0	0	100	0
12 ‡	Clean air project	0	0	80	0
13	Internal project	0	0	30	30
14	Internal project	0	0	2	70
15*	Foreign advertising project	0	0	0	100

* Project 15 is conditioned upon acceptance of Projects 2 and 6.
† Projects 3 and 4 are mutually exclusive.
‡ Projects 9 and 12 are mutually exclusive.

---

5. This example is adapted from Lee and Lerro (1974), Run 11; however, their estimated NPV objective (D in their appendix) has been eliminated and an ordinal priority goal substituted (since here the interest rate is unknown due to imperfect capital market conditions). Also, we have reordered some priorities; therefore, our solution differs.

10.4 Project Selection by Goal Programming Methods

The undiscounted costs of the 15 projects are given in Table 10.2, along with brief goal-related descriptions. The estimated annual income and cash flows from each project are given in Table 10.3. The annual incomes are displayed because one management goal is a 50% increase in annual income.

The following verbal description (the complete model is in Appendix 10A) will assist in understanding the GP formulation. First, the firm's capital budget for each year is $B_1 = \$200$; $B_2 = \$250$; $B_3 = \$260$; $B_4 = \$120$ (all amounts in tens of thousands of dollars). For strict control over expenditures the four budget constraints are:

$$\sum_{j=1}^{15} C_{tj} x_j \leq B_t \qquad (10.34)$$

where C_{tj} = the cost of the jth project in the tth year ($t = 1, 2, 3, 4$)
 x_j = the proportion of the jth period accepted; that is, $0 \leq x_j \leq 1$

Table 10.3 Estimated annual income and cash

Project		Estimated Annual Income ($ = \$10^4$)				Estimated Annual Cash Inflow			
	j	t = 1	t = 2	t = 3	t = 4	t = 1	t = 2	t = 3	t = 4
1		$0	$25	$30	$0	$75	$100	$100	$100
2		10	100	70	105	150	150	150	300
3		0	1	5	5	5	10	20	40
4		−100	5	0	−2	0	140	50	40
5		−10	65	−15	26	200	200	50	200
6		0	−115	0	40	0	60	70	80
7		0	−75	−45	385	0	30	0	450
8		0	−140	97	58	0	0	120	150
9		0	95	−10	0	0	160	0	0
10		0	−13	−12	80	0	0	100	110
11		0	0	50	200	0	0	250	295
12		0	0	−100	190	0	0	0	240
13		0	0	30	-55	0	0	114	187
14		0	0	6	62	0	0	38	194
15		0	0	0	290	0	0	0	550
Σ =		−$100	−$52	$106	$1384	$430	$820	$1062	$2936

Casting these constraints into goal form, we rewrite Equation (10.34) as:

$$\sum_{j=1}^{15} C_{tj}x_j + d_t^- - d_t^+ = B_t \qquad (10.35)$$

where d_t^- = an underachievement (underspending) of the budget goal.
d_t^+ = an overachievement (overspending) of the budget goal.

For this goal, the positive or overspending deviation is minimized.

Second, a corollary of the budget limits allows for transfer of funds between periods. The *cumulative* amount available for capital allocation is simply the sum of the unspent budgets. Thus, for the second year, if $50 is spent in the first year, there would be available $150 + $250 = $400 for allocation. Likewise, cumulative allocations are subjected to these limits. This allows the firm some flexibility to take advantage of opportunities regardless of their timing. The generalized goal expression for funds mobility between years is:

$$\sum_{j=1}^{15}\sum_{t=1}^{t'} C_{tj}x_j + d_k^- - d_k^+ = \sum_{t=1}^{t'} B_t \qquad (10.36)$$

where t' = 2, 3, 4 and k = 5, 6, 7. These three goals express management's goal that the cumulative expenditures for accepted projects not exceed the cumulative budgeted amounts, for t = 2, 3, or 4.

Third, the firm faces public regulatory pressure to control the noise from one plant and to abate some serious air pollution problems. Projects 3 and 4 are mutually exclusive methods of solving the noise problem, and projects 9 and 12 are the mutually exclusive alternatives for abating the air pollution problem. Normally, these objectives would be formulated as constraints, such as

Solve noise pollution: $x_3 + x_4 = 1$
Air pollution: $x_9 + x_{12} = 1$.

However, in the present formulation the combination (x_3 = 1, x_9 = 1) is better and corresponding deviations are minimized. The deviation variables for accepting projects 3 and 9 are (d_8^- and d_8^+), and (d_9^- and d_9^+), respectively; and for rejecting projects 4 and 12 they are (d_{40}^- and d_{40}^+) and (d_{41}^- and d_{41}^+).

Foreign-based manufacturing and distribution operations have become more profitable and attractive, and project 2 is a new captive manufacturing plant in a foreign country, which will open up new markets there and in adjacent countries. Project 6 proposes to select and establish new distribution channels in the foreign countries, and project 15 is an advertising campaign. Neither the distribution plan nor the advertising campaign should be undertaken without the plant. These limitations can be expressed by conditional constraints:

$$\left.\begin{array}{l} x_{15} - x_6 - d_{23}^+ + d_{23}^- = 0 \\ x_6 - x_2 - d_{24}^+ + d_{24}^- = 0 \end{array}\right\} \qquad (10.37)$$

10.4 Project Selection by Goal Programming Methods

whose positive deviations (d⁺) would be minimized. Thus, with sufficiently high priority, the d⁺ values would be zero, with the following results:

x_2	x_6	x_{15}	Decision
0	0	0	Execute nothing
1	0	0	Build plant
1	1	0	Build plant; distribute
1	1	1	Build plant; distribute, advertise

(This assumes that $d^- = 0$ also.)

The remaining objectives relate to the firm's cash flow, growth, and total value. Since a discount rate for the intertemporal transfer of funds is not previously known in imperfect capital markets, a workable substitute for discounting is to place a higher utility on the earlier flow of funds.[6] The GP format supports this by assigning arbitrary weights to deviations within a priority class. To incorporate this concept, one objective is written for each period. For example, for $t = 1$, the cash inflow objective is:

$$75X_1 + 150X_2 + 5X_3 + 200X_5 + d_{19}^- - d_{19}^+ = 430 \tag{10.38}$$

where the coefficients of the X_j are the cash flows for $t = 1$ in Table 10.3 and the RHS value, 430, is simply the sum of the coefficients on the LHS. In the achievement function of the GP problem, however, one wishes to minimize the weighted negative deviations; so,

$$\text{Min } S = P_i(W_1 d_{19}^- + W_2 d_{20}^- + W_3 d_{21}^- + W_4 d_{22}^-) \tag{10.39}$$

where P_i = priority level of the funds flow objectives, and $W_1, \ldots, W_4$ = arbitrary weights (utilities) assigned to the negative deviations d_t^-, where $W_1 > W_2 > W_3 > W_4$ to imply greater utility attachment to early funds flow. So goals like Equation (10.37) express the objective of achieving cash inflow goals, and the weighted negative deviations [Equation (10.39)] assure incorporation of a value weighting on each cash inflow. A similar weighting process expresses a time-value preference for cash outflows.

To maximize the chances for market growth, the firm desires a 50% growth rate in its income (not cash flows). In detail the planning staff has established the following annual income goals: $15, $22.50, $34, and $51. Four constraints, corresponding to each year ($t = 1, 2, 3, 4$) and ($k = 10, 11, 12, 13$) in the planning horizon, express this growth requisite:

$$\sum_{j=1}^{15} (NI)_{tj} X_j + d_k^- - d_k^+ \tag{10.40}$$

where $(NI)_{tj}$ = net income from project j in period t (Table 10.3)
         X_j = decision variable, $0 \le X_j \le 1$
         $(TI)_t$ = total income for year t ($15, $22.50, $34, $51).

---

6. See Baumol and Quandt (1965) and Bernhard (1971).

Minimizing the negative deviations, d_k^-, approaches the objective as closely as possible on the underachievement side; if a d_k^+ comes into a solution, the corresponding growth goal is exceeded.

The firm expects a large cash outflow near the end of year 3 to meet a contractual obligation with its union. To minimize external financing, management desires the internal cash inflow during year 3 to be at least $5 million. This goal is formulated as:

$$\sum_{j=1}^{15} Y_{tj}X_j + d_{14}^- - d_{14}^+ = 500 \qquad (10.41)$$

where Y_{3j} = cash inflow from project j, accepted in year 3. Management also desires to maintain a minimum liquidity level for each period, as follows: $300, $500, $500, and $1,700. Just as the cash inflow was maximized for year 3 in Equation (10.41), four additional goals (k = 15, 16, 17, 18) can be written for the minimum liquidity requirements, as follows:

$$\sum_{j=1}^{15} Y_{tj}X_j + d_k^- - d_k^+ = (TCF)_t \qquad (10.42)$$

where $(TCF)_t$ = the minimum liquidity levels (total cash flows). Minimizing the negative deviations most closely approaches the liquidity goals.

Last, no multiples of projects should be permitted. The 15 constraints (k = 25, ... , 39) that prevent multiple projects are of the form

$$X_j + d_k^- - d_k^+ = 1$$

where the positive deviation d_k^+, are minimized with a high priority.

The formulation of the GP objectives is now complete. An annotated summary of the model can be found in Appendix 10A. The GP achievement function is formulated as follows. Management's consensus on relative priorities ($P_1 > P_2 > \ldots\ldots > P_k$) for the objectives and within a priority class the arbitrary weights (that is, $W_1 > W_2 > W_3 > W_4$) follow this rationale:

Priority 1
   A. Prohibit multiples of projects.
   B. Minimize annual budget overruns, incorporating a utility of money for time preference.

Priority 2
   A. Accept pollution control projects 3 and 9.
   B. Reject pollution control projects 4 and 12.
   C. Maximize cash inflow in year 3 (for union contract obligations).
   D. Value A and B twice as much as C.

Priority 3
      Achieve early funds (cash) inflow, incorporating a utility of money for time preference.

10.4 Project Selection by Goal Programming Methods

Priority 4

Maintain 50% growth rate on total income.

Priority 5

A. Minimize the shifting of projects from one period to another [that is, minimize intertemporal transfer of funds via Equation (10.36)].
B. Assure foreign investment in new foreign plant, distribution system, and/or advertising.
C. The foreign investments (B) are three times as important as the shifting of funds objective (A).

Priority 6

Minimize the firm's excess liquidity in each period.

The foregoing considerations result in the following achievement function for the GP problem:

GP problem:

$$\text{Min } Z = P_1 \underbrace{\{10(d_{25}^+ + d_{26}^+ + \cdots + d_{39}^+)}_{\text{Deviations permitting multiples of projects}} + \underbrace{4d_1^+ + 3d_2^+ + 2d_3^+ + 1d_4^+\}}_{\substack{\text{Excess budget overruns} \\ \text{incorporating utility-of-} \\ \text{money weights}}}$$

$$+ P_2 \underbrace{\{2(d_8^- + d_8^- + d_9^- + d_9^-)}_{\substack{\text{Accept pollution control} \\ \text{Projects 3 and 9}}} + \underbrace{d_{40}^+ + d_{40}^+ + d_{41}^- + d_{41}^+}_{\substack{\text{Reject pollution control} \\ \text{Projects 4 and 12}}}$$

$$+ \underbrace{d_{14}^-\}}_{\substack{\text{Underachievement} \\ \text{of Year 3 cash flow goal}}} \quad (10.43)$$

$$+ P_3 \underbrace{\{4d_{19}^- + 3d_{20}^- + 2d_{21}^- + 1d_{22}^-\}}_{\substack{\text{Underachievement of funds} \\ \text{inflow, incorporating utility-of} \\ \text{-money function}}} + P_4 \underbrace{\{d_{10}^+ + d_{11}^+ + d_{12}^+ + d_{13}^+\}}_{\substack{\text{Maintain 50\% income} \\ \text{growth rate}}}$$

$$+ P_5 \underbrace{\{3d_{23}^+ + 3d_{24}^+}_{\substack{\text{Assue foreign} \\ \text{investment at} \\ \text{weight = 3}}} + \underbrace{d_5^+ + d_6^+ + d_7^+\}}_{\substack{\text{Prevent excess} \\ \text{transfer of funds} \\ \text{from period to} \\ \text{period}}} + P_6 \underbrace{\{d_{15}^+ + d_{16}^+ + d_{17}^+ + d_{18}^+\}}_{\substack{\text{Minimize excess} \\ \text{liquidity in each of} \\ \text{the 4 years}}}$$

This project selection problem was originally solved by a modified linear goal programming code developed by Bershader (1975) and reported in Ignizio (1976). (The modifications consisted principally of accommodating 60 objectives and 10 priority levels.) Bershader's coding follows the modified simplex procedure, so that a stagewise optimization procedure seeks successive suboptima in decreasing priority order. The indicated solution with some fractional acceptance levels is:

Project	Description	Acceptance Level
1	Internal project	0.224
2	Foreign factory	1.000
3	Noise pollution control	1.000
5	Internal project	0.344
8	Internal project	0.600
9	Clean air project	1.000
13	Internal project	1.000
14	Internal project	1.000

The highest priority goals were satisfied exactly, or exceeded. No multiples of projects were accepted and no budget overruns were allowed (budgets in years 1, 2, and 3 were used exactly, and year 4's was under-utilized by $2.21). Under priority 2, the designated pollution control projects were selected,[7] which were projects 3 and 9. Also under priority 2, year 3's cash inflow goal was exceeded by $143.94.

Beginning with priority 3, some goals were partially achieved. The priority 3 goals of maximizing early funds inflows, weighted by an assumed utility ordering, were all underachieved. For year 1, the under-achievement was $279.40; for year 2, $418.80; for year 3, $342; and for year 4, $1390.70. Therefore, the utility equivalent of maximum net present value was not achieved. The priority 4 goal of maintaining a 50% growth rate in income was underachieved in years 1 and 2 ($−9.2 and $−35.2) but overachieved in years 3 and 4 ($181.1 and $424.2). The priority 5 goal of building the foreign factory was achieved, but the proposed distribution system and advertising program were rejected. Also under priority 5, the minimization of intertemporal transfer of funds was achieved for years 1–2 and 2–3 but failed for years 3–4. Finally the liquidity goal (priority 6) was underachieved in years 1, 2, and 4 but overachieved in year 3. The complete numerical solution is presented in Appendix 10A.

Obviously, the foregoing solution would change with a simple reordering of priorities or other modifications in the conceptual formulation. Since this kind of problem restructuring is equivalent to restating the decision-maker's preferences, it simply demonstrates the principle enunciated by Hirshleifer, Bernhard, and others: In an imperfect capital market, the optimal project selection results from the decision-maker's choices. If these can be rationally ordered, consistent decisions can be made. In Lee and Lerro's article (1974), the next step is made by examining eight decisional permutations of a capital budgeting problem, which permits management to ask what if questions beforehand.

10.5 SUMMARY

This chapter has covered three areas. First, using a simple case developed by Hirshleifer in which capital market (interest) rates differ for borrowing and lending, we demonstrated

---

7. The other pollution control projects could just as well have been eliminated for our purposes. The entire problem, however, was adapted from one reported by Lee and Lerro (1974), who permitted the alternative pollution control projects.

that maximizing NPV collapses and the separation theorem is invalidated. No longer can the optimal set of projects be chosen separately from the method of financing them. Second, using the Bernhard generalized horizon model, we showed some alternative formulations of the selection problem, when interest rates are not procurable internally or externally. These models are basically undiscounted flow-of-funds models of the firm that comply with some terminal wealth restriction. Then we presented some of Bernhard's results, obtained by deriving the Kuhn-Tucker conditions, for capital budgeting models. One of the most important is a substantiation of Baumol and Quandt's finding that the interest rate used to find NPV is not internally procurable from the model. In other words, the source of funds and its cost must be known to use NPV.

Finally, to demonstrate that the project selection problem can be made operational in imperfect capital markets, we summarized the goal programming approach and then illustrated how a 15-project problem could be formulated, subjected to prioritized goals, and solved.

APPENDIX 10.A COMPILATION OF PROJECT SELECTION PROBLEM

The model used here is Lee and Lerro's (1974) model 4, modified to eliminate the net present value objective and substitution of an ordered utility-of-money function.

Formulations of the Objectives

A. Budget objectives (priority 1)

(First year): $50X_1 + 100X_2 + 20X_3 + 75X_4 + 200X_3 + d_1^- - d_1^+$ = 200.0

(Second year): $10X_1 + 20X_3 + 20X_4 + 75X_6 + 200X_7 + 100X_8 +$
$60X_9 + 10X_{10} + d_2^- - d_2^+$ = 250.0

(Third year): $5X_1 + 2X_4 + 55X_5 + 30X_7 + 5X_9 + 86X_{10} + 180X_{11} +$
$80X_{12} + 30X_{13} + 2X_{14} + d_3^- - d_3^+$ = 260.0

(Fourth year): $8X_2 + X_4 + 10X_8 + 4X_{10} + 30X_{13} + 70X_{14} + 100X_{15} +$
$d_4^- - d_4^+$ = 120.0

B. Permissible interperiod transfer of funds (priority 5)

(Year 2): $60X_1 + 100X_2 + 40X_3 + 77X_4 + 200X_5 + 150X_6 +$
$90X_7 + 100X_8 + 60X_9 + 10X_{10} + d_5^- - d_5^+$ = 450.0

(Year 3): $65X_1 + 100X_2 + 40X_3 + 79X_4 + 255X_5 + 150X_6 +$
$120X_7 + 100X_8 + 65X_9 + 96X_{10} + 180X_{11} +$
$80X_{12} + 30X_{13} + 2X_{14} + d_6^- - d_6^+$ = 710.0

(Year 4): $65X_1 + 108X_2 + 40X_3 + 80X_4 + 255X_5 + 150X_6 +$
$120X_7 + 110X_8 + 65X_9 + 100X_{10} + 180X_{11} +$
$80X_{12} + 60X_{13} + 72X_{14} + 100X_{15} + d_7^- - d_7^+$ = 830.0

C. Accept pollution control projects 3 and 9

$$X_9 + d_8^- - d_8^+ = 1$$
$$X_3 + d_9^- - d_9^+ = 1$$

D. Maintain 50% income growth (priority 4)

$$10X_2 - 100X_4 - 10X_5 + d_{10}^- - d_{10}^+ = 15.0$$

$$25X_1 + 100X_2 + X_3 + 5X_4 + 65X_5 - 115X_6 - 75X_7 - 140X_8 + 95X_9 - 13X_{10} + d_{11}^- - d_{11}^+ = 22.5$$

$$30X_1 + 70X_2 + 5X_3 - 15X_5 - 45X_7 + 97X_8 - 10X_9 - 12X_{10} + 50X_{11} - 100X_{12} + 30X_{13} + 6X_{14} + d_{12}^- - d_{12}^+ = 34.0$$

$$105X_2 + 5X_3 - 2X_4 + 26X_5 + 40X_6 + 385X_7 + 58X_8 + 80X_{10} + 200X_{11} + 190X_{12} - 55X_{13} + 62X_{14} + 290X_{15} + d_{13}^- - d_{13}^+ = 51.0$$

E. Maximize cash flow in year 3 (priority 2)

$$100X_1 + 150X_2 + 20X_3 + 50X_4 + 50X_5 + 70X_6 + 120X_8 + 100X_{10} + 250X_{11} + 114X_{13} - 38X_{14} + d_{14}^- - d_{14}^+ = 500.0$$

F. Minimize excess liquidity (priority 6)

(Year 1): $\quad 75X_1 + 150X_2 + 5X_3 + 200X_5 + d_{15}^- - d_{15}^+ = 300.0$

(Year 2): $\quad 100X_1 + 150X_2 + 10X_3 + 140X_4 + 200X_5 + 60X_6 + 160X_9 + d_{16}^- - d_{16}^+ = 500.0$

(Year 3): $\quad 100X_1 + 150X_2 + 20X_3 + 50X_4 + 50X_5 + 70X_6 + 120X_8 + 100X_{10} + 250X_{11} + 114X_{13} + 38X_{14} + d_{17}^- - d_{17}^+ = 500.0$

(Year 4): $\quad 100X_1 + 300X_2 + 40X_3 + 40X_4 + 200X_5 + 80X_6 + 450X_7 + 150X_8 + 110X_{10} + 295X_{11} + 240X_{12} + 187X_{13} + 194X_{14} + 550X_{15} + d_{18}^- - d_{18}^+ = 1{,}700.0$

G. Place utility ranking on funds flow (priority 3)

(Year 1, $W=4$): $\quad 75X_1 + 150X_2 + 5X_3 + 200X_5 + d_{19}^- - d_{19}^+ = 430.0$

(Year 2, $W=3$): $\quad 100X_1 + 150X_2 + 10X_3 + 140X_4 + 200X_5 + 60X_6 + 160X_9 + d_{20}^- - d_{20}^+ = 820.0$

(Year 3, $W=2$): $\quad 100X_1 + 150X_2 + 20X_3 + 50X_4 + 50X_5 + 70X_6 + 120X_8 + 100X_{10} + 250X_{11} + 114X_{13} + 38X_{14} + d_{21}^- - d_{21}^+ = 1{,}062.0$

(Year 4, $W=1$): $\quad 100X_1 + 300X_2 + 40X_3 + 40X_4 + 200X_5 + 80X_6 + 450X_7 + 150X_8 + 110X_{10} + 295X_{11} + 240X_{12} + 187X_{13} + 194X_{14} + 550X_{15} + d_{22}^- - d_{22}^+ = 2{,}936.0$

H. Increase foreign investment (priority 5)

$$X_{15} - X_6 + d_{23}^- - d_{23}^+ = 0$$
$$X_{15} - X_2 + d_{24}^- - d_{24}^+ = 0$$

I. Define range of project acceptance (priority 1)

$$X_1 + d_{25}^- - d_{25}^+ = 1$$
$$X_2 + d_{26}^- - d_{26}^+ = 1$$

$$\vdots \quad \vdots \quad \vdots$$
$$\vdots \quad \vdots \quad \vdots$$
$$X_{15} + d_{39}^- - d_{39}^+ = 1$$

(Min d_j^+ forces nonmultiples of projects.)

J. Reject two pollution control projects (priority 2)

$$X_4 + d_{40}^- - d_{40}^+ = 0$$
$$X_{12} + d_{41}^- - d_{41}^+ = 0$$

Solution to the GP problem, using the objective in Equation (10.43)

Projects:

$X_1 = 0.2237$ $X_9 = 1.0000$
$X_2 = 1.0000$ $X_{10} = 0$
$X_3 = 1.0000$ $X_{11} = 0$
$X_4 = 0$ $X_{12} = 0$
$X_5 = 0.3441$ $X_{13} = 1.0000$
$X_6 = 0$ $X_{14} = 1.0000$
$X_7 = 0$ $X_{15} = 0$
$X_8 = 0.6008$

Deviations (listed as they appear in the objective, Equation (10.43)
1. A. Multiples of projects:

$$d_{25}^+, d_{26}^+, \ldots, -d_{39}^+ = [0]$$

B. Budget goals:

$$d_1^+, \ldots, -d_4^+ = [0] \quad d_1^-, \ldots, -d_3^- = [0] \quad d_4^- = 2.2095$$

2. A. Accept pollution control projects 3, 9:

$$d_8^+, d_8^- = 0 \quad d_9^+, d_9^- = 0$$

B. Reject pollution control projects 4, 12:

$$d_{40}^+, d_{40}^- = 0 \quad d_{41}^+, d_{41}^- = 0$$

C. Achieve cash inflow in year 3:

$$d_{14}^+ = 143.9418 \quad d_{14}^- = 0.$$

3. A. Achieve early cash inflows:

$$d^+_{19}, \ldots\ldots, -d^+_{22} = [0] \quad d^-_{19} = 279.3921 \quad d^-_{20} = 418.8652$$
$$d^-_{21} = 342.0579 \quad d^-_{22} = 1{,}390.7168$$

4. A. Maintain 50% growth rate on income:

$$d^+_{10} = 0, d^-_{10} = 9.1168 \quad d^+_{11} = 0, d^-_{11} = 35.1549$$
$$d^+_{12} = 181.0716, d^-_{12} = 0 \quad d^+_{13} = 424.1506, d^-_{13} = 0$$

5. A. Minimize intertemporal transfer of funds:

$$d^+_5 = 0, d^-_5 = 0 \quad d^+_6 = 0, d^-_6 = 0 \quad d^+_7 = 0, d^-_7 = 2.2105$$

B. Assure foreign investment:

$$d^+_{23} = 0, d^-_{23} = 0 \quad d^+_{24} = 0, d^-_{24} = 1.0000$$

6. A. Minimize excess liquidity:

$$d^+_{15} = 0, d^-_{15} = 66.2921 \quad d^+_{16} = 0, d^-_{16} = 98.8659$$
$$d^+_{17} = 219.9418, d^-_{17} = 0 \quad d^+_{18} = 0, d^-_{18} = 154.7171$$

Summation of terms in the achievement function

P_1:	$\sum Wd_t =$	0
P_2:	$\sum Wd_t =$	0
P_3:	$\sum Wd_t =$	4449 ($\Rightarrow$ underachievement)
P_4:	$\sum Wd_t =$	605.22 ($\Rightarrow$ underachievement)
P_5:	$\sum Wd_t =$	0
P_6:	$\sum Wd_t =$	219.94 ($\Rightarrow$ underachievement)

REFERENCES

BAUM, SANFORD, "Observations on 'Discount Rates in Linear Programming Formulations of the Capital Budgeting Problem'," *The Engineering Economist*, **30**(3) (Spring 1985), pp. 292–293.

BAUMGARTEN, EDWIN O., "An Investigation of the Goal Programming Method" (unpublished MS Thesis, Kansas State University, 1973).

BAUMOL, WILLIAM J., and RICHARD E. QUANDT, "Investment and Discount Rates Under Capital Rationing–A Programming Approach," *The Economic Journal*, **75**(298) (June 1965), pp. 317–329.

BERNHARD, RICHARD H., "Mathematical Programming Models for Capital Budgeting–A Survey, Generalization, and Critique," *Journal of Financial and Quantitative Analysis*, **4**(2) (June 1969), pp. 111–158.

BERNHARD, RICHARD H., "Some Problems in the Use of a Discount Rate for Constrained Capital Budgeting," *AIIE Transactions*, **3**(3) (September 1971), pp. 180–184.

BERSHADER, PAULA S., "Linear Goal Programming Package," University Park, PA: The Pennsylvania State University (Spring 1975) (reproduced in Ignizio (1985), pp. 227–242).

BYRNE, R., A. CHARNES, W. W. COOPER, and K. KORTANEK, "A Chance-Constrained Approach to Capital Budgeting with Portfolio Type Payback and Liquidity and Horizon Posture Controls," *Journal of Financial and Quantitative Analysis*, **2**(4) (December 1967), pp. 339–364.

CARRAWAY, ROBERT L., and ROBERT L. SCHMIDT, "An Improved Discrete Dynamic Programming Algorithm for Allocating Resources Among Interdependent Projects," *Management Science*, **37**(9) (September 1991), pp. 1195–1200.

CHARNES, A., and W. W. COOPER, *Management Models and Industrial Applications of Linear Programming* (Wiley, 1961).

CHARNES, A., W. W. COOPER, and M. H. MILLER, "Application of Linear Programming to Financial Budgeting and the Costing of Funds," *Journal of Business*, **32**(1) (January 1959), pp. 20–46.

CHUI, YU, and CHAN S. PARK, "Capital Budgeting Decisions with Fuzzy Projects," *The Engineering Economist*, **43**(2) (Winter 1998), pp. 125–150.

FREELAND, JAMES R., and MEIR J. ROSENBLATT, "An Analysis of Linear Programming Formulations for the Capital Rationing Problem," *The Engineering Economist*, **24**(1) (Fall 1978), pp. 49–61.

FREELAND, JAMES R., and MEIR J. ROSENBLATT, "A Note on Discount Rates in Linear Programming," *The Engineering Economist*, **30**(2) (Winter 1985), pp. 191–192.

GLOVER, FRED, "Improved Linear Integer Programming Formulations of Nonlinear Integer Problems," *Management Science*, **22**(4) (December 1975), pp. 455–460.

GOLDSTEIN, PAULA M., and HOWARD M. SINGER, "A Note on Economic Models for R&D Project Selection in the Presence of Project Interactions," *Management Science*, **32**(10) (October 1986), pp. 1356–1360.

GURGUR, CIGDEM Z., and JAMES T. LUXHOJ, "Application of Chance-Contrained Programming to Capital Rationing Problems with Asymmetrically Distributed Cash Flows and Available Budget," *The Engineering Economist*, **48**(3) (2003), pp. 241–258.

HAYES, JAMES, "Discount Rates in Linear Programming Formulations of the Capital Budgeting Problem," *The Engineering Economist*, **29**(2) (Winter 1984), pp. 113–126. [See comments by Freeland and Rosenblatt (1985) with author's reply in *The Engineering Economist*, **30**(3) (Spring 1985), pp. 294–295, and comments by Baum (1985) with author's reply in *The Engineering Economist*, **30**(4) (Summer 1985), pp. 397–398.]

HAYES, JAMES, "Dual Variables in Pure Capital Rationing Linear Programming Formulations," *The Engineering Economist*, **34**(3) (Spring 1989), pp. 255–260.

HEIDENBERGER, KURT, and CHRISTIAN STUMMER, "Research and Development Project Selection and Resource Allocation: A Review of Quantitative Modeling Approaches," *International Journal of Management Reviews*, **1**(2) (June 1999), 197–224.

HIRSHLEIFER, J., "On the Theory of Optimal Investment Decision," *Journal of Political Economy*, **66**(5) (August 1958), pp. 329–352; reprinted in Ezra Solomon, ed., *The Management of Corporate Capital* (The Free Press, 1959).

IGNIZIO, JAMES P., *Goal Programming and Extensions* (Lexington Books, Heath, 1976).

IGNIZIO, JAMES P., "An Approach to the Capital Budgeting Problem with Multiple Objectives," *The Engineering Economist*, **21**(4) (Summer 1976), pp. 259–272.

IJIRI, YUJI, *Management Goals and Accounting for Control* (North-Holland, 1965).

KACHANI, SOULAYMANE, and JEROME LANGELLA, "A Robust Optimization Approach to Capital Rationing and Capital Budgeting," *The Engineering Economist* **50**(3), 2005, pp. 195–229.

KIRA, DENNIS, MARTIN KUSY, and IAN RAKITA, "The Effect of Project Risk on Capital Rationing Under Uncertainty," *The Engineering Economist*, **45**(1) (2000), pp. 37–55.

KUMAR, P. C., and TRAMI LU, "Capital Budgeting Decision in Large Scale, Integrated Projects: Case Study of a Mathematical Programming Application," *The Engineering Economist*, **36**(2) (Winter 1991), pp. 127–150.

LEE, SANG M., *Goal Programming for Decision Analysis* (Auerbach, 1972).

LEE, SANG M., *Goal Programming Methods for Multiple Objective Integer Programs* (Institute of Industrial Engineering, 1985).

LEE, SANG M., and A. J. LERRO, "Capital Budgeting for Multiple Objectives," *Financial Management*, **3**(1) (Spring 1974), pp. 58–66.

MANNE, ALAN S., "Optimal Dividend Policies for a Self–Financing Business Enterprise," *Management Science*, **15**(3) (November 1968), pp. 119–129.

MEIER, HELGA, NICOS CHRISTOFIDES, and GERRY SALKIN, "Capital Budgeting Under Uncertainty—An Integrated Approach Using Contingent Claims Analysis and Integer Programming," *Operations Research*, **49**(2) (March–April 2001), pp. 196–206.

NÄSLUND, BERTH. "A Model of Capital Budgeting Under Risk," *Journal of Business*, **39**(20) (April 1966), pp. 257–271.

NEMHAUSER, G. L., and Z. ULLMANN, "Discrete Dynamic Programming and Capital Allocation," *Management Science*, **15**(9) (May 1969), pp. 494–505.

PARK, CHAN S., and GUNTER P. SHARP-BETTE, *Advanced Engineering Economics* (Wiley, 1990).

RAVID, S. ABRAHAM, "On Interactions of Production and Financial Decisions," *Financial Management*, **17**(3) (Autumn 1988), pp. 87–99.

ROSENBLATT, MEIR J., "On the Relationship Between Discount Factors and Dual Variables in the Formulation of the Pure Capital Rationing Problem," *The Engineering Economist*, **30**(2) (Winter 1985), pp. 173–181.

SARPER, HUSEYIN, "Capital Rationing Under Risk: A Chance Constrained Approach Using Uniformly Distributed Cash Flows," *The Engineering Economist*, **39**(1) (Fall 1993), pp. 49–76.

TAYLOR, BERNARD W. III, LAURENCE J. MOORE, and EDWARD R. CLAYTON, "R&D Project Selection and Manpower Allocation with Integer Nonlinear Goal Programming," *Management Science*, **28**(10) (October 1982), pp. 1149–1158.

WEINGARTNER, H. MARTIN, *Mathematical Programming and the Analysis of Capital Budgeting Problems* (Prentice-Hall, 1963; reprinted by Markham, 1967).

WEINGARTNER, H. MARTIN, "Capital Budgeting of Interrelated Projects: Survey and Synthesis," Management Science, **12**(7) (March 1966) pp. 485–516.

WHITE, D. J., "A Bibliography on the Applications of Mathematical Programming Multiple–objective Methods," *The Journal of the Operational Research Society*, **41**(8) (August 1990), pp. 669–691.

PROBLEMS

10-1. How is the separation theorem invalidated by a difference in the borrowing and lending rates of interest in the capital market?

10-2. What other factors in the real capital market can cause the separation theorem to be invalidated?

10-3. What are some possible consequences of imperfect capital market conditions on the project selection problem?

10-4. Using the necessary variables defined in Section 10.3.3 and the additional variables defined in Section 10.3.5 (1. Cash balance constraints, etc.), write the firm's cash inflow and outflow expressions, and then derive Equation (10.2b) as a cash balance equation at t.

10-5. (Bernhard Generalized Model.) A firm is considering four new projects whose cash flow streams are independent. The projects have various lives, but the planning horizon is 5 years. The availability of the firm's managerial resources is a limiting factor, and at any time the total managerial resources available to all new projects is limited to 30 "executive units." Projects 1-4 will use 10, 8, 5 and 15 executive units, respectively. The expected cash flows (in $1,000s) to and from the projects are:

Time, $t =$	Project 1	Project 2	Project 3	Project 4
1	$-500	$-300	$-300	$-500
2	+50	+100	+100	-50
3	100	100	150	+50
4	100	100	100	100
5	100	-50	100	300
Present worth (time $T = 5$) of post-T cash flows:	$ 400	$ 200	-0-	$ 400

Cash throw-offs from other projects in which the firm has already invested will be available in the following amounts:

Year	Amount
1	$100,000
2	25,000
3	50,000
4	50,000
5	100,000

These other projects will continue to generate cash inflows after the planning horizon and the present value of these cash inflows at time $t = 5$ will be $2,000,000.

The firm's borrowing and lending rates of interest are 12% and 6% per year, respectively. The maximum debt in any year is $300,000. Also, the firm must maintain a cash balance at all times of at least $5,000 plus 10% of its outstanding debt. To attract new shareholders, the Board of Directors has decided on the following minimum dividends over the next 5 years: $50,000; $55,000; $61,000; $67,000; and $74,000, respectively.

(a) Formulate the project selection problem using the Bernhard generalized model. In doing so, maximize terminal wealth plus the discounted dividend stream (use a discount rate of 6% on dividend amounts), and include the following constraints: (1) cash balance restrictions, (2) scarce resource restriction, (3) terminal wealth restriction, (4) upper bound(s) on borrowing, (5) lower bound(s) on payment of dividends, (6) prohibition of multiple projects.

(b) Solve (a) by a computer-based linear programming code, such as Solver. Use the model from (a). Nonnegativity constraints will be required.

(c) Write the Lagrangian corresponding to (a), and derive the properties of the optimal solution by taking partial derivatives and setting the results equal to zero.

10-6. (Bernhard special case 2b, in which the borrowing and lending rates are equal and timewise invariant.) Consider four independent projects, not mutually exclusive, with expected cash flows as follows:

Expected Cash Flows, Y_{tj}

End of Period, t=	Project Identification, j =			
	1	2	3	4
1	$-20,000	$-25,000	$-45,000	$-35,000
2	-1,500	+2,000	+3,000	-25,000
3	+1,000	5,000	5,000	10,000
4	1,500	15,000	10,000	7,500
5	1,500	15,000	15,000	7,500
6	750	4,000	2,000	5,000
7	250	2,000	2,000	1,000
8	-0-	500	-0-	-0-

Other assumptions are that the planning horizon is $T = 3$, borrowing and lending rates of interest are constant at 10% per year, and the upper limits on borrowing for periods 1 and 2 are $B_1 = \$20,000$ and $B_2 = \$25,000$. Cash inflows to the firm from other projects outside this study are:

Time, t =	1	2	3	4	5	6	7
Amount, $	15,000	10,000	5000	5000	3000	2000	-0-

There is a 100-person manpower restriction on the total number of employees available for the new projects. The number of employees required for each project is:

Project, j=	1	2	3	4
People	20	25	45	30

A constant cash balance requirement of $5000 is necessary in all periods (except $C_0 = 0$). In addition, 50% of the outstanding debt at any time must be kept liquid (as cash). A minimum dividend of $30 must be paid during each period up to the planning horizon.

(a) Formulate this problem as a mathematical program which maximizes the sum of the annual dividend payments and the terminal wealth at time t = 3, subject to the following constraints: (i) cash balance restrictions including liquidity constraints, (ii) scarce resource restriction(s), (ii) terminal wealth restriction, (iv) terminal wealth posture restriction, (v) upper bound(s) on borrowing, (vi) prohibition of multiple projects, (vii) lower bound(s) on dividends, (viii) nonnegativity restrictions.

(b) Solve the problem by a computer-based linear programming code and give economic interpretations of the optimal solution.

10-7. The A-B-C Company is studying the desirability of adopting one or more of four independent projects. The expected cash flows from the projects and budget limits (in $million) are:

	Time, t =		
Project	0	1	2
1	$-1	$ 3	0
2	-1	2	1.5
3	0	-1	2
4	0	-2	3.5
Budget limit	$ 1	$ 2	0

For simplicity, assume that the cash flows are zero after $t = 2$ and that the market borrowing and lending rates of interest are equal and constant. Amounts available for investment, however, are from external sources and are constrained as shown in the table. The firm must retain, at all times, 20% of its borrowings in liquid form (cash or quick, marketable securities). The firm can borrow only at $t = 1$, up to a maximum of $1; but it may lend money at any time. There are no other scarce resources. The expected terminal wealth at $t = 2$ is $2 more than the sum of the dividends paid in periods 1 and 2. Neither multiple nor partial projects are acceptable.

(a) Formulate the problem as a mathematical program, writing the objective function as Max $F = f(W_1, W_2, G)$, where W_1, W_2 = dividends to be paid in periods 1 and 2, respectively, and G = terminal wealth of the firm at $t = 2$.

(b) Solve the problem using computer based linear programming software such as Solver. Assume an interest rate of 10%.

(c) How does the value of the terminal wealth, G, change as the interest rate changes between 0% and 15%? Does the project selection change as the interest rate changes within this range?

10-8. Redo problem 10-5 parts (a) and (b) assuming that projects 1 and 2 are synergistic. If both projects are undertaken, then the net returns from year 1 on are increased by 25%.

10-9. Redo problem 10-5 parts (a) and (b) assuming that projects 1 and 2 are competitive. If both projects are undertaken, then the net returns from year 1 on are reduced by 15%. (See Chapter 9 or (Goldstein and Singer, 1986) on interacting projects.)

10-10. Redo problem 10-5 part (a) assuming that the first costs of projects 1 and 2 can be reduced by 20% if they are both undertaken. Net annual returns in later years are not changed.

10-11. (Goal Programming Problem). EMPCO Corporation is an automotive equipment manufacturer. The State Air Pollution Control Board has recently issued regulations for more stringent emission controls. EMPCO has designed an emission control device, No. EC-1, to meet these requirements. Because of the Federal governments energy policy, however, an additional project concerning dual-fuel injection for small cars is also under consideration. This project is independent of the EC-1 project. The U.S. market for fuel injection equipment is still somewhat limited; hence, EMPCO plans to market its products in foreign countries also. This is the third project under consideration. The undiscounted cash outflows due to the three projects are tabulated below.

Project	$t = 1$	$t = 2$	$t = 3$
Dual-fuel project	$50M*	$10M	$0
Foreign Project	10M	15M	20M
EC-1 emission project	50M	0	0

*M = $1,000,000.

The cash inflows from these projects are:

Project	$t = 1$	$t = 2$	$t = 3$
Dual-fuel project	$10M	$50M	$60M
Foreign Project	5M	20M	40M
EC-1 emission project	5M	30M	45M

The company has allocated a capital budget of $110M for the first year, $30M for the second year, and $15M for the third year. Management has assigned the following priorities:

Priority 1
Minimize annual budget overruns (incorporate an ordinal weighting to express a higher preference for early funds flow.)
Priority 2-Subpriorities with Priority 2
(i) Accept EC-l emission control project (weight = 3).
(ii) Assure foreign marketing campaign (weight = 2).
(iii) Accept the dual-fuel project (weight = I).

Priority 3
Assure early cash inflows (incorporate an ordinal weighting to express a higher preference for early funds flow).
(i) Formulate this problem as a goal program, using the priorities and weights to establish the multiple-term achievement function.
(ii) Obtain a solution to this GP problem.

10-12. (Goal Programming Problem.) A company is considering three projects, which must be either selected or rejected, as a partially funded project is infeasible. The expected net profits, personnel needed, and project costs are tabulated below.

Project	Net Profit (Units)	Number of Personnel (Units)	Cost in Year 1 (Units)	Cost in Year 2 (Units)
A	3	12	8	6
B	4	15	12	10
C	2	10	5	6

The allowable total budget for year 1 is 20 units and 18 units in year 2. Total number of employees in that company is currently 25 units. The desired priorities are as follows:

With Priority 1, the total costs should not exceed the total budget each year.
With Priority 2, maximize net profit.
With Priority 3, minimize new employment (weighting = 2) and maximize utilization of employees (weighting = 1) in order to minimize the labor turnover.

10-13. A company is considering three projects, but the capital available over the 3-year life of these projects cannot fully support all three; however, a project can be partially accepted. The expected profits, minimum dividends, and projected costs for these projects are tabulated below.

Project	Net Profit in Year t ($ million)			Dividend in Year t ($ million)			Cost in Year t ($ million)		
	$t=1$	$t=2$	$t=3$	$t=1$	$t=2$	$t=3$	$t=1$	$t=2$	$t=3$
1	$70	40	20	$6	9	10	$60	30	10
2	6	60	80	5	4	10	50	40	10
3	20	40	60	3	6	9	30	30	30
Total	$96	140	160	$14	19	29	$140	100	50

The capital availability for investment in year 1: $100,000,000; year 2: $50,000,000; and year 3: $20,000,000. The company's goals are prioritized below:

P_1:   (a) Multiple units of the same project are prohibited.
        (b) Capital allocation for each year cannot be exceeded.
P_2:   Maximize profit.
P_3:   Maximize dividends paid.
P_4:   Minimize the shifting of project funds from one period to another.
        (a) Formulate the problem; (b) Find the optimal solution.
            (This problem was contributed by Joe M. Taio.)

10-14. Solve problem 10-11 using a weighted model for the multiple goals. Assign a weight of 5 to the priority 1 goals, a weight of 3 to the priority 2 goals, and a weight of 1 to the priority 3 goals. Apply weights within categories and adjust the units so that the weights on each category represent their relative importance.

10-15. Solve problem 10-12 using a weighted model for the multiple goals. Assign a weight of 5 to the priority 1 goals, a weight of 3 to the priority 2 goals, and a weight of 1 to the priority 3 goals. Apply weights within categories, and adjust the units so that the weights on each category represent their relative importance.

10-16. Solve problem 10-13 using a weighted model for the multiple goals. Assume that the priority 1 goals must be met completely, before any other goal may be considered. Assign a weight of 5 to the priority 2 goals, a weight of 3 to the priority 3 goals, and a weight of 1 to the priority 4 goals. Apply weights within categories and adjust the units so that the weights on each category represent their relative importance.

11

Utility Theory

11.1 INTRODUCTION

Now we relax the certainty conditions of earlier chapters by considering the uncertainty that is inherent in all project planning and selection methodology. Obviously, in reality all possible outcomes of a project(s) cannot be known with certainty. We simply cannot know all future values of sales, expenses, labor to be utilized, machinery needed, power consumed, and raw materials converted; in fact, we may not have *any* exact data. We also know that in reality a perfect capital market does not exist—a firm's borrowing rate of interest generally exceeds its lending rate and both interest rates are uncertain and may vary over time.

As Haley and Schall (1973) point out when we depart from assumed certainty and perfect capital markets into the world of uncertainty, we can think of these departures as being of two kinds: (1) departures that result from *changes in the actual cash flow stream* from the predicted values (the cash flow effect) and (2) departures due to *changes in the discount rates* (the rate effect) that are applied. Under perfect capital markets, the cash flow effect (as we saw in Chapter 9) may not invalidate the net present value method. The basis is the existence and additive properties of the parameter, NPV. However, capital market imperfections introduce rate effects (for example, simply a difference in borrowing and lending interest rates will suffice), which immediately invalidate the NPV criterion since the Separation Theorem (Chapter 10) is thereby destroyed.

We now propose to depart from certainty via the cash flow effect concept. We shall no longer assume that future conditions are known with certainty, even though we shall continue (for a time) with the assumption of a perfect capital market. We shall look at the occurrence and magnitudes of the cash flows themselves as being uncertain.

This chapter will focus on the normative development of utility theory, on its theoretical application to project selection, and on its empirical support. We will then focus on the certainty equivalent approach, which has been operationalized to a much larger extent. Then later chapters will address decision–making with probability distributions including the consideration of risk. Any discussion of risk must have its limits, because the literature is extremely broad. These chapters are no exception. For example, as pointed out by Hodder and Riggs (1985), many analyses ignore inflation, differing levels of risk in different project phases, and management's ability to mitigate risk. Because our emphasis is more theoretical, these elements are not included here.

11.1.1 Definitions of Probability. Project selection under uncertainty requires a basic knowledge of statistical theory. For example, the concepts of expected value, variation, random variables, sampling, and other topics are used repeatedly in this book. We expect that students who use this text will have had one or more courses in statistical methods.

In some contexts, *risk* is used to describe consequences and probabilities. In other contexts *risk* is used for situations with explicit probabilities and *uncertainty* for situations with no probabilistic data. Since we do not cover the latter problem, we will use the term that is closest to everyday terminology, *uncertainty*.

Historically, we are taught that an objective probability is the *limit* of the relative frequency of an event in a series of repeated trials. Thus, if we have a black bag containing 60 white balls and 40 red balls and if we repeatedly draw one ball from the bag with replacement between draws, the probability of drawing a white ball should equal six-tenths *in the limit*. Thus, objective probabilities depend on the existence of two conditions for a model where randomness prevails:

1. There are a finite number of outcomes to an experiment
2. The experiment is repeatable a large number (perhaps an infinite) of times under *identical conditions*.

The other approach to probability is that of the *subjective* or *personalistic* school. The subjectivist regards probability similarly to the man in the street: a measure of personal belief in a particular proposition, such as will it rain tomorrow. Jakob Bernoulli, in his *Ars Conjectandi* (1713), suggested that probability is a "degree of confidence" (later writers state "degree of belief") that an individual attaches to an uncertain event and that this degree depends on the individual's knowledge and differs between individuals. This theme was further developed by DeMorgan (1847) and Laplace (1952), but subjective probability as an operational theory was first formulated by Ramsey (1931) according to Raiffa (1968). Today this "degree of belief" may also be used to express the accuracy or reliability of the estimated probability, rather than the value of the probability. Work by Kahneman, Slovic, and Tverskey (1982) has pointed out many ways in which typical behavior and subjective estimates depart from the rules of objective probability.

While we face many unique events in solving project selection problems, we often have data on similar past events. This data may be the basis of an objective model, or it may be the foundation for subjectively estimated probabilities. At times, the subjectively estimated probabilities may have their basis entirely in the decision-maker's judgment. Indeed, Ackoff, Gupta, and Minas (1962) maintain that the decision-maker possesses more information regarding the decision environment than an assumption of outright uncertainty would require, merely by being able to specify the subjectively probable outcomes of a prospective action. The succeeding chapters are thus based on the generality and legitimacy of the different bases for probabilities.

11.2. CHOICES UNDER UNCERTAINTY: THE ST. PETERSBURG PARADOX

For numerical simplicity and consistency with the numbers in the cited papers, the following examples are stated in terms of $10 or $20. Realistically, for students on a limited budget today these values might be raised to $100 or $200, or for students with more money to $1000 or $2000, or for small firms to $10,000 or $20,000, or for large firms to $10M or $20M. Scale does matter; once the values are large enough, utility theory is needed to explain how industrial projects are evaluated.

Under conditions of absolute certainty and one objective, choosing between two alternatives is usually trivial. For example, suppose that

> Alternative A pays us $10, and
> Alternative B pays us $20.

We obviously prefer B, since common sense tells us that more is better. (The economist takes this rationale for granted.)

Suppose instead, that

> A pays us $10 with certainty, and
> B pays us $50 if the flip of a fair coin turns heads up and nothing if it is tails.

Now it is not so obvious that B is preferred. Obviously, receiving $10 is good but is it better (or worse) than a 50–50 chance of getting $50? We don't know. There is no *objective* way to make this choice—each individual's choice may differ.

Karl Borch (1968) constructed an ingenious example. Suppose we must decide on one action $A_0, A_1, ..., A_n ...$, in an experiment. If we decide on A_n, we shall receive

> either S_n dollars with probability P_n,
> or nothing with probability $(1 - P_n)$.

S_n and P_n are given by the relationships

$$P_n = \frac{10}{10+n} \qquad S_n = 10\left(\frac{n}{10}\right)^{0.9}$$

and tabulated in Table 11.1 (we can interpolate and extrapolate as needed). Clearly A_0 should not be chosen, but how far down the table do we go? As the payoffs become greater, the probabilities of success become smaller. Another question is what criterion do we use to judge the proper action?

Table 11.1 Hypothetical Payoffs

Action	P_n	S_n
A_0	1.0	$ 0
A_1	0.9	2
A_{10}	0.5	10
A_{20}	0.33	19
A_{30}	0.25	27
A_{40}	0.20	35
A_{90}	0.10	72
A_{190}	0.05	140
A_{990}	0.01	625
A_{9990}	0.001	5,000
A_{99990}	0.0001	40,000
A_{999990}	0.00001	250,000

The answer to the latter question is that, if we expect the experiment to be repeated many times then the correct criterion is to maximize the expected gain. In Borch's example, the expected gain of each action is

$$E_n = P_n S_n = \frac{100}{10+n}\left(\frac{n}{10}\right)^{0.9} = \frac{10^{1.1} n^{0.9}}{10+n}$$

and taking this as a continuous function, then $(E_n)_{max}$ is found where

$$\frac{dE_n}{dn} = \frac{10^{1.1}}{n^{0.1}(10+n)^2}(9 - 0.1n) = 0,$$

from which $(E_n)_{max}$ occurs at $n = 90$. This corresponds to action A_{90}—receive $72 with a chance of 1 in 10.

This criterion arises from the *law of large numbers* in statistics. If we repeat the experiment many times, then A_{90} is very close to the average gain per trial over the whole number of trials that will give us (with almost certainty) a gain of $P_{90}S_{90}$ = $7.20, per trial. Thus, in the long run, we do best by choosing the action with the greatest expected gain.

However, applying this principle of maximizing expected gain to the capital allocation problem runs into a serious theoretical problem. If we could undertake the same project over and over, then the law of large numbers would apply, but we usually do a project only once. Therefore, the law of large numbers rarely applies. So, one–shot decisions are made on some basis other than the law of large numbers.

The following discussion explores the theoretical foundations for decisions. However, a possible practical perspective is that firms typically make a large number of these one–shot decisions, so that large number statistics applies over the large number of projects—not the one time each project happens.

Daniel Bernoulli (1954) pointed out as early as 1738 that the maximum expected gain criterion is not always followed by all decision–makers, especially if the law of large numbers is not expected to apply. In a classic counter example, known as the St. Petersburg paradox, Bernoulli reported the shortcomings of the expected gain criterion by citing a game devised by his cousin, Nicolas Bernoulli. The game goes as follows. A coin is tossed until it falls heads. If heads occurs for the first time on the nth toss, the player gets a price of 2^{n-1} ducats (dollars) and the game is over. The probability of a heads for the first time on the nth toss is $(2)^n$ since the trials (tosses) are independent and Prob(tails) = Prob(heads) = ½ for a fair coin. Hence, the expected gain in this game is

$$E = \sum_{n=1}^{\infty} 2^{n-1} \left(\frac{1}{2}\right)^n = \frac{1}{2} + \frac{1}{2} + \frac{1}{2} + \cdots = \infty$$

since it is theoretically possible for the game to go on forever.

Now, since this game has a theoretical expectation of an infinite gain, then a maximizer of expected gain should be willing to pay any finite amount, for the opportunity of playing it. Yet we know people do not decide things this way. Indeed, Levy and Sarnat (1984) report that this experiment, conducted with students, revealed that most would pay only two or three dollars to play, and a few would pay as much as eight dollars, but no one offered more. This contradiction between the games is *infinite mathematical expectation* and what reasonable people are willing to pay to play is the St. Petersburg paradox. People apparently do not always decide on the basis of maximizing expected gain.

11.3 THE BERNOULLI PRINCIPLE: EXPECTED UTILITY

11.3.1 The Bernoulli solution. Daniel Bernoulli (1954) [also described in Cramer and Smith (1964) and Hammond (1998)], proposed two explanations for why people do not always follow the expected gain principle. (The two solutions differ principally in the mathematical function that describes the decision–maker's choices.) We shall describe the Bernoulli solution, which leads to the generalized *Bernoulli principle* or expected utility.

Bernoulli's solution assumes that an individual values the usefulness, or *utility value*, rather than the actual *money value* of the alternative prizes and that the incremental utility afforded by incremental money decreases as the prize's money value increases. (The latter assumption is the modern economist's principle of diminishing marginal utility of money.) Therefore, while total utility increases as monetary gain increases, it does so at a diminishing rate. Bernoulli assumed that the money's utility is a logarithmic function of the money prize:

$$U(x) = b \log (x/a) \qquad (11.1)$$

Where $U(x)$ = utility assigned to a money prize of x dollars, and a,b = positive coefficients.

Equation (11.1) is non–decreasing and its second derivative is negative.[1] Hence, the marginal utility is always positive but decreases with increasing wealth (x)—thus satisfying the modern concept of positive but diminishing marginal returns.

Bernoulli asserted that an individual would consider the prize's utilities rather than their monetary amounts; so the amount of money paid to play the game would correspond to the game's expected utility rather than its expected monetary value. If n = the number of tosses of the coin until heads first appears and $U(n)$ = the utility derived from the prize awarded after n tosses, the money prize will be 2^{n-1}, and the prize's utility is given by Equation (11.1); so

$$U(n) = b \log \frac{2^{n-1}}{a} = b \log 2^{n-1} - b \log a = b[(n-1) \log 2 - \log a] \quad (11.2)$$

According to Bernoulli's assertion that the individual's willingness to pay is based on the game's expected utility and that $P(n)$ is $(2)^n$, we calculate the expected value of Equation (11.2):

$$E[U(n)] = \sum_{n=1}^{\infty} p(n)U(n) = \sum_{n=1}^{\infty} p(n)b[(n-1) \log 2 - \log a] \quad (11.3)$$

Now, $P(n)$ is $(½)^n$, so

$$E[U(n)]\} = \sum_{n=1}^{\infty} \left(\frac{1}{2}\right)^n b[(n-1) \log 2 - \log a] \quad (11.4)$$

$$= b \log 2 \sum_{n=1}^{\infty} (n-1) \left(\frac{1}{2}\right)^n - b \log a \sum_{n-1}^{\infty} \left(\frac{1}{2}\right)^n.$$

However, since both summations equal one,[2] the game's expected utility is

$$E[U(x)] = b \log 2 - b \log a = b \log (2/a). \quad (11.5)$$

According to Equation (11.1) this is also the utility of a game whose certain money prize is $2.

So, Bernoulli's *expected utility criterion* values the game at $2 rather than the

---

1. Since both $a, b > 0$, then the first derivative
$$U'(x) = (b/x) \geq 0,$$
thus, the function $U(x)$ is nondecreasing and has positive marginal utility. Also, the second derivative is
$$U''(x) = -(b/x^2) \leq 0,$$
which is negative, thus the marginal utility decreases as wealth (x) increases.

2. The first series is an infinite geometric series that converges to [(first term)/(1 − r)] when the common ratio, r, is fractional (i.e., $0 < r < 1$). Since $r = 2$ here, the sum is unity. Similarly, the second series
$$\sum_{n=1}^{\infty}(n-1)\left(\frac{1}{2}\right)^n$$
is equivalent to
$$\sum_{n=2}^{\infty}\left(\frac{1}{2}\right)^n + \sum_{n=3}^{\infty}\left(\frac{1}{2}\right)^n + \sum_{n=4}^{\infty}\left(\frac{1}{2}\right)^n + \cdots = \frac{1}{2} + \frac{1}{4} + \frac{1}{8} + \ldots = \sum_{n=1}^{\infty}\left(\frac{1}{2}\right)^n = 1$$

infinity of the *expected monetary criterion*. An individual who chooses according to Bernoulli's logarithmic utility function will pay at most $2 to play the coin game, or the player is *indifferent* between playing the game and a certain $2. [Cramer used the utility function $U(x) = \sqrt{x}$ and the game's expected utility is $2.93.] Bernoulli simply wanted to answer the paradox and demonstrate the superiority of the expected utility criterion over the simple expected monetary criterion; the economic consequences of the Bernoulli–Cramer result were not appreciated until 200 years later.

11.3.2 Preference theory: the Neumann–Morgenstern hypothesis. The Bernoulli–Cramer solutions to the paradox were proved behaviorally by Frank Ramsey (1931). The paper laid the axiomatic foundations and derived the expected utility principle as a preference criterion. Ramsey's paper went unnoticed, apparently because no economist of that day thought the results important. Then, in 1944, the British mathematician, Jon von Neumann, and the economist Oskar Morgenstern published their famous collaborative work, *Theory of Games and Economic Behavior*. In an appendix to the book's second edition (1947) they derived the expected utility hypothesis directly from a set of behavioral axioms. While the expected utility hypothesis was important for the theory of games, the criterion is of much more general application now, and it is somewhat unfortunate that the Neumann–Morgenstern (N–M) development is intuitively associated with games.

In any event, Neumann–Morgenstern provided us with a method of ordering choices over a set of prospects. In traditional utility theory, a set of finite *commodities* makes up a *market basket*, to which we assign a relative value, or *utility*. In N–M preference theory, however, we need to pass from the world of finite things into a world that is infinitely divisible. We must *convolute* the probability functions of compound events and assign utility values to the convoluted probabilities. In the N–M framework, there is a *preference ordering over a set of probabilistic money returns*. Since the expected utility hypothesis of preference ordering is so important to understanding the economics of uncertainty, we shall outline below one method of showing that the expected utility criterion derives directly from some rather simple axioms of rational human behavior. Our previous edition included an appendix with a more complete proof based on Luce and Raiffa (1957).

11.3.3 The axiomatic basis of expected utility. The fundamental contribution of Neumann and Morgenstern is that utility functions can be used in decision problems so that an individual who acts solely in accord with the criterion of expected utility also acts in accord with his true preferences.

Assume that an individual faces three risky options (alternative lotteries A, B, and C) from which one must be chosen. The individual is assumed to make choices in accord with the following six behavioral axioms:

Axiom 1 (Comparability): Any two alternatives may be compared, and either the individual prefers one or is indifferent. Thus, a person can arrive at a choice between two alternative options, or he is indifferent between them.

Axiom 2 (Transitivity): Both the indifference and preference relations in Axiom 1 are transitive. That is, if the individual prefers A to B and B to C, then A is preferred to C

also. Likewise, indifference between A and B and also between B and C implies indifference between A and C.

Axiom 3 (Decomposition Axiom): Where a risky option has as a prize another risky option, the first option is *decomposable* into its more basic elements. (An example would be the French roulette wheels that yield tickets in the French National Lottery as prizes.) To illustrate the decomposition axiom let G be a lottery that includes two other lotteries, L_1 and L_2, as prizes. In abstract notation,

$$G = [qL_1, (1-q)L_2] \tag{11.6}$$

and

$$L_1 = [P_1 A_1, (1-P_1)A_2] \tag{11.7}$$
$$L_2 = [P_2 A_1, (1-P_2)A_2]$$

where q = probability of winning lottery L_1
L_1 = a lottery with probability P_1 of winning prize A_1 and $(1-P_1)$ for A_2
L_2 = a lottery with probability P_2 for winning A_1 and $(1-P_2)$ for A_2.

Axiom 3 makes it possible for us to decompose the compound lottery, G, into its more basic form. If we take the sign $\sim$ to mean *equivalent to*, then we have from axiom 3 that

$$G = [qL_1, (1-q)L_2] \sim \{q[P_1 A_1, (1-P_1)A_2], (1-q)[P_2 A_1, (1-P_2)A_2]\}$$

or

$$G \sim \{[qP_1 + (1-q)P_2]A_1, [q(1-P_1) + (1-q)(1-P_2)]A_2\}$$

or

$$G \sim [P^* A_1, (1-P^*)A_2] \tag{11.8}$$

where $P^* = qP_1 + (1-q)P_2$. Thus, the compound lottery G can be decomposed (or reduced) to a simple lottery, whose outcomes, A_1 and A_2, have compound probabilities P^* and $(1-P^*)$.

Axiom 4 (Substitution Axiom): If an individual is indifferent between two risky options, they may be interchanged in any compound option. Thus, equally preferable payoffs may be substituted for each other in any problem. Specifically, nonmonetary consequences (payoffs) can be expressed as their *certainty* or *cash equivalents*, where the decision-maker is indifferent between the nonmonetary payoff and the cash equivalent.

Axiom 5 (Monotonicity Axiom): If two risky options have the same two alternatives, then the option in which the more preferred outcome as the higher probability is itself preferred. This axiom is the one that expresses a desire for a high probability of success. For example, if $Q_1 = (\frac{1}{3} \cdot 5, \frac{2}{3} \cdot 12)$ and $Q_2 = (\frac{1}{2} \cdot 5, \frac{1}{2} \cdot 12)$ and if the individual prefers getting 12 to 5, Q_1 is preferred to Q_2 since Q_1 has the higher probability of getting 12.

Axiom 6 (Continuity Axiom): If A is preferred to B and B to C, then there exists a lottery involving A and C, for which the individual expresses indifference to B. For example, if the individual prefers \$20 to \$10, and \$10 to \$1, then some value of probability, P, can be such that, $P \cdot 20 + (1 - P) \cdot 1 \sim 10$.

The foregoing six axioms are sometimes called the N–M *axioms of rational behavior* since they formally express the approach of a rational or consistent person to decisions about risky options. If we accept these axioms, *the optimal investment policy under conditions of uncertainty is the policy that maximizes expected utility*. This is the *Bernoulli principle*. In other words, the selection criterion of maximum expected utility can be derived directly from the axioms.

While the N–M axioms provide a basis for an interval–scale preference–ordering function (the utility function), the class of admissible functions comprises those functions that provide the *same ranking* for a given set of risky alternatives. Beyond the sole necessity for ranking, the absolute levels of the utilities have no significance. Those utility functions that keep the rank order intact are *linear* transformations of one another. So the N–M utility function has neither an absolute (true) zero nor a unique unit of measure and can, therefore, be transformed by any linear transformation. For example, consider three utility functions

$$U_1(x) = x^{1/2}$$

$$U_2(x) = 100 x^{1/2}$$

$$U_3(x) = 10 + 100 x^{1/2}.$$

For several values of x, the corresponding utilities are

x:	1	49	81	400
$U_1(x)$:	1	7	9	20
$U_2(x)$:	100	700	900	2000
$U_3(x)$:	110	710	910	2010

While the values of $U_j(x)$ change, all three utility functions rank the utilities in the same order as x itself, which preserves the rank–ordering of x.

Stronger measures of preferential utility (e.g., a proportional scale with rational zero) have been proposed, for example by Restle (1961). However, the capital budgeting problem requires only that one can determine which project (or set of projects) has a greater utility; it is not necessary to know how much greater. Thus, a proportional (or ratio) scale utility with rational zero is a refinement not required by the selection model.

11.4 PROCURING A NEUMANN–MORGENSTERN UTILITY FUNCTION

11.4.1 The standard lottery method. Ascertaining a person's utility function typically relies on a series of hypothetical lotteries. The first step is to arbitrarily choose two widely separated points to fix the range of the scale. We might choose the utility of zero dollars, $U(0)$, as being zero *utiles* (the unit *utiles* describes the fact that the functional has units of value or worth and is not simply dimensionless). We also choose an upper scale value, say $U(\$100{,}000) = 10$ utiles.

The second step is to present a *standard lottery* to the decision–maker:

Alternative A	Alternative B
Payoff = $25,000 with probability $p_1 = 1.0$	(a) Payoff of $100,000 with probability p_2 (b) Payoff of $0 with probability $(1 - p_2)$

The decision–maker responds with the value of p for indifference between alternatives A and B. This is the *indifference probability*.

The third step is to calculate the lottery's value, $U(L)$, at the indifference probability, using axiom 6. For example, suppose that the decision–maker's indifference probability is $p = 0.5$ for alternatives A and B, above, then the lottery's utility would be $U(A) = U(B)$ at $p = 0.5$, or

$$p_1 U[(\$25{,}000)] = p_2[U(\$100{,}000)] + (1 - p_2)[U(0)]$$

$$1.0 U(\$25{,}000) = 0.5(10 \text{ utiles}) + (1 - 0.5)(0 \text{ utiles})$$

Therefore, $U(\$25{,}000) = 5.0$ utiles.

Negative utilities are also needed, so the decision–maker is presented with lotteries involving a probable loss:

Alternative C	Alternative D
Payoff = $0 with $p_1 = 1.0$	(a) Payoff = $ + 100,000 with p_2 (b) Payoff = $ −30,000 with $(1 - p_2)$

Suppose his indifference probability is $p_2 = 0.5$. Then,

$$(1.0)U(0) = 0.5 U(\$100{,}000) + (1 - 0.5)U(\$-30{,}000)$$

or

$$0 = 0.5(10) + 0.5 U(\$-30{,}000)$$

Therefore, $U(\$-30{,}000) = [-0.5(10)]/0.5 = -10.0$ utiles.

Repeating this procedure maps $U(x)$ versus x, and describes the decision–maker's utility function for money increments. Our example data are plotted in Figure 11.1.

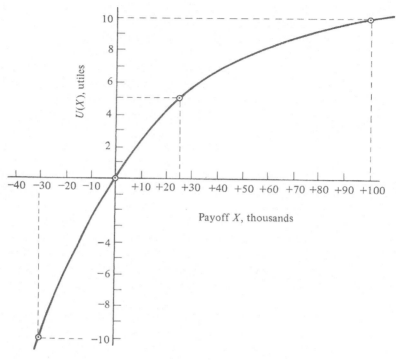

Figure 11.1 Typical utility of money function

11.4.2 Empirical determinations of utility functions. The Neumann–Morgenstern utility theory has been used to empirically determine the utility functions of *individuals*. Mosteller and Nogee (1951) established utility functions for individuals, using overt betting behavior in a gambling game. Davidson, Suppes, and Siegel (1957) made additional determinations. Grayson (1960), using the standard lottery technique, determined the utility functions of 11 principal decision–makers in the petroleum exploration and development business. Green (1963) reported on the utility functions of 16 middle management personnel in a large chemical company, representing the four major divisions of the firm (production, sales, finance, and research). Spetzler (1968) reported on 36 executives from one company. This article is part of a special issue on decision theory edited by Howard (1968) that puts utility curves in a broader historical context. Swalm (1966) empirically determined the utility functions of 13 executives. Cramer and Smith (1964), also using the standard lottery method, determined the utility functions of 8 R&D and manufacturing executives of a leading U.S. corporation. Hershey and Schoemaker (1980) showed that utilities for losses cannot be characterized as purely risk averse or risk seeking. Laughhunn et al. (1980) studied 224 managers in the United States, Canada, and Europe, where 71% were risk seeking for below target returns. Busche and Hall (1988) represent one of numerous studies concerning betting behavior on sporting events by individuals. Unlike many other betting studies, this work focused on behavior that was *not* risk preferent. A comprehensive review of utility theory can be found in Barbara, Hammond, and Seidl (1998 and 2004).

Several of the derived utility functions were depicted in this book's previous edition. While they were generally of the shape of Figure 11.1, individual decision-makers often displayed a unique function. Obviously, a utility function expresses a ranking of personal choices unique to the individual. From a different standpoint, a utility function cannot always be determined for a group. Arrow (1951) demonstrated by his *voting paradox* that the transitivity axiom can be violated by *circular preferences* in a group, which can prevent ranking by group choice. Thus the curve's individuality and the inability to generalize to a group are barriers to use of the derived curves for decision-making by the firm.

Second, we note that most of the utility functions were concave downward (like Figure 11.1), while some were convex, at least over a portion of the first quadrant. As we shall see later, the concavity or convexity of the utility function is indicative of risk-avoiding or risk-seeking behavior by the decision-maker. Risk avoiders have concave-downward utility functions, while risk seekers display some convexity in the first quadrant.

Third, the utility functions are generally concave downward in the third quadrant (where payoff is negative—that is a loss). In fact, most functions become rather steep at small negative values of return, thereby expressing a pronounced dislike for large losses. This, as noted above, has been challenged for below-target returns by Laughhunn et al. (1980).

Fourth, there is no one form of mathematical function that will fit an individual's observed data points exactly. One cannot specify from theory any particular mathematical function that is a utility function for an individual. Theory may disqualify certain functions (such as those that do not provide for the required linear transformation property—see Section 11.3.3), but one is never assured that any chosen mathematical function is theoretically correct. The best we can do is to assume one, although the choice of function can be narrowed by using regression and analysis of variance techniques [for example, see Bussey (1970)].

11.5 RISK AVERSION AND UTILITY FUNCTIONS

11.5.1 Risk aversion as a function of wealth. The wealth of the firm or individual is defined as, w, given that $0 \leq w < \infty$. As noted by Pratt (1964) the local *risk aversion function*, $r(w)$, is defined as

$$r(w) = \frac{-U''(w)}{U'(w)} = \frac{-\partial^2 U(w)/\partial w}{\partial U(w)/\partial w}. \tag{11.9}$$

We would expect that in most cases risk aversion would be a decreasing function of wealth. This is known as the Arrow–Pratt characterization [see Arrow (1951), Pratt (1964) and Machina (1982, 1987)]. A wealthier firm is better able to absorb a fixed level of loss. In fact, it has been postulated [see for example, Thompson and Thuesen (1985)] that marginal utility is a function of the ratio of the potential equity change to the firm's initial equity position.

Risk aversion behavior can also be described by a *risk premium*, which decreases as wealth increases. As described in Levy and Sarnat (1984), the risk premium, RP, is defined as follows:

$$E[U(w+x)] = U(w + E(x) - \text{RP}), \quad (11.10)$$

where x is the random variable that describes the potential change in equity position. Thus, the risk premium is an amount the individual will give up to avoid a risky option.

While this discussion has been in terms of a decreasing risk aversion with increasing wealth, not all utility functions have this property. In fact, this is the problem with the quadratic utility function [see Rabin (2000)], which is why the quadratic utility function is not discussed further in this edition.

11.5.2 Other risk–avoiding utility functions. Other mathematical functions can be used to overcome the limits of the quadratic utility function. For risk–avoiding utility functions (1) the function must be continuous, (2) the first derivative must be everywhere positive (thus increasing returns increase utility), (3) the second derivative must be everywhere negative (decreasing marginal utility), and (4) linear transformations of the function must not destroy its relative ordering properties. In essence, the risk–avoiding utility function, $U(X)$, must be continuous, monotonic increasing, and concave downward in X.

Hillier (1969) specifies coefficients and models so that certain classes of *hyperbolic* and *exponential* functions can be used as risk–avoiding utility functions. Expected utility can be approximated from the Hillier models, and if normality is assumed in the exponential model, then expected utility can be derived directly. For both hyperbolic and exponential models, subjectively estimating the utility function reduces to one of subjectively estimating the asymptotes of $U(X)$.

Freund (1956) develops the model $U(X) = 1 - \exp(-BX)$, which is a risk–avoiding utility function when X is normally distributed, plus only the risk–avoidance constant, B, need be estimated from an individual's responses to the standard lottery. Park and Sharp-Bette (1990) note that the negative exponential function (see problem 10–12) has a constant risk aversion function and that a logarithmic function (see problem 10–13) has decreasing risk aversion. Both of these utility functions are risk–averse utility functions, so they retain that strength of the quadratic function without at least one of the quadratic function's weaknesses. Brockett and Golden (1987) develop a general mathematical class of utility functions that contains those utility functions whose derivatives alternate in sign.

11.5.3 Linear utility functions: Expected monetary value. The indifference to risk of a linear utility function is a special case of the risk–avoiding utility functions: *Expected* monetary value (EMV) is maximized not expected utility. The expected monetary value principle is a special case of the Bernoulli principle, and it does not contradict the St. Petersburg paradox (maximizes expected utility).

The linear utility function, does not consider risk and $r(w) = 0$ in Equation 11.9. The expected utility criterion reduces to

$$E[U(X)] = A\mu = AE[X]. \quad (11.10)$$

This is merely a straight line passing through the origin $\{X = 0, E[U(X)] = 0\}$ with a slope, A, times the expected monetary value of the return, $E[X]$. Hence, a person who is indifferent to risk will maximize expected monetary value.

This allows the firm to decentralize the process of decision–making. Even utility functions that are nonlinear (e.g., risk avoiding) can be represented as a *piecewise–linear* model. Then, if the linear segments "predict" the nonlinear decision results without serious error, the EMV principle can be used for each given range. Thus, expected net present values and equivalent annual amounts are used in engineering economy studies as criteria for comparing investment alternatives.

The delegation process involves simply determining a range of returns over which a linear approximation to the utility function can be used satisfactorily and then delegating the appropriate decision–making. For example in a reasonably large firm, a piece–wise policy might be established that "go or no go" is decided for investments involving a net return of less than $10,000 at the work–unit level; for investments involving net returns of about $100,000 at the superintendent level; and for investments under $10M at the vice–president level. These limits would depend on the scale of the firm. In each case, EMV criteria could be used but perhaps with different discount rates to reflect the different trade–off rates between expected return and risk, which are implied by the nonlinear parent utility function.

Another way to understand trade–offs between expected value decision–making by the firm and the risk preferent and risk averse behavior described in the next section is to focus on the number of decisions being made by the firm. Note that large projects such as a Trans Alaska Pipeline or opening a firm's *first* manufacturing facility cannot be "lumped together" with a firm's list of typically sized projects. If a firm is making a decision between

- Hundreds of projects where 100 or more of those will be implemented ➔ expected monetary value decision–making is expected.
- Tens of projects where 10 or more will be implemented ➔ expected monetary value decision–making *may be* the norm.
- One project to implement from three choices ➔ expected monetary value decision–making is less likely. Issues involving business strategy tend to dominate.

11.5.4 Complex utility functions: Risk seekers and insurance buyers. Decision–makers sometimes violate the rational risk–avoiding assumptions of concave–downward utility functions (Section 11.5.1). For example, a business executive may prospect for petroleum and simultaneously buy fire and casualty insurance. A risk avoider would not engage in a business activity where the uncertainties are great (the ratio of dry holes to successful wells varies from about 10:1 to perhaps 30:1 in unproved territory) and where payoffs, if attained, sometimes may be disappointingly small. On the other hand, a risk seeker would not purchase insurance against a loss of other physical assets that could very well result in financial ruin.

Similar examples were used to discredit the Neumann–Morgenstern utility theory shortly after it was proposed. Critics said that such irrational behavior shows that the theory is sterile and incapable of predicting the choices of a decision–maker. In defense, Friedman and Savage (1948) demonstrated how a decision–maker would explain the simultaneous acceptance of an unfair gamble and the purchase of insurance to avoid loss in another quarter with a complex utility function that exhibited both risk preferent and risk averse behavior in different ranges.

Markowitz (1959) points out that a complex utility function is consistent with the investor's behavior in diversifying his investments when important money is at stake. Such an investor will insure against large losses, take small bets, and diversify his portfolio. Evidence for complex utility functions in another context was shown by Siegel (1957) who focused on a psychological *level of aspiration* that equates to an inflection point (*A* in Figure 11.2) and that the individual will take risks to obtain a return of at least *A*'s utility. Siegel substantiated his derivation by correlating the risk preferences of 20 students with their levels of aspiration for their semester grades. At a more personal level you may demonstrate both risk avoiding and risk seeking behavior by purchasing insurance on your life, car, house, or professional performance, and by buying chances in roulette or a lottery.

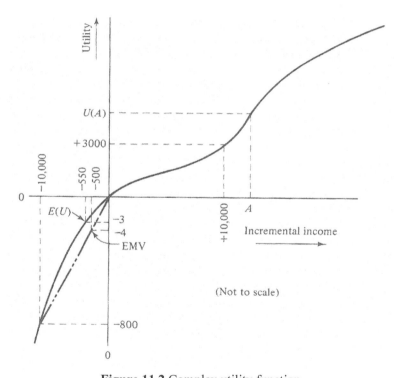

Figure 11.2 Complex utility function

In fact, as well explained in Park and Sharp–Bette (1990) but known for over 40 years (Friedman and Savage, 1948) these are rational explanations for risk–preferent behavior. First, the risk may be an *end* as well as a *means*. For individuals, gambling is often classed as entertainment. This is somewhat far–fetched if applied to firms. However, a second explanation might be extrapolated from the fact that poor people buy disproportionately more lottery tickets than middle–class or wealthy people. The rationale is that a poor person (or perhaps a small firm), to change his standing will take risks that would be shunned by those in wealthier positions. It is worth noting that the observation of risk–preferent behavior by horse track bettors is so common, that the converse is an anomaly (Busche and Hall, 1988).

11.5.5. Reconciling firm's utility and behavior by employees and managers. Even when a firm has decreed that expected value decision–making (Section 11.5.3) shall govern, decisions are influenced and made by employees and managers that exhibit risk averse and risk seeking behavior. Implementing industrial projects often involves a series of decisions within a firm. These decisions are not made by isolated individuals thinking in terms of utility functions. Rather a firm is a hierarchical network where it is not possible to establish a utility function for the firm.

While a firm may be implementing hundreds of projects so that it wants expected value decision–making, it is likely that each project has someone or many people whose annual evaluation, salary, bonus, and promotion prospects depend on that *one* project. Not surprisingly, such individuals have been known to allow personal considerations to influence their decisions and what numbers and analyses are presented to higher–level managers. It is well–documented that employees and managers will take short–term actions to meet a quarterly target, to protect their or a colleague's job, and to support other objectives. At the lowest level, this can consist of shipping an empty box on Friday afternoon to meet a production quota. At the highest level, there are many corporate mergers that have benefited the managers far more than the firms' stockholders.

To reliably connect employee behavior to a desired emphasis on expected value decision–making, firms would have to consistently audit projects (and their rejected alternatives) over their lives to see if initial estimates and decisions were correct. *Then* they would have to connect that audited performance to employee rewards. We know of *no* firm that conducts such audits for all successful and unsuccessful projects. More importantly, in many firms employee evaluations often focus on the last month or the last quarter, not the last year and certainly not longer periods.

Examples of the employee–firm disconnect include JPMorgan Chase where traders tried to hide losses of nearly $6 billion (Gogoi, 2012). Traders at the London office were speculating in credit default swaps, a form of derivative. Aggressive risky trading grew out of proportion, causing the huge losses. The traders and their management were later fired. The incident also caused a $20 billion loss in value of the company's stock. While not publicly detailed, industrial projects have been started, delayed, expanded, revised, and stopped to support a manager's desires for promotion, salary and bonuses, and job retention. In other cases, employees have selected the "safe" choice, so that their job would not be at risk.

11.6 SUMMARY

When certainty assumptions are relaxed for project selection, two departures from certainty occur— (1) *changes in the cash flow stream* away from predicted values and (2) *changes in the discount rates* that apply to the cash flows. The first is the *cash flow effect*, and the latter is the *rate effect*. Under conditions of uncertainty, such as estimating the worth of a future project, both effects usually occur. This chapter has introduced a theoretical means of handling the cash flow effect, the theory of preference ordering, which is based upon the expected utility principle (the Bernoulli principle), derived from the Neumann–Morgenstern axioms of rational behavior. The rate problem has not been considered here.

We note particularly that this development of the expected utility criterion results in a *present–time* device. The Neumann–Morgenstern behavioral axioms describe a rational decision–maker's behavior when he considers a lottery, which is a set of probabilistic returns in present time with no delay in payoff once a lottery is selected. The N–M axioms do not consider an investor who purchases a lottery with a present–time outlay and then waits for one or more periods for the payoff, during which the time value of money ought to be considered.

Furthermore, most authorities recognize that a decision–maker's utility function for money returns is not generally stable along a time axis. Firms and people become older, their level of wealth changes, and their risk attitudes change.

For these reasons, N–M utility methods cannot be applied directly to future money returns. To do so would apply the present–time utility function to future sums (returns) of money, which is not logical. Hence, the multi–period project selection problem, under conditions of uncertainty, cannot be solved by directly applying N–M utility theory to the future cash flow streams. Nevertheless, considerable insight can be gained by the limited application of expected utility, and some simpler models are discussed in Chapter 12.

The special case of a linear utility function leads to the criterion of expected monetary value (EMV), thus resolving any possible inconsistency between maximizing EMV and maximizing expected utility via the Bernoulli principle. They are simply different utility functions. While the expected value criterion is theoretically clear and easy to apply, utility theory helps explain why employees and managers will sometimes allow other criteria to apply—even when the firm specifies expected value as *the* criterion.

Quantitative application of utility theory is difficult to impossible for many industrial scale decisions. It is simply impossible to establish which utility curve should be used for a firm. However, utility theory often *explains* why employees, managers, and firms act as they do. Thus, utility theory is *very* important. The decreasing utility of additional returns (doubling returns does not typically double their utility) is the underlying reason behind the risk aversion that is common in economic decision–making.

REFERENCES

ACKOFF, RUSSELL L., SHIV GUPTA, and J. SAYER MINAS, *Scientific Method: Optimizing Applied Research Decisions* (Wiley, 1962).

ARROW, KENNETH, J., *Social Choice and Individual Values* (Wiley, 1951).

BARBERA, SALVADOR, PETER J. HAMMOND, and CHRISTIAN SEIDL, *Handbook of Utility Theory, Volume 1, Principles,* (Kluwer Academic Publishers, 1998).

BARBERA, SALVADOR, PETER J. HAMMOND, and CHRISTIAN SEIDL, *Handbook of Utility Theory, Volume 2, Extensions,* (Kluwer Academic Publishers, 2004).

BERNOULLI, DANIEL, "Exposition of a New Theory of the Measurement of Risk," *Econometrica,* **22**(1) (1954), pp. 23–36. Translation of a paper "Specimen Theoriae Novae de Mensura Sortis," *Papers of the Imperial Academy of Sciences in Petersburg,* **V**, 1738.

BORCH, KARL HENRIK, *The Economics of Uncertainty* (Princeton University Press, 1968).
BROCKETT, P. L., and L. L. GOLDEN, "A Class of Utility Functions Containing All the Common Utility Functions," *Management Science*, **33**(8) (August 1987), pp. 955–964.
BUSCHE, KELLEY, and CHRISTOPHER D. HALL, "An Exception to the Risk Preference Anomaly," *Journal of Business*, **61**(3) (July 1988), pp. 337–346.
BUSSEY, LYNN E., "Capital Budgeting Project Analysis and Selection with Complex Utility Functions." (Unpublished PhD Dissertation, Oklahoma State University, 1970).
CRAMER, ROBERT H., and BARNARD E. SMITH, "Decision Models for the Selection of Research Projects," *The Engineering Economist*, **9**(2) (1964), pp. 1–20.
DAVIDSON, DONALD, PATRICK SUPPES, and SIDNEY SIEGEL, *Decision–Making: an Experimental Approach* (Stanford University Press, 1957).
DE MORGAN, AUGUSTUS, *Formal Logic, or The Calculus of Inference, Necessary and Probable* (Taylor & Walton, 1847).
FREUND, RUDOLF, J., "The Introduction of Risk into a Programming Model," *Econometrica*, **24**(3) (July 1956), pp. 253–263.
FRIEDMAN, MILTON, and L. J. SAVAGE, "The Utility Analysis of Choices Involving Risk," *Journal of Political Economy* (April 1948), pp. 279–304.
GOGOI, PALLAVI, "JPMorgan Chase Trading Loss Closer to $6 Billion Than $2 billion CEO had Reported," *Detroit Free Press* (July 14, 2012).
GRAYSON, C. JACKSON, JR., *Decisions under Uncertainty: Drilling Decisions by Oil and Gas Operators* (Harvard University, Division of Research, Graduate School of Business Administration, 1960).
GREEN, PAUL E., "Risk Attitudes and Chemical Investments Decisions," *Chemical Engineering Progress*, **59**(1) (January 1963), pp. 35–40.
HALEY, CHARLES W., and LAWRENCE D. SCHALL, *The Theory of Financial Decisions* (McGraw-Hill, 1973).
HAMMOND, PETER J., "Objective Expected Utility: A Consequentialist Perspective," in Salvador Barbera, Peter J. Hammond, and Christian Seidl, *Handbook of Utility Theory, Volume 1, Principles*, (Kluwer Academic Publishers, 1998).
HERSHEY, JOHN C., and PAUL J. H. SCHOEMAKER, "Risk Taking and Problem Context in the Domain of Losses: An Expected Utility Analysis," *Journal of Risk and Insurance*, **47**(1) (March 1980), pp. 111–132.
HILLIER, F. S., *The Evaluation of Risky, Interrelated Investments* (North-Holland, 1969).
HODDER, JAMES E., and HENRY E. RIGGS, "Pitfalls in Evaluating Risky Projects," *Harvard Business Review*, **85**(1) (January–February 1985), pp. 128–135.
HOWARD, RONALD, ed. "Special Issue on Decision Analysis," *IEEE Transactions on Systems Science and Cybernetics*, **SSC–4**(3) (September 1968), pp. 199–366.
KAHNEMEN, DANIEL, PAUL SLOVIC, and AMOS TVERSKY, eds., *Judgement under Uncertainty: Heuristics and Biases* (Cambridge University Press, 1982).
LAPLACE, PIERRE SIMON DE, *A Philosophical Essay on Probabilities,* English translation of 5th (1825) ed. (Truscott & Emory, 1952).
LAUGHHUNN, D. J., J. W. PAYNE, and R. CRUM, "Managerial Risk References for Below–Target Returns," *Management Science*, **26**(12) (December 1980), pp. 1238–1249.

LEVY, HAIM, and MARSHALL SARNAT, *Portfolio and Investment Selection* (Prentice-Hall, 1984) p. 111.

LUCE, R. D., and H. RAIFFA, *Games and Decisions* (Wiley, 1957).

MACHINA, MARK J., "A Stronger Characterization of Declining Risk Aversion," *Econometrica*, **50**(4) (July 1982), pp. 1069–1079.

MARKOWITZ, HARRY M., *Portfolio Selection: Efficient Diversification of Investments.* Cowles Foundation Monograph 16 (Wiley, 1959).

MOSTELLER, FREDERICK, and PHILIP NOGEE, "An Experimental Measurement of Utility," *The Journal of Political Economy,* **59**(5) (October 1951), pp. 371–404.

NEUMANN, JOHN VON, and OSKAR MORGENSTERN, *Theory of Games and Economic Behavior,* 2nd ed. (Princeton University Press, 1947).

PARK, CHAN S., and GUNTER P. SHARP–BETTE, *Advanced Engineering Economy* (Wiley, 1990).

PRATT, J. W., "Risk Aversion in the Small and in the Large," *Econometrica*, **32**(1–2) (January–April 1964), pp. 122–136.

RABIN, MATTHEW, "Risk Aversion and Expected-Utility Theory: A Calibration Theorem," *Econometrica*, **68**(5) (September 2000), pp. 1281–1292.

RAIFFA, HOWARD, *Decision Analysis* (Addison-Wesley, 1968).

RAMSEY, F. P., "Truth and Probability," in *The Foundations of Mathematics and Other Logical Essays* (London: Kegan, Paul, Trench, Trusner & Co., 1931).

RESTLE, F., *Psychology of Judgement and Choice: A Theoretical Essay* (Wiley, 1961).

SIEGEL, SIDNEY, "Level of Aspiration and Decision Making," Psychological Review, 64(4) (July 1957), pp. 253–262.

SPETZLER, CARL S., "The Development of a Corporate Risk Policy for Capital Investment Decisions," *IEEE Transactions on Systems Science and Cybernetics*, **SSC–4**(3) (September 1968), pp. 279–300.

SWALM, RALPH O., "Utility Theory: Insights into Risk Taking," *Harvard Business Review* **44**(6) (November–December 1966), pp. 123–136.

THOMPSON, ROBERT A., and GERALD J. THUESEN, "Dynamic Investment Criteria for Capital Budgeting Decisions," *The Engineering Economist*, **31**(1) (Fall 1985), pp. 1–26.

PROBLEMS

11-1. What is meant mathematically, when one states that a utility function is defined *up to a linear transformation?*

11-2. The expected utility criterion was derived in Section 11.3.1 using Bernoulli's assumed utility function, $U(x) = b \log (x/a)$. Compare Bernoulli's solution to Cramer's solution, which uses $U(x) = x^{1/2}$.

11-3. Define mathematically a utility function that expresses risk aversion and that applies over the range of the argument from $x = -1$ to $x = +4$. The function passes through the point $[(U(x), x)] = (0, 0)$. What is the expectation of this function if the probability of any value is 0.5?

11-4. A person may estimate the expected cash flow in a particular year to be $10,000. If this figure is used in project analysis, what assumption(s) is (are) being made?

11-5. The following describes a decision–maker's utility function:

Dollars	Utility Measure
$-3,000	-3,000
-1,000	-1,000
-600	-500
-500	-350
0	0
100	100
500	150
1,000	200
2,000	350
4,000	500
10,000	11,000

(a) What is the utility of a gamble in which the decision–maker would win $10,000 or nothing with probabilities of 0.5?
(b) What amount would the decision–maker accept *for certain*, for indifference between the two choices?
(c) What is the gamble's expected monetary value?
(d) What risk premium is being paid?

11-6. Use the utility function defined in problem 11–5.
(a) What is the utility of a gamble in which the decision–maker would win $10,000 or lose $3000 with probabilities of 0.5?
(b) What amount would the decision–maker accept *for certain*, for indifference between the two choices?
(c) What is the gamble's expected monetary value?
(d) What risk premium is being paid?

11-7. Use the utility function defined in problem 11–5.
(a) What is the utility of a gamble in which the decision–maker would win with probability 0.6 or lose with probability 0.4? Winning will earn $10,000 40% of the time and $4000 the other 60% of the time. Losing will lose $1000 with probability 0.7 and $3000 with probability of 0.3.
(b) What amount would the decision–maker accept *for certain*, for indifference between the four choices?
(c) What is the gamble's expected monetary value?
(d) What risk premium is being paid?

11-8. Assume an investor has the following utility function:

Dollars	Utility Measure
$−20,000	−400
−10,000	−100
0	0
7,200	80
8,600	90
10,000	100
18,600	140
20,000	150
30,000	190
35,800	200
40,000	220
60,000	240

(a) Should the investor accept an investment whose initial outlay is $10,000 and which will result in either complete loss of investment or an immediate cash inflow of $30,000 with probabilities of 0.5?

(b) Should the investor undertake the following investment? Assume that a 5% discount rate for money applies.

End of Period	Cash Flow	
0	$−10,000	
1	0	with probability 0.5
	+30,000	with probability 0.5

(c) What would be your recommendation if the probabilities were 0.6 for failure (cash flow = 0 at end of period 1)?

11-9. Use the utility function defined in problem 11–8, assume that the initial investment is $10,000, and assume that the return is in period 1 (with a 5% discount rate)?

(a) What is the utility of a gamble in which the returns are a win $60,000 or a loss of $20,000 with probabilities of 0.5?

(b) What amount would the decision–maker accept *for certain*, for indifference between the two choices?

(c) What is the gamble's expected monetary value?

(d) What risk premium is being paid?

11-10. Use the utility function defined in problem 11–8, assume that the initial investment is $10,000, and assume that the return is in period 1 (with a 5% discount rate)?

(a) What is the utility of a gamble in which the firm would win with probability 0.6 or lose with probability 0.4? Winning will earn $60,000 40% of the time and $30,000 the other 60% of the time. Losing will lose $10,000 with probability 0.7 and $20,000 with probability of 0.3.

(b) What amount would the firm accept *for certain*, for indifference between the four choices?
(c) What is the gamble's expected monetary value?
(d) What risk premium is being paid?

11-11. Which of the Neumann–Morgenstern axioms is most likely to fail under conditions of reality?

11-12. For the assumed utility function $U(x) = x^{1/2}$, fit a series of piecewise–linear functions $U_1(x) = A_1 + B_1 x$ (A and B are constants), so that the maximum relative error, $|UB_1(x) - U(x)| \div U(x)$, is 5% or less, over the interval $+1 \leq x \leq +4$.

11-13. What is the significance of each of the two inflection points in a quartic utility function $U(x) = Ax - Bx^2 + Cx^3 - Dx^4$?

11-14. For a negative exponential utility function show that the risk aversion function, Equation (11.9), is a constant. Also show that a firm with this utility function is risk averse: $U(w) = 1 - e^{-cw}$ where $c > 0$.

11-15. Show that a logarithmic utility function has a decreasing risk aversion function, Equation (11.9), with w. Also show that a firm with this utility function is risk averse: $U(w) = \ln(w + d)$ where $d \geq 0$.

11-16. Using a series of standard gambles construct a utility function for a key decision–maker in the firm you work(ed) for (or if none available, a classmate).

12
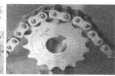

Stochastic Cash Flows

12.1 INTRODUCTION

In Chapter 11 the Bernoulli, or expected utility, principle was applied to lotteries when the outcomes were uncertain—as a present-time decision criterion. So the expected utility principle is used to judge stream effects of uncertainty rather than rate effects. When we consider the value of the cash flow stream, we will discount it at some interest rate, i_t. For the present, *we will assume a perfect capital market*, so that we may know i_t. We will also assume that i_t *is constant at i*, the known market rate of interest. Under these conditions, we may discount a present value stream to $t = 0$ and obtain its net present value.

Under conditions of uncertainty in cash flows, each of the cash flows may be considered as a *random variable*, with mean, variance, and possibly higher moments. The stream of cash flows into and out of the project is just a time sequence of random variables. The idea of a mean cash flow, or some linear combination of mean cash flows, conveys the concept of expected value of the project's net cash flow(s). However, the variance of the project's cash flows or some function of the individual cash flows variances complicates the meaning of value to a decision-maker. We associate variation in the cash flows with risk or uncertainty. Therefore, this and subsequent chapters rely on the standard deviations or variances of the cash flows or some function based on these to measure uncertainty about future cash flows.

In this chapter, we discuss the random variable basis of single, risky projects and how we may obtain and use a probability distribution, in present time to judge a project's worth. This is the *net present value* method. For multiple projects we look at combining the moments of the probability distributions. For further reading, a literature survey was conducted by Carmichael and Balatbat (2008) that identifies much of the significant literature in the field of probabilistic discounted cash flow analysis.

The previous edition also included material which used the expected utility concept and certainty equivalents to evaluate single projects and mutually exclusive sets of projects. In the last chapter we concluded that utility functions were theoretically very important in explaining the actions of individuals and firms. However, we also concluded that it was not possible to derive the utility functions that firms were actually using. Thus, we have omitted the material on certainty equivalents from this chapter.

We also want to note that the next chapter expands on the treatment of uncertainty in the economic analysis of projects by focusing on the tools of simulation and decision trees.

12.2 SINGLE RISKY PROJECTS—RANDOM CASH FLOWS

12.2.1 Estimates of cash flows. Project j, under conditions of uncertainty, is characterized by its random cash flows, ($t = 0, 1, 2, \ldots, N$), where N is the project's life. Y_{tj} may be positive (an inflow to the firm), negative (an outflow from the firm, that is, an investment in the project), or zero. Y_{tj} is the net cash flow, resulting from all possible inflows and outflows at t, and N may also be a random variable. For the present, we assume that N is fixed and known.

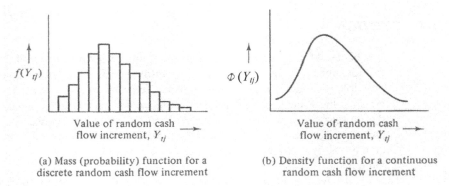

(a) Mass (probability) function for a discrete random cash flow increment

(b) Density function for a continuous random cash flow increment

Figure 12.1 Distributions of random cash flow increments

The relative frequencies of each cash flow's values form probability or density functions, such as Figure 12.1. In Figure 12.1a, the relative frequency of each discrete value forms the mass function $f(Y_{tj})$, while in Fig. 12.1b the relative frequency over a range of values for a continuously distributed Y_{tj} is

$$p(Y_{tj})_{a,b} = \int_{Y=a}^{b} \Phi(Y_{tj}) dY. \tag{12.1}$$

where $\Phi(Y_{tj})$ is a probability density function. For the discrete case, the mass functions values are in the range $0 \leq f(Y_{tj}) \leq 1$ and they sum to 1; for the continuous case, the density functions values are nonnegative and the integral equals 1. Thus, for the probabilistic case each net cash flow is a random variable, either discretely or continuously distributed, rather than a known constant value. Consequently, each random cash flow will have at least a mean and a variance, and possibly higher central moments as well. For now we consider the case that has only means and variances of the cash flows.

How then, does one determine the probability distribution (or at least the mean and variance) of each cash flow? In general, future data must be estimated. By applying subjective (or predictive) estimates to each source element, the analyst can develop a mean (expected value) estimate for each net cash flow. However, the probabilistic model requires more. The analyst also must realistically evaluate the variance of each net cash flow or specify its probability or density function.

One common approach is based on the properties of the beta distribution, used extensively in PERT and critical path methodology in project management. Use of the beta distribution, proposed by Wagle (1967) and summarized by Hillier (1969, pp. 87–89), requires that the analyst make three estimates—*optimistic, pessimistic,* and *most likely* for each cash flow. These three estimates are assumed, respectively, to be the upper bound, the lower bound, and the mode of the β distribution.

A typical β distribution is bell-shaped, but (1) the β-density function is truncated in the tails (not continuing indefinitely like the normal) and (2) it is skewed right or left whenever the mode (most likely value) is not midway between the extreme bounds.

Each optimistic estimate should describe the cash flow that would occur if everything goes as well as reasonably possible (the pessimistic, as poorly as reasonably possible) with each having one chance in several hundred. Thus, the optimistic and pessimistic values are taken to be the slightly probable extremes—that is, approximately six standard deviations should exist between the optimistic and pessimistic cash flow estimates.

By assuming that the β distribution represents the underlying cash flow density function with six standard deviations between the bounds, the mean and variance of each cash flow can be found by

$$E[Y_{tj}] = \frac{[Est\,(Y_{tp}) + 4\,Est\,(Y_t) + Est\,(Y_{to})]}{6} \quad (12.2)$$

and

$$V[Y_{tj}] = \left\{\frac{[Est\,(Y_{to}) - Est\,(Y_{tp})]}{6}\right\}^2 \quad (12.3)$$

where $E[Y_{tj}]$ = mean cash flow for period t
$V[Y_{tj}]$ = variance of the cash flow for period t
Est (Y_t) = most likely estimate of cash flow in period t
Est (Y_{tp}) = pessimistic estimate of cash flow in period t
Est (Y_{to}) = optimistic estimate of cash flow in period t.

To demonstrate this technique we assume that expert subjective analysis has yielded the optimistic, pessimistic, and most likely cash flow estimates shown in Table 12.1. One can verify, by applying Equations (12.2) and (12.3), that the mean cash flows and the variances for each year are those given in Table 12.1.

Table 12.1 Means and variances of β distributed cash flows

Year t	Pessimistic Estimate	Most Likely Estimate	Optimistic Estimate	Mean Cash Flow, $E[Y_{tj}]$	Cash Flow Variance, $V[Y_{tj}]$
0	$-20,000	$-10,000	$-6,000	$-11,000	5.44 x 10^6 (^{2})
1	-6,000	+2,000	+6,000	+1,333	4.00 x 10^6
2	+6,000	+12,000	+18,000	+12,000	4.00 x 10^6
3	+12,000	+18,000	+28,000	+18,667	7.11 x 10^6

These basic measures of central tendency and dispersion do not discriminate between "upside potential" and "downside risk." Thus, measures such as the variance have been applied by Hoskins (1978), Porter et al. (1975), and others to the project selection problem. These measures can be applied to present worth sums of cash flows (as is done in the next section for means and variances). However, there are difficulties as shown by Eldred and Barnes (1979) for the semivariance. Even for two independent variables, if either has an asymmetric distribution, then the sum's distribution must be derived and the semivariance calculated.

Buck and Askin (1986) have developed the *partial means* measure of economic risk. This measure can be used to discretize a continuous distribution for use in decision trees, as well as for calculating other risk measures. Transforms, both Z and Laplace, have also been used to construct the probability density functions. Buck, with Hill and Tanchoco (1971, 1975, and 1977), have applied transform theory to both deterministic and stochastic problems. Stochastic analysis has been furthered by Barnes, Zinn, and Eldred (1978). A good textbook presentation can be found in Park and Sharp-Bette (1990). This body of work also includes discussions of stochastic dominance, which can be used to compare the risk profiles of projects.

12.2.2 Expectation and variance of project net present value. A project's net present value is simply the sum of the discounted periodic cash flows. In the deterministic case, each periodic cash flow is assumed to be known with certainty. In contrast, the assumption in the probabilistic case is that each periodic cash flow takes on an unknown value, Y_{tj}, but the effect of the randomness can be expressed through the mean and variance of the distribution Y_{tj}. Thus, in the probabilistic case, the summation of the discounted periodic random cash flows implies that the project's net present value is also a random variable. That is,

$$P_j = Y_{0j} + \frac{Y_{1j}}{1+i} + \frac{Y_{2j}}{(1+i)^2} + \cdots + \frac{Y_{ij}}{(1+i)^t} + \frac{Y_{Nj}}{(1+i)^N}, \qquad (12.4)$$

where P_j is the random net present value for project j; Y_{tj} is the random cash flow in the tth period for the jth project; and i is the known minimum attractive rate of return. So in the probabilistic case, P_j is a random variable, with some mean, $E[P_j]$, and some variance, $V[P_j]$.

The random net present value for the project will possess a *mean net present value*, $E[P_j]$, *and a variance of net present value*, $V[P_j]$. These values permit us to relate the unknown P_j to the project's random cash flows. The project's mean net present value is simply the sum of the discounted mean cash flows,

$$E[P_j] = \sum_{t=0}^{n} \frac{E[Y_{tj}]}{(1+i)^t}. \qquad (12.5)$$

So the mean values of the cash flows (for example, those in Table 12.1) and a fixed discount rate, i, lead directly to the project's mean net present value via Equation (12.5).

It is not so easy to establish the variance of net present value. Essentially, one of three relationships may exist between cash flows in different periods: (1) complete independence, (2) complete dependence, or (3) partial dependence. Cash flows for the entire project are completely independent if there is no relationship between any two cash flows. That is, any cash flow can be determined solely from events occurring within that period and these events bear no relationship to or dependency on events in prior or subsequent periods. If, on the other hand, there is some relationship between events in two periods, then the cash flows are dependent. For example, operating expenses may trend upward in succeeding years as a result of prior year actions or sales may exceed estimates in early years, which imply they are likely to exceed estimates in succeeding years. These are fairly common phenomena and when such conditions exist the cash flows are not independent.

Complete dependence between cash flows exists if there is a one-to-one relationship among events in succeeding periods, and partial dependence exists if the relationship is less than one-to-one. In evaluating the variance of project net present value the exact relationship among cash flow relationships must be specified (whether known or assumed).

For the *independent* case, the variance of project net present value is found by letting $(1+i)^{-t} = k_t$. Since the cash flows, Y_{tj}, are random variables, then:

$$V\left[\frac{Y_{tj}}{(1+i)^t}\right] = V[k_t Y_{tj}] = k_t^2 V[Y_{tj}] \tag{12.6}$$

and these variance for the N periods are additive, so that:

$$V[P_j] = \sum_{t=0}^{n} V\left[\frac{Y_{tj}}{(1+i)^t}\right] = \sum_{t=0}^{n} k_t^2 V[Y_{tj}] = \sum_{t=0}^{n}\left[\frac{\sigma_{Y_{tj}}^2}{(1+i)^{2t}}\right], \tag{12.7}$$

where σ_{tj} is the variance of the tth cash flow. Note that the exponent for the discounting factor in the variance formulation is $2t$, compared with simply t for the mean in Equation (12.5).

If the cash flows are not completely independent, then the cash flows are statistically correlated, either perfectly or partially, due to causative relationships. Perfect correlation results in the one-to-one complete dependency mentioned earlier, and partial correlation results in cases intermediate between perfectly correlated and mutually independent cash flow streams.

To determine the variance of project net present value in dependent cases, a useful approach is to assume that cash flows form an autocorrelated time series whose autocorrelation coefficient matrix or autocovariance matrix can be obtained. Then one can derive an expression for the variance of net present value, as described below.

Dropping the subscript j for clarity but recognizing that the development is for a single project, let Y_t be the distributed random cash flow for period t with mean $E[Y_t]$ and variance $V[Y_t] = \sigma_t$. Let P be the project net present value with mean $E[P]$ and variance $V[P]$. Let Y_τ and Y_θ ($\tau, \theta \in t$; $\tau \neq \theta$) be correlated cash flows, such that the covariance between Y_τ and Y_θ is defined by:

$$\text{Cov}(Y_\tau Y_\theta) = \rho_{\tau\theta} \sigma_\tau \sigma_\theta \qquad (\tau \neq \theta), \tag{12.8}$$

where $\rho_{\tau\theta}$ is the simple correlation coefficient ($-1 \leq \rho_{\tau\theta} \leq +1$) and σ_τ and σ_θ are the standard deviations in the τth and θth periods. Symmetry among the correlation coefficients is assumed ($\rho_{\tau\theta} = \rho_{\theta\tau}$). Then, the variance of project net present value is:

$$V[P] = V[Y_0] + \frac{V[Y_1]}{(1+i)^2} + \frac{V[Y_2]}{(1+i)^4} + \cdots + \frac{V[Y_N]}{(1+i)^{2N}}$$

$$\frac{2\,Cov[Y_0 Y_1]}{(1+i)} + \frac{2\,Cov[Y_0 Y_2]}{(1+i)^2} + \cdots \frac{2\,Cov[Y_1 Y_2]}{(1+i)^3} \quad (12.9)$$

$$+ \cdots \frac{2\,Cov[Y_\tau Y_\theta]}{(1+i)^{\tau+\theta}} + \cdots + \frac{2\,Cov[Y_{N-1} Y_N]}{(1+i)^{2N-1}}$$

or, on substituting Equation (12.8) and $\sigma = V[Y_t]$,

$$V[P] = \sum_{t=0}^{N} \frac{\sigma_t^2}{(1+i)^{2t}} + 2 \sum_{\tau=0}^{N-1} \sum_{\theta=1}^{N} \frac{\rho_{\tau\theta}\sigma_\tau\sigma_\theta}{(1+i)^{\tau+\theta}} \quad \text{for } \tau < \theta. \quad (12.10)$$

An alternative form of Eq. (12.10) exists for the covariance matrix:

$$V[P] = \sum_{t=0}^{N} \frac{\sigma_t^2}{(1+i)^{2t}} + 2 \sum_{\tau=0}^{N-1} \sum_{\theta=1}^{N} \frac{Cov(Y_\tau Y_\theta)}{(1+i)^{\tau+\theta}} \quad \text{for } \tau < \theta. \quad (12.11)$$

One may now define what is meant statistically by *complete dependence* and *partial dependence* in the cash flow stream. When cash flows are completely dependent, the correlation is perfect and all values of $\rho_{\tau\theta}$ are either -1 or $+1$. Values intermediate between -1 or $+1$ (excepting zero), imply partial correlation and partial dependence. Note that $\rho_{\tau\theta} = 0$ implies that the second summation term in Equation (12.10) equals zero, which is the case for an *in*dependent cash flow stream.

When either partial or complete dependencies exist among the periodic cash flow increments, the variance of project net present value can be found by either Equation (12.10) or Equation (12.11). Both require a mathematical statement of the relationships that exist among the cash flows by specification of the autocorrelation coefficients or the autocovariances.

12.2.3 Autocorrelations among cash flows (same project). A project's cash flows are often correlated across time (autocorrelated). Successful sales can lead to continued success, high gasoline prices tend to remain high, and sales declines due to obsolescence tend to continue their decline. Autocorrelated cash flows can have a significant impact on future cash flows, and where they are present, they need to be incorporated into any forecast. Where correlation exists, the analyst can: (1) decompose each cash flow, Y_t, into its contributing elements, such as gross sales, operating expense, and depreciation, and determine the autocorrelations between these elements; (2) examine the cash flow stream as a time series and statistically test a correlative model for goodness of fit; (3) construct decision-tree like scenarios to simulate the possible outcomes (Bonini, 1975), or (4) rely on expert estimates of end-of-period cash flows (Leung, et al., 1989).

Giaccotto (1984) used a Markov model of the correlation between successive periods. While the approach is interesting itself, the most interesting point is that he found that serial correlation increased his example's NPV faster than it increased the example's variance. Thus, the higher the serial correlation, the higher the probability of a positive present value (or the lower the risk). This is quite surprising due to the many examples where serial correlation increases risk [see, for example, Hillier (1963)].

12.2.4 Probability statements about net present value. In deriving Equations (12.5) and (12.10), no specific form of the resulting statistical distribution of P was defined, except to require that it have finite mean and variance. So, $E[P]$ and $V[P]$ do not depend on any assumptions about the form of the distributions of the random cash flows, Y_{tj}. The equations are obtained by the application of the expected value operator, which is a distribution-free operator. As Hillier (1963) states, these equations estimate $E[P]$ and $V[P]$ for any *linear combination* of Y_{tj}, regardless of distribution. We can find the mean and variance of net present value, but now we must use them.

If we can make distributional assumptions concerning net present value, then probability statements about NPV are possible. For example, if the cash flows are *normally* distributed, then the central limit theorem implies that net present value (P_j) is also normally distributed with mean $E[P_j]$, and variance $V[P_j]$. Moreover, even if the Y_{tj} are not normally distributed, P_j might still be approximated by a normal distribution if one of the several versions of the central limit theorem is applicable [see Hillier (1969), pp. 24–29]. Finally, if no distributional assumptions concerning the random Y_{tj} can be made, certain weak probability statements can be made via the Chebyshev inequality or via a stronger form of this inequality investigated by Grosh and Bussey (1973). Numerical examples will illustrate these principles.

Example 12.1 A proposed project has the following cash flows:

End of Year	Mean Net Cash Flow, Y_t	Standard Deviation Of Cash Flow, $\sigma_{Y(t)}$
0	$- 800,000	$ 250,000
1	+1,000,000	450,000
2	+1,000,000	600,000

Assuming independent cash flows, a normally distributed net present value, $N(\overline{P}, \sigma)$, and a minimum attractive rate of return of 15%, what is the probability that the net present value will be
 a. $210,000 or less?
 b. Zero or less (i.e., that the IRR is < 0.15 rather than its mean of 0.9059)?
 c. $1,000,000 or greater?

Solution

$\overline{P} = -800K + 1000K(P/F, 15\%, 1) + 1000K(P/F, 15\%, 2) = 826K$

$$\sigma_P^2 = (250K)^2 + (450K)^2(P/F, 15\%, 2) + (600K)^2(P/F, 15\%, 4) = 422{,}000K^2$$

$$\sigma_P = \sqrt{422{,}000K^2} = 650K$$

Assuming that $P \sim N(\bar{P}, \sigma_P^2)$; we find

a. $Z_\alpha = \dfrac{P - \bar{P}}{\sigma_P} = \dfrac{210K - 826K}{650K} = -0.947$ ➔ $\alpha = 0.173$ (from Normal tables).

b. $Z_\alpha = \dfrac{0 - 826K}{650K} = -1.27$ ➔ $\alpha = 0.102$, or 10.2%.

c. $Z_\alpha = \dfrac{1{,}000K - 856K}{650K} = +0.268$

So, $\alpha = 0.605 = \text{Prob}[P \leq 1{,}000{,}000]$.

Therefore, $\text{Prob}[P \geq 1{,}000{,}000] = 1 - \alpha = 0.395$

Example 12.2 Use the data of Example 12.1 without distribution assumptions.
a. What is the probability that NPV is zero or negative?[1]

Solution
From Grosh and Bussey (1973), we have a strong form of the Chebyshev inequality applying to *single-tail* probabilities, so that

---

1. Note: some difficulties attach to equating the probabilities that NPV ≤ 0 and IRR $\leq i'$. Originally, Hillier (1963) asserted that the two probabilities are identical. As pointed out by Kaplan and Bernhard, however, Hillier erred in this assumption and corrected his error (Hillier, 1965). Bernhard (1967) commented further. In general, the identity holds only if the following relationship is satisfied with probability one or nearly one:

$$\text{Prob}[i^* < i'] = \text{Prob}[NPV < 0 | i']$$

For example, the relationship holds only if the investment is a simple one (only one root, i^*, and initial cash flow(s) < 0; hence, $NPV(i)$ is a monotonic decreasing function of i). If the cash flows, Y_t, are normally distributed, then *some* of the Y_t could become negative which might invalidate the monotonic decreasing requirement for $NPV(i)$, and invalidate the rule given in Bernhard (1967). The remedy is to assume that all Y_t have small variances compared to their means, so that the probability of a sign change in Y_t is very small. Then the equation will be satisfied (nearly). Additional information on multiple roots may be found in Chapter 7.

Note: In the next chapter we will note that E(NPV) and the E(IRR) may indicate different decisions.

$$\text{Prob}[(P - \bar{P}) \leq -t\sigma_P] \leq \frac{1}{(1+t^2)}$$

or, equivalently, that $\alpha = \frac{1}{(1+t^2)}$, where $t = \frac{(P-\bar{P})}{\sigma_P}$. Therefore,

$$t = \frac{0 - 826K}{650K} = -1.27; \quad \rightarrow \quad t^2 = 1.61, \text{ and}$$

$$\text{Prob}[P \leq 0] \leq \frac{1}{1 + t^2} = \frac{1}{2.61} = 0.383.$$

b. If it is known that the net present value, P, has a unimodal and symmetrical distribution, then the Camp-Meidell inequality (Grant and Leavenworth, 1972) applies:

$$[(\bar{P} - t\sigma) \leq P \leq (\bar{P} + t\sigma)] \leq 1 - \frac{1}{2.25t^2}$$

So that, with $t = -1.27$ [as in (a) above], then (assuming symmetry):

$$2\alpha = 1 - \text{Prob}[(\bar{P} - t\sigma) \leq P(\bar{P} + t\sigma)] = \frac{1}{2.25t^2}$$

This is obvious from the following sketch:

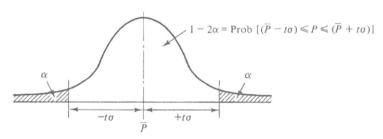

Therefore,

$$\alpha \leq \frac{1}{(2)(2.25)t^2} = 0.138$$

and since $\text{Prob}[P \leq 0] = \alpha$, we have that $\text{Prob}[P \leq 0] \leq 0.138$.

12.3 MULTIPLE RISKY PROJECTS AND CONSTRAINTS

In Chapter 9, to allocate resources optimally, we followed Fleischer (1966) to create *mutually exclusive* bundles of candidate projects and then maximized net present value to obtain the single optimal subset of feasible projects. We also demonstrated that Weingartner's mathematical programs did this for the constrained budget model under certainty conditions.

The same general approach is valid for projects whose cash flows are random variables. For example, Peterson (1975) uses nonlinear chance constraints within mathematical programming to account for the possibility that the random cash flows will cause a violation of a budget constraint. However, there is an extremely important difference between the deterministic and random cases. In the deterministic cases in Chapter 9, we only had to consider synergistic or competitive interactions between the cash flows of different projects. In the stochastic case, however, we need to consider possible *correlative effects* between the cash flow streams of two or more projects. If correlative effects exist, then the variances (and hence, the risk) of the cash flow distributions for the combination of cash flows can be seriously affected by the covariance terms. Practically, the right project combination can reduce risk, whereas the wrong combination can increase it. For this reason, we begin with methods of estimating the variance of combinations of random cash flows and then proceed to possible selection methods for multiple risky projects.

The risk of project portfolios is reexamined in Chapter 16.

12.3.1 Variance of cross-correlated cash flow streams. Sometimes economic factors may affect the cash flows of more than one project. When the cash flows Y_{tj} and Y_{tk}, of two projects in the same period are affected, then the projects net present values will be cross-correlated. In the absence of cross-correlations between their cash flows, P_j and P_k are independent of each other.

When the cash flows of two or more projects are correlated, then we compose one set of statistical parameters (one mean and one variance of net present value) representing each *pair* of correlated cash flows. The procedure is as follows. Obtain the expected net present value for each pairwise overall project by taking the sum of the net present values:

$$E(P_{j,k}) = E(P_j) + E(P_k). \tag{12.12}$$

The method of obtaining the variance of the pairwise overall project, $V[P_{j,k}]$, depends on the relationship between their cash flows. If the two cash flow streams are perfectly correlated year by year, merely add them together and calculate the resulting variance.

If the two cash flow streams are only partially correlated between projects, then we must specify the nature of the relationship. The development is considerably simplified if we can assume that the correlation is only between cash flow elements Y_{tj} and Y_{tk} in the *same* time period. This describes the usual situation in which a common causative economic circumstance tends to push the two cash flow elements up or down simultaneously rather than in different time periods. The development is further simplified if we assume that the partial correlation coefficient relating projects j and k, $\rho_{jk}(t)$ can be estimated.

This estimate cannot be created from the estimated means and variances of the cash flows in the two projects. The correlation coefficient(s) must be separately estimated. This could be done by looking at similar historical projects, which were likely part of the process in estimating the data in Table 12.2. In most cases, the correlation coefficient $\rho_{jk}(t)$ will be assumed to be constant over time so that it becomes ρ_{jk}.

The data in Table 12.2 will be used to illustrate the method. Suppose that the mean cash flows were estimated from common economic considerations that imply a year by year correlation. Equation (12.13) provides the estimated variance of NPV for the combined project, AB, analogously to Equation (12.10). However, the cross-correlation coefficient, $\rho_{AB}(t)$, is now constant among periods.

$$V(P_{AB}) = \sum_{t=0}^{n} \frac{\sigma_{tA}^2}{(1+i)^{2t}} + \sum_{t=0}^{n} \frac{\sigma_{tB}^2}{(1+i)^{2t}} + 2\rho_{AB} \sum_{t=0}^{n} \frac{\sigma_{tA}\sigma_{tB}}{(1+i)^{2t}} \qquad (12.13)$$

Example 12.3 Multiple correlated projects
Use the data of Table 12.2 and assume that $i = 15\%$ and $\rho_{AB} = 0.75$.

Table 12.2 Cash flow streams for two hypothetical projects with $i = 15\%$ and $\rho_{AB} = 0.75$

Year (t)	Cash Flow for Project A (Y_{tA})	Variance, $V[Y_{tA}]$	Cash Flow for Project B (Y_{tB})	Variance $V[Y_{tB}]$
1	2.0	0.20	19.0	2.0
2	2.0	0.24	21.0	2.2
3	3.0	0.31	24.0	2.6
4	3.0	0.35	27.0	3.0
5	3.0	0.40	27.0	3.2
6	4.0	0.42	29.0	3.3
7	4.0	0.46	31.0	3.2
8	6.0	0.62	35.0	3.6
9	6.0	0.64	36.0	3.8
10	6.0	0.68	37.0	4.0

Solution
As summarized in Figure 12.2, this is most easily solved by using the SUMPRODUCT function. If the two projects had been independent, the resulting variance of NPV would have excluded the covariance terms, and the variance would only have been 8.61.
 The positive correlation ($\rho_{AB} = 0.75$) between the two cash flow streams has increased the NPV variance of the project pair, $V(\text{NPV}_{AB})$, from 8.61 to 12.62, indicating an increased risk of undertaking the two correlated projects simultaneously. Similarly, if the correlation coefficient, ρ_{AB}, were negative, then the risk of the joint project, AB, would be lower than if the projects were independent.

362 Chapter 12, Stochastic Cash Flows

	A	B	C	D	E(Yt)	F	G	H
1		15% interest			0.75 correlation coreffcient			
2	Year		Project A			Project B		
3	(t)	E(Y$_t$)	V[Y$_t$]	Stdev	E(Y$_t$)	V[Y$_t$]	Stdev	1/(1+i)^2t
4	1	2	0.20	0.45	19	2.0	1.41	0.7561
5	2	2	0.24	0.49	21	2.2	1.48	0.5718
6	3	3	0.31	0.56	24	2.6	1.61	0.4323
7	4	3	0.35	0.59	27	3.0	1.73	0.3269
8	5	3	0.40	0.63	27	3.2	1.79	0.2472
9	6	4	0.42	0.65	29	3.3	1.82	0.1869
10	7	4	0.46	0.68	31	3.2	1.79	0.1413
11	8	6	0.62	0.79	35	3.6	1.90	0.1069
12	9	6	0.64	0.80	36	3.8	1.95	0.0808
13	10	6	0.68	0.82	37	4.0	2.00	0.0611
14								
15	Indep. Variance	8.61	=SUMPRODUCT(C4:C13,H4:H13)+SUMPRODUCT(F4:F13,H4:H13)					
16	Covariance	4.01	=2*D1*SUMPRODUCT(D4:D13,G4:G13,H4:H13)					
17	Total variance	12.62	=B15+B16					

Figure 12.2 Spreadsheet for Example 12.3

12.3.2 The candidate set of projects. Obviously correlational effects among the future cash flow streams of candidate projects have important implications for the firm's decision policies, and for composing the candidate sets of projects. For example, suppose an expansion of existing facilities is a candidate project. Suppose also that the firm's analysts predict a high degree of positive correlation between the cash flows of the expansion and of the existing facilities—that is, the unchanged "going concern." The adoption of the expansion would thereby increase the firm's risk because of the additive property of the positive correlational effect. Therefore, even though the expected return would be increased by the expansion, a disproportionate increase in the firm's risk might result in rejecting the project.

For the going concern, this has two implications. First, correlated cash flows mandate that existing undertakings be included in the candidate set of investments. That is, the project selection process must evaluate future cash flows both of additional projects and of present projects. In essence, then, the decision set covers all future correlated cash flows, in all feasible combinations, *including the future cash flows anticipated from presently existing projects and investments.*

Second, in the absence of synergistic effects, the firm ought to undertake future projects that complement its existing investments—that is, projects that are negatively correlated with present activities. This implies diversifying the firm's activities, which reduces the firm's overall risk position. However, often investments that reinforce the firm's current activities will have larger relative returns than complementary projects, since the reinforcing projects often have synergistic relationships with existing activities.

Consequently, a merger with or the acquisition of another firm can be viewed conceptually as a project. In either case, a present outlay of funds results in future cash inflows to the firm. The advantage of mergers and acquisitions lies in the opportunity for the acquiring firm to reduce its overall risk, while increasing its overall return. This area for the application of the theory of investment of the firm is beyond the scope of this book. However, we should note that *under the assumptions of a perfect capital market*, Haley and Schall (1973) demonstrate that the stochastic relationships (that is, correlational effects) between cash flow streams are irrelevant—the firm's risk stance is not affected by correlational affects. Consequently, under perfect capital market assumptions we can judge candidate projects on an *incremental* basis with respect to the firm's present investments—independently of the firm's present activities. However, if borrowing and lending rates differ (as they normally do) or if any other factor destroys the perfect capital market assumptions, then the entire scenario, *including the firm's present investments*, must be evaluated for their future consequences.

12.3.3 Multiple project selection by maximizing expected net present value. Selecting a set of multiple projects by maximizing the expected NPV is a straightforward extension of the methodology of Sections 12.2.1–12.2.4 for single projects which relies on the means and variances of the candidate project NPVs.

Once these values are procured, generally by a method outlined in Section 12.2, then candidate projects are combined into mutually exclusive subsets. If correlational effects are present, then the variance of the NPV of affected subsets must be adjusted. Constraints are also applied to eliminate infeasible subsets.

When the appropriate means and variances of NPV are established for each feasible subset (*bundle*), then each paired value, [E(NPV), σ(NPV)], is converted into its expected utility, and the expected utility is maximized over the feasible set of bundles. The single, mutually exclusive subset of projects that maximizes expected utility is the optimal solution. If no utility function can be established, then expected value decision-making is assumed and only the expected value is maximized.

There are some fundamental objections to this method for finding an optimal set of projects. First, net present value, even as a random variable, can exist only under perfect capital market conditions. Alternatively, we can assume that the NPV for candidate projects exists if we hypothesize the firm's marginal cost-of-capital rate, to be used as the interest or discount rate; and if we assume that additional new projects in the future will yield a rate at least equal to the assumed cost of capital, to provide a sink for recovered funds. This latter requirement stems from the reinvestment assumption underlying NPV, which was developed in Chapter 9.

Second, the method assumes implicitly that an expected utility indifference function is obtainable for the firm. This may or may not be true.

Third, as with other constrained budget investment problems, the method does not consider intertemporal transfer and/or reinvestment of idle funds at rates other than the assumed discount rate. Hence, it has limited applicability in practice.

12.4 ACCOUNTING FOR UNCERTAIN FUTURE STATES

Theoretically, both rate and stream effects can enter the same model. Then, not only might the discounting rate vary between periods, but also the decision-maker would have different preferences (different utility functions) about future cash flows. These define the time-state-preference (TSP) model, where the utility of money at a future time is a function of the time from "now" to "then" and the state of the decision-maker when the money is received or paid. For example, 10 years from now the receipt of $1000 would be viewed differently by a poor man than by a wealthy one, even though at the present time both men have similar means. Thus, the TSP model emphasizes the state of nature and the decision-maker at a given time.

Briefly, the model assumes a finite set of states, which occur in a time sequence with one, two, several, or many states at any time. However, all states must be specified, which includes a particular *sequence of events*, assumed to occur from now to the time when the state is defined. The number of possible states increases rapidly as the model becomes more complex. Furthermore, all possible states must be identified—all possible combinations of prior states and paths must be defined—and then only one state can occur at one point in time (states are mutually exclusive and exhaustive and all possible combinations must be accounted for).

In this model, cash flow streams and utility functions may be state-dependent, and probabilities of occurrence are defined for each state. If necessary, the time value of money can be made state dependent; this model can include intertemporal transfer of funds at different interest rates (for borrowing and lending purposes) and with differing *utilities* for borrowing and lending. As a result, the TSP approach to modeling the capital investment decision is a powerful one. Very sophisticated models can be built using the TSP concept; however, its extreme generality is also its practical downfall. Because it is so general, it is difficult to formulate the models so that solutions can be obtained.

Other approaches that rely on examination of the state that the firm is in over time have been described by Park and Thuesen (1979) and Thompson and Thuesen (1987). The first combines the idea of project balance with the certainty equivalent approach discussed in this chapter. The second approach develops criteria that focus on how the firm's health (equity position) influences and is influenced by project selection criteria. Three new criteria (discounted certainty equivalent, time-weighted utility, and horizon utility) were contrasted with the expected present value and mean-variance criteria. In a simulation experiment the "time weighted utility" produced only 4–14% as many bankruptcies as the other criteria.

12.5 SUMMARY

This chapter examined models based on the Bernoulli principle of expected utility for handling uncertainty connected with the project cash flows—maximizing the expected net present value. In addition, we examined timewise autocorrelations within a project, and cross-correlations between projects for estimating future cash flow and their stochastic relationships.

All of the models rely on perfect capital market assumptions—or at least, under conditions with known or assumed rate effects. Finally, a time-state-preference model was briefly described, in which it is theoretically possible to allow varying uncertainty and rate effects. This model's generality makes it extremely difficult to construct and solve.

Several applications of this material will be found in the following chapter.

REFERENCES

BARNES, J. WESLEY, C. DALE ZINN, and BARRY S. ELDRED, "A Methodology for Obtaining the Probability Density Function of the Present Worth of Probabilistic Cash Flow Profiles," *AIIE Transactions*, **10**(3) (September 1978), pp. 226–236.

BERNHARD, RICHARD H., "Probability and Rates of Return: Some Critical Comments," *Management Science*, **13**(7) (March 1967), pp. 598–600.

BONINI, CHARLES P., "Comment on 'Formulating Correlated Cash Flow Streams'," *The Engineering Economist*, **20**(3) (Spring 1975), pp. 209–214.

BUCK, JAMES R., and RONALD G. ASKIN, "Partial Means in the Economic Risk Analysis of Projects," *The Engineering Economist*, **31**(3) (Spring 1986), pp. 189–212.

BUCK, JAMES R., and T. J. HILL, "Laplace Transforms for the Economic Analysis of Deterministic Problems in Engineering," *The Engineering Economist*, **16**(4) (July–August 1971), pp. 247–263.

BUCK, JAMES R., and T. J. HILL, "Additions to the Laplace Transform Methodology for Economic Analysis," *The Engineering Economist*, **20**(3) (Spring 1975), pp. 197–208.

BUCK, JAMES R., and JOSE M. A. TANCHOCO, "A Closed–Form Methodology for Computing Present Worth Statistics of Risky Discrete Cash–Flows," *AIIE Transactions*, **9**(3) (September 1977), pp. 278–287.

BUSSEY, LYNN E., and G. T. STEVENS, JR., "Formulating Correlated Cash Flow Streams," *The Engineering Economist*, **18**(1) (Fall 1972), pp. 1–30.

BUSSEY, LYNN E., and G. T. STEVENS, JR., "Reply to 'Comment' on 'Formulating Correlated Cash Flow Streams'," *The Engineering Economist*, **20**(3) (Spring 1975), pp. 215–221.

BUSSEY, LYNN E., and G. T. STEVENS, JR., "Net Present Value from Complex Cash Flow Streams by Simulation," *AIIE Transactions*, **3**(1) (March 1971), pp. 81–89.

CARMICHAEL, DAVID G. and MARIA C. A. BALATBAT, "Probabilistic DCF Analysis and Capital Budgeting and Investment—A Survey," The Engineering Economist, **53**(1) (January–March 2008), pp. 84–102.

ELDRED, BARRY S., and J. WESLEY BARNES, "The Semi–Variance of a Sum of Independent Random Variables," *The Engineering Economist*, **24**(2) (Winter 1979), pp. 129–131.

FLEISCHER, GERALD A., "Two Major Issues Associated with the Rate of Return Method for Capital Allocation: The 'Ranking Error' and 'Preliminary Selection,'" *The Journal of Industrial Engineering,* **17**(4) (April 1966), pp. 202–208.

GIACCOTTO, CARMELO, "A Simplified Approach to Risk Analysis in Capital Budgeting with Serially Correlated Cash Flows," *The Engineering Economist*, **29**(4) (Summer 1984), pp. 273–286.

GRANT, EUGENE L., and RICHARD S. LEAVENWORTH, *Statistical Quality Control*, 4th ed. (McGraw–Hill, 1972).

GROSH, DORIS, L., and L. E. BUSSEY, "A Generalized Cramer Inequality," *Special Report No 115, Kansas Engineering Experiment Station* (Kansas State University, November 1973).

HALEY, CHARLES W., and LAWRENCE D. SCHALL, *The Theory of Financial Decisions* (McGraw–Hill, 1973).

HILLIER, FREDERICK, S., "The Derivation of Probabilistic Information for the Evaluation of Risky Investments," *Management Science* **9**(3) (April 1963), pp. 443–457.

HILLIER, FREDERICK, S., "Supplement to 'The Derivation of Probabilistic Information for the Evaluation of Risky Investments,'" *Management Science,* **11**(3) (January 1965), pp. 485–487.

HILLIER, FREDERICK, S., *The Evaluation of Risky Interrelated Investments* (North Holland, 1969).

HOSKINS, COLIN G., "Capital Budgeting Decision Rules for Risky Projects Derived from a Capital Market Model Based on Semivariance," *The Engineering Economist,* **23**(4) (Summer 1978), pp. 211–222.

LEUNG, L. C., Y. V. HUI, and G. A. FLEISCHER, "On the Present–Worth Moments of Serially Correlated Cash Flows," *Engineering Costs and Production Economics,* **16**(4) (July 1989), pp. 281–289.

LEVY, HAIM, and MARSHALL SARNAT, *Capital Investment and Financial Decisions,* 4th ed. (Prentice–Hall, 1990).

PARK, CHAN S., and GUNTER P. SHARP-BETTE, *Advanced Engineering Economics* (Wiley, 1990).

PARK, CHAN S., and GERALD J. THUESEN, "Combining the Concepts of Uncertainty Resolution and Project Balance for Capital Allocation Decisions," *The Engineering Economist,* **24**(2) (Winter 1979), pp. 109–127.

PETERSEN, CLIFFORD C., "Solution of Capital Budgeting Problems Having Chance Constraints: Heuristic and Exact Methods," *AIIE Transactions,* **7**(2) (June 1975), pp. 153–158.

PORTER, R. B., R. P. BEY, and D. C. LEWIS, "The Development of a Mean–Semivariance Approach to Capital Budgeting," *Journal of Financial and Quantitative Analysis* **10**(4) (November 1975), pp. 639–649.

SASIENI, M. W., "A Note on PERT Times," *Management Science,* **32**(12) (December 1986), pp. 1652–1653. See also T. K. LITTLEFIELD, JR., and P. H. RANDOLPH, "An Answer to Sasieni's Question on PERT Time," *Management Science,* **33**(10) (October 1987), pp. 1357–1359; C. GALLEGHER, "A Note on PERT Assumptions," *Management Science,* **33**(10) (October 1987), p. 1360, and KYUNG C. CHAE and SEHUN KIM, "Estimating the Mean and Variance of PERT Activity Time Using Likelihood-Ratio of the Mode and the Midpoint," *IIE Transactions,* **22**(3) (September 1990), pp. 198–203.

THOMPSON, ROBERT A., and GERALD J. THUESEN, "Applications of Dynamic Investment Criteria for Capital Budgeting Decisions," *Engineering Economist,* **33**(1) (Fall 1987), pp. 59–86.

VAN HORNE, JAMES C., *Financial Management and Policy,* 8th ed. (Prentice-Hall, 1989).

WAGLE, B., "A Statistical Analysis of Risk in Investment Projects," *Operational Research Quarterly,* **18**(1) (March 1967), pp. 13–33.

PROBLEMS

12-1. Dauntless Electric Company is deciding whether to invest $1,800,000 in a new plant to expand their output of stereo equipment. Management forecasts indicate that the new plant will generate independent random net cash flows of Y_t dollars at the end of each year. The means and standard deviations of the cash flow are estimated as follows:

End of Year, t	E(NCF) (in $ thousands)	σ(NCF) (in $ thousands)
0	$-1800	$ 0
1	+ 100	10
2	+ 300	20
3	+ 500	50
4	+ 800	60
5	+ 800	80

Assume a marginal investment rate of 15%.
(a) Calculate the mean and variance of net present value.
(b) If net present value is normally distributed, what is the probability that the firm will fail to recover its investment plus its required return?
(c) Assuming normality, what is the probability that the project's rate of return will be less than 15%? What assumptions are implicit in your answer?
(d) If the annual net cash flow amounts are normally and independently distributed, what is the probability that *no* return will be realized on the investment (i.e., that only the investment itself will be recovered)? (*Hint*: Find E[net cash flows] at an interest rate $i = 0$ and then compare this mean value with net present value = 0).
(e) If the distributional form of net present value is unknown, but it can be assumed unimodal, what is the probability that the company will fail to recover its investment plus the required return of 15%?

12-2. A firm is analyzing a project with a "known" first cost of $2,000,000. The annual returns at the end of each year for 10 years are independent random cash flows. Assume a marginal investment rate of 10% and a constant expected value for the annual returns of $400,000 per year. Calculate the mean and standard deviation of net present value, and assuming normality—the probability that the firm will not recover its investment plus its required rate of return. Assume that the standard deviation of the cash flows is:
(a) Constant at 10% of the expected annual cash flows.
(b) Increasing from 10% of the expected annual cash flows in year 1 to 28% of the expected annual cash flows in year 10 (+2% per year).
(c) Constant at 28% of the expected annual cash flows.

12-3. Redo problem 12-2 assuming that the expected value for the annual returns is a gradient that begins at $250,000 per year and increases at $50,000 per year. Assume that the standard deviation of the cash flow in each year is:
(a) 10% of the expected value for the first year.
(b) 10% of the expected value for each year.
(c) Increasing from 10% of the expected value for the first year to 28% of the expected value for year 10 (+2% per year).

12-4. Considering the results of problems 12-2 and 12-3, what suggestions about the sensitivity of the results can you make?

12-5. In a cash flow stream that is autocorrelated across time, what is the effect on the variance of net present value if the correlation coefficients are positive (i.e., $0 \le \rho \le 1$)? What is the effect of cash flow increments that are negatively correlated?

12-6. Develop an economic interpretation of a cash flow stream in which the annual cash flows are strongly and negatively correlated.

12-7. Identify some of the possible economic and physical causes of strong, positive autocorrelations in a project's cash flow stream.

12-8. Assume that the cash flow stream for a particular project has been synthesized from relevant economic and technical data, with the following estimates of the parameters of the cash flows:

End of Period	Mean Cash Flow, $E(Y_t)$	Standard Deviation of Cash Flow, $\sigma(Y_t)$
0	$-5.0	$0.750
1	+2.0	0.500
2	4.0	1.000
3	3.0	1.732
4	3.0	2.449

(a) If the cash flows, Y_t, are normally and independently distributed and if the firm's marginal investment rate is assumed to be 15%, what is the probability that the net present value of the project will be zero or negative?

(b) If the cash flows, Y_t, are normally distributed and autocorrelated with the autocorrelation coefficients shown in the table below and if the firm's marginal investment rate is assumed to be 15%, what is the probability that the project will fail to return its investment plus the required return rate? (Use the net present value method.)

T	Correlation Coefficients, $\rho_{T\theta}$, for $\theta =$				
	0	1	2	3	4
0	+1.00	0	0	0	0
1		+1.00	+0.10	+0.05	+0.02
2			+1.00	+0.15	+0.12
3	(Symmetrical)			+1.00	−0.16
4					+1.00

(c) Assuming that the cash flows, Y_t, are normally and independently distributed and using discount rates of 15, 30, 45, 60, 75, and 100%, calculate the probabilities that NPV ≤ 0. Plot these probabilities as a function of the discount rate, i, and sketch a continuous function of NPV(i) versus i. What relation does this function bear to the function $P(i^* \leq 0)$ versus i, where i^* is the internal rate of return of the project?

12-9. For problem 12-2 assume that the annual returns are autocorrelated and that the standard deviation of the cash flows is 10% of the expected value of the cash flow. Calculate the mean and standard deviation of net present value and, assuming normality, the probability that the firm will not recover its investment plus its required rate of return. Assume that $\rho_{t+2,t} = 0.5\ \rho_{t+1,t}$ and that $\rho_{t+m,t} = 0$ for $m \geq 3$. Assume that the autocorrelation coefficient, $\rho_{t+1,t}$, is:

(a) 0.25
(b) 0.50
(c) 0.75
(d) 1.00

(Note that the fixed autocorrelation relationship between 1, 2, and 3 or more years is questionable as ρ changes, especially for $\rho = 1.00$.)

12-10. Redo problem 12-9 assuming that $\rho_{t+2,t} = -0.5\ \rho_{t+1,t}$ and that $\rho_{t+m,t} = 0$ for $m \geq 3$. Assume that the autocorrelation coefficient, $\rho_{t+1,t}$, is:

(a) −0.25
(b) −0.50
(c) −0.75
(d) −1.00

12-11. Considering problem 12.2 and problems 12.9 and 12.10, how important is the assumed autocorrelation value for this problem? What tentative comments can you make about other applications?

12-12. The cash flows from a contemplated project are assumed to have a β distribution with the following estimated values:

End of Year, $t =$	Cash Flows		
	Pessimistic, Y_p	Modal	Optimistic, Y_o
0	$−14,000	$−12,000	$−10,000
1	0	2,000	4,000
2	5,000	8,000	11,000
3	8,000	12,000	16,000

(a) Calculate the means and standard deviations of the cash flows using PERT estimating techniques.
(b) Using a marginal attractive rate of return of 15%, formulate the equations for mean and standard deviation of net present value, and find the expected net present value and the standard deviation of net present value.

13
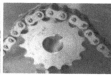

Decision Making Under Risk

13.1 INTRODUCTION

Many industrial problems involve making decisions over a period of time. Decisions lead to possible outcomes, generally involving chance, and often more decisions. The decisions and probabilistic outcomes for a project may range from fairly simple and straightforward to alternatives that may become quite complex. Decision trees help to organize the information and aid the decision making process by laying out the alternatives in a clear manner. The decision trees may then be used to determine expected NPV or other valuations.

This chapter explores creating decision networks and trees to organize decisions, options, and chance events. We begin with general purpose decision networks and how they evolve into decision trees when information becomes stochastic. Determining project value of both concurrent and sequential trees is demonstrated, as well as the use of decision trees to determine the expected value of perfect information.

Standard sensitivity analysis tools, such as tornado diagrams and spiderplots, illustrate the impact of uncertainty when a single input is varied. Monte Carlo analysis is a widely used and flexible simulation tool that provides a range of valuations as multiple inputs are varied simultaneously. The use of Monte Carlo analysis is shown to provide greater insight into a project than performing simple expected value analysis.

13.2 DECISION NETWORKS

Decision networks have been used for many years to describe the logical analysis of choice among various courses of action (Raiffa and Schlaifer, 1961). In their simplest forms, decision networks are simply chronological diagrams which show all of the possible alternatives that are available to the decision maker (Holloway, 1979). A high degree of certainty in the information is assumed, and there are no chance events. The diagrams help to structure problems, and are especially helpful when there are many future decisions.

Decision nodes, which are shown by a square, represent points in time where the decision maker must choose between alternative courses of action. The various alternatives, sometimes called option arcs, are represented by lines (usually arrowed) to the right of the square. The decision network links together the decision nodes and all of their possible outcomes.

Example 13.1 Cookie brand
A major food company was marketing a brand of cookies. Research determined that they could create a sandwich cookie (two cookie wafers with a layer of sweet icing in the middle) that was highly preferred over the market leader. Management conducted a product and economic evaluation to see if this would be a worthwhile business venture. Initial estimates showed that the product should sell at least $100 million per year. Management needed to make a series of decisions, each requiring increasing investment, as follows:

    Decision 1. Whether to complete product development? $500,000
    Decision 2. Design the facility for national production. $1 million.
    Decision 3. Build national production capacity. $10 million
    Decision 4. Market introduction. $100 million

Each step required an increasing financial commitment. Each early step had a relatively low risk in itself; the project was largely dependent on the risk of being successful in the marketplace. Each decision node offered alternatives. At any point, the project could be delayed, funded further, or abandoned, as shown in Figure 13.1.

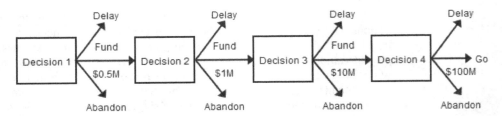

Figure 13.1 Decision network

The decision network illustrates the decisions to be made over time, and each decision node represents a point in time where a decision must be made. The arcs show the decision options (alternatives) that are available at each decision node, and each arc has its own potential financial value.

13.3 DECISION TREES

The term "decision network" accurately describes the connections between choices under a high degree of certainty. However, in many projects, there is a high degree of uncertainty regarding future information and events. When information is probabilistic, we add a chance node (sometimes called an event node) to represent the various outcomes that may result—shown on outcome arcs. The "decision network" becomes a "decision tree" with the addition of chance nodes and outcome arcs.

The format for drawing decision trees dates back at least to the early 1960s (Magee, 1964). Decision points (or nodes) are again represented by a square. These are structured so that all possible decisions are identified, but the decision maker may choose only one option arc. Chance nodes are drawn with a circle. Outcome arcs must be created so that each is mutually exclusive; only one outcome may occur (Clemen and Reilly, 2001). The set of chance nodes must be collectively exhaustive—including all possibilities. In probability terms, the sum of the probabilities for all arcs must equal 1.

Decision trees can become very complex; the more complex the problem, the more helpful decision trees become in organizing the information. Probabilities may be shown next to a given chance node. The values of various outcomes can also be shown on the far right of the decision tree, as shown in Figure 13.2. Given the values of the outcomes and probabilities, the value of each decision in the tree may be determined.

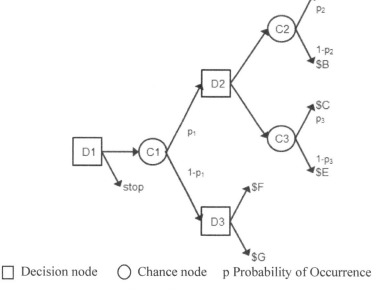

☐ Decision node   ◯ Chance node   p Probability of Occurrence

Figure 13.2 Decision tree

The structuring and understanding of a problem is coupled to drawing the decision tree. This begins by drawing, from left to right in chronological order, the various decisions and chances that occur. This stage often involves multiple iterations as different parts of the organization and different individuals contribute what they know and ensure that the tree is complete with good estimated values. At each chance node, it is critical to show the probabilities that are associated with each possible outcome and verify that the sum of these probabilities is 1.0. The value of each final outcome, on the right side of the decision tree, may include all costs and revenues, or these values may be linked with specific decision or chance arcs.

Once the tree has been completed, the second stage is to use this information to determine the value of the decision options and to recommend choices. The value of the project may be determined using the decision tree, working from right to left. The process has been called a variety of names, including "Rollback" (Magee, 1964), "Backwards Induction" (Raiffa & Schlaifer, 1961) and "Averaging Out and Folding Back" (Raiffa, 1968). Each decision node and chance node has a value, with each decision node value based on the decision made, and the chance node value based on the probabilities involved. In Figure 13.2, chance node 2 (C2) are calculated based on the weighted average of the expected values. So the value of C2 is

$$E(C2) = p_2(\$A) + (1 - p_2)(\$B)$$

Likewise, the value of chance node 3 is

$$E(C3) = p_3(\$C) + (1 - p_3)(\$E)$$

Decision nodes are valued on the worth of each decision. Again in Figure 13.2, the expected value of D2 is the larger of the C2 or C3 chance node expected values. If E(C2) has a higher value than E(C3), then E(D2) has a value equal to E(C2).

Example 13.2 Chemical process
A chemical company is planning to add a new specialty chemical to its manufacturing plant. There are two different technologies in development, and each has its own costs and potential revenue.

Process A has a 90% probability of succeeding. If successful, the project will have an expected NPV of $20 million. If unsuccessful, the project will have spent development funds and will have an expected NPV of −$0.5 million. Process B has an 80% probability of succeeding; it has an expected NPV of $30 million if successful, and −$0.7 million if unsuccessful. Which process should be pursued? What are the relative risks of each alternative?

Solution
Figure 13.3 shows the decision tree for the process options. The expected value of chance node A is:

$$E(A) = (0.9)(\$20M) + (0.1)(-\$0.5M) = \$17.95 \text{ million}$$

The expected value of chance node B is:

$$E(B) = (0.8)(\$30M) + (0.2)(-\$0.7M) = \$23.86 \text{ million}$$

There is a third option: to not do the project at all, with a value of zero.

Standard deviation can be used as a measure of risk:

$$\sigma = \sqrt{E(X^2) - [E(X)]^2}$$

If the project is not pursued, the standard deviation is zero. For Process A:

$$E(PW_A^2) = 0.9(20)^2 + 0.1(-0.5)^2 = 360.025$$

$$\sigma_A = \sqrt{360.025 - 17.95^2} = \$6.15 \text{ million}$$

For Process B:

13.3 Decision Trees

$$E(PW_B^2) = 0.8(30)^2 + 0.2(-0.7)^2 = 720.098$$

$$\sigma_B = \sqrt{720.098 - 23.86^2} = \$12.28 \text{ million}$$

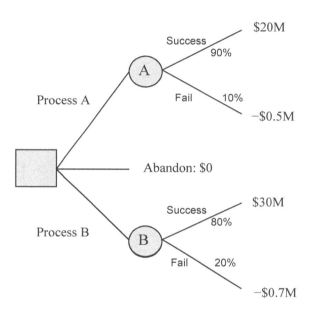

Figure 13.3 Chemical process decision tree

If the decision criterion is maximize expected PW, then Project B is the best alternative with a value of $23.86 million. However, the increased project value comes with an increased risk.

Example 13.3 Dementia drug
A drug company is seeking approval for a new drug product. The company is hoping for approval from the Food and Drug Administration (FDA) two years from now. The drug will have patent protection for 10 years after FDA approval. Once on the market, year-one net cash from sales is expected to be $8M (million), year two net revenues are expected to be $15 million, and years three through ten are expected to be $22 million.

The facility to produce the new drug will take two years to build at a cost of $38 million with a $5 million salvage value at the project horizon. There is a 90% probability that the FDA will approve the new drug. The minimum attractive rate of return (MARR) for the project is 25%.

If facility construction begins after FDA approval, initial sales will be delayed two years. Because the patent limits the time horizon for sales, two years of revenues would be lost. If facility construction begins now, it will be available to produce the drug upon

FDA approval. However, if the FDA does not approve the drug, the unused $38M facility will have a salvage value of only $9 million at the end of year 2. The question facing the firm is whether the facility should be built now, delayed pending FDA approval, or not be built? The alternative chosen should maximize expected value.

Solution
The decision tree is shown in Figure 13.4. The firm has three alternatives: (1) build the facility now, (2) build the facility later (delay), or (3) do not build the facility.

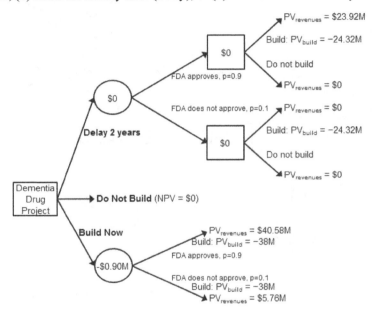

Figure 13.4 Decision tree, dementia drug problem

The facility can be built now, in expectation of FDA approval. The cost to build is $38 million. If the drug is approved, it is expected to generate a present value of $40.58 million in revenue. If the facility is built now and the FDA does not approve the drug, the facility has a value of only $9 million two years from now, or a present value of $5.76 million. The value of building now is

$$(0.9)(40.58M - 38M) + (0.1)(5.76M - 38M) = -\$0.90 \text{ million}$$

The project can be delayed, with the decision to build being made after the FDA approval decision. The FDA may approve (90% probability) or may not approve. If the FDA approves, the company can decide to either build or not. If they build, revenues will have a present value of $23.92 million, and the present value of the $38 million cost is −$24.32 million, for an NPV of −$0.40 million. If the plant is not built, there is no cost and no revenue. The best decision is to not build, for a value of $0.

If the FDA does not approve the project, the firm still has the option to build, but this would not be economical. The best decision is to not build, for a value of $0. The delay option therefore has a value of:

$$(0.9)(0) + (0.1)(0) = \$0$$

The project has three alternatives. The firm can build now, with an estimated NPV of −$0.90 million, and they can delay the decision, with an estimated NPV of $0 million, or they can decide to not pursue the project, with an NPV of $0. They should pick the most economic option, which is to not build now. The decision node has a value of zero, the larger of the two values, $0 and −$0.9M.

Computer software is available to build and analyze decision trees. Figure 13.5 shows the dementia drug decision tree, similar to Figure 13.4, built in Precision Tree software (Palisade Corporation).

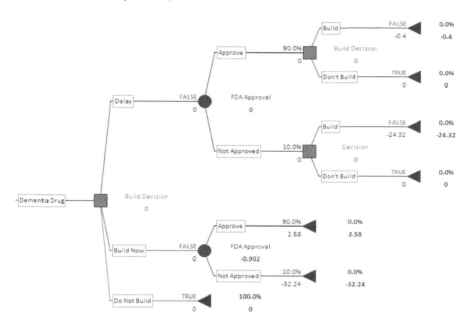

Figure 13.5 Dementia drug decision tree in Precision Tree software

13.4 SEQUENTIAL DECISION TREES

Decision trees are often used to organize and evaluate multi-stage projects that take place over a period of time. Firms often undertake large investments in stages. Rather than invest all required funds at the beginning of a large project, many firms will make a series of sequential investments based on the success of previous investments. Many projects involve multiple stages, and multi-stage economic analysis is often necessary.

Project management courses teach that large projects are often carried out as a series of stages or phases. The project manager often completes one stage, and then must obtain management approval and funding for the next stage. Decision trees can assist in identifying the value of such projects.

An example of a multi-stage project is pharmaceutical drug development, where new drug products must pass a series of clinical trials, and where successive clinical trials are performed (or not) depending on the success of the previous clinical tests. That is, Phase II tests are only performed if Phase I tests are successful, and Phase III testing is conducted only if Phase II is successful. Staged funding also occurs in many other large projects, where new ideas pass from concept development to product design and development to engineering, creation of manufacturing capacity, and product introduction into the marketplace. Each stage usually involves rapidly increasing monetary commitments, and each stage is funded only if the previous stage is successful. Staged funding is a method of managing the investment risk. While at each *passed* stage the probability of a product reaching the market increases, the increasing financial stakes imply increasing amounts of risk.

Example 13.4 Hypertension drug development
A drug candidate for treating hypertension (high blood pressure) has been identified and has completed initial testing. As with the dementia drug example, a series of clinical tests needs to be conducted following established FDA rules. Three clinical trials need to be successfully completed, followed by FDA approval and a launch phase. Each phase is dependent on the success of the previous phase.

As summarized in Figure 13.6, the testing and approval process is expected to take ten years. If all goes according to plan, the drug would have 10 years of exclusive marketing rights upon FDA approval. In Phase I testing, the drug would be given to 20–80 healthy people to determine human safety. The testing is expected to cost $8 million (assumed in year 2 for simplicity) and take two years to complete, with an estimated 70% chance of success. In Phase II testing, the drug would be given to 100–300 people to determine the efficacy for treating hypertension. The probability of success is estimated at 30%. Phase II testing is expected to require 2 years to complete, and cost $30 million (in year 4). In Phase III clinical testing, the drug would be given to 1000–5000 people to determine safety and efficacy in a broad spectrum of the population. This testing is expected to take three years to complete and would start pending successful results from Phase II. The Phase III trials would cost $300 million (in year 7) and have an 80% chance of success. To obtain FDA approval, a new drug application (NDA) would need to be written; this will require $10 million (in year 8) and one year to complete. FDA approval is expected to take two years, and there is a 90% probability of obtaining the needed approval. Successfully launching the product would require $350 million (primarily marketing costs) in year 10.

The hypertension drug has the potential of generating large profits, with net revenue of $450 million per year for ten years, starting in year 11. While the development costs are high and the chances of success are low, the potential payout is high if success can be achieved. The question therefore becomes: should the drug be developed? The MARR is 20%. This information is summarized in Table 13.1 and Figure 13.6.

Solution
The costs and revenues of the project are examined assuming that payments are made at the conclusion of each phase. Product launch and market introduction costs $350 million in year 10. The MARR is 20%.

Cost at year 10 = $350M

Revenue at year 10 = 450M(*P/A*, 20%, 10) = $1886.4M
Conditional Expected Value (EV) = 1886.4M − 350M = $1536.4M

Table 13.1 Hypertension drug cash flows

Year	Investment Required ($ millions)	Conditional Probability of Success
2	8	70 %
4	30	30
7	300	80
8	10	90
10	350	

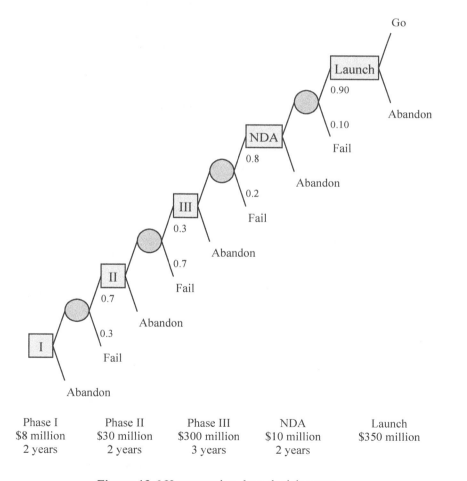

Phase I	Phase II	Phase III	NDA	Launch
$8 million	$30 million	$300 million	$10 million	$350 million
2 years	2 years	3 years	2 years	

Figure 13.6 Hypertension drug decision tree

The NDA will cost $10 million in year 8. To determine the likely costs in year 8, we add the NDA preparation cost to the probability weighted discounted cost from year 10.
Cost at year 8 = 10M + (0.9)(350M)(P/F, 20%, 2) = $228.75M
Revenue at year 8 = 1886.4M(0.9)(P/F, 20%, 2) = $1179M
Conditional EV = 1179M − 228.75M = $950.25M

In year 7, we need to pay for the Phase III testing and need to assume probability-adjusted costs of further testing.

Cost at year 7 = 300M + (.80)(228.75M)(P/F, 20%, 1) = $452.5 million
Revenue at year 7 = 1179M(0.8)(P/F, 20%, 1) = $786.0M
Conditional EV = 786.0M − 452.5M = $333.5M

In year 4, we need to pay $30 million for Phase II tests.
Cost at year 4 = 30M + (0.3)(452.5M)(P/F, 20%, 3) = $108.6 million
Revenue at year 4 = 786.0M(0.3)(P/F, 20%, 3) = $136.5M
Conditional EV = 136.5M − 108.6M = $27.9M

In year 2, we need to pay $8 million for Phase I tests.
Cost at year 2 = 8M + (0.7)(108.56)(P/F, 20%, 2) = $60.77 million
Revenue at year 2 = 136.5M(0.7)(P/F, 20%, 2) = 66.35
Conditional EV = 66.35−60.77 = $5.58M

Discounting this to year 0, we have
Conditional EV = 5.58(P/F, 20%, 2) = $3.88M

The expected NPV of the project is $3.88 million. The project is worth pursuing, but the NPV is not particularly high given the investment, the long timeline and the many hurdles that need to be overcome. While the project is risky, staged funding is a form of hedge; the firm will not commit the entire $698 million because they do not need to; they only need to justify spending $8 million for Phase I testing. They do not need to commit to a 10-year project, only a 2-year study. Staged funding is a hedge that protects the upside revenue potential while minimizing the cost and risk exposure of the firm.

13.5 DECISION TREES AND OUTCOME VARIABILITY

Solving a decision tree based on expected value is straightforward, and is useful in many situations. However, there are many situations where the outcome contains a significant degree of risk. Decision trees can be used to help quantify the risk profile of a problem. Alternatives can be assigned probabilities based on expert judgment. Using the rollback method along with these probabilities, uncertain alternatives can be evaluated, and strategies can be compared.

Example 13.5 Improve sales
Figure 13.7 contains a decision tree for a company considering alternatives to increase sales. There are three alternatives: (1) create a new product improvement, (2) increase

13.5 Decision Trees and Outcome Variability

marketing, and (3) hope for improvement. The company is willing to invest $50 million dollars in one of the first two strategies.

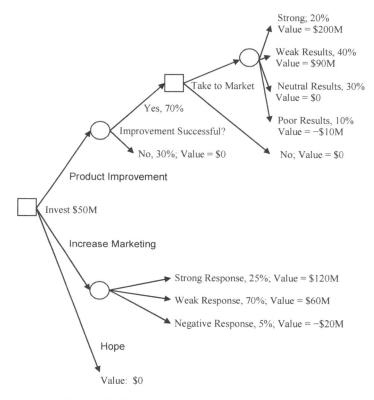

Figure 13.7 Decision tree for sales improvement

The first strategy involves creating a product improvement, which has a 70% chance of success. If the development project is successful, the company can decide whether to commercialize the product. If they do, there are four possibilities regarding success in the marketplace, all having different probabilities of success. These are also shown in Table 13.2. If the product is a strong success, significant returns could be made on the investment.

TABLE 13.2 Alternative results for product improvement

Resulting Sales	Probability	PV of Improvement
Strong	20%	$200M
Weak	40%	90M
Neutral	30%	0M
Decline	10%	−10M

As an alternative to creating a product improvement, the company could use the $50 million to increase marketing efforts. Three possible outcomes could result, each with different probabilities, as shown in Table 13.3. Which strategy is recommended?

TABLE 13.3 Alternative results for increased marketing

Resulting Sales	Probability	PV of Improvement
Strong	25%	$120M
Weak	70%	70M
Negative	5%	−20M

Solution

The expected value of the product improvement, if it goes to market, is the weighted average based on its probabilities:

$E(PV) = (0.20)(200M) + (0.40)(90M) + (0.30)(0M) + (0.10)(-10M) = \$75$ million

However, there is only a 70% probability that the product will be commercialized, so

$E(PV) = 0.7(75M) = \$52.5$ million

With a cost of development of $50 million, the resulting NPV of the product improvement strategy is $2.5 million.

Another way to view the possibilities is to roll back the market possibilities to the product improvement decision node, including the 70% chance of getting to the market. Each possibility is then:

Strong = (0.70)(0.20)(200M) = $28.0 million
Weak   = (0.70)(0.40)(90M) = $25.2 million
Neutral = (0.70)(0.30)(0)   = $0
Decline = (0.70)(0.10)(-10M) = −$0.70 million
Total   = $52.5 million

The expected value of the increased marketing strategy is also the weighted average based on its probabilities:

$E(PV) = (0.25)(120M) + (0.70)(60M) + (0.05)(-20M) = \$71$ million

The potential revenues are not as high, but there is an increased probability of achieving improvement. Because $50 million are spent on marketing, the NPV of this strategy is $21 million and is the preferred strategy.

13.5.1 Stochastic decision trees. In Example 13.5, chance nodes had up to four different possible outcomes. When the number of possible outcomes is increased, and probabilities are associated with each outcome, the tree becomes a stochastic decision tree. Any decision tree can be made to mirror a probability distribution by adding an arbitrarily large number of branches at each chance node. The addition of a large number of branches can be used to represent any probability distribution (Hespos and Strassmann, 1965). However, this makes the tree very complex and cumbersome, and becomes impractical.

An alternative is to replace the many branches at each chance node by a single distribution. This requires simulation in order to solve, and requires computer software to solve it efficiently. At each chance node, a random selection is made consistent with the probability distribution, and the value of the decision tree is determined. This represents one iteration. Many iterations are run using computer simulation. This technique is explored in greater detail in following sections of this chapter.

13.5.2 Applications. There are a number of examples of decision trees being used to aid decision making in a wide variety of industries. Hespos and Strassmann (1965) is an early example of using stochastic decision trees for investment decisions. They found that decision trees allowed the evaluation of nearly all feasible combinations of decisions, including expected and risk aversion, allowing an improved set of decisions.

Decision trees have been used to evaluate the use of prescribed burning in forest management (Cohan, et al., 1984). Uncertainties concerning weather conditions and fire behavior were used to make decisions on whether to initiate controlled burns. The technique was also used to document the rationale of how decisions were made, which was used to communicate within the forest management organization.

These techniques have been used to choose which of two locations for drilling an oil well (Hosseini, 1986). Decision trees were used to create a framework for systematically analyzing geological and engineering problems in order to decide the site for an exploratory well. The tools were used to quantify uncertainties in terms of subjective probabilities, and incorporated risk and time value of money.

The U.S. Postal Service used decision analysis to decide to use nine digit zip codes (ZIP+4) and to buy additional equipment as part of its ongoing automation efforts (Ulvila, 1987). Automation offered the potential of huge labor savings, but these savings were highly uncertain. Decision analysis was the basis for the Office of Technology Assessment to go forward with the capital investment.

Decision trees have begun to be used in medical treatment decisions (Hazen, Pellissier, and Sounderpandian, 1998). An example of analyzing a decision of whether a patient should undergo total hip replacement was performed from the perspectives of the patient (using utility analysis) and of total cost. Risk assessment, including human mortality, was included in the analysis.

13.6 EXPECTED VALUE OF PERFECT INFORMATION

Most projects face uncertainty in one or more issues. We often deal with this uncertainty by assuming the underlying parameters follow a known distribution in order to put limits on the risk involved. Additional work, such as market testing or analysis, is sometimes performed to increase knowledge and to decrease the level of risk and uncertainty. However, extra testing has its costs, including the time involved and the resulting delays in getting good ideas to market. The cost of decreasing risk needs to be balanced with the value of the information received.

One way of placing limits on the value of information is to determine the value of perfect information. Is it still worth pursuing an idea if all things go well? If the answer is no, then it is in the best interest of the firm to quickly abandon that idea.

Project selection often involves making decisions between alternative choices. The choices can be visualized in a decision tree with a mix of decision and chance nodes. Decisions must often be made before the uncertainty is resolved. If perfect information could be forecast on which chance branches would occur, then optimal decisions could be made. These optimum choices present the case where the maximum project value can be achieved and there is no loss of opportunity. This optimum represents the expected value with perfect information (EVwPI). The difference between the expected NPV and the expected value with perfect information (EVwPI) is the expected value *of* perfect information (EVPI). The expected value of perfect information is the amount of money that would be gained, over and above the expected NPV, if all information were known and the best possible decisions were made. Another term for EVPI is the expected opportunity loss, or EOL (Pratt, Raiffa, and Schlaifer, 1965).

Let's return to the Dementia project described previously. The project has many variables. To simplify the problem, assume that all values are known with certainty except for the FDA approval decision. The FDA approval is the project's primary source of risk. With this simplified problem all uncertainty is shown in the decision tree in Figure 13.4.

With perfect information, we will not delay the decision. If our perfect forecast is that the FDA approves, then we build now. If our perfect forecast is that the FDA does not approve, then we don't build. With perfect information, we cannot change the 90% probability of FDA approval, but we do get correct predictions of what will happen. Thus, the expected value with perfect information (EVwPI) equals 90% of the NPV of the "build now" option plus 10% of zero, or

$$EVwPI = (0.9)(2.58M) + (0.1)(0) = \$2.3M$$

As stated above, the expected value *of* perfect information (EVPI) is the difference between the expected NPV and the expected value *with* perfect information (EVwPI). EVPI is calculated according to equation (1). In this case the best decision without the perfect information has a present value of $0 since the decision is *not* to proceed; the decision tree in Figure 13.3 has an NPV of zero.

$$\begin{aligned}EVPI &= EVwPI - NPV \\ &= 2.3M - 0 = 2.3M\end{aligned} \tag{13.1}$$

The EVPI of the dementia drug project is $2.3 million.

13.7 SIMULATION

Standard sensitivity techniques such as tornado diagrams and spiderplots do an excellent job of demonstrating how an outcome, such as NPV, changes as individual inputs change. The graphical tools convey a wealth of information that is difficult to visualize without their aid. However, these tools are limited in that they demonstrate how the outcome changes as one, and only one, input changes over the course of its range, thus missing two important points: (1) input variables tend to have a higher probability of occurrence near the center of their range than at their extreme points, and (2) more than one input often changes at any given time. Also, the NPV distribution may be skewed, due to the interaction of economic factors that establish the cash flows, and then probability statements about NPV using normal distribution tables will be in error. These limitations can be partially overcome by using Monte Carlo simulation.

Imagine a computer performing a simulation for a new product. Market share and price will be limited in their range and likelihood. The new product's present worth will depend on the market share and the price, among other factors. A simulation begins with the computer randomly choosing a market share, consistent with the limits imposed. The price is then chosen randomly following its limits. The computer then calculates the present worth, using these new pieces of information as inputs. The process is repeated many times and the computer keeps track of the results. After many iterations, the computer can create a graph of the distribution of the market share, the price, and the resulting present worth. The computer can also describe each distribution in terms of mean, standard deviation, and other descriptive variables. This process is known as Monte Carlo simulation; as in gambling, the results depend on the random selections of values that are made (Clemen and Reilly, 2001).

Monte Carlo simulation is used to deal with problems with a high level of uncertainty. Monte Carlo programs can efficiently handle problems where there are multiple sources of risk. After running a simulation many times (a thousand times or more is common), a distribution of likely outcomes for any alternative is created. When all of the sources of uncertainty are included, the simulation model captures the relevant uncertainties and provides a distribution of likely outcomes in the form of a graph (Clemen and Reilly, 2001). This graph can provide a likely outcome (mean), risk (standard distribution), and likelihood of failure (percent above or below a given value).

Hess and Quigley (1963) and Hertz (1964, 1968) were among the first to use Monte Carlo simulation for the construction of NPV distributions. Hess and Quigley assumed complete independence among all cash flows, and while Hertz stated that some cash flows might be correlated, he did not report a method for considering correlated effects. Bussey and Stevens (1971) reported on a simulation model that considered some complex internal dependencies and interactions, resulting in correlative effects. Cobb and Charnes (2004) used correlated inputs using Crystal Ball software to determine real option volatility (which is explored in more detail in Chapter 14).

Simulation tools became more widely available with the advent and advancement of personal computers. This increased computing power has made fairly advanced simulation tools available. Several programs are available that perform Monte Carlo simulations in conjunction with spreadsheets. The simulations contained in Chapters 13 and 14 were performed using @Risk software (Palisade Corporation).

The concept of simulation is to derive an outcome based on sampling from inputs, which are defined as probability distributions. When this process is repeated, characteristics of the outcome, such as an expected NPV, can be derived and information can be gleaned.

To build a simulation, a spreadsheet is written containing the inputs and outputs of interest, along with the needed equations or algorithms linking the two. A distribution is needed for each uncertain input. Depending on the distribution, other information may be needed. Quite often, the actual distribution is not known, in which case using the triangular distribution is recommended. This is simply a triangle with three known points: the minimum, the maximum, and the mode (the most likely point in the distribution). Triangular distributions need not be symmetric; in fact, triangular distributions in real projects are often not symmetric. Take project costs for example. Many projects exceed their budgets, sometimes by a large factor (the Big Dig in Boston is an example). Yet projects usually do not underspend their budgets by a large amount; available money tends to be spent on something. A project manager is likely to overspend the budget by more than they are to underspend the budget.

Normal distributions are also possible, but they are not common in engineering projects. If a normal distribution is present, then the mean and the standard deviation are required to define the input distribution. Stock and commodity prices often follow lognormal distributions where prices are nonnegative. Project schedules often use beta distributions when determining the probability of project completion times. Many other predetermined distributions are available, depending on the software, to mimic the input variability. (Many advanced simulation software packages have data assessment tools to help identify likely distributions given available sample data.)

When defining a project, some input variables are known with certainty, and others exist as a range; they vary and have risk. All risky variables can be identified with their means (or most likely values), ranges, and distributions. Simulation programs generate random variates (sample from distributions). This number (or set of numbers if there is more than one risky variable) is then used to calculate the output variable, such as NPV. This is considered one iteration. The process of sampling is repeated, the output is determined a second time, and a second iteration is completed. This process is repeated again and again. While this can be done by hand, it can be enormously time consuming. Computer programs can cycle through thousands of iterations in a few seconds, identifying the range and distribution of all inputs as well as the resulting variable output. The result is an output range which occurs as the inputs change. The output is available as a histogram with the accompanying statistical analysis (mean, mode, standard deviation, etc.).

Example 13.6 Monte Carlo simulation

As a very simple example, we will have two variables, the expected price of a product, a, and its expected sales volume, b. The mean of a is $20, with a lower limit of $10 and an upper limit of $30 following a triangular distribution. The input a is shown in Figure 13.8. Variable b also follows a triangular distribution but is not symmetric; it has a mode of 30 thousand units, a minimum of 10 thousand units, and a maximum of 40 thousand units. Variable b is shown in Figure 13.9.

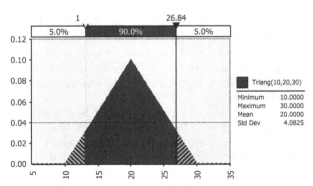

Figure 13.8 Input variable a for Monte Carlo simulation

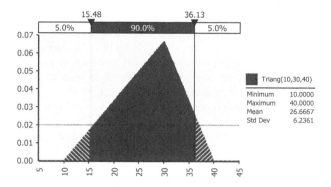

Figure 13.9 Input variable *b* for Monte Carlo simulation

The output for this example is the revenue, *a* times *b*. You would expect the mean of *ab* to be about 533, since the mean of *a* is 20 and the mean of *b* is 26.67. Figure 13.10 shows the results if 1000 iterations are run. Every simulation outcome will be slightly different due to the use of random number generation.

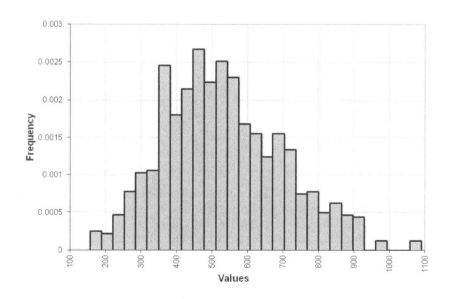

Figure 13.10 Output *ab* for Monte Carlo simulation.

Example 13.7 Dementia drug project

We return to the Dementia drug project that was introduced in Example 13.3. The decision tree recommended that the facility not be built, but many of the project's inputs are not known with certainty. As mentioned earlier, if facility construction for this project begins after FDA approval, it will take two years to build the plant, and the start of sales will be delayed two years. If facility construction begins now, it will be available to produce the drug given FDA approval. However, if the FDA does not approve the drug, the facility will have limited use and will be worth far less. The cash flows are shown in Table 13.4.

The question facing the firm is whether the facility should be built now or delayed pending FDA approval. The estimated base case values are summarized in Table 13.5 along with the lower and upper limits for each parameter. Expected values were calculated for these parameters using the assumption of triangular distributions, using the Base Case values as the mode.

Table 13.4 Dementia drug cash flows

Year	Build Now FDA Approves	Build Now FDA Disapproves	Build Later FDA Approves
0	$-38	$-38	$0
1	0	0	0
2	0	9	-38
3	8		0
4	15		0
5	22		8
6	22		15
7	22		22
8	22		22
9	22		22
10	22		22
11	22		22
12	27		27

Several important input variables (investment, first year revenue, and second year revenue) have asymmetric distributions. It is more likely that the facility will cost more than the estimate, and thus the distribution is skewed toward the higher cost. The FDA approval is a binomial distribution; 90% of the time it is approved (a value of 1.0) and 10% of the time it is rejected (value of zero). When calculating the resulting NPV and IRR, the expected values (means) were used, not the modes, as this treats the base case decision tree and simulations more consistently.

A Monte Carlo simulation using triangular distributions (binomial in the case of the FDA approval) and the parameters from Table 13.4 provide a distribution of the expected NPV. Each year's cash flows are correlated to the previous year's cash flows, with an r^2 of 0.90 (see Cobb and Charnes, 2004). The output of the Monte Carlo simulation includes the NPV of the "Build Now" option and the IRR of the "Build Now" and "Delay" options. Figures 13.11, 13.12, and 13.13 show the output distributions using @Risk software.

Table 13.5 Dementia drug variables

Parameter	Base Case	Low	High	Distribution
First Cost	$38 M	−5%	+15%	Triangular
Salvage Value: new plant	$ 9 M	−100%	+100%	Triangular
Salvage Value (terminal)	$ 5 M	−100%	+100%	Triangular
P(FDA approval)	0.9			Binomial
Revenue 1st year	$ 8 M	−40%	+20%	Triangular
Revenue 2nd year	$15 M	−40%	+30%	Triangular
Revenue > 2 years	$22 M	−40%	+40%	Triangular
Hurdle Rate (discrete)	25%	−20%	+20%	Triangular
Risk-free Rate	5%	−40%	+40%	Triangular
Time to FDA decision	2 yr	1 yr	3 yr	Triangular
Time to build facility	2 yr	1 yr	3 yr	Triangular

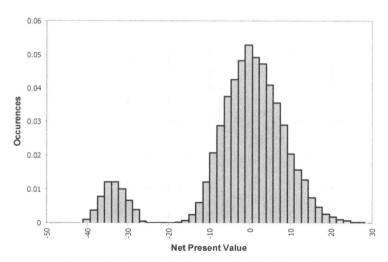

Figure 13.11 NPV distribution for "Build Now"

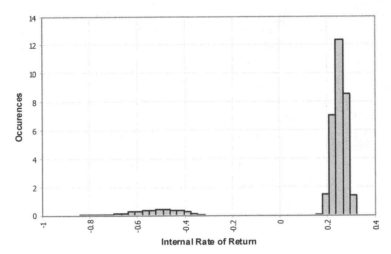

Figure 13.12 IRR distribution for "Build Now"

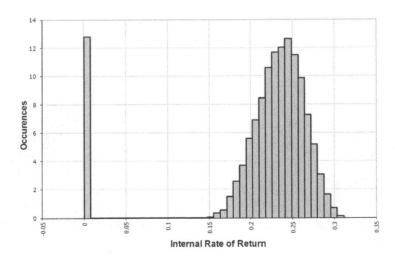

Figure 13.13 IRR distribution for "Delay"

Figure 13.11 shows a bimodal distribution of the NPV due to the bimodal nature of the FDA decision. If the FDA approves the project, the right hand distribution occurs. If the FDA does not approve the project, then the left distribution occurs, and the project loses a significant amount of money. Figure 13.12 shows the bimodal distribution of the IRR. Again, the bimodal nature of the distribution is caused by the risk of the FDA decision.

The expected NPV for the project is less than zero, and the average IRR is less than the MARR. Using traditional NPV or IRR analysis, the project would not normally be funded. The simulation shows that there is an opportunity to make significant profit, but the project is risky. Because the risk and the resulting low return is driven by the FDA decision, an option to wait until after the FDA decision could be made. The change in the resulting risk and returns is demonstrated using the Monte Carlo simulation of the delayed project IRR, shown in Figure 13.13. If the FDA approves the project, a positive IRR is highly likely, but it may not meet the required hurdle rate of the firm. If the FDA does not approve, then the facility is not built and the return on the project is zero.

Building a facility now is risky and has a negative average NPV. We would not recommend building the facility now. The company could wait for 2 years until after the FDA makes a decision. There is an opportunity for a positive NPV if one or more of the variables trend positive, but this appears unlikely. Simulation recommends that the project be abandoned.

Example 13.8 The biodiesel initiative (Rangarajan et al., 2012)
A project was being planned to design a mid-size biodiesel plant in northern Illinois. Used cooking grease was to be collected from regional restaurants and converted into biodiesel fuel. Monte Carlo simulation was used to evaluate the risk due to uncertainties in the project. Table 13.6 shows the project variables, along with their most likely values (modes) and ranges of uncertainty.

Table 13.6 Input variables and uncertainties

Input Variables	Base Case	Lower Limits	Upper Limits
Initial Investment	$6,630,000	95%	110%
Quantity of Cooking Grease (lbs.)	33,750,000	80%	110%
Cost of Cooking Grease	$0.40/lb.	75%	130%
Transportation (# of Rail Cars)	160	80%	110%
Cost per Rail Car	$1200	95%	120%
Quantity of Methanol (gal)	560,000	80%	110%
Cost of Methanol	$0.84/gal.	70%	125%
Quantity of Catalyst (lbs.)	320,000	80%	110%
Cost of Catalyst	$0.40/lb.	85%	130%
Total Fixed Expenses	$1,186,796	90%	110%
Quantity of Biodiesel Produced (gal) (dependent on amount of grease collected)	4,500,000	93%	110%
Sale Price of Biodiesel	$3.65/gal.	85%	125%
Revenue from sale of glycerin	$40,000		
After-tax NPV ($i = 5\%$)	$911,372		
After-tax IRR	7.5%		

For the base case, the initial investment for the biodiesel project is $6,630,000. The annual cash flows are found by taking the annual revenues ($16,465,000), and subtracting annual fixed costs ($1,186,796) and annual variable costs ($14,290,400). Cash flows are discounted using an interest rate of 5% to arrive at a before-tax NPV of $2,125,155 and an after-tax NPV of $911,372 (assuming MACRS depreciation, a 10-year property class, and a 35% tax rate). The before-tax IRR is 10.3% and the after-tax IRR is 7.5%.

Figure 13.14 shows the output obtained for the after-tax NPV from the simulation using the variables from Table 13.6. Triangular distributions were used for all variables. The NPV histogram resembles a normal distribution in spite of the fact that none of the inputs were normally distributed, with a mean after-tax NPV of $6.4 million. (This is often the result of the Central Limit Theorem.) Note that the mean NPV from the simulation is significantly larger than the base case. The fact that the inputs are not symmetrical (the revenue is positively skewed), improves the economics. Based on the statistics of the distribution, there is approximately a 68% probability of achieving an after-tax NPV greater than zero. This was calculated by determining the ratio of iterations with positive NPV values versus total iterations.

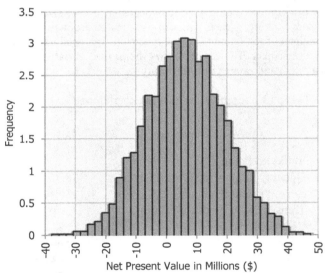

Figure 13.14 After-tax NPV for biodiesel project

The IRR analysis in Figure 13.15 also resembles a normal distribution with a mean after-tax IRR of 26.2%. This is again significantly higher than the base case value. IRR analysis again shows that the project will likely achieve the required rate of return, with an 84% probability of achieving an IRR greater than the required 5% and an 89% probability of achieving a positive return.

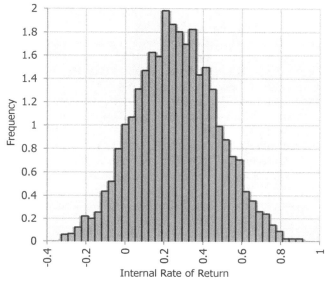

Figure 13.15 After-tax IRR for biodiesel project

13.8 SUMMARY

Decision trees help to organize information and aid the decision making process by displaying the alternatives in a clear manner. Where there are multiple decisions to be made, decision trees are often needed to bring order to the decision-making process. Decision trees are helpful in determining NPV, IRR, or other valuations for complex decisions, including those carried out over a period of time.

Simulation can aid in determining the range of outcomes when project inputs are uncertain. Especially when there are multiple points of uncertainty, simulation models can provide a range of project valuations as multiple inputs are varied concurrently. If input variables are skewed, mean output results can be significantly different than base case values.

REFERENCES

BUSSEY, LYNN E., and G. T. STEVENS, JR., "Net Present Value from Complex Cash Flow Streams by Simulation," *AIIE Transactions,* **3**(1) (March 1971), pp. 81–89.

CLEMEN, ROBERT T., and TERENCE REILLY, *Making Hard Decisions*, (South-Western, 2001)

COBB, BARRY R., and JOHN M. CHARNES, "Real Options Volatility Estimation with Correlated Inputs," *The Engineering Economist,* **49**(2) (April–June 2004), pp. 119–137.

COHAN, DAVID, STEPHEN M. HAAS, DAVID L. RADLOFF, and RICHARD F. YANCIK, "Using Fire in Forest Management: Decision Making under Uncertainty," *Interfaces,* **14**(5), (Sept.–Oct. 1984), pp. 8–19.

Hazen, Gordon B., James M. Pellissier, and Jayavel Sounderpandian, "Stochastic–Tree Models in Medical Decision Making," *Interfaces*, **28**(4),(July–August 1998), pp. 64–80.

Hertz, David B., "Risk Analysis in Capital Investment," *Harvard Business Review* **42**(1) (January–February 1964), pp. 95–106.

Hertz, David B., "Investment Policies That Pay Off," *Harvard Business Review* **46**(1) ((January–February 1968), pp. 96–108.

Hespos, Richard F., and Paul A. Strassmann, "Stochastic Decision Trees for the Analysis of Investment Decisions," *Management Science*, **11**(10), (August 1965), pp. B244–B259.

Hess, Sidney W., and Harry A. Quigley, "Analysis of Risk in Investments Using Monte Carlo Techniques," *Chemical Engineering Progress Symposium Series 42: Statistics and Numerical Methods in Chemical Engineering* (American Institute of Chemical Engineering, 1963).

Holloway, Charles A., *Decision Making Under Uncertainty: Models and Choices*, (Prentice-Hall, 1979).

Hosseini, Jinoos, and Thomas E. Blair, "Decision Analysis and Its Application in the Choice between Two Wildcat Oil Ventures," *Interfaces*, **16**(2), (March–April 1986), pp. 75–85.

Magee, John F., "Decision Trees for Decision Making," *Harvard Business Review*, **42**, (July–August 1964), pp. 126–138.

Pratt, John W., Howard Raiffa, and Robert Schlaifer, *Introduction to Statistical Decision Theory*, (McGraw-Hill, 1965); expanded and published (The MIT Press, 1995).

Rangarajan, Kiran, Suzanna Long, Norbert Ziemer, and Neal Lewis, "An Evaluative Economic Development Typology for Sustainable Rural Economic Development," *Community Development*, **43**(3), (July 2012), pp. 320–332.

Raiffa, Howard, *Decision Analysis, Introductory Lectures on Choices under Uncertainty*, (Addison-Wesley Publishing Company, 1968).

Raiffa, Howard, and Robert Schalifer, *Applied Statistical Decision Theory*, (Harvard University Press, 1961).

Ulvila, Jacob W., "Postal Automation (ZIP+4) Technology: A Decision Analysis," *Interfaces*, **17**(2), (March–April 1987), pp. 1–12.

PROBLEMS

13-1. Century Motorworks needs to decide between two proposed projects. Both projects will cost the same, and both have risks. The expected benefits and risks of both projects are shown below. What is the annual value of the best decision?

Proposal 1		Proposal 2	
Annual Benefit	Probability	Annual Benefit	Probability
$20,000	0.25	$18,000	0.20
30,000	0.5	28,000	0.45
40,000	0.25	44,000	0.35

13-2. The University of New London is planning on an expansion of its engineering building. Due to the uncertainty of future enrollments and the uncertainty of current funding, three scenarios are being investigated. The school's cost of capital is 5%.

1. Build a 2-story building now, and build another 2-story building in four years when another expansion is expected to be needed. Land is available. The first 2-story building will cost $3 million now, and the second building will cost $4 million in four years. There is a 60% probability of needing the second building.
2. Build a 2-story building now, with a structure suitable for adding an additional 2 floors later. This will cost $4 million now and $2.5 million in four years for the additional 2 floors. There is a 60% probability of needing the additional 2 floors.
3. Build a 4-story building now at a cost of $6 million. There is a 60% chance of needing the addition floors over the next 5 years.

(a) Draw a decision tree showing current and future decisions.
(b) What is the NPV of each option?
(c) Which option do you recommend? Why?

13-3. What is the value of decision A in the following decision tree? What is the value of decision B? What is the best decision? Values shown represent net present values.

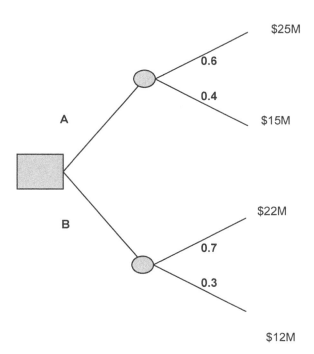

13-4. For problem 13-3, determine the standard deviation of each alternative. What does this say about the relative risk of the alternatives?

396   Chapter 13, Decision Making Under Risk

13-5. What is the NPV for the following decision tree? Probabilities are shown in bold.

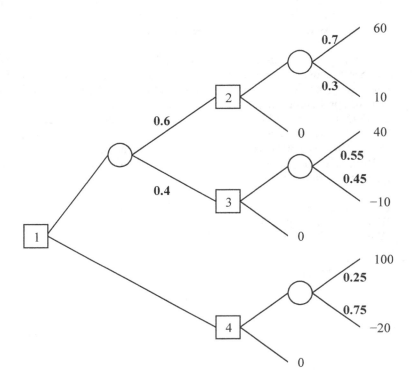

13-6. A company is looking at automating a process in its factory. This is being done on a part-time basis by one of the engineering staff.

Stage 1 of the project is to study the process and make recommendations. This work will cost $40,000 and take one year. The cost is payable at the end of the year, when the operating budget will be reimbursed. The company expects the chances of success (a positive recommendation) to be 80%.

Stage 2 involves creating a prototype to prove the concept. This prototype will cost $100,000 (payable at the end of year 2) and will take one year to build and test. The probability of success is 70%.

Stage 3 involves building production scale equipment. This will cost $1 million, take one year to build, and is payable at the end of year 3.

Stage 4 involves the savings over a three year period. The savings is dependent on market volumes, which are not certain. The following shows three scenarios of possible savings over years 4, 5, and 6.

Probability	Savings
0.30	$2,000,000 / year
0.50	750,000 / year
0.20	250,000 / year

(a) Draw the decision tree supporting this set of decisions and risks
(b) What is the NPV of the potential savings as of the beginning of year 4 if the MARR is 15%.
(c) What is the expected value of the project as of the beginning of year 3?
(d) What is the expected value of the project as of year 0?

13-7. A consumer products company is readying a new sunscreen. Launching the product requires a current investment of $11.5 million. The company's hurdle rate for this type of project is 20%.

The research department has recently identified a new sunscreen active ingredient, which is not yet available. Including it would delay the product's launch by a year. However, it would improve product efficacy and increase cash flows if it were used. The investment would be 5% higher if the project is delayed one year because the project would need to adopt a crash schedule. The life of the equipment and the new formulation technology is ten years in either scenario. The expected cash flows and costs, along with their anticipated ranges, are shown below.

(a) Draw a decision tree showing the decisions to be made.
(b) What is the NPV of the decision to launch the product now?
(c) What is the NPV of the decision to delay launching?
(d) What is the best decision?

Sunscreen Cash Flows

Year	Cash Flows, $million		Lower Limit	Upper Limit
	Launch now	Delay 1 year		
1	1.0	0.0	−40%	+20%
2	2.0	1.0	−40%	+30%
3	2.5	2.5		
4	3.0	3.5		
5	3.0	3.5		
6	3.0	3.5		
7	3.0	3.5	−40%	+30%
8	3.0	3.5		
9	3.0	3.5		
10	3.0	3.5		
11	0.0	3.5		
Initial investment	$11.50 million	$12.08 million	−5%	+15%
Salvage value	$0.75 million	$0.75 million	−100%	+100%
Hurdle rate	20%	20%	−20%	+20%
Extra cost for delaying construction		$0.575 million	−40%	+20%

13-8. What is the approximate range of NPV for the sunscreen project if all input variables are triangular distributions? Use Monte Carlo simulation.

13-9. What is the approximate range of IRR for the sunscreen project if all input variables are triangular distributions? Use Monte Carlo simulation.

13-10. Using the cash flow information in Example 12.1, and assuming a normal distribution for each cash flow, create a Monte Carlo simulation determining the range of expected NPV.

13-11. Using the output from problem 13-10, find the standard deviation of the output histogram. From the output, find the probability that the NPV will be
(a) $200,000 or less?
(b) Zero or less
(c) $1,000,000 or greater?
(d) Compare your answers with those in Example 12.1.

13-12. The values in the following decision tree are net present values. What is the expected value of each alternative, and which decision should be made?

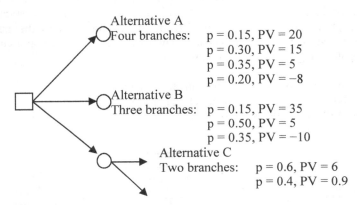

13-13. Identify the standard deviations of A, B, and C in problem 13-12. What are the risk/return ratios of the alternatives?

13-14. Two products are being considered by a company, A and B. Predicted sales volumes are as follows.

	Low	Medium	High
Probability	0.3	0.5	0.2
A (million units)	3.0	4.0	4.5
B (million units)	2.0	8.0	16.0

(a) What is the expected sales volume of each product?
(b) What is the standard deviation of each product's sales?

13-15. A new product is in development, but annual revenues are considered risky. To date, $250K has been spent, but an additional $400k will need to be spent to launch the product. What is the expected NPV of the project, using the following probabilities, if the firm's MARR is 12%? Draw a decision tree.

Probability	0.30	0.45	0.25
Annual Revenues	$150k	$300k	$500k
Life expectancy	3 years	4 years	5 years

13-16. In problem 13-15, describe the risk to the firm to proceed.

13-17. In problem 13-15, instead of probabilities, the annual revenues are represented by a triangular distribution where the minimum annual revenue is $20K, the most likely is $125K and the maximum is $250K. The life expectancy is also represented by a triangular distribution where the minimum is 2 years, the most likely is 4 years, and the maximum is 7 years. Perform a Monte Carlo simulation to determine the most likely project NPV. Describe the distribution of the resulting NPV.

13-18. Two process modifications are being considered. The initial costs and savings are shown below. The savings below depends on future sales volumes and product lifetimes. The firm's MARR is 14%.

Project	Cost	Initial Savings per year	Probability	Project Life	Probability
A	$160K	$30K	0.25	3	0.60
		60K	0.55	5	0.40
		90K	0.20		
B	$200K	$50K	0.35	2	0.70
		80K	0.50	4	0.30
		120K	0.15		

Create a decision tree showing expected NPV.

13-19. In problem 13-18, determine the standard deviation of each alternative, and the probability of losing money.

13-20. If a process change is made, product yield will increase from 92% to 94%. Of the material that does not become finished product, 60% is recycled at a cost of $0.70 per pound, with the balance being waste. The product has a value of $4.50 per pound. Future prices are uncertain, but are expected to be normally distributed with a mean of $4.50 per pound with a standard deviation of $0.40. Product sales volumes are also uncertain, and normally distributed with a mean of 60 million pounds per year and a standard deviation of 10 million pounds. The process change will cost $5 million. MARR is 12%. The project has an expected lifetime of 4 years. What is the probability of the project having a positive present value?

13-21. A project's initial cost is $450,000 with an interest rate of 10%. The project's expected life is 8, 9, 10, 11, or 12 years, with equal probabilities of each. The annual benefit follows a normal distribution with a mean of $85,000 and a standard deviation of $14,000. Using Monte Carlo analysis and 1000 iterations, what is the expected value and the standard deviation of the net present worth? How does this compare with the NPV using the expected life and the expected annual benefit? Which method provides more information?
(b) Repeat the Monte Carlo analysis five times with 5000 iterations each. Is there a change in values with each replication? Are the changes significant?

13-22. The mechanical vapor recompression evaporator has just failed. The large motor has burned out and needs to be replaced. One spare motor is in stock, but the placement of the motor and the alignment of the high speed shafts must be done with perfection. The service contractor can arrive in two days, and will require one day to perform the installation and alignment. The cost will be $15,000. Alternatively, your own crew can try to do the work in 2 days. They can start immediately, but they only have a 60% probability of performing the work correctly. If they do not do the work right, then you will need to contact the service representative, and lose an additional 2 days of production. There is also a 10% risk that they will damage the replacement motor which would cost $24,000 and one additional day. The value of lost production is $45,000 per day. Draw a decision tree and determine the right course of action.

13-23. A new furnace is required for the office building. An oil fired furnace can be purchased for $4100. This will require $6600 of oil the first year, and oil prices are expected to rise 4% per year for the next 5 years. A propane system can be installed for $7000 (including the tank). This will require $6500 of propane the first year, and propane prices are expected to rise 2% per year for the next 5 years. An electric system can be installed for $2400. This is expected to cost $7900 in electricity the first year, and electric prices are expected to rise 3% for the next 5 years. Natural gas is not available in the area. The MARR is 5%. Determine the right equipment to install.

14

Real Options Analysis

14.1 INTRODUCTION

Discounted cash flow (DCF) techniques have been criticized for biasing evaluators toward conservative conclusions. Good ideas are sometimes not pursued because DCF techniques assume that expenses and cash flows occur without the possibility of being changed. In reality, management has options of making changes during the project's life, especially during the early stages, which can improve the chance that the project succeeds. Management can delay a project if there is value in waiting, they can expand a project if it is successful, they can decrease the size of a project if market estimates prove to be too optimistic, and they can abandon a project if it seems likely to be unprofitable. Management has flexibility, and that flexibility has value.

Real options analysis is a tool intended to place a monetary value on the managerial flexibility in future choices. The theoretical foundation for real options comes from options on financial securities. A typical financial option is an opportunity to buy or sell a company's stock at a fixed price at a future time. For example, you may want the option to buy 100 shares of Google 6 months from now at a set price. You will need to pay for this option, because the option has value. In 6 months, if the price of Google is above your set price (called the strike price), then you can buy the 100 shares for the strike price, and you have made money. If the price of Google is below the strike price, then you can simply let the option expire and do nothing. An option is a right, but not an obligation, to do something in the future.

Real options take a similar approach. Delaying a decision until there is additional information may have value. Having the option to abandon a project and collect its salvage value may also have value. Options analysis uses the traditional DCF inputs of benefits, costs, time horizon, and interest rates, and then adds a fifth variable, volatility. This volatility reflects the uncertainty that is present, and incorporates risk into the option models. The intent of options analysis is to identify the value of such options, preserving the future upside potential of a project while minimizing the downside risk.

When a project's NPV is large, there is no need to determine an option value—do it. When the NPV is highly negative, the project should be abandoned; no option value will justify the project. Real options have their application only in those projects where the NPV is close to zero, where there is uncertainty, and where management has the ability to exercise their managerial options.

In this chapter, we will investigate the background of real options analysis, including the methods and problems related to the volatility parameter. The Black-Scholes model, traditionally used to price financial options, will be discussed, as well as the use of binomial lattices to solve real options problems. The methods of solving deferral, abandonment, and staging options will be detailed. We conclude with a discussion of the problems facing real options analysis, and why these issues prevent widespread adoption of these methods.

14.2 FINANCIAL OPTIONS

Real options have their basis in financial options. A financial option is an asset that gives the owner the right, without an obligation, to buy or sell another asset (such as common stock) for a specified price at or before some specified time in the future. For example, suppose a person wants to opportunity to buy Microsoft stock at a certain price six months in the future. Assume that Microsoft is currently selling for $28 per share, and the person thinks the price will increase. An option to buy the stock at $30 in 6 months may be purchased, giving the owner the right to purchase Microsoft at $30. If the price increases beyond $30, then the owner of the option may exercise the option (buy the stock at $30) and possibly make a profit. If the stock does not increase to $30, then the option is allowed to expire. The option has a value, and must be paid for.

The option to buy, such as the Microsoft example, is called a call option. The option to sell is called a put option. Options are continuously bought and sold through markets such as the Chicago Board Options Exchange (CBOE). Stock option prices are listed much like common stock prices. Options are considered derivative securities because their value is derived from the value of another asset. The trading of standardized options contracts in the United States started in 1973 when the CBOE began listing call options.

The value of a financial option is related to several variables, including
- The current stock price. In the Microsoft example, this is $28 per share.
- The strike price (the price a person is willing to pay), which is $30 in the Microsoft example.
- The time to expiration. This is 6 months in the Microsoft example.
- The risk-free rate of return.
- The volatility of the stock's rate of return. This is based on historical stock price fluctuations.

Some options may only be exercised on the date that they mature. These are known as European options. If the Microsoft stock example were a European option, then it can only be exercised six months in the future on the day that the option matures. Financial options mature on the third Friday of the month. Options that can be exercised at any time up to their date of maturity are known as American options.

Fisher Black and Myron Scholes (1973), along with Robert Merton (1973), were the first to publish an accepted pricing model for financial options. Scholes and Merton later shared the 1997 Nobel Prize in Economics for this work (Fisher Black died in 1995). The Black-Scholes model continues to be one of the most widely used methods of calculating option prices. Equation (14.1) approximates the value of a European call option, an option to buy which can be exercised only at the option's maturity date. The Black-Scholes equation is:

$$C = S_0 \Phi(d_1) - X e^{-rT} \Phi(d_2) \tag{14.1}$$

where

$$d_1 = \frac{\left(\ln {S_0}/{X}\right) + \left(r + {\sigma^2}/{2}\right)T}{\sigma\sqrt{T}}$$

$$d_2 = d_1 - \sigma\sqrt{T}$$

and

$\Phi(d_x)$ = the cumulative standard normal distribution of the variable d_x
S_0 = the current stock price
X = the strike price
σ = the volatility, the standard deviation of the past rate of return
r = the risk-free interest rate, and
T = the time to expiration of the option

$\Phi(d_x)$ is the area under a normal curve from negative infinity to d_x, and may be found in a standard normal distribution table (z-table) in most statistics books. The standard normal distribution may also be found in most spreadsheets; in Excel, this function is known as NORMDIST (NORMSDIST may also be used).

The derivation of the Black-Scholes equation is shown in Appendix 14.A. There are limitations to the model, which were clearly stated in the original article, including:
1. The short-term interest rate is known and constant.
2. The stock price follows a random walk in time and follows a lognormal distribution over time. The variance of the return on the stock is constant.
3. The stock pays no dividends (there is no cost for waiting).
4. The option can only be exercised at maturity.
5. There are no transaction costs in buying or selling the stock or option.

Allowing for dividends was one of the first modifications to the Black-Scholes model (Black, 1975). An extension is also widely used to determine the value of a put option (an option to sell), using the same variables:

$$P = Xe^{-rT}[1 - \Phi(d_2)] - S[1 - \Phi(d_1)] \tag{14.2}$$

Cox, Ross, and Rubenstein (1979) published the binomial option pricing method, an alternative technique to the Black-Scholes model which is more flexible. The binomial model is a discrete time model, while Black-Scholes is a continuous time model. Binomial lattices use simple mathematics to price options, and can be applied to a variety of problems where closed form equations are not available.

14.3 REAL OPTIONS

14.3.1 Historical development. During the 1970s, academic researchers started viewing corporate securities as options on the firm's assets. Stewart Myers (1977) explored the concept that financial investments generate options, and it was he who created the term "real options." Myers argued that valuing financial investments using standard DCF methods ignores the value of options that occur in projects that have a high degree of uncertainty.

Dixit and Pindyck (1994) explored the concept of the value in keeping options open until further information can be obtained and uncertainty can be minimized. Many early real option applications occurred in industries closely aligned with commodities markets: mining, oil and gas, and electrical power generation. The reason was that price volatility could be determined through futures markets. Merck in the pharmaceutical industry began using options analysis in the early 1990s in part due to the long R&D time required for new products, the high risk of project failure in the R&D stages, and the high potential payout for successful new drugs.

Real options analysis was initially developed with stochastic differential equations. This work evolved into proving that the Black-Scholes and related pricing models could be used for what was believed to be reasonably accurate estimates of option values. As will be discussed in this chapter, most of the calculus-based models contain assumptions that simply do not fit real projects. These assumptions include the lognormal distribution of the underlying asset's value and geometric Brownian motion, among others. These assumptions fit many financial options for which they were designed, but often do not fit their real option counterpart. What is more, these assumptions are important.

Calculus was used to explain the mathematics of real options when academics first approached industry about using options analysis. Most managers, whose responsibilities included large capital investments, simply did not have the mathematics background to understand the complex models that were proposed. One exception was Merck, who used real options for selecting research projects (Nichols, 1994). However, Merck's project Gamma, which has been used as an example of real options in action, has also been criticized because it failed to consider patent expiration, which led to overestimating the option value. This resulted in an incorrect decision to license another company's technology (Bowman and Moskowitz, 2001).

Practitioner guides attempted to simplify the complex mathematics, and to support the use of options analysis in industry (see Amram and Kulatilaka, 1999; Copeland and Antikarov, 2001 and 2003; Mun, 2002 and 2006, and Brach, 2003). Much of the application literature focuses on the use of binomial lattices which are easier to understand, though cumbersome to construct.

Binomial lattices make use of risk-neutral probabilities. As explored in Chapter 11, people tend to be risk averse; so are companies. Utility theory demonstrates that risk aversion decreases with wealth; a poor person might not risk losing $1000 while a rich person would not be so concerned. Firms are much the same way. They will tend to be risk neutral if the investment cost is small, but will become risk averse if the investment is a significant percentage of the firm's value.

As will be demonstrated in Chapter 16, portfolio diversification is virtually always used to reduce the risk of owning financial securities—which includes financial options. With a diversified portfolio of financial options, each security will represent a small enough share that risk neutrality can be expected. Because real projects represent larger investments and because real options analysis is generally reserved for larger projects where the investment size justifies the additional analysis and complexity, we cannot casually accept the idea of risk-neutrality being applied to all projects.

Projects with a high NPV or IRR are considered good investments, and investment is justified without further analysis. Projects with NPVs that are highly negative are usually abandoned because they will not deliver the required return. Projects with an NPV close to zero require additional effort to determine whether they should be funded or abandoned. The decision often takes one of three forms: fund the project, abandon the project, or keep the "option" open, keeping the project alive without funding it. In real options analysis, the option creates an expanded net present value (Trigeorgis, 1996), defined as:

$$\text{ENPV} = \text{NPV} + \text{Option Value} \qquad (14.3)$$

Management has options. One of those options is to wait on the funding decision while additional information becomes available. The option to wait has a value, especially when there are many sources of uncertainty, but there is also a cost of waiting. If the ENPV is positive, then it will be worth keeping the option open, which signals that the project should be kept alive; it does not signal that the project should be funded.

The use of options analysis can be used as more than a valuation tool; options can help identify business strategies. Luehrman (1998) said that business strategy is much more like a portfolio of options than it is like a series of static cash flows. McGrath and MacMillan (2000) also pointed out that real option thinking is a logic for funding projects that maximizes learning and supports the access to upside potentials while containing costs and risks.

Copeland and Antikarov (2001) developed a four-step process to carry out a real options analysis. The steps include:
1. Compute the present value without flexibility using discounted cash flow.
2. Model the uncertainty using decision trees.
3. Identify and incorporate the managerial flexibilities.
4. Conduct real options analysis.

These steps have much in common with the flow of our presentation in Chapters 12 and 13 on dealing with uncertainty.

14.3.2 The real option model. The five primary variables involved in the Black-Scholes calculation for financial assets are related to real assets as shown in Table 14.1 (Trigeorgis, 1996). There is a large gap between the original financial option theory and its application in real options.

Table 14.1 Black-Scholes option variables

Variable	Financial Options	Real Options
S	Stock price	PV of future cash flows
X	Exercise price	Implementation cost
T	Time to expiration	Time to expiration
r	Risk-free interest rate	Risk-free interest rate
σ	Volatility of stock price movement	Volatility of future returns

The value of simple options can be quickly estimated using the Black-Scholes model, but the original model is limited in the types of problems where it can be applied. The alternative method using binomial lattices will be discussed later in the chapter.

The following sections discuss some of the questions in the literature regarding these variables and their values. However, a larger problem is that the forecasts used to determine the inputs tend to be far less accurate for real options. Challenges in forecasting future events, costs, and revenues cause real option analysis to be less precise than financial option analysis. It is not worth refining an analysis technique to create a theoretically exact option price when the inputs are inherently inaccurate.

14.3.3 Interest rates. In real option analysis, the risk-free rate is often used. This is generally considered to be the rate on U.S. Treasury bills and bonds with a duration that matches the option. For example, if an option has a time horizon of one year, then a one year Treasury bill rate may be used as the risk-free rate.

Net present value is typically determined using a single hurdle rate and discrete compounding. Financial options are typically calculated using a risk-free rate with continuous compounding. Barton and Lawryshyn (2011) recommend using only risk-free rates for real options. The literature on real options usually uses a mixture of these, with the method depending on the author.

Luehrman (1998) first proposed that costs are known with greater certainty than future revenues are, and that the risk surrounding project costs are less than the risks of the forecast revenue stream. For this reason, he suggested that costs be discounted using risk-free interest rates when determining the NPV for a real option project. This idea has been supported by Mun (2006) and others. The idea is controversial, and has not gained wide support. This practice contradicts standard engineering economy methods of using a single interest rate. The use of the risk-free rate to discount the investment will provide a larger present value of the cost, thereby decreasing the NPV and providing a conservative recommendation.

For option valuation, interest rates may be applied using either discrete or continuous compounding. There is no uniform approach to using either discounting method when applied to real options. The finance literature consistently uses continuous discounting, in keeping with their calculus-based approach to options analysis. Practitioner books are consistent in continuous discounting of future costs while using discrete discounting of future net revenues (Copeland and Antikarov, 2001; Mun, 2006). The engineering economy literature uses both methods, and is not consistent.

The use of discrete or continuous interest rates, for example, can have a significant effect on the value of the option, and together with the choice of compounding method can change the recommended decision. Lack of consistency has been identified in the literature with at least one attempt to unify the techniques (Copeland and Antikarov, 2005), but this has not met with success. The assumptions of the practitioner books will be followed in this chapter's examples.

14.3.4 Time. The time involved is the time span of the option. For a deferral option, it is the proposed time of the delay. In finance, an option that can be exercised only on its maturity date is known as a European option. An option that can be exercised at any time up to its maturity date is known as an American option. In a real options context, an option that is waiting for an event or decision to occur is a European option. Waiting for conditions to change, such as prices to increase, is an American option. Merton (1973) pointed out that exercising an option early has no value unless there is a dividend (a loss in value due to waiting).

14.3.5 Present value of future cash flows. Option values are very sensitive to changes in the forecasted net revenues. Many authors have pointed out that there is often value in delaying a decision, hence the value of a deferral option. What few authors point out is that there is always a cost involved in the delay of a real engineering project. If nothing else, projected revenues will be delayed, causing a decrease in their present value due to discounting. The value of delaying will often be larger than the cost of waiting, but deferral costs must not be ignored as they are in much of the literature.

The traditional view of a delay cost is to model it after dividends, as is done in financial options. However, the dividend model is rarely the correct model because it fails to accurately describe the nature of lost cash flows. Delay models must be matched to the details of the case being analyzed (Eschenbach, Lewis, and Hartman, 2009). Including waiting costs is virtually a requirement for realistic industrial projects.

The Black-Scholes equation can be modified to include the cost of waiting, W. This modification can be expressed is several forms, including Equation (14.4). See Eschenbach, Lewis, and Hartman (2009) for a continued discussion on the cost of waiting.

$$C = (S_0 - W)\Phi(d_1) - Xe^{-rt}\Phi(d_2) \tag{14.4}$$

$$d_1 = \frac{\ln\left((S_0 - W)/X\right) + \left(r + \sigma^2/2\right)t}{\sigma\sqrt{t}} \qquad d_2 = d_1 - \sigma\sqrt{t}$$

14.4 REAL OPTION VOLATILITY

14.4.1 Actionable volatility. Volatility is the one parameter added to the discounted cash flow information set in order to calculate a real option's value. Determining a financial option's volatility is not a problem because it is based on historical prices. This is known information, with the only question being whether historical volatility will hold in the future. Volatility for real options is based on the standard deviation of projected cash flows. This is not historical information, but forecasted data. The only way to determine the standard deviation of forecasted data is with simulation or by using management estimates of possible future scenarios.

Unlike financial options, there is no single, theoretically justified approach for calculating the volatility coefficient in real options. There is also no well-defined approach to the question of which sources of variability should be included in determining the volatility coefficient.

The concept of *actionable volatility* (Lewis, Eschenbach, and Hartman, 2008) comes from examining the analysis of financial options. A stock option's value can be realized when the stock's price enters a range that makes the option worthwhile. The option's owner has the ability to take advantage of the stock volatility by exercising the option and immediately reaping its benefits. However, real options are more complex. There are a number of sources of variability which occur well beyond the time an option must be exercised. With real options, much of the variability is the randomness of each year's product mix, demand, prices, costs, and so forth. Moreover, the value of this variability cannot be "seized" by exercising the option, because at that moment the variability still lies in the future beyond the option's exercise date. Correctly valuing the real option requires identifying the volatility that is dependent on the option decision.

If volatility is overstated by including variability that cannot be captured by the option, then the option value is also overstated. Because the option value increases with increasing volatility in deferral options (and related growth options), the value of delay is overstated. This can lead to inappropriate delays in funding and executing projects. It can also lead to projects being kept "alive" when they should rightfully be abandoned. Actionable volatility relates not only to what variables are included in the volatility, but also how long these variables are included. If an option is kept open for two years, then it makes no sense to include variability that occurs after the option is mature. This proves to be a problem with most of the existing volatility estimation methods in the literature.

Keeping a real option open costs money and thus carries risk. For instance, delaying a decision to invest is usually the risk-averse decision, but a more volatile future increases the real option valuation of delaying the decision to invest. One reason that practitioners do not use real options analysis is because higher volatility means a higher option value, leading to keeping the option open, despite higher risk (Block, 2007).

The techniques used to estimate project volatility are summarized in the following subsections. The logarithmic cash flow method that is used for financial options is presented first.

14.4.2 Logarithmic cash flow method. For financial options, volatility is the natural log of the rate of return. The annual return may be found by dividing the price at the current time by the price from one year earlier as shown in Equation (14.5).

$$r = ln(p_t/p_{t-1}) \tag{14.5}$$

A series of returns is tabulated. The volatility is the standard deviation of the resulting r values, following Equation (14.6).

$$volatility = \sqrt{\left(\frac{1}{n-1}\right)\sum_{i=1}^{n}(r-\bar{r})^2} \tag{14.6}$$

While this technique is widely used for financial options, it does not work well with real options. The primary drawback is that the return may be negative during some periods for a real project, and there is no natural log of a negative number that can be used.

14.4.3 Stock proxy method. Trigeorgis (1996) and Amram and Kulatilaka (1999) recommended finding a traded security that is correlated with the project's asset value, known as a twin security. This approach received widespread support during the 1990s and early 2000s by a wide array of authors. This twin security may be a stock or an index that parallels the project's volatility , such as the implied volatility of short term call options on related companies (Brach and Paxson, 2001). A range of stock proxies has also been used, identifying a conservative range of companies operating in similar businesses. This was practiced by Merck in the early 1990s (Nichols, 1994).

The primary problem with the stock proxy approach is that the variables that are included in a stock's variability are usually completely different than the many variables involved in the project. Some innovations are not correlated to any other market traded asset, especially a new product that is unrelated to any existing ones (Pindyck, 1991). The volatility of a technical project is not the same as the volatility of the company's equity (Copeland and Antikarov, 2001).

The central limit theorem and related work implies that the standard deviation of returns for a firm performing many projects is linked to the standard deviation of each project. The number of projects makes the firm less risky than each individual project, especially since project diversification makes it unlikely for all projects to be dependent. Stock price volatility and project volatility are not closely linked, and finding a stock proxy that accurately matches the risk of any project is highly unlikely. Typically, the volatility of a stock will be lower than the volatility of an individual project.

14.4.4 Management estimates method. Management estimates are sometimes used, employing standard statistics with optimistic and pessimistic estimates (Mun, 2006). For example, in project management, the variance of a project's critical path completion time can be estimated from the beta distribution. Each critical path task's time variance is estimated from the optimistic and pessimistic duration estimates, as shown in Equation (1).

$$\sigma^2 = \left[(b-a)/6 \right]^2 \tag{14.7}$$

where a is the optimistic duration
b is the pessimistic duration

The variance is dependent on the accuracy of the estimates. Similar approaches can be taken using a variety of statistical methods, applied to expert forecasts.

14.4.5 Logarithmic present value returns method (CA method). A method of identifying the volatility for real options using the logarithmic present value returns was offered by Copeland and Antikarov (2001) and reviewed by Herath and Park (2002). In this method, the *estimated* future cash flows are discounted (using the MARR) to two present values: one for time period 0 and another for time period 1. In Copeland and Antikarov's (CA) approach, the time 0 value is treated as a static value, while the time 1 value is varied through Monte Carlo simulation. The present value at the present time (t = 0) is treated as the static denominator, as shown in Equation (14.8).

$$\sum_{t=0}^{N} PVCF_i = \frac{CF_0}{(1+i)^0} + \frac{CF_1}{(1+i)^1} + \frac{CF_2}{(1+i)^2} + \ldots + \frac{CF_N}{(1+i)^N} \tag{14.8}$$

The present value at year 1 omits CF_0 and discounts later cash flows by one less period, calculated as shown in Equation (14.9).

$$\sum_{t=1}^{N} PVCF_i = \frac{CF_1}{(1+i)^0} + \frac{CF_2}{(1+i)^1} + \frac{CF_3}{(1+i)^2} + \ldots + \frac{CF_N}{(1+i)^{N-1}} \tag{14.9}$$

A logarithmic ratio of the present values of the cash flows is calculated as:

$$z = \ln \left(\frac{\sum_{t=1}^{N} PVCF_i}{\sum_{t=0}^{N} PVCF_i} \right) \tag{14.10}$$

Because the investment cost is not included in this equation, these present worth values of net returns are typically large; we do not have to worry about the inability to calculate the log values.

The volatility is the standard deviation of this z ratio which must be estimated by performing a Monte Carlo simulation of the future cash flows. The cash flows can be correlated (see Cobb and Charnes, 2004), and often should be. In the CA method, only the numerator in Equation (14.10) is varied during the simulation; the denominator is held constant. Monte Carlo simulations in this chapter are performed using @Risk®, but can be also be performed using Excel® or Crystal Ball® software. This technique has been widely supported since its introduction.

Herath and Park (2002) describe the same technique, but they take a different approach in performing the Monte Carlo simulation. Rather than holding the denominator static, they allow both the numerator and the denominator to vary. They point out that the numerator and denominator are independent random variables; that a different set of random numbers must be generated for each. Herath and Park's method provides a volatility that is approximately 1.4 times larger than the CA method over a variety of cases.

14.4.6 Standard deviation of cash flows. Van Putten and MacMillan (2004) and MacMillan et al. (2006) recommend using the annualized standard deviation of the cash flows, divided by the mean of the cash flows derived from the Monte Carlo simulation. They add a correction when the volatility of the costs (essentially manufacturing costs, not the investment costs) exceeds the volatility of the revenues. In this case, they recommend dividing the revenue volatility by the cost volatility, then multiplying this ratio by the project volatility. This adjusts the project volatility down, because cost volatility should not increase the option's value, according to the authors. MacMillan and van Putten have been issued a U.S. patent on a real option valuation method that is based on this principle (MacMillan and van Putten, 2005).

Brandão, Dyer, and Hahn (2005a) used the CA method in the use of binomial decision trees. This was critiqued by Smith (2005), demonstrating that the CA method created too large a volatility, resulting in an overvalued option. Smith used the standard deviation of the cash flows to estimate the range of a project's present value of net revenues. Calculating these same ranges using the CA volatility resulted in much wider ranges. The conclusion was that the CA method overstated the volatility, resulting in an inflated option value. Brandão, Dyer, and Hahn (2005b) responded with a modified CA technique, where only CF_1 (in the numerator) in Equation (14.9) was allowed to vary during the Monte Carlo simulation; all other cash flows were held static. In their case, this modification provided a result quite close to Smith's.

14.4.7 Internal Rate of Return. In some applications, the uncertainty in the investment cost is the most likely concern to be resolved by waiting. Volatility in financial options is defined as the standard deviation of the rate of return (Black and Scholes, 1973). Measuring volatility in real options in a similar way, using the internal rate of return (IRR) has been proposed (Lewis, Eschenbach, and Hartman, 2008).

In financial options, volatility is measured in fractional changes in the rate of return, and often expressed as a percentage. In real options, the same is done; when using the logarithmic present value method, volatility is calculated as a fraction of the average return. Volatility can be expressed as the standard deviation of the IRR divided by the mean IRR, as shown in Equation (14.11)

$$CV = \sigma_{IRR}/\mu_{IRR} \qquad (14.11)$$

The coefficient of variability provides unreasonable results when the mean IRR approaches zero. This is generally not a concern because when the IRR approaches zero, the project is clearly not worthy of investment, and real options would not be of value. Likewise, if the IRR were negative, the CV would also be unreasonable; however, such a project would generally not be viable and a real option calculation would again be of no value.

14.4.8. Actionable volatility revisited. Most of the above volatility methods use projected estimates over the course of the entire project. Using years of data would assume that exercising the option enabled the owner to gain value from volatility years in the future. In fact, most future variability is simply random and not influenced by exercising the option. Correctly valuing the option requires identifying the volatility that is dependent on the option decision, including the time increment involved. Project volatility should be based on very short time increments, often one year, rather than using the volatility generated over the project's life. By limiting the time horizon that is included, the volatility parameter will be smaller and more realistic.

There are examples of uncertainties that may be resolved in the short-run that may reduce the long-term variability of a project's cash flows. For example, a pilot plant might more precisely define long-term operating costs, efficiencies, and yields. Delineation drilling will better describe the long-term limits of mineral resources, just as a test market will better define long-term product sales. In each case, the uncertainties could represent *actionable volatility*.

14.5 BINOMIAL LATTICES

Binomial lattices were created to provide a simplified approach to option valuation (Cox, Ross, and Rubinstein, 1979). Lattices are more flexible and can be used to calculate a variety of option values. The literature sometimes claims that Black-Scholes is wrong and that lattices are right (Copeland and Tufano, 2004, for example). This is not quite true. As previously mentioned, Black-Scholes is limited in how it can be applied, while lattices are more flexible. However, the proof that lattices work was based on the fact that they would provide the same answer as Black-Scholes in the limit (Cox, Ross, and Rubinstein, 1979). So in their simplest form, such as deferral and abandonment options—simple calls and puts—the methods are essentially the same. In these cases, lattices are a discrete form of the continuous Black-Scholes model. While lattices are flexible and the math is straightforward, it takes a significant amount of time to set them up.

To determine the value of a simple option, there are always at least two lattices involved in the calculation. The first is the evolution of the underlying value. The lattice normally starts with the value of S_0 on the left, and expands the value over one time step to one of two values. In the case of a delay option, there is a cost of waiting, and the underlying value needs to be modified by subtracting the waiting cost, obtaining $(S_0 - W)$ as is done in the Black-Scholes model.

The binomial lattice is constructed from left to right. The first item is the adjusted present value of the project's net revenues, $(S_0 - W)$. During the next step in time, conditions may improve (go up in value) to the state $(S_0 - W)u$. However, conditions may deteriorate to the state $(S_0 - W)d$. The "up" factor u is defined as

$$u = e^{\sigma\sqrt{dt}} \qquad (14.12)$$

where dt is the length of each time-step in the lattice. The "down" factor d is defined as

$$d = e^{-\sigma\sqrt{dt}} = 1/u \qquad (14.13)$$

Over time, a series of time steps are made, left to right, each of which has two resulting states. This first theoretical lattice, using 5 time steps, is shown in Figure 14.1.

$(S_0-W)u^5$

$(S_0-W)u^4$

$(S_0-W)u^3 \qquad (S_0-W)u^4 d$

$(S_0-W)u^2 \qquad (S_0-W)u^3 d$

$(S_0-W)u \qquad (S_0-W)u^2 d \qquad (S_0-W)u^3 d^2$

$(S_0-W) \qquad (S_0-W)ud \qquad (S_0-W)u^2 d^2$

$(S_0-W)d \qquad (S_0-W)d^2 u \qquad (S_0-W)d^3 u^2$

$(S_0-W)d^2 \qquad (S_0-W)d^3 u$

$(S_0-W)d^3 \qquad (S_0-W)d^4 u$

$(S_0-W)d^4$

$(S_0-W)d^5$

Figure 14.1 Generic lattice, evolution of the underlying project value

For a simple problem, the second lattice determines the value of the option. For more complex options, there may be a series of lattices that follow. The construction of lattices for many types of options is explained by Mun (2006) and the following examples also include lattices.

14.6 THE DEFERRAL OPTION: DEMENTIA DRUG EXAMPLE

14.6.1 Definition and NPV calculation. Management often has the option of delaying (or deferring) a decision, such as delaying the funding for a project until more information becomes available. There tend to be two types of deferral options: (1) waiting for an event to occur, and (2) waiting for conditions to improve. In the first, waiting for an event, the option cannot be exercised until the event occurs; this follows the logic and the mathematics of a European option. In the second, waiting for conditions to improve, the option may be exercised at any time that the situation warrants. This is an example of an American option.

14.6 The Deferral Option: Dementia Drug Example

A simple deferral option follows the mathematics of a financial call option (the option to buy). For very simple problems, the Black-Scholes equation may be used. For more complex problems, a two-stage binomial lattice may be a better choice.

Example 14.1 A dementia drug

The dementia drug example was introduced in Chapter 13 as Example 13.3. The details of the case are repeated here for clarity.

A drug company is seeking approval for a new drug product. The company is hoping for approval from the Food and Drug Administration (FDA) 2 years from now. The drug will have patent protection for 10 years after FDA approval (the 20-year patent was applied for 8 years ago at an earlier stage of the development process). Once on the market, year 1 net cash from sales is expected to be $8M (million), year 2 net revenues are expected to be $15M, and years 3 through 10 are expected to be $22M.

The facility to produce the new drug will take 2 years to build at a cost of $38M with a $5M salvage value at the project horizon. There is a 90% chance that the FDA will approve the new drug. The hurdle rate for this project is 25%.

If facility construction begins after FDA approval, initial sales will be delayed 2 years. The patent limits the horizon for sales, so 2 years of revenues are lost. If facility construction begins now, it will be available to produce the drug upon FDA approval. However, if the FDA does not approve the drug, the unused $38M facility will have a salvage value of only $9M at the end of year 2. The cash flows are shown in Table 14.2.

The question facing the firm is whether the facility should be built now or delayed until after FDA approval. The estimated base case values are summarized in Table 14.3 along with the lower and upper limits for each parameter (the cash flows are uncertain). Delays in approval can lead to shorter periods of product exclusivity. The NPV of the "build now" option is −$0.90 million, and the NPV of the "build later" option is −$0.36 million. Traditional discounted cash flow methods would indicate that the project should not be funded now or 2 years from now.

TABLE 14.2 Dementia drug investment options and cash flows

	Cash Flows		
Year	Build Now FDA Approves	Build Now FDA Rejects	Build Later FDA Approves
0	$−38	$−38	$ 0
1	0	0	0
2	0	9	−38
3	8		0
4	15		0
5	22		8
6	22		15
7	22		22
8	22		22
9	22		22
10	22		22
11	22		22
12	27		27

TABLE 14.3 Dementia drug variables

Parameter	Base Case	Low	High
First Cost	$38M	−5%	+15%
Salvage Value of new plant	9M	−100%	+100%
Salvage Value (terminal)	5M	−100%	+100%
Probability of FDA approval	0.9	−20%	+20%
Revenue 1st year	8M	−40%	+20%
Revenue 2nd year	15M	−40%	+30%
Revenue > 2 years	22M	−40%	+40%
Hurdle Rate	25%	−20%	+20%
Risk-free Rate	5%	−40%	+40%
Time to FDA decision, yr.	2 yrs	−50%	+50%
Time to build facility, yr.	2 yrs	−50%	+50%
Time to patent expiration, yr.	10 yrs	−40%	+20%

Solution
The first step in solving the problem is to organize the input variables shown in Table 14.1. The resulting cash flows are shown in Table 14.4. Future net revenues are discounted using the hurdle rate and discrete (annual) compounding, providing a value for S_0. Costs for determining the NPV are also discounted using the hurdle rate and discrete compounding, consistent with traditional engineering economics. Costs for determining the option value are discounted using the risk-free interest rate and continuous compounding, consistent with the Black-Scholes model (and consistent with most users of binomial lattices).

TABLE 14.4. Incremental cash flows and the cost of waiting

	Cash Flows		
Year	Build Now, 90% Chance of Approval	Build Later if FDA Approves	Incremental Revenues
0	$−39.3	$ 0	
1	0	0	$ 0
2	0.9	−39.3	0.9
3	6.7	0	6.7
4	13.1	0	13.1
5	19.8	7.5	12.3
6	19.8	14.5	5.3
7	19.8	22.0	−2.2
8	19.8	22.0	−2.2
9	19.8	22.0	−2.2
10	19.8	22.0	−2.2
11	19.8	22.0	−2.2
12	24.3	27.0	−2.7
NPV	$−2.60	$−1.52	$13.06

Some of the input variables have been defined as having asymmetric uncertainty. For example, the first cost has an expected value of $38 million, with uncertainties of −5% or +15%. This means that the minimum will be $36.1 million, the maximum will be $43.7 million, and the mode will be $38 million. Because of the asymmetric uncertainty, the mean will be $39.3 million, not $38 million; the mean is higher than the mode because the upside uncertainty is greater than the downside (you are likely to overspend the budget by more than you are likely to underspend the budget). Asymmetric uncertainties are common in real engineering projects. Revenues in years 1 and 2 will average $7.5 and $14.5 million, somewhat smaller than their modes. Year 2 cash flows reflect the 10% probability of not receiving FDA approval at the end of year 2, recognizing the $9 million salvage value at this time.

The mean investment and the mean net revenues need to be used to determine the cost of waiting. The waiting cost is determined by calculating the NPV of the incremental value between the "build now" and the "build later" alternatives. Discounting these differences using the 25% hurdle rate provides a total cost of waiting, W, of $13.06 million. This greatly affects the value of the net revenues, significantly decreasing the resulting option value. Without the cost of waiting, the PV of future net revenues, S_0, would be $36.67 million. By subtracting the cost of waiting, $(S_0 - W)$ becomes $23.61 million. The project NPV for the "build now" option is −$2.60 million using the mean cash flows.

14.6.2 Volatility. Before the option value can be determined, the volatility must be estimated. Three methods are used: logarithmic present value of returns (the CA method), the standard deviation of cash flows, and the IRR method. All three methods require the use of Monte Carlo simulation. This technique requires that input variables be given a mean, a distribution, and an indication of variation for each. High and low values are based on the information from Table 14.3.

The actionable inputs to the Dementia problem include the capital investment, the FDA decision, and the salvage value of the new plant (the value if the FDA decision is negative). All other input variables are beyond the time to execute the option, where the value of the volatility cannot be seized, and should not be used in the calculation for volatility. Note that the FDA decision is a binomial distribution (90% of the time having a value of 1, and 10% of the time having a value of 0). This binomial distribution has a profound effect on the potential outcomes.

The histogram of the NPV of the project (if it is built now) is shown in Figure 14.2. Note the bimodal nature of the distribution: the NPV is near zero about 90% of the time, when the FDA approves the project, but the NPV is highly negative about 10% of the time if the FDA does not approve. The simulation was carried out using 5000 iterations. Because this is a simulation, results will change slightly each time the simulation is performed.

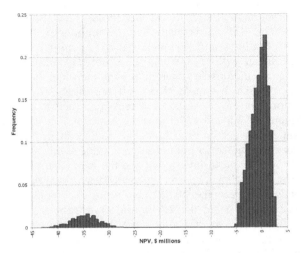

Figure 14.2 NPV of the dementia drug project (Build Now) using actionable variables

Logarithmic Present Value Returns Method (CA Method). Volatility for the dementia drug problem can be determined using the logarithmic present value returns method. Using Monte Carlo simulation, the histogram of the z value (from Equation 14.10) is shown in Figure 14.3; the standard deviation is 0.62, providing a volatility of 62% for the "Build Now" option.

Standard Deviation of Cash Flows Method. Using the same simulation, the present values of the net revenue are shown in Figure 14.4; the standard deviation is 10.02, with a mean of 36.89. The volatility is the coefficient of variation, or 10.02/36.89, resulting in a volatility of 0.27 or 27%. As in previous studies, this is significantly smaller than the CA method.

Internal Rate of Return Method. The IRR of the "Build Now" option has a mean IRR of 17.47%, which is below the hurdle rate of 25%. The standard deviation is 0.230, producing a coefficient of variation of 1.32 and an actionable volatility of 132%. One reason that this method produces a higher volatility is that it is the only one to include the investment risk.

The above techniques were calculated using traditional methods as they appear in the literature. However, the concept of actionable volatility dictates that we need to limit the time involved to the period that is influenced by executing the option. Rather than using the entire project time horizon to determine the above volatility parameters, the simulations can be performed, using only the first year of cash flows (occurring in year 2 shown in Table 14.4) as inputs. When this is done, the value of the volatility decreases, as expected. The results are shown in Table 14.5.

14.6 The Deferral Option: Dementia Drug Example

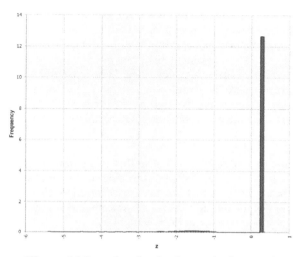

Figure 14.3 z value for the dementia drug project using actionable variables

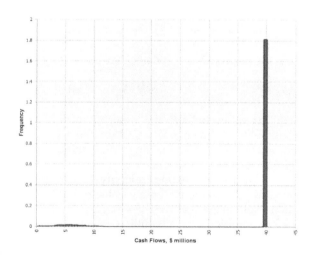

Figure 14.4 Present Value of Cash Flows for the dementia drug project using actionable variables

Table 14.5 Dementia project volatility

Method:	PV of future cash flows (CA)	Std. Deviation of Cash Flows	Internal Rate of Return
Traditional	62%	27%	132%
Actionable time	4.6	5.9	6.4

The literature has not settled on any one method for determining project volatility. The PV of future cash flows (CA method) has received much attention and criticism. The standard deviation of cash flows is the method that was used to demonstrate that the CA method overstated volatility (Smith, 2005). The IRR method was proposed as a method that includes investment risk; however, investment risk should lower a project's value, not raise it as project volatility. When the volatility methods are limited to the time horizon where the results are actionable, the three methods arrive at essentially the same answer. This is not typical, as different methods rarely provide the same answer. The results are also far smaller, indicating that the volatility that might be "captured" is limited.

14.6.3 Black-Scholes results. Because this is a simple delay option, we can substitute the input values into the Black-Scholes pricing model. We will first use the volatility using the CA method, 62%, as that has been used multiple times in the literature. Applying our input variables, including the cost of waiting and using the risk-free interest rate, d_1 and d_2 are determined first using Equation (14.4).

$$d_1 = \frac{\ln\left((36.67 - 13.06)/39.27\right) + \left(0.05 + 0.62^2/2\right)2}{0.62\sqrt{2}} = -0.028$$

$$d_2 = -0.028 - 0.62\sqrt{2} = -0.904$$

Using Equation (14.4), the option value is

$$C = (36.67 - 13.06)\Phi(-0.028) - 39.27e^{-(0.05)(2)}\Phi(-0.904) = 5.05$$

$$ENPV = -2.60 + 5.05 = 2.45$$

The option to delay the project 2 years is worth $5.05 million. The expanded net present value of the project is $2.45 million. In other words, we should keep the option open, and be willing to spend up to $2.45 million in order to keep the project alive but not fully funded.

If we use the option volatility based on the standard deviation of cash flows method, we determine the option value to be $0.79 million and the ENPV to be −$1.81 million. Note that the change in volatility from 62% to 27% decreased the option value from $5.05 million to $0.79 million, and decreased the ENPV to below zero, making the project not worth keeping open. Which is correct? This remains an open question; methods of validating the volatility do not currently exist. This is a key issue with real option methodology; different volatility methods provide different results, and there is no way to verify which, if any, are correct.

If any of the actionable volatility results are used, the option value falls to zero. This again makes the project unworthy of keeping open. A conservative approach to the use of options would indicate that the option is not worth keeping open.

14.6.4 Binomial Lattices. The first lattice is the evolution of the underlying value. The lattice begins with the value of $(S_0 - W)$ on the left, and expands the value to the right. The first theoretical lattice, using 5 time steps, was shown in Figure 14.1. Substituting the values

14.6 The Deferral Option: Dementia Drug Example

for the dementia drug project, the first lattice is shown in Figure 14.5. The delay time t is 2 years and there are 5 time steps, so dt is 2/5 or 0.4 years. Using 0.27 for the volatility,

$$u = e^{\sigma\sqrt{dt}} = e^{0.27\sqrt{.4}} = 1.19 \qquad d = 1/u = 0.843$$

```
                                                55.45
                                         46.75
                                  39.41         39.41
                           33.22         33.22
                    28.01         28.01         28.01
             23.61         23.61         23.61
                    19.90         19.90         19.90
                           16.78         16.78
                                  14.15         14.15
                                         11.92
                                                10.05
```

Figure 14.5 Evolution of the Underlying Project Value

The second lattice begins at the top of the right hand column in position A. To determine the value of position A, take the value of the same location in the first lattice (55.45M) and subtract the cost (39.27M). If the cost exceeds the project value at any point in time, then the project would not be pursued, leaving an effective value of zero. This recognizes management's flexibility to not fund a money-losing project. So the value of position A is max[0,$(S_0 - W)u^5 - X$], or \$16.18 million. This approach is continued down the right hand column, as shown in Figure 14.6. The right hand column represents the range of values that the option is expected to have at the maturity date of the option. If the value is above zero, the project should be executed. If the value is zero, then the project should be abandoned.

```
                                                16.18  A
                                         8.26  B
                                  4.21  C       0.14
                           2.15          0.07
                    1.10          0.04          0.00
             0.56          0.02          0.00
                    0.01          0.00          0.00
                           0.00          0.00
                                  0.00          0.00
                                         0.00
                                                0.00
```

Figure 14.6 Option Valuation Lattice

The column to the left is calculated next, starting with position B. There are three options for each cell in this column. The project could be executed (demonstrating that this is an American option), the option can be left open, or the project can be abandoned. In this project, we will not have the FDA decision for about 2 years, but we can begin construction (executing the project) at any time.

To determine the value of executing the option, we take the corresponding cell in the first lattice (46.75 for the top cell) and subtract the value of X, as was done in the right column. The value of executing the project at time step B is (46.75 − 39.27) = 7.48.

The value of keeping the option open can be determined using the risk-neutral probability approach (Mun, 2006).

$$Value = [pV^+ + (1-p)V^-]e^{-r(dt)} \tag{14.14}$$

where

$$p = \frac{e^{r(dt)} - d}{u - d} \tag{14.15}$$

The risk-neutral probability (p) is 0.516 according to equation (14.15). The value of keeping the option open for position B in Figure 14.6 then becomes [(0.516)(16.18) + (1 − 0.516)(0.14)]exp(−0.05)(0.4) or 8.26.

The value of the cell will not be less than zero, since the project would be abandoned in this case. So the value of the cell is the maximum of zero, the value of keeping the option open (8.26), and the value of executing the option (7.48). The highest value is in keeping the option open. This process is continued down the column.

The column to the left, starting with position C, is calculated in the same way, and this process is repeated in all cells to the left side of the lattice. The final cell is the value of the delay option, $0.56 million. This is similar to the $0.79 million option value found by the Black-Scholes equation. The option value obtained by using a lattice will approach the Black-Scholes value as more time steps are used (as the change in time approaches zero, the discrete method approaches the value of the continuous method).

14.7 THE DEFERRAL OPTION: OIL WELL EXAMPLE

Example 14.2 Oil well
An oil exploration company is considering leasing a plot of land and drilling an oil well. The well will have characteristics that are typical of this company's projects, and the equipment sizing, outputs, and costs are well understood. The characteristics of the project are shown in Table 14.6.

TABLE 14.6 Oil well project information

First cost	$75 million	Current oil price	$90/bbl
Salvage cost	$10 million	Oil price volatility/yr	35%
Output, yr 1, 10^6 bbls/yr	0.60	Project volatility, %	To be determined
Well depletion rate/yr	15%	Project delay	Up to 2 years
Operating costs	$40/bbl	Risk-free rate	5%
Hurdle rate, %	16%		

14.7 The Deferral Option: Oil Well Example

14.7.1 NPV. The first step is to determine the project's NPV. Note that in addition to the initial investment of $75 million, there is an additional salvage cost of $10 million at the end to return the site to its original condition. The well's output begins in Year 2, and then declines at 15% per year. The well produces for 7 years before it is shut down. Cash flows are shown in Table 14.7. The NPV, using the hurdle rate of 16%, is −$3.67 million if the project were started now. The well is not worth starting given the current price of oil.

14.7.2 Delay Option Formulation.

The company has several options. First, it could invest in the project; however, the NPV is negative, making this unattractive. Second, the company can abandon the project, as suggested by the NPV. Third, the company could keep the project open by paying a lease on the property, investing later only if prices increase in the future to make investment economical. By paying the lease, the company can pay a relatively small amount to preserve the option of future investment. The question is: how much should the company be willing to spend on the lease to keep the option (the project) open?

TABLE 14.7 Project Cash Flows ($M)

Year	Invest Now	Delay (up to 2 years)
0	−75.00	0
1	0	0
2	30.00	−75.00
3	25.50	0
4	21.68	30.00
5	18.42	25.50
6	15.66	21.68
7	13.31	18.42
8	11.31	15.66
9	−10.00	13.31
10		11.31
11		−10.00

The company should be willing to pay up to the expanded net present value (ENPV) to keep the option open. A simple delay option can be solved using the Black-Scholes model. In this case, we can delay the project up to two years and execute the option (implement the project) at any time up to the two-year expiration of the lease. In finance, this is known as an American option. Equation (14.1) is used to determine the value of the delay option (although the Black-Scholes equation models a European option, it will be used for now). The value for S_0, the present value of the net cash flows, can be found with our existing data. The first cost is known with certainty.

In this case, there is only one source of volatility: the price of oil. It is important to realize that the project's volatility is not the same as the volatility of the input variables; the project volatility is not necessarily 35% simply because we have one input whose volatility is 35%.

The first cost is known to be $75 million. For simplification of the problem, we assume that this is known with certainty, and that the value does not change if the project is delayed. However, if the project is delayed, then the first cost must be discounted to the present time.

Waiting costs can be modeled for a European option without too much difficulty. Delays are known, and lost or deferred revenues can be modeled based on the length of the delay. In the present case, we have an American option, where the firm is waiting for the price of oil to increase to the point where it is economic to pursue the project. What is the waiting cost in this case? It depends on how long the project is delayed, at what price the project is implemented, and what occurs in the future. If the current project were not delayed, there would be no cost of waiting (there would also be a negative NPV). If the current case were delayed the full two years, there would be a cost of waiting of $18.32 million, and there would still be a negative NPV. A smaller delay results in a smaller cost of waiting, but also a smaller option value. The traditional approach is to model the delay to the time horizon of the option.

Solution

Before the option value can be determined, the volatility must be estimated. The method of logarithmic present value of returns (CA method) and the standard deviation of cash flows is used. The key inputs are the future cash flows, determined by the price of oil, the operating cost, and the production of the well. The price of oil is determined to be the only input having volatility (having a standard deviation of 35%). The oil price is assumed to follow a lognormal distribution (which is supported by historical data). The oil prices are correlated 90% from year to year; the price in one year will tend to follow the price of the preceding year (see Cobb and Charnes, 2004). The correlation is important; without it, each year's price would vary independently of preceding prices. Without correlated oil prices, there is no actionable volatility due to oil prices, and project volatility would be meaningless. The NPV histogram for the project is shown in Figure 14.7

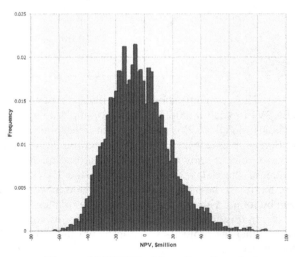

Figure 14.7 NPV of the oil well project

14.7 The Deferral Option: Oil Well Example

Most of the real option literature assumes that the underlying asset price follows geometric Brownian motion. In the case of oil prices, and indeed many commodities, this is not true. Oil prices tend to be mean reverting (Smith and McCardle, 1999). When prices are high, new production capacity tends to come online and prices tend to decrease to their long term mean. When prices are low, many marginal wells are shut down, and prices tend to increase to their long term mean. Oil prices (West Texas Intermediate Crude Oil) over a 5 year period are shown in Figure 14.8. This affects the true volatility of oil prices, because high prices do not increase indefinitely, and low prices do not continue to decrease. Actual volatility is significantly less than predicted by geometric Brownian motion. Before we determine the project volatility, oil price volatility must be adjusted for mean reversion.

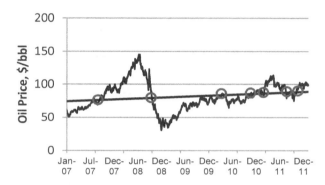

Figure 14.8 Mean reversion of WTI crude oil, 2007-2012

Hafner (2003) demonstrated several models using approximations for use in determining the pricing of options under stochastic volatility and mean reversion. This work included models of adjusting the volatility to take into account the mean reverting nature of a commodity. The resulting adjusted volatility could be used in standard models, including Black-Scholes. Hafner's adjusted volatility is shown in Equation (14.16).

$$V_{t,T} = \frac{1}{T-t} \int_t^T e^{-2\kappa(T-s)} \sigma_s^2 ds \qquad (14.16)$$

where $V_{t,T}$ = variance with mean reversion
t = beginning time period, usually 0
T = time horizon
κ = mean reversion speed (approximately 1.0 year^{-1}
s = incremental time, left as a variable
σ = is the standard deviation (price volatility) without mean reversion

The effect of the variables κ and σ are shown in Figure 14.9. In this graph, the time horizon is two years, consistent with our oil well project. As an example, if an oil development lease were held for 2 years with a mean reversion speed of 1.0 and an oil price volatility of 35%, the effective (adjusted) price volatility would be 17.3%, or half of the original volatility. Because options and other derivative prices are highly dependent on volatility, this would make a dramatic effect on the option's value.

For the oil well case, price volatility is adjusted from 35% to 17.3%.

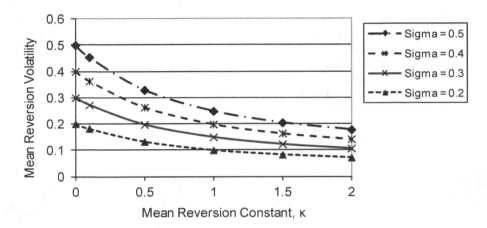

Figure 14.9 Effect of Reversion Constant κ and Volatility σ when T = 2 years.

Logarithmic Present Value Returns Method (CA Method). The variables used for determining the z value are shown in Table 14.8. Using Monte Carlo simulation, the histogram of the z value (from Equation 14.10) is shown in Figure 14.10; the standard deviation is 0.31, providing a volatility of 31%.

Standard Deviation of Cash Flows Method. The present values of the net revenue (the variable CF column in Table 14.8) are shown in Figure 14.11; the standard deviation is 20.91, with a mean of 71.33. The volatility is the coefficient of variation, 20.91/71.33, or 0.29.

TABLE 14.8 Monte Carlo variables for the z calculation

Year	Static CF	Variable CF	$(1+i)^n$	(variable) $CF_n/(1+i)^{n-1}$	(static) $CF_n/(1+i)^n$
0	0	0	1.00		0.00
1	0	0	1.16	0.00	0.00
2	30.00	30.0	1.35	25.86	22.29
3	25.50	25.5	1.56	18.95	16.34
4	21.68	21.7	1.81	13.89	11.97
5	18.42	18.4	2.10	10.18	8.77
6	15.66	15.7	2.44	7.46	6.43
7	13.31	13.3	2.83	5.46	4.71
8	11.31	11.3	3.28	4.00	3.45
9	-10.00	-10.0	3.80	-3.05	-2.63
10	0.00	0.0	4.41	0.00	0.00
11	0.00	0.0	5.12	0.00	0.00
	NPV =	71.33			
			Sum	82.75	71.33
			z		0.1484

Actionable volatility. If the simulations are performed using only the first year of cash flows (occurring in year 2 shown in Table 14.7) as inputs, the value of the volatility decreases. The volatility from the logarithmic present value returns method decreases from 0.31 to 0.096. The volatility from the standard deviation of cash flows method decreases from 0.29 to 0.097. As in the dementia drug project, the actionable volatility values from the two methods agree.

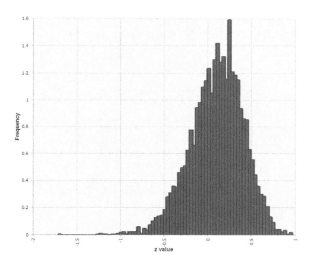

Figure 14.10 z value for oil well project

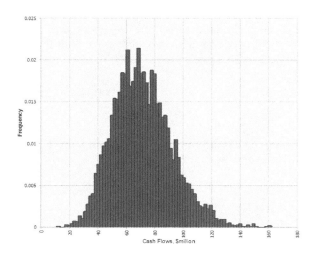

Figure 14.11 Present value of cash flows for oil well project

14.7.3 Black-Scholes results. The option value will be determined by using both the Black-Scholes model and binomial lattices. The Black-Scholes model determines the value of a European option. As the oil well project is an American option, the Black-Scholes provides a lower bound for the option value. Because the American option can be exercised at any time, it will have a slightly higher value than one that can only be exercised at maturity. Rather than use a closed form equation, binomial lattices are used to determine the American option value. Applying the input variables, using a volatility of 29%, we find an option value of $4.09 million. Adding the option value to the NPV of drilling a well now provides an ENPV of $0.42 million. The firm should be willing to spend up to $0.42 million in order to keep the option open.

$$d_1 = \frac{\ln\left((S_0 - W)/X\right) + \left(r + \sigma^2/2\right)t}{\sigma\sqrt{t}}$$

$$= \frac{\ln\left((71.33 - 18.32)/75\right) + \left(0.05 + 0.29^2/2\right)2}{0.29\sqrt{2}} = -0.397$$

$$d_2 = d_1 - \sigma\sqrt{t} = -0.397 - 0.29\sqrt{2} = -0.807$$

$$C = (S_0 - W)\phi(d_1) - Xe^{-rt}\phi(d_2)$$
$$C = (71.33 - 18.32)\phi(-0.397) - 75e^{-0.05(2)}\phi(-0.807) = 4.09$$

Using the actionable volatility of 0.096 yields an option value of $0.11 million and an ENPV of −$3.56 million. In this case, a lease should not be paid.

14.7.4 Binomial lattices. The option value may be calculated using binomial lattices using the same methods as used in the dementia drug problem. First, the evolution of the underlying value lattice is created, expanding the value of $(S_0 - W)$ over the two-year time period. This is shown in Figure 14.12 using 5 time steps.

$$u = e^{\sigma\sqrt{dt}} = e^{0.29\sqrt{0.4}} = 1.201$$
$$d = 1/u = 1/1.201 = 0.832$$

Each up-step multiplies the previous value by 1.201, and each down-step multiplies the previous value by 0.832.

14.7 The Deferral Option: Oil Well Example

```
                                            132.63
                                    110.40
                            91.90            91.90
                    76.50            76.50
            63.68            63.68            63.68
    53.01            53.01            53.01
            44.13            44.13            44.13
                    36.73            36.73
                            30.58            30.58
                                    25.45
                                            21.19
```

Figure 14.12 Evolution of the underlying value, oil well project

The second lattice is the option valuation lattice, shown in Figure 14.13. The right hand column shows the value of exercising the option at the end of 2 years. Each cell in the right column is determined by subtracting the cost of implementation ($75 million) from the equivalent node in the first lattice. For example, node A will have the value (132.63 − 75.00) or $57.63 million. This is repeated down the column. If any value is below zero, then the project will not be implemented, and will have a value of zero.

```
                                            57.63   A
                                    36.89 B
                            22.46            16.90
                    13.23            8.43
            7.61             4.21             0.00
    4.30             2.10             0.00
            1.05             0.00             0.00
                    0.00             0.00
                             0.00            0.00
                                     0.00
                                             0.00
```

Figure 14.13 Option valuation lattice, oil well project

Columns to the left have three possibilities; the project may be exercised (because this is an American option), the option may be left open, or the project may be abandoned. The value for exercising the option is found exactly the same way it was in the right-most column. The value of exercising the option at node B is (110.40 − 75.0) or $35.4 million. The value of keeping the option open is determined using risk neutral probabilities, equations (14.14) and (14.15).

$$p = \frac{e^{r(dt)} - d}{u - d} = \frac{e^{0.05(0.4)} - 0.832}{1.201 - 0.832} = 0.509$$

The value of keeping the option open (node B) is

$$Value = [pV^+ + (1-p)V^-]e^{-r(dt)}$$
$$= [(0.509)(57.63) + (1-0.509)(16.90)]e^{-0.05(0.4)} = 36.89$$

Note that the values are discounted by the risk-free rate for one time-step using continuous compounding. The value of node B is max (35.4, 36.89, 0) or 36.89. This method is repeated for all remaining cells, first completing each column, then moving from right to left.

The option value using binomial lattices is $4.30 million, compared to $4.09 million using Black-Scholes. If the number of time steps were increased, the results from the lattice method will approach the results from Black-Scholes, although in theory, the lattice should produce a slightly higher number reflecting the ability to implement the project at any time.

14.8 THE ABANDONMENT OPTION

A project may be abandoned if the salvage value of the project's assets exceeds the future income potential. The abandonment option assumes that you can abandon the project and obtain the salvage value of the assets if it is in your best interest, but you are not obligated to do so. The abandonment option is intended to identify the value of management flexibility where there is a likelihood that the firm will abandon the project. Abandonment options are common, and may occur at the same time as other options.

An abandonment option is a form of a put option. A modification of the call option model is available for determining the value of a put option (Black and Scholes, 1973), as shown earlier in the chapter.

$$P = Xe^{-rT}[1 - \Phi(d_2)] - S[1 - \Phi(d_1)] \qquad (14.2)$$

Example 14.3 Abandonment option

A consumer products company is developing a new product. The company is not yet sure that the product will be economically viable, and has been performing a financial feasibility study. The present value of the future revenues has been estimated to be $100 million. The company has also found another firm that is interested in the new technology, and has identified that the project could sell all of its assets during the next three years for about $80 million. The option to abandon the project and sell its assets has value.

In discounted cash flow analysis, the salvage value would be discounted to the present at the firm's working MARR. If the assets are sold and the salvage value taken, the revenues would cease. In real options analysis, the option to abandon the project and obtain the salvage value is simply an option; it is not an obligation.

Assume the following:
- S_0 = $100 million ($PV$ of the future revenues)
- X = $80 million salvage value
- t = 3 years
- N = 3 time steps
- dt = 1 year (T/N)

14.8 The Abandonment Option

σ = 30% annual volatility
r = 5% risk-free rate (Treasury rate for a 3-year bond)

Solution

The abandonment option can be solved using either the Black-Scholes method or binomial lattices; both will be demonstrated.

Black-Scholes method

As in the deferral option, the values for d_1 and d_2 must be determined first. Then the option value is found using Equation (14.2):

$$d_1 = \frac{\ln(S_0/X) + (r + \sigma^2/2)t}{\sigma\sqrt{t}} = \frac{\ln(100/80) + (0.05 + 0.30^2/2)3}{0.30\sqrt{3}} = 0.978$$

$$d_2 = d_1 - \sigma\sqrt{t} = 0.978 - 0.30\sqrt{3} = 0.458$$

$$P = 80e^{-(0.05)(3)}[1 - \Phi(0.458)] - 100[1 - \Phi(0.978)] = \$5.84 \text{ million}$$

Binomial Lattice Method

To calculate the binomial lattices, the up-step, the down-step, and the risk neutral probabilities must first be identified. These are calculated as before, based on the project volatility and the length of a time step. The lattice is brief this time.

$$u = e^{\sigma\sqrt{dt}} = e^{0.3\sqrt{1}} = 1.35$$

$$d = 1/u = 1/1.35 = 0.741$$

$$p = \frac{e^{r(dt)} - d}{u - d} = \frac{e^{0.05(1)} - 0.741}{1.35 - 0.741} = 0.509$$

The first lattice, shown in Figure 14.14, shows the evolution of the underlying asset, similar to the method used in the deferral option. The lattice starts on the left and moves one time step to the right, multiplying the original value by the up factor and the down factor, creating two new nodes. Each of these nodes is again multiplied by the up and down factors, creating new nodes to the right, until the required number of steps is performed.

```
                          246.0
                 182.2
        135.0            135.0
100.0            100.0
        74.1             74.1
                 54.9
                          40.7
```

Figure 14.14 Lattice of the underlying asset

The option valuation lattice is shown in Figure 14.15. Calculations begin on the right side of the lattice, identifying the value of the option at that point in time. The ENPV at time t is calculated at each node; it is either the evolved value of the underlying asset (from the first lattice) or the salvage value, whichever is greater. Node A is either $S_0 u^3$ (246.0) or the salvage value (80). Since the asset value is greater, the node value is 246.0. Under these conditions, the project should be continued and not abandoned. This same procedure is continued down the column. Node C has a value of 80; the salvage value is greater than the asset value, so it is worthwhile to exercise the abandonment option and to collect the salvage value.

```
                                          246.0 A
                         182.2 D
              136.3                       135.0 B
    106.7                102.8
              87.1                        80.0 C
                         80.0
                                          80.0
```

Figure 14.15 Lattice of the expanded NPV

Internal points on the lattice are calculated using the backward induction technique (Mun, 2006). Any node value is determined based on the maximum value of either keeping the option open or exercising the option and abandoning the project. The method to determine the value of keeping the option open is the same as that used for the deferral option. For node D, the value of keeping the option open is:

$$Value = [pV^+ + (1-p)V^-]e^{-r(dt)}$$
$$= [(0.509)(246.0) + (1-0.509)(135.0)]e^{-0.05(1)} = 182.2$$

The minimum value of any node is the salvage value (80).

At the extreme left side, the final value is the expanded net present value (ENPV), 106.7. The option value can be determined by subtracting the PV of future revenues (100.0); the abandonment option is worth \$6.7 million. The minimum value of any abandonment option is zero, and the maximum value of an abandonment option is its salvage value.

The binomial lattice and the Black-Scholes equation provide similar, but not identical, results. For the example problem, the value of the Black-Scholes put option is 5.84. The result of the 3-step binomial lattice is 6.7. A three-step lattice is very short, and the lattice result will approach the Black-Scholes result as the number of time steps increases.

A number of problems regarding the deferral option were explored in the previous sections. Most of the same problems facing the deferral option are faced in the abandonment option. The example above demonstrates the standard methods by which an abandonment option can be calculated. Like the deferral option, it is very difficult to accurately identify project volatility, and this has a significant impact on the results.

Other theoretical issues continue to be present, such as the assumption that revenues are lognormally distributed, that cash flows follow Brownian motion, and that the Black-Scholes equation applies only to European options. While all of these issues are true, the inability to verify an accurate volatility is the most important.

14.9 COMPOUND OPTIONS

Compound options depend on the value of other options. Black, Scholes, and Merton recognized that the equity of a leveraged (partially debt financed) firm is actually an option on the firm's value (Black and Scholes, 1973; Merton, 1973). An option written on the firm's equity can therefore be considered an option on an option. Geske (1979) identified a closed form solution for compound options, and identified that the Black-Scholes pricing model is a special case of the compound option model.

14.9.1 Multi-stage Options Modeling. Current methods for determining the value of staged projects use NPV analysis based on expected costs, expected revenues, and the probabilities of passing from one stage to the next. Decision trees are often used to organize the information and to calculate project value, as addressed in Chapter 13. Real options analysis can use compound options to determine an expanded net present value of a staged project. In high risk, high payoff projects, such as drug development, where the probabilities of moving forward are fairly low, options analysis may provide a very different, and possibly more positive, project assessment. There is a possibility that options analysis will provide a more accurate project valuation than traditional methods if existing problems and concerns that exist with real options can be overcome.

As with single stage analysis, an option value for a multiple stage project can be computed. However, there is significantly more complexity in evaluating a multi-stage project when compared to a single-stage project. Because of this complexity, closed form solutions are generally not possible. Also, due to the complexity, many questions about multi-stage analysis remain open.

The investment at each project stage can be thought of as an option. At the conclusion of stage 1, the company can choose to continue to invest or not, depending on the results of this first stage. The decision of whether to continue to invest in later stages depends on the success of each incremental stage. Succeeding investments (known as staged funding) are a series of options that are dependent on the outcome of previous options. In options analysis, this is known as a sequential compound option, also known as a staging option. In theory, this approach accurately follows the actual decision making process regarding multi-stage project funding. Some real options proponents (Mun, 2006; Copeland and Keenan, 1998) have suggested that sequential compound options is a preferred method over decision tree analysis for determining the value of multi-stage projects. Some authors feel that the sequential compound option captures uncertainty, using the volatility parameter, better than decision trees which use specific probabilities. Determining the value of the option is complex, and state of the art techniques contain open, unresolved questions.

The most common tool for modeling multi-stage projects uses binomial lattices to simulate possible outcomes of the multiple investments. This is best described using an example. We return to the Hypertension drug example that was introduced in Chapter 13.

14.9.2 Multi-stage Option Example.

Example 14.4 Multi-Stage Options

A drug candidate for treating hypertension (high blood pressure) has been identified and has completed initial (animal) testing. In order to develop the drug candidate for market, a series of clinical tests would need to be conducted, following established Food and Drug Administration (FDA) rules. Three clinical trials would be needed, followed by FDA approval and a launch phase. Each phase is increasingly more expensive, and each is dependent on the success of the previous phase.

The testing and approval process is expected to take ten years. If all goes according to plan, the drug would have 10 years of exclusive marketing rights, beginning with FDA approval. In Phase I testing, the drug would be given to 20–80 healthy people to determine human safety. The testing is expected to cost $8 million (in year 2) and take 2 years to complete, with an estimated 70% chance of success. In Phase II testing, the drug would be given to 100–300 people to determine the efficacy for treating hypertension. The probability of success is estimated at 30%. Phase II testing is expected to require 2 years to complete, and would cost $30 million (in year 4). In Phase III clinical testing, the drug would be given to 1000–5000 people to determine safety and efficacy in a broad spectrum of the population. This testing is expected to take 3 years to complete and would start pending successful results from Phase II. The Phase III trials would cost $300 million (in year 7) and have an 80% chance of success. To obtain FDA approval, a new drug application would need to be written; this will require $10 million (in year 8) and 1 year to complete. FDA approval is expected to take 2 years, and there is a 90% probability of obtaining the needed approval. Successfully launching the product would require $350 million (primarily marketing costs) in year 10.

The hypertension drug has the potential of generating large profits, with net revenue of $450 million per year for ten years, starting in year 11. While the development costs are high and the chances of success are low, the potential payout is high if success can be achieved. The question therefore becomes: should the drug be developed? The MARR is 20%, and the costs and success probabilities for the stages are summarized in Table 14.9.

TABLE 14.9 Hypertension product costs

Year	Investment Required ($ million)	Probability of Success
2	8	70%
4	30	30
7	300	80
8	10	90
10	350	100

We determined in Chapter 13 that the NPV of the project is $3.88 million.

Binomial Lattices. The up-step, the down-step, and the risk-neutral probability are calculated as before. The volatility parameter is estimated to be 0.40 *at each stage* and

14.9 Compound Options

the risk-free rate is 5%. Note: a constant volatility parameter is typical in published work (see Section 14.9.4). We build the lattices so that each time step is one year.

$$u = e^{\sigma\sqrt{dt}} = 1.4918$$
$$d = \frac{1}{u} = 0.6703$$

$$p = \frac{e^{r(dt)} - d}{u - d} = 0.4637$$

Because there are five project stages, the option calculation consists of six lattices, each related to the previous one. The first lattice is the underlying lattice, starting with the present value of the predicted net revenues on the extreme left side. This is shown in Figure 14.16, and follows the same technique described in previous lattice techniques.

The first lattice is carried out over the time horizon of the project of 10 years. The left column shows the present worth of the net revenues, $S_0 = \$305$ million. For year 1, the up-step is equal to $S_0 u$, and the down-step is $S_0 d$. This is repeated across the lattice. The right hand column shows the possible range of the project's value at the time horizon. This is informative, because it places limits on what future costs can be economically incurred.

Year:	0	1	2	3	4	5	6	7	8	9	10
											16634
										11150	
									7474		7474
								5010		5010	
							3358		3358		3358
						2251		2251		2251	
					1509		1509		1509		1509
				1012		1012		1012		1012	
			678		678		678		678		678
		455		455		455		455		455	
	305		305		305		305		305		305
		204		204		204		204		204	
			137		137		137		137		137
				92		92		92		92	
					62		62		62		62
						41		41		41	
							28		28		28
								19		19	
									12		12
										8	
											6

Figure 14.16 Evolution of the underlying project

The second lattice is the equity lattice for the project's execution, shown in Figure 14.17. To determine the value of executing the project, the cost incurred for the final launch phase of the project is subtracted from the value of the first lattice in year 10 (from Figure 14.16), creating a new right hand column for the new lattice in Figure 14.17. The successive columns to the left are then discounted based on risk-neutral probabilities and the time value of money. The value of each point in the lattice is the maximum of three values: (1) the value of executing the project (the value from the same location in the preceding lattice minus the cost), (2) the value of keeping the option open (discounting the column to the right based on risk-neutral probabilities and time), and (3) zero, representing management's flexibility to not execute a money losing project.

	0	1	2	3	4	5	6	7	8	9	10
											16284
										10817	
									7157		7124
								4709		4677	
							3072		3042		3008
						1980		1950		1918	
					1258		1226		1192		1159
				785		753		716		679	
			482		452		417		373		328
		291		266		235		197		145	
	172		153		130		101		64		0
		87		70		51		28		0	
			37		25		12		0		0
				12		5		0		0	
					2		0		0		0
						0		0		0	
							0		0		0
								0		0	
									0		0
										0	
											0

Figure 14.17 Equity lattice, launch phase

The next lattice is the equity lattice for the project's New Drug Application (NDA) phase, shown in Figure 14.18. For this lattice, the cost incurred for the NDA ($10M) is subtracted from the value at year 8 in the previous lattice. This creates the new right hand column in Figure 14.18. Note that there are no values for years 9 and 10. The successive columns to the left are then discounted as before, determining the maximum of executing the project, keeping the option open (using risk-neutral probabilities), or zero.

	0	1	2	3	4	5	6	7	8
									7147
								4699	
							3063		3032
						1972		1940	
					1249		1216		1182
				778		744		707	
			476		445		407		363
		286		260		228		188	
	169		149		125		95		54
		84		67		47		24	
			35		23		10		0
				11		5		0	
					2		0		0
						0		0	
							0		0
								0	
									0

Figure 14.18 Equity lattice, NDA

The next lattice is the equity lattice for the project's Phase III, shown in Figure 14.19. For this lattice, the cost incurred for the third clinical phase ($300M) is subtracted from the value at year 7 in the previous lattice with 0's replacing any negative values. This creates the new right hand column in Figure 14.19. The successive columns to the left are then discounted as before, with each cell being the maximum of executing the project, keeping the option open, or zero.

14.9 Compound Options

```
   0      1      2      3      4      5      6      7
                                                    4399
                                             2777
                                      1700         1640
                               1006          931
                        577           502           407
                 323           262           179
          176           133            79             0
   95             67            35             0
          33             15             0             0
                   7             0             0      0
                          0             0      0
                                        0      0
                                               0
                                                      0
```

Figure 14.19 Equity lattice, Phase III

The next lattice is the equity lattice for Phase II of the project, shown in Figure 14.20. For this lattice, the cost incurred for the second clinical phase ($30M), which occurs in year 4, is subtracted from the value at year 4 in the previous lattice. A new year-4 column is created, and columns to the left are again discounted as before.

```
   0      1      2      3      4
                                 976
                          549
                   296           232
          155             105
   79             47              5
          21              2       0
                   1      0
                          0
```

Figure 14.20 Equity lattice, Phase II

The final lattice is the equity lattice for Phase 1 of the project, shown in Figure 14.21. A new year-2 column is created by subtracting the $8M for the first trial, and the columns to the left are calculated as before. The far left hand column of this last lattice is the value of the option in year 0. The option value is $73.66 million. The NPV for the project was $3.88 million, so the ENPV for this project is $77.54 million.

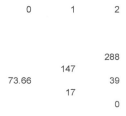

Figure 14.21 Equity lattice, Phase I

436 Chapter 14, Real Options Analysis

The large option value relative to the NPV is theoretically due to several factors (Copeland and Keenan, 1998). First, the option approach identifies all of the opportunities that are available to the firm, including those that are not obvious from the decision tree (such as uncertainties regarding price, sales volume, and market conditions). Second, the option includes a value for managerial flexibility where uncertainty exists.

The five equity lattices may be combined into a single consolidated lattice. The last two columns are from Figure 14.17, the column for year 8 is from Figure 14.18, the columns for years 5 to 7 are from Figure 14.19, the columns for years 3 and 4 are from Figure 14.20, and the first three columns are from Figure 14.21. This consolidated lattice is shown in Figure 14.22. The option value is highly dependent on the volatility, as shown in Figure 14.23.

0	1	2	3	4	5	6	7	8	9	10
										16284
									10817	
								7147		7124
							4399		4677	
						2777		3032		3008
					1700		1640		1918	
				976		931		1182		1159
			549		502		407		679	
		288		232		179		363		328
	147		105		79		0		145	
74		39		5		0		54		0
	17		2		0		0		0	
		0		0		0		0		0
			0		0		0		0	
				0		0		0		0
					0		0		0	
						0		0		0
							0		0	
								0		0
									0	
										0

Figure 14.22 Consolidated equity lattice, hypertension problem

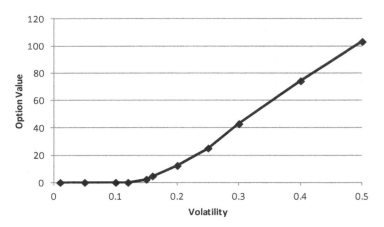

Figure 14.23 Effect of volatility on option value

14.9.3 Closed form solution. Geske (1979) derived a mathematical model for valuing corporate liabilities as compound options. As the number of nested options increase, so does the complexity of the solution. If the compound option contains two options, the solution contains bivariate normal distributions which require the solution of double integrals. Higher order nested options require multi-variate normal distributions. The complexity of the closed form model increases dramatically as additional stages are included.

14.9.4 Volatility Issues in Multi-stage Modeling. Volatility is a key issue in multi-stage options as it is in other types of real options. As with most options, increased volatility will generally increase the real option value. However, for multi-stage options, this conclusion is not straightforward. In real multi-stage projects, the volatility is not constant. Each stage of a multi-stage project has a different level of risk, and in general, project risks decrease over the life of the project as facts become better known, and early hurdles are overcome. As volatility decreases over time, project costs often increase dramatically, so the highest cost stages have lower probabilities of failure. Risk, however, is often viewed as the probability of failure times the potential loss, so risk does not necessarily decrease as the project progresses. This is not captured by traditional techniques nor is this addressed in most real options work. However, it is possible to capture this effect using real options tools. Changing volatility over time complicates the lattice, because each stage will require new calculations of d, u, and risk neutral probabilities. This makes valuation difficult but not impossible.

If each stage of a project has its own volatility, how can we determine stage volatility? This is very difficult; estimating the volatility of a simple option is challenging. Uncertain variables contribute to project volatility. Only the volatility that can be captured by exercising the option should be included in the volatility coefficient. Accurately determining volatility for each stage of a multi-stage project has not been addressed by those who propose the use of compound options. The standard (published) compound option method uses the average project volatility at each stage. More work is needed to understand the impact of volatility on multi-stage projects.

14.10 CURRENT ISSUES WITH REAL OPTIONS

When authors complain about the shortcomings of discounted cash flow analysis, they are usually referring to simple NPV and IRR. When dealing with traditional discounted cash flow tools, decision trees, sensitivity analysis, and simulation need to be included. These are all part of the traditional toolkit, and a wealth of information is provided by these techniques.

There are a number of reasons that companies do not adopt real option analysis. One of the primary reasons is the feeling that options analysis overestimates the value of projects (Block, 2007). We have found at least two major reasons why this is likely. First, it is difficult to estimate the volatility, and it is easy to overstate the volatility. Second, while many authors point out the value of waiting in deferral options, few include the fact that there is always a cost of waiting that must be considered.

Volatility estimates must be limited to examining only those items the firm can seize by exercising the option, which we call *actionable* volatility. Many inputs are independent variables and will not change whether or not the option is exercised. Including these independent variables in the option volatility calculation can overstate the volatility. This in turn can overstate the value of the option, leading a decision-maker to continually hold an option open when the investment should either be exercised or abandoned. Given the number of inputs that define the volatility of a capital investment, it is unlikely that a unifying theory will eventually define a single measure of volatility to be utilized in real options analysis. However, it should be clear that the analysis should include only those risky parameters that impact the value of an option. Miller (2011) recommends the use of separate volatility coefficients for the investment and future cash flows in order to address this concern.

Options analysis of even simple problems is a complex, time consuming tool that provides a slightly different perspective than traditional tools. It is interesting to note that whether traditional methods and real options analysis lead to the same recommendation will depend on the value of the volatility coefficient used in the real option calculation—and there are questions about the theoretically correct way to estimate this coefficient. We suggest that traditional methods, especially simulation, provide somewhat more information. Real options analysis requires significantly more time and analysis, and provides an imprecise answer due to the difficulty in estimating forecasted parameters. However, real options may provide value when experience with other similar projects provides a volatility coefficient of sufficient reliability. Then real options may be substituted for detailed decision tree and simulation analysis.

The problems and the perceptions regarding real options need to be overcome before the tools can become widely accepted in practice. The problems are quite real, and the negative perceptions are directly related to these problems. The issues plaguing simple options are also present in compound options, and the parameters under which compound options can be successfully applied are not yet understood. Carmichael, Hersh, and Parasu (2011) recommend using a probabilistic present value approach because of the concerns regarding volatility estimation. We suggest continued use of the "entire" traditional toolkit until questions concerning real options analysis are answered.

There are a variety of real options types and applications. We have explored the deferral (delay), abandonment, and staging options here. There are others. Baldi (2010) explored the application of switching options as one example. Others include expansion, contraction, and chooser options (see Mun, 2006).

14.11 SUMMARY

Real option analysis has been presented as an alternative to traditional discounted cash flow tools for identifying a project's value. The mathematics of real options is based on financial options, which have proven useful for a number of years. However, there are problems with the translation from financial options to real options, and these problems cannot be overlooked. The issues lead us to question the accuracy of real options valuation.

Standard methods were used to illustrate methodologies for valuating deferral, abandonment, and compound options. Several methods for calculating the option volatility were also discussed. Finally, a concern regarding real options analysis closed the chapter. Current concerns regarding options analysis will need to be resolved before these methods are widely adopted.

APPENDIX 14.A DERIVATION OF THE BLACK-SCHOLES EQUATION

14.A.1 The Black-Scholes Differential Equation. We will define an asset price as S(t), and a call option price as C(S, t) because the value of the asset is dependent on time, and the option is dependent on both the asset value and time. We will also assume that S(t) follows a lognormal distribution. The price of the asset will vary with time and also display volatility, following what is widely known as the "random walk." So

$$dS = \mu S dt + \sigma S dx \qquad (14.A.1)$$

where μSdt is the drift with time, and σSdx is the volatility, which will vary randomly within a normal distribution with a mean of zero. This is often referred to as Brownian motion.

Construct a portfolio having one option and some portion, −Δ of the underlying asset. In this way the portfolio is "long" one option and "short" a portion of the asset. The value of the portfolio is

$$\Pi = C - \Delta S \qquad (14.A.2)$$

In one time step, the value of the portfolio will change as follows:

$$d\Pi = dC - \Delta dS \qquad (14.A.3)$$

Ito's lemma (Ito and McKean, 1965; McKean, 1969) states

$$dC = \sigma S \frac{\partial C}{\partial S} dx + \left(\mu S \frac{\partial C}{\partial S} + \frac{1}{2} \sigma^2 S^2 \frac{\partial^2 C}{\partial S^2} + \frac{\partial C}{\partial t} \right) dt \qquad (14.A.4)$$

Modifying Equation (14.A.1),

$$\Delta dS = \Delta \mu S dt + \Delta \sigma S dx \qquad (14.A.5)$$

Combining equations (14.A.3), (14.A.4), and (14.A.5),

$$d\Pi = \sigma S \frac{\partial C}{\partial S} dx + \left(\mu S \frac{\partial C}{\partial S} + \frac{1}{2} \sigma^2 S^2 \frac{\partial^2 C}{\partial S^2} + \frac{\partial C}{\partial t} \right) dt - \Delta \mu S dt - \Delta \sigma S dx \qquad (14.A.6)$$

Combining terms,

$$\begin{aligned} d\Pi &= \left(\sigma S \frac{\partial C}{\partial S} - \Delta \sigma S \right) dx + \left(\mu S \frac{\partial C}{\partial S} + \frac{1}{2} \sigma^2 S^2 \frac{\partial^2 C}{\partial S^2} + \frac{\partial C}{\partial t} - \Delta \mu S \right) dt \\ &= \sigma S \left(\frac{\partial C}{\partial S} - \Delta \right) dx + \left(\mu S \frac{\partial C}{\partial S} + \frac{1}{2} \sigma^2 S^2 \frac{\partial^2 C}{\partial S^2} + \frac{\partial C}{\partial t} - \Delta \mu S \right) dt \end{aligned} \qquad (14.A.7)$$

We choose that

so
$$\Delta = \frac{\partial C}{\partial S}$$

$$d\Pi = \left(\frac{\partial C}{\partial t} + \frac{1}{2}\sigma^2 S^2 \frac{\partial^2 C}{\partial S^2}\right) dt \qquad (14.A.8)$$

This portfolio contains no risk because there is no Brownian motion term, dx. Therefore, the value of the portfolio changes based only on the interest rate *r* over time.

$$\begin{aligned} d\Pi &= r\Pi dt \\ &= r(C - \Delta S)dt \\ &= r\left(C - \frac{\partial C}{\partial S}S\right) dt \end{aligned} \qquad (14.A.9)$$

Combining Equations (14.A.8) and (14.A.9),

$$r\left(C - \frac{\partial C}{\partial S}S\right) dt = \left(\frac{\partial C}{\partial t} + \frac{1}{2}\sigma^2 S^2 \frac{\partial^2 C}{\partial S^2}\right) dt \qquad (14.A.10)$$

Dividing both sides by dt,

$$rC - rS\frac{\partial C}{\partial S} = \frac{\partial C}{\partial t} + \frac{1}{2}\sigma^2 S^2 \frac{\partial^2 C}{\partial S^2} \qquad (14.A.11)$$

or

$$\frac{\partial C}{\partial t} + \frac{1}{2}\sigma^2 S^2 \frac{\partial^2 C}{\partial S^2} + rS\frac{\partial C}{\partial S} - rC = 0 \qquad (14.A.12)$$

This is the *Black-Scholes partial differential equation* (based on Wilmott, Howison, and DeWynne, 1995, Chapter 3).

14.A.2. Deriving the Black-Scholes equation. The following derives the Black-Scholes equation for a European Call option, where the option can only be exercised at maturity. As S approaches a value of zero, then C will also approach a value of zero. As S approaches infinity, then C approaches S. So

$$C(0,t) = 0 \qquad C(S,t) \sim S \text{ as } S \to \infty$$

and
$$C(S,T) = \max(S - E, 0)$$

where E is the strike price of the option. We will use E as the strike price to avoid confusion between X (as the strike price) and the variable x defined in Equation (14.A.13).

We want to eliminate the S and S^2 terms in Equation (14.A.12), so we set

$$S = Ee^x \qquad (14.A.13)$$

Appendix 14.A Derivation of the Black-Scholes Equation

$$t = T - \frac{\tau}{\frac{1}{2}\sigma^2} \tag{14.A.14}$$

$$C = Ev(x,\tau). \tag{14.A.15}$$

Applying the chain rule to the partial derivatives, we can state

$$\frac{\partial C}{\partial t} = E\frac{\partial v}{\partial \tau}\frac{\partial \tau}{\partial t} = \frac{-E\sigma^2}{2}\frac{\partial v}{\partial \tau} \tag{14.A.16}$$

$$\frac{\partial C}{\partial S} = E\frac{\partial v}{\partial x}\frac{\partial x}{\partial S} = \frac{E}{S}\frac{\partial v}{\partial x} = e^{-x}\frac{\partial v}{\partial x} \tag{14.A.17}$$

$$\frac{\partial^2 C}{\partial S^2} = \frac{E}{S^2}\frac{\partial^2 v}{\partial x^2} - \frac{E}{S^2}\frac{\partial v}{\partial x} = \frac{e^{-2x}}{E}\left(\frac{\partial^2 v}{\partial x^2} - \frac{\partial v}{\partial x}\right) \tag{14.A.18}$$

Substituting the partial differentials from Equations (14.A.16 – 14.A.18) into the Black-Scholes partial differential equation (14.A.12),

$$-\frac{1}{2}E\sigma^2 S^2 \frac{\partial v}{\partial t} + \frac{1}{2}\sigma^2 S^2 \left[\frac{e^{-2x}}{E}\left(\frac{\partial^2 v}{\partial S^2} - \frac{\partial v}{\partial x}\right)\right] + rSe^{-x}\frac{\partial v}{\partial x} - rC = 0 \tag{14.A.19}$$

This eventually simplifies to

$$\frac{\partial v}{\partial \tau} = \frac{\partial^2 v}{\partial x^2} + \left(\frac{2r}{\sigma^2} - 1\right)\frac{\partial v}{\partial x} - \frac{2r}{\sigma^2}v$$

Letting $k = 2r/\sigma^2$

$$\frac{\partial v}{\partial \tau} = \frac{\partial^2 v}{\partial x^2} + (k-1)\frac{\partial v}{\partial x} - kv \tag{14.A.20}$$

The initial condition $C(S,T) = \max(S - E, 0)$ becomes

$$v(x, 0) = \max(e^x - 1, 0).$$

We will apply a change of variables, and let

$$v = e^{\alpha x + \beta \tau}u(x,\tau)$$

where we need to determine the constants α and β. Differentiation gives

$$\beta u + \frac{\partial u}{\partial \tau} = \alpha^2 u + 2\alpha \frac{\partial u}{\partial x} + \frac{\partial^2 u}{\partial x^2} + (k-1)\left(\alpha u + \frac{\partial u}{\partial x}\right) - ku \tag{14.A.21}$$

We can obtain an equation with no u term by defining β:

$$\beta = \alpha^2 + (k-1)\alpha - k$$

In order to eliminate the $\partial u / \partial x$ term, α will be defined as

$$0 = 2\alpha + (k-1)$$

So α and β become

$$\alpha = -\frac{1}{2}(k-1) \qquad \beta = -\frac{1}{4}(k+1)^2 \qquad (14.\text{A}.22)$$

The transformation from v to u is

$$v(x,\tau) = e^{-\frac{1}{2}(k-1)x - \frac{1}{4}(k+1)^2 \tau} u(x,\tau) \qquad (14.\text{A}.23)$$

where

$$\frac{\partial u}{\partial \tau} = \frac{\partial^2 u}{\partial x^2} \qquad (14.\text{A}.24)$$

for $-\infty < x < \infty, \tau > 0$

which is the classic heat diffusion partial differential equation.

$$u(x,0) = u_0(x) = \max\left(e^{\frac{1}{2}(k+1)x} - e^{\frac{1}{2}(k-1)x}, 0\right) \qquad (14.\text{A}.25)$$

The solution to the diffusion equation is:

$$u(x,\tau) = \frac{1}{2\sqrt{\pi\tau}} \int_{-\infty}^{\infty} u_0(s) e^{-(x-s)^2/4\tau} \, ds \qquad (14.\text{A}.26)$$

where $u_0(x)$ is defined in Equation (14.A.25).

Solving the integral is simplified if we allow $x' = (s-x)/\sqrt{2\tau}$. Equation (14.A.23) becomes

$$\begin{aligned} u(x,\tau) &= \frac{1}{2\sqrt{\pi\tau}} \int_{-\infty}^{\infty} u_0(x'\sqrt{2\tau} + x) e^{-\frac{1}{2}x'^2} \, dx' \\ &= \frac{1}{2\sqrt{\pi}} \int_{-x/\sqrt{2\tau}}^{\infty} e^{-\frac{1}{2}(k+1)(x+x'\sqrt{2\tau})} e^{-\frac{1}{2}x'^2} \, dx' \\ &\quad - \frac{1}{2\sqrt{\pi}} \int_{-x/\sqrt{2\tau}}^{\infty} e^{-\frac{1}{2}(k-1)(x+x'\sqrt{2\tau})} e^{-\frac{1}{2}x'^2} \, dx' \end{aligned} \qquad (14.\text{A}.27)$$

We will break this into two equations, by naming the integrals:

$$u(x,\tau) = I_1 - I_2 \qquad (14.\text{A}.28)$$

We will evaluate I_1 first. Simplifying the exponent,

Appendix 14.A Derivation of the Black-Scholes Equation

$$I_1 = \frac{1}{2\sqrt{\pi}} \int_{-x/\sqrt{2\tau}}^{\infty} e^{-\frac{1}{2}(k+1)(x+x'\sqrt{2\tau}) - \frac{1}{2}x'^2} dx'$$

$$= \frac{e^{\frac{1}{2}(k+1)x}}{2\sqrt{\pi}} \int_{-x/\sqrt{2\tau}}^{\infty} e^{\frac{1}{4}(k+1)^2 \tau} e^{-\frac{1}{2}\left(x' - \frac{1}{2}(k+1)\sqrt{2\tau}\right)^2} dx'$$

$$= \frac{e^{\frac{1}{2}(k+1)x + \frac{1}{4}(k+1)^2 \tau}}{2\sqrt{\pi}} \int_{-x/\sqrt{2\tau} - \frac{1}{2}(k+1)\sqrt{2\tau}}^{\infty} e^{-\frac{1}{2}\rho} d\rho$$

$$I_1 = e^{\frac{1}{2}(k+1)x + \frac{1}{4}(k+1)^2 \tau} \Phi(d_1) \tag{14.A.29}$$

where

$$d_1 = \frac{x}{\sqrt{2\tau}} + \frac{1}{2}(k+1)\sqrt{2\tau} \tag{14.A.30}$$

and

$$\Phi(d_1) = \frac{1}{\sqrt{2\pi}} \int_{-\infty}^{d_1} e^{\frac{1}{2}s^2} ds \tag{14.A.31}$$

which is the cumulative distribution function for the normal distribution.

The calculation of I_2 is identical to that of I_1 except that $(k+1)$ is replaced by $(k-1)$. Finally, recalling Equation (14.A.23),

$$v = e^{-\frac{1}{2}(k-1)x - \frac{1}{4}(k+1)^2 \tau} u(x, \tau) \tag{14.A.23}$$

Also, recalling Equations (14.A.13 – 14.A.15),

$$S = E e^x \qquad t = T - \frac{\tau}{\frac{1}{2}\sigma^2} \qquad C = Ev(x,\tau).$$

We can rewrite these as

$$x = \ln(S/E) \tag{14.A.32}$$

$$\tau = \frac{1}{2}\sigma^2 (T - t) \tag{14.A.33}$$

$$C = Ev(x, \tau) \tag{14.A.34}$$

So

$$C = Ev(x,\tau) = E e^{-\frac{1}{2}(k-1)x - \frac{1}{4}(k+1)^2 \tau} u(x,\tau) \tag{14.A.35}$$

$$C = Ev(x,\tau) = E e^{-\frac{1}{2}(k-1)x - \frac{1}{4}(k+1)^2 \tau} (I_1 - I_2) \tag{14.A.36}$$

$$I_1 = e^{\frac{1}{2}(k+1)x + \frac{1}{4}(k+1)^2 \tau} \Phi(d_1)$$

$$I_2 = e^{\frac{1}{2}(k-1)x + \frac{1}{4}(k-1)^2 \tau} \Phi(d_2)$$

Substituting,

$$C = Ev(x,\tau) = E e^{-\frac{1}{2}(k-1)x - \frac{1}{4}(k+1)^2 \tau} [e^{\frac{1}{2}(k+1)x + \frac{1}{4}(k+1)^2 \tau} \Phi(d_1) - e^{\frac{1}{2}(k-1)x + \frac{1}{4}(k-1)^2 \tau} \Phi(d_2)]$$

This eventually simplifies to

$$C = Ee^x\Phi(d_1) - Ee^{-r(T-t)}\Phi(d_2) \tag{14.A.37}$$

Because $S = Ee^x$ and $t = 0$

$$C = S\Phi(d_1) - Ee^{-r(T-t)}\Phi(d_2) \tag{14.A.38}$$

Substituting Equations (14.A.32-14.A.34) into the equations for d_1 and d_2, d_1 becomes

$$d_1 = \frac{\ln(S/E) + (r + \tfrac{1}{2}\sigma^2)T}{\sigma\sqrt{T}} \tag{14.A.39}$$

and

$$d_2 = \frac{\ln(S/E) + (r - \tfrac{1}{2}\sigma^2)T}{\sigma\sqrt{T}} \tag{14.A.40}$$

which is usually written as

$$d_2 = d_1 - \sigma\sqrt{T}.$$

Replacing the term E with X in order to match the existing Chapter 14 terms, and setting t equal to zero,

$$C = S_0\Phi(d_1) - Xe^{-rT}\Phi(d_2) \tag{14.1}$$

$$d_1 = \frac{\ln(S_0/X) + (r + \sigma^2/2)T}{\sigma\sqrt{T}}$$

$$d_2 = d_1 - \sigma\sqrt{T}$$

(Based on Wilmott, Howison, and DeWynne, 1995, Chapter 5.)

REFERENCES

AMRAM, MARTHA, and NALIN KULATILAKA, *Real Options* (Harvard Business School Press, 1999).

BALDI, FRANCESCO, "Switch, Switch, Switch! A Regime-Switching Option-Based Model for Valuing a Tolling Agreement," *The Engineering Economist*, **55**(3) (July–September 2010), pp. 268–304.

BARTON, KELSEY, and YURI LAWRYSHYN, "Integrating Real Options with Managerial Cash Flow Estimates," *The Engineering Economist*, **56**(3) (July–September 2011), pp. 254–273.

BLACK, FISHER, and MYRON SCHOLES, "The Pricing of Options and Corporate Liabilities," *Journal of Political Economy*, **81**(3) (May/June 1973), pp. 637–654.

BLACK, FISHER, "Fact and Fantasy in the Use of Options," *Financial Analysis Journal*, **31**(4) (July–August 1975), pp. 36–41; 61–72.

BLOCK, STANLEY, "Are 'Real Options' Actually Used in the Real World?" *The Engineering Economist*, **52**(3) (July–September 2007), pp. 255–267.

Bowman, Edward H., and Gary T. Moskowitz, "Real Options Analysis and Strategic Decision Making," *Organization Science*, **12**(6) (Nov/Dec 2001), pp. 772–777.

Brach, Marion A., and Dean A. Paxson, "A Gene to Drug Venture: Poisson Options Analysis," *R&D Management*, **31**(2) (April 2001), pp. 203–214.

Brach, Marion A., *Real Options in Practice* (John Wiley & Sons, 2003).

Brandão, Luiz E., James S. Dyer, and Warren J. Hahn, "Using Binomial Decision Trees to Solve Real–Option Valuation Problems," Decision Analysis, **2**(2) (June 2005a), pp. 69–88.

Brandão, Luiz E., James S. Dyer, and Warren J. Hahn, "Response to Comments on Brandão et al. (2005)," Decision Analysis, **2**(2) (June 2005b), pp. 103–109.

Carmichael, David G., Ariel M. Hersh, and Praneet Parasu, "Real Options Estimate Using Probabilistic Present Worth Analysis," *The Engineering Economist*, **56**(4) (October–December 2011), pp. 295–320.

Cobb, Barry R., and John M. Charnes, "Real Options Volatility Estimation with Correlated Inputs," *The Engineering Economist*, **49**(2) (April–June 2004), pp. 119–137.

Copeland, Thomas E., and Keenan, Phillip T., "Making Real Options Real," *The McKinsey Quarterly*, **1**(3) (Summer 1998), pp. 128–141.

Copeland, Tom, and Vladimir Antikarov, *Real Options: A Practitioner's Guide* (Texere LLC, 2001). Revised edition, 2003.

Copeland, Tom, and Peter Tufano, "A Real–World Way to Manage Real Options," *Harvard Business Review*, **82**(3) (March 2004), pp. 90–99.

Copeland, Thomas E., and Vladimir Antikarov, "Real Options: Meeting the Georgetown Challenge," Journal of Applied Corporate Finance, **12**(2) (Spring 2005), pp. 32–51.

Cox, John C., Stephen A. Ross, and Mark Rubinstein, "Option Pricing: A Simplified Approach," *Journal of Financial Economics*, **7**(3) (September 1979), pp. 229–263.

Dixit, Avinash K., and Pindyck, Robert S., *Investment Under Uncertainty* (Princeton University Press, 1994).

Eschenbach, Ted G., Neal A. Lewis, and Joseph C. Hartman, "Technical Note: Waiting Cost Models for Real Options," *The Engineering Economist*, **54**(1) (January–March 2009), pp. 1–21.

Geske, Robert, "The Valuation of Compound Options," *Journal of Financial Economics*, **7**(1) (March 1979), pp. 63–81.

Herath, Hemantha S. B., and Chan S. Park, "Multi–Stage Capital Investment Opportunities as Compound Real Options," *The Engineering Economist*, **47**(1) (January–March 2002), pp. 1–27.

Ito, Kiyosi, and Henry P. McKean, Jr., *Diffusion Processes and Their Simple Paths*, (Springer, 1965).

Lewis, Neal A., Ted G. Eschenbach, and Joseph C. Hartman, "Can We Capture the Value of Option Volatility," *The Engineering Economist*, **53**(3) (July–September 2008), pp. 230–258.

Luehrman, Timothy A., "Investment Opportunities as Real Options: Getting Started on the Numbers," *Harvard Business Review*, **76**(4), (July/August 1998), pp. 51–64.

MacMillan, Ian C., and Alexander B. van Putten, "Cost Option Based Real Options Investment Valuation, U.S. Patent #US 2005/0131791A1 (June 16, 2005).

MacMillan, Ian C., Alexander B. van Putten, Rita Gunther McGrath, and James D. Thompson, "Using Real Options Discipline for Highly Uncertain

Technology Investments," *Research Technology Management*, **49**(1) (Jan/Feb 2006), pp. 29–37.

McGrath, Rita Gunther, and Ian C. MacMillan, "Assessing Technology Projects Using Real Options Reasoning," *Research–Technology Management*, **43**(4) (July–August 2000), pp. 35–49.

McKean, Henry P., *Stochastic Integrals*, (American Mathematical Society, 1969).

Merton, Robert C., "Theory of Rational Option Pricing," *Bell Journal of Economics & Management*, **4** (June 1973), pp. 141–183.

Miller, Tom W., "Active Management of Real Options," *The Engineering Economist*, **56**(3) (July–September 2011), pp. 205–230.

Mun, Johnathan, *Real Options Analysis, 2nd ed.* (John Wiley & Sons, 2006).

Myers, Stewart C., "Determinants of Corporate Borrowing," *Journal of Financial Economics*, **5**(7) (1977), pp. 147–175.

Nichols, Nancy A., "Scientific Management at Merck: An Interview with CFP Judy Lewent," *Harvard Business Review*, **72**(1) (month 1994), pp. 88–99.

Pindyck, Robert S., "Irreversibility, Uncertainty, and Investment," *Journal of Economic Literature*, **29**(3) (September 1991), pp. 1110–1148.

Smith, James E., and Kevin F. McCardle, "Options in the Real World: Lessons Learned in Evaluating Oil and Gas Investments," *Operations Research*, **47**(1) (Jan.–Feb. 1999), pp. 1–15.

Smith, James E., "Alternative Approaches for Solving Real–Options Problems (Comment on Brandão et al., 2005)," *Decision Analysis*, **2**(2) (June 2005), pp. 89–102.

Trigeorgis, Lenos, *Real Options: Managerial Flexibility and Strategy in Resource Allocation* (The MIT Press, 1996).

van Putten, Alexander B., and Ian C. MacMillan, "Making Real Options Really Work," *Harvard Business Review*, **82**(12) (December 2004), pp. 134–141.

Wilmott, Paul, Sam Howison, and Jeff DeWynne, *The Mathematics of Financial Derivatives*, (Cambridge University Press, 1995).

PROBLEMS

14-1. A new computer installation is under consideration that should save the firm operating expenses if the new technology works as planned. An investment of $120,000 is needed, but the project should save $40,000 the first year, increasing 10% per year. The project has an expected life of 3 years. The savings are uncertain, and could be 10% higher or 15% lower than expected. Assume a MARR of 12%. Apply the uncertainty to the first year's savings using a triangular distribution and determine the project volatility.
  (a) Using the standard deviation of cash flows method
  (b) Using the logarithmic present value returns method

14-2. A new product is planned having the following forecast (volumes and costs are in thousands).

Year	Price per unit	Sales volume	Costs
0			$2000
1	$14.00	50	700
2	14.00	70	950
3	13.50	80	1050
4	13.50	85	1050
5	13.00	80	1050

The sales volumes are uncertain, and could be 20% higher or 30% lower for each year and follow triangular distributions. The MARR is 15%. Determine the project volatility based on:
(a) Logarithmic present value returns method
(b) Standard deviation of cash flows method

14-3. Repeat problem 14-2, but make the sales volume for each year 90% correlated to the previous year's sales volume. Determine the project volatility based on the following methods.
(a) Logarithmic present value returns method
(b) Standard deviation of cash flows method
(c) What happened to the volatility when cash flows became correlated?

14-4. A project has the following attributes for a possible deferral option.

Present value of net revenues, S_0	$15 million
Cost of waiting, W	$ 2.5 million
Investment cost, year 0, X	$12 million
Risk free interest rate, r_f	4%
Total delay time, T	1 year
Volatility, σ	15%

Using binomial lattices,
(a) Determine the option value without the cost of waiting. Use four time steps.
(b) Determine the option value including the cost of waiting. Use four time steps. What is the impact of including the cost of waiting?
(c) Determine the option value including the cost of waiting. Use eight time steps. What is the impact of using more time steps?

14-5. Repeat problem 14-4 part b using the Black-Scholes equations.

14-6. A project has the following attributes for a possible deferral option.

Present value of net revenues, S_0	$80 million
Cost of waiting, W	$ 4.0 million
Investment cost, year 0, X	$80 million
risk free interest rate, r_f	4%
Total delay time, T	2 years
volatility, σ	20%

Using binomial lattices,
(a) Determine the option value without the cost of waiting. Use four time steps.
(b) Determine the option value including the cost of waiting. Use four time steps. What is the impact of including the cost of waiting?

(c) Determine the option value including the cost of waiting. Use eight time steps. What is the impact of using more time steps?

14-7. Repeat problem 14-6 part b using the Black-Scholes equations.

14-8. A project has the following attributes for a possible abandonment option.
Present value of net revenues, S_0 $50 million
Salvage value, X $45 million
risk free interest rate, r_f 5%
Total delay time, T   1 year
volatility, σ 25%

Using binomial lattices, determine the option value, P.
(a) Use four time steps.
(b) Use eight time steps.

14-9. Repeat problem 14-8 using the Black-Scholes equations.

14-10. A project has the following attributes for a possible abandonment option.
Present value of net revenues, S_0 $10 million
Salvage value, X $8 million
Risk free interest rate, r_f 5%
Total delay time, T   3 years
Volatility, σ 30%

Using binomial lattices, determine the option value, P.
(a) Use three time steps.
(b) Use six time steps.

14-11. Repeat problem 14-10 using the Black-Scholes equations.

14-12. A proposed project has three stages: concept development, engineering, and commercialization.
Stage 1 will cost $1 million, take 2 years to complete, and has a 40% chance of success.
Stage 2 will cost $10 million, take 1 year to complete, and has a 50% chance of success.
Stage 3 will cost $50 million and have a 35% chance of success. Revenues will be received over 5 years, and are expected to be as follows:
   Year 1 $ 20 million
   Year 2   60 million
   Year 3   80 million
   Year 4   100 million
   Year 5   110 million
The hurdle rate is 15%, risk-free rate 4%, and project volatility 20%.
Determine the ENPV of the project. Should the project be undertaken? Use binomial lattices to solve a sequential compound option.

14-13. A company is looking at adding a new product. The investment required to start the project is $100 million. The present value of the net benefits is $95 million. The cost structure might change in the future, making the project look better. The project can be delayed for 2 years. The volatility of the project is 25%. The project's minimum acceptable rate of return is 20% and the risk-free rate is 5%. If the project is delayed, the present value of the net

benefits will be reduced by $30 million.
(a) What is the project's traditional NPV if it is started now?
(b) What is the project's NPV if it is delayed two years?
(c) Using the Black-Scholes model, what is the project's ENPV with the flexibility to delay?

14-14. Solve problem 14-13 using binomial lattices.
(a) Use three time steps to solve the problem.
(b) Use seven time steps to solve the problem. Why is your answer different from (a)?

14-15. A project has the following characteristics, and management is considering deferring the funding decision until more information becomes available.

S_0	$10 million	PV of the future revenues
X	$10 million	project cost
t	1 year	potential delay time
σ	0.30	project volatility
r	0.05	risk-free rate of return
i	16%	MARR

The cost of waiting is due to revenues being postponed by one year.
(a) What is the project's NPV?
(b) What is the option value for the delay option using the Black-Scholes method?
(c) What is the project's ENPV?
(d) Should the project be funded now?
(e) Should the project be delayed with the option kept open?

14-16. Repeat problem 14-15 using binomial lattices, using 3 time steps.

14-17. A new technology is being tried for an existing product. The project requires an investment of $2.60 million. The present value of total benefits is $2.40 million, with a project volatility of 25%. At any time during the next two years, the project's assets can be sold for $2.30 million. The risk-free rate is 5%.
(a) Should the project be funded based on NPV?
(b) What is the value of the option to abandon the project?
(c) Should be project be done with the option to abandon?

14-18. A new technology has been created and patented, and the company is planning on licensing the new technology. The company expects revenues of $300,000 the first year, increasing $100,000 each year for four years (for a total of five years of revenues). The project required an investment of $1.5 million. The amount of increase is uncertain, and could be as low as zero and as high as $200,000 in any given year. The company may want to delay the project one year. Assume that the increase in revenues follows a triangular distribution. The MARR is 14%. Use 1000 iterations in the simulation. This requires a spreadsheet with Monte Carlo simulation software (such as @Risk or Crystal Ball).
(a) Calculate the project's NPV.
(b) Calculate the project's IRR.
(c) Determine the project's volatility using the standard deviation of cash flows method.
(d) Determine the project's volatility using the logarithmic present value return method.

14-19. Massive Coal Company is planning a new mine. Capital costs, including the rights to mine the land, infrastructure, and equipment cost a total of $30 million. The mine will produce $8 million worth of coal each year for 10 years. When the mine is closed, $15 million will be spent to return the land to its original grade. The MARR is 15%. The income is uncertain because of the fluctuating price of coal. The annual revenue can be considered to be a normal distribution with a mean of $8 million per year and a standard deviation of 20%. Use 1000 iterations in the simulation. This requires a spreadsheet with Monte Carlo simulation software (such as @Risk or Crystal Ball).

(a) Calculate the project's NPV.
(b) Calculate the project's IRR.
(c) Determine the project volatility based on
    (1) The standard deviation of cash flows method,
    (2) The logarithmic present value returns method,
    (3) The IRR method, what is the project's volatility
Only the first year's annual revenue is considered to be an actionable variable, so only that value is varied to determine the volatility
(d) Calculate the ENPV, using the Black-Scholes model and the volatility from the standard deviation of cash flows method in (c1).
(e) Repeat (d) using binomial lattices.

Mini-Case 1: Sunscreen

A consumer products company is readying a new sunscreen which blocks the sun's ultraviolet rays. The product has been developed and the size of the market has been estimated. Launching the product requires a current investment of $9.0 million. The company's MARR for this type of project is 20%.

The research department has recently identified a new sunscreen active ingredient, which is not yet available. Including it would delay the product's launch by a year. However, it would improve product efficacy and increase cash flows if it were used. The investment would be 5% higher if the project is delayed one year because the project would need to adopt a crash schedule. The equipment's life and the new formulation technology is 10 years in either scenario.

	Cash Flows, $million		Lower Limit	Upper Limit
	Launch now	Delay 1 year		
1	1.0	0.0	−40%	+20%
2	2.0	1.0	−40%	+30%
3	2.5	2.5		
4	3.0	3.5		
5	3.0	3.5		
6	3.0	3.5		
7	3.0	3.5	−40%	+30%
8	3.0	3.5		
9	3.0	3.5		
10	3.0	3.5		
11	0.0	3.5		
Initial investment	9.00 million	9.45 million	−5%	+15%
Salvage value	0.75 million	0.75 million	−100%	+100%
MARR	20%	20%	−20%	+20%
Extra cost for delaying construction		0.45 million	−40%	+20%

Most of the variables involved in the sunscreen project are realized after the time when the option to delay has expired. The investment occurs within the option timeline and the first year revenues may be estimated with good accuracy by tracking the market. Only these two variables are considered actionable variables for a true volatility calculation.

Use the average values for cash flows for each variable rather the most likely (use the simulation to determine the average) for a more accurate evaluation.

(a) Determine the sunscreen project volatility using triangular distributions of the two variables and using the following volatility methods:
(1) The standard deviation of cash flows method
(2) The logarithmic present value returns method
(3) The internal rate of return method.

(b) Determine the ENPV for the sunscreen case using
(1) Black-Scholes method
(2) Binomial lattices. Use 5 time steps.

Mini-Case 2: Steel mill

A firm is considering building a new steel mill that would use an innovative mini-mill iron smelting process. Building the plant would require spending $12M over two years with $7M at t_0 and $5M at t_1. The MARR for the project is 14%.

The project's cash flows were calculated by subtracting the annual fixed and variable costs from the gross revenues in each year. The following table shows the detailed net cash flows. The risk-free interest rate is 4%. The project is estimated to have a 10-year lifespan including the 2-year construction phase.

Year	Build Now Net Cash Flow	Delay Net Cash Flow	Waiting cost
0	$-7.0	$0	$0
1	-5.0	-7.0	0
2	0.2	-6.0	0.2
3	2.8	0.2	2.6
4	3.4	2.8	0.6
5	3.4	3.4	0
6	3.4	3.4	0
7	3.4	3.4	0
8	3.4	3.4	0
9	3.4	3.4	0
10		3.4	-3.4

There are many potential risks involved in the steel industry including operational hazards; disruptions in supply, fluctuations in demand, changes in the legal, tax, and environmental regulatory climate, and technology obsolescence. One method for dealing with these risks is to defer the project by one year to see if circumstances are more advantageous to proceed at that time. The deferral option would push the initial construction costs to time t_1. Deferring the project would increase the investment cost in the second year from $5 million to $6 million. The estimated net cash flows would remain the same, but the delay postpones all cash flows, decreasing the NPV from −$0.37M to −$1.00M.

What level of volatility is needed to justify keeping the project alive? The ENPV would need to be 0 if the project is to break even. Because the NPV is negative, an option value equal to the NPV is needed to provide a minimum ENPV of zero. What volatility will create this option value? Use Goal Seek to determine the breakeven volatility.

Mini-Case #3: Abandonment option

A high-tech firm is planning a new product that is highly dependent on rare earth elements. While it is not certain that the technology will work, they feel confident that the project will have enough valuable equipment and materials that it will have a high salvage value. The project is expected to have net revenues of $45M for three years, at which time the assets will either be turned over to a second generation project or sold. The equipment is expected to have a salvage value of $120M at any time during the next three years. However, this salvage value is highly uncertain, and could be anywhere between 10% and 190% of this estimate. This is the only source of actionable volatility.

The MARR for the project is 25% and the risk-free rate is 5%.

Because the salvage value may be obtained at any time during the next three years, this is an American option and is best calculated using binomial lattices. Determine the following.

(a) Determine the volatility using the standard deviation of cash flow method.
(b) Determine the abandonment option value, P, using binomial lattices and five time steps.
(c) Determine the sensitivity of the option value to the project volatility by graphing the option value (y axis) vs. the volatility (x axis) over a volatility range of 0.01 to 0.4.

15
Capacity Expansion and Planning

15.1 INTRODUCTION

Expansion opportunities can be analyzed with the tools described in previous chapters. That is, the cash flows associated with the expansion, including the additional investment, expected returns, and costs associated with the additional capacity, can be estimated. This option should include only the incremental costs and returns associated with the additional investment and not the total returns from the complete project (including previous investments and expansions. Once the associated cash flow diagram has been defined, analysis of the project merits can commence, along with evaluating all associated risks and noneconomic factors. The decision to expand or not can be soundly made. This decision set may also include the delay option.

We take our analysis one step further in this chapter because a company may want to look farther into the future. That is, they would like to plan for future expansions. This can be very useful when determining future capital budgeting plans, as money can be made available and funds can even be sought (which can take considerable time) if deemed necessary. These plans are often referred to as capacity plans, developments from capacity planning, or capital expenditure (capex) plans. Public companies often divulge their planned capital expenditures, including methods to finance the projects, in annual Securities and Exchange Commission (SEC) filings.

Capacity planning is a bit different than the decisions that we have evaluated thus far in this text. Previously we evaluated a number of possible decisions that had already been identified. That is, we laid out possible capital expenditures and evaluated whether they should be pursued or not. In this chapter, we present a general approach which allows for the amount and timing of expansions to be decision variables. Specifically, the models will determine (1) when and (2) how much to expand without us explicitly identifying all of the options. This is a much more difficult problem, so we will need to make some simplifying assumptions. Nevertheless, the solutions will provide valuable insight into future expansions and capacity plans.

Because both the timing and amount to expand are decision variables, it is clearly not possible to define all possibilities, let alone evaluate all of them. Further, these options can become quite complicated depending on assumptions made with respect to demand. Thus, we take a simpler approach and assume that expansions will follow a strategy. We define these strategies in the next section and then illustrate how to implement these strategies given the investment climate.

15.2 EXPANSION ANALYSIS

Firms expand operations for a number of reasons, but mainly it is to meet expected demand—demand that is usually growing. The decision to expand is risky, because demand may not materialize and the investment is lost. Furthermore, it often takes a long time to expand. Numerous examples exist where companies take months, and often years, to increase capacity, which can include adding a production line or building a new plant. When one considers high technology firms, such as semiconductor producers, it is understandable that expansion can take time. Electric power generation plants are likely to take over a decade to build and put online, as another example.

Clearly a lot can happen in the time between when an expansion is approved and when the project becomes operational. The realities of this were illustrated in Chapter 14 as potential delays in investments were investigated due to potential changes market conditions, including prices and demand. Our analysis in the following two sections will illustrate deterministic and probabilistic analyses in dynamic settings.

15.2.1 Dynamic deterministic evaluation. Our analysis of an expansion problem in a dynamic setting addresses the questions of: (1) when to expand and (2) how much capacity to add. Figure 15.1 illustrates a decision network where an expansion can take place in each period (over three periods) and there are two levels of expansion. Each path from the initial decision node to a terminal node represents a possible sequence of decisions with an associated cash flow diagram. The decision not to expand generally results in a present worth of zero, as only revenues and costs associated with the expansion are considered.

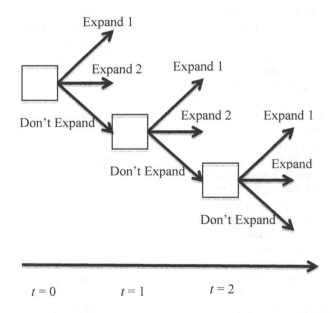

Figure 15.1 Decision network for expansion options over time

Figure 15.1 illustrates the difficulty with expansion analysis in that the number of options can grow considerably as the time horizon increases. The decision-maker must determine which set of decisions is most viable for analyses in order to pare down the feasible set of options. We illustrate the procedure in Example 15.1.

Example 15.1 Expansion decision network
An automobile production facility, with capacity to build 400,000 vehicles per year, can be expanded to a capacity of 600,000 vehicles per year at a cost of $500M. Assume per unit costs are $33,000 per vehicle with per vehicle revenues, from all sales, of $35,000 due to increased supply to the system. (If the expansion is not pursued, revenues will total $35,500 per vehicle.) Annual fixed costs are $200M without the expansion and increase by $7.5M with the expansion. The plant is expected to be in operation, with or without the expansion, for the next 8 years, with a salvage value of $50M without the expansion and $75M with it. The MARR is 20%. Finally note that the expansion takes 1 year to complete but does not interrupt current production.

The expansion can be considered immediately or at the end of either of the next two periods. Should the expansion commence, and if so, when?

Solution
The decisions available are drawn in the decision network in Figure 15.2. Note that the option to do nothing results in operations continuing with current capacity and revenue levels.

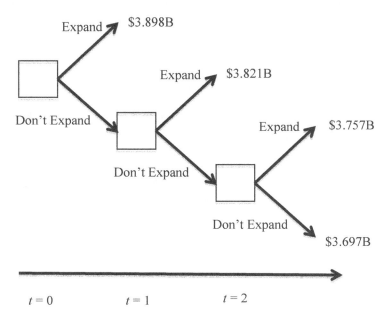

Figure 15.2 Expansion options for automobile assembly plant

The net cash flows and present worth of each alternative are given in Table 15.1. Note that $800M is received in each year until the expansion is complete.

456   Chapter 15, Capacity Expansion and Planning

Table 15.1 Cash flows and net present values ($M) for do-nothing and expansion alternatives

Year	Do-Nothing	Expand (0)	Expand (1)	Expand (2)
0	$800	$300	$800	$800
1	800	993	300	800
2	800	993	993	300
3	800	993	993	993
4	800	993	993	993
5	800	993	993	993
6	800	993	993	993
7	850	1068	1068	1068
PW(20%)	$3698	$3898	$3821	$3757

The option to expand immediately has the highest present worth and is the recommended choice. In fact, each expansion option is preferred to the do-nothing option, despite the lower per vehicle revenues.

15.2.2 Dynamic probabilistic evaluation. The extension to a probabilistic setting adds a chance node after each decision alternative. These chance nodes are followed by a number of possible outcomes with their associated probabilities (see Figure 15.3), which define an expected return. For capacity planning problems, typically the largest uncertainty is demand.

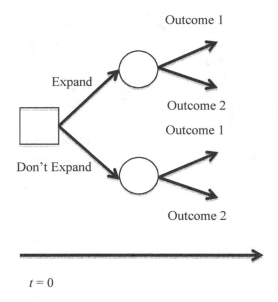

Figure 15.3 Decision tree for expansion option at time zero

15.2 Expansion Analysis

Example 15.2 Expansion decision tree
We continue our previous example but assume that either one of three scenarios will occur after the expansion: demand increases either by 100,000, 150,000, or 200,000 vehicles with estimated probabilities as given in Table 15.2. Note that without any expansion, demand is expected to meet current capacity of 400,000 vehicles per year. The probabilities of having better sales increase with time.

Table 15.2 Probabilities of annual sales levels given expansion time.

Expansion Time	Annual Sales		
	100,000	150,000	200,000
0	0.50	0.25	0.25
1	0.25	0.50	0.25
2	0.25	0.25	0.50

What is the best strategy for expansion, if any, given this added new data?

Solution
The decision tree is given in Figure 15.4. The decision to expand leads to a chance node with three arcs defining the level of sales (H for high, M for medium, and L for low).

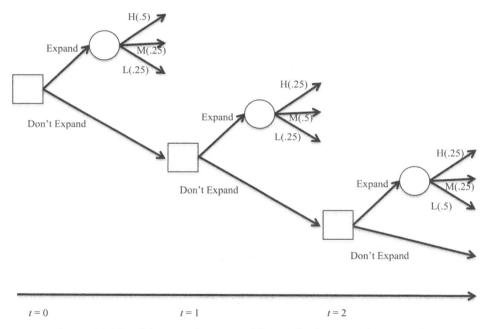

Figure 15.4 Decision tree for automobile production capacity expansions

As this is a decision tree with probabilistic arcs, we must traverse backwards in the tree. We begin with the final decision node of whether to expand at time period two or continue with operations (which assumes no expansions occurred earlier). If the expansion is to be taken, then the following present worth (valued at time period two) is received:

E(PW(20%)) = $800M − $500M + 0.25[$793M(*P/A*, 20%, 5)] + 0.25[$893M(*P/A*, 20%, 5)] + 0.50[$993M(*P/A*, 20%, 5)] + $75M(*P/F*, 20%, 5) = $3104M.

The present worth of the do-nothing alternative (at time period two) is computed as follows:

PW(20%) = $800M + $800M(*P/A*, 20%, 5) + $50M(*P/F*, 20%, 5) = $3213 M.

At this stage in the tree, the best decision is to do-nothing as it carries a larger present worth. Peeling back this value through the tree to time period one, the decisions are evaluated as follows:

E(PW(20%)) = $800M − $500M + 0.25[$793M(*P/A*, 20%, 6)] + 0.5[$893M(*P/A*, 20%, 6)] + 0.25[$993M(*P/A*, 20%, 6)] + $75M(*P/F*, 20%, 6) = $3318M.

For the do-nothing decision at time period one:

PW(20%) = $800M + $3213M/(1+0.20) = $3477M.

Again, the decision is to do-nothing as it leads to a higher present worth. Finally, the decision to expand at time period zero carries an expected present worth of:

E(PW(20%)) = $800M − $500M + 0.50[$793M(*P/A*, 20%, 7)] + 0.25[$893M(*P/A*, 20%, 7)] + 0.25[$993M(*P/A*, 20%, 7)] + $75M(*P/F*, 20%, 7) = $3469M.

For the do-nothing alternative:

PW(20%) = $800M + $3477M/(1+0.20) = $3698M.

The decision at time zero, and each ensuing period, is to do-nothing. Thus, the plant would be expected to continue to work at full capacity because an expansion is not justified at this time.

15.3 CAPACITY PLANNING STRATEGIES

We now take a more proactive approach to expansion decisions, as opposed to always reacting to demand. We do this by looking at expansion *strategies* which look forward in time based on expected demand patterns.

Recall that the risk associated with an expansion is that demand will not materialize and the investment could be lost. The risk associated with expansions can also be described by market timing. Consider demand that is expected to increase with time.

Demand generally builds continuously with time. However, capacity does not build continuously. This is due to the lead time to complete an expansion and capacity is added in discrete quantities. Figure 15.5 illustrates this point.

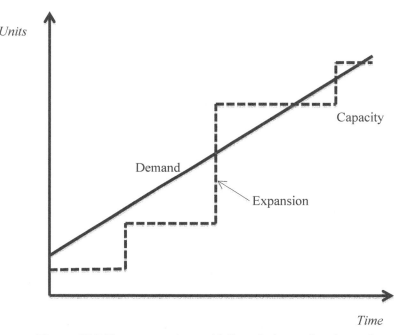

Figure 15.5 Three expansions with linearly increasing demand

The figure shows demand growing linearly with time. Over that time, the company expands its production capacity three times. The capacity added in the first and last expansion is the same, but the middle expansion is much larger. The figure illustrates the difficulty of the decision and its impact on the wealth generated, namely:

1. *There are numerous possible decisions.* The figure illustrates a few of the many possibilities with respect to the timing of and size of expansions.

2. *The difference between capacity and demand defines profitability.* If there is excess demand, then potential revenues are lost. If there is excess capacity, then investment is wasted. Risk can be measured as the difference between the demand and capacity curves in Figure 15.5 in that there are risks associated with lost returns and those with having excess capacity.

Figure 15.5 clearly simplifies reality, as firms can also use inventories and outsourcing in order to meet demand. However, it does provide a nice overview of options and decisions that they face.

We examine two general strategies defined by the firm's goal of wanting to either maximize *market share* or maximize *capacity utilization*. Figure 15.6 depicts these strategies. The left figure maximizes market share by ensuring there is always capacity to meet demand. The right figure maximizes capacity utilization by ensuring that capacity never exceeds demand. Clearly, more strategies can be defined but most other strategies can be viewed as some combination of these two strategies, as they represent the extreme approaches to capacity planning in that demand is always met or capacity is always fully utilized.

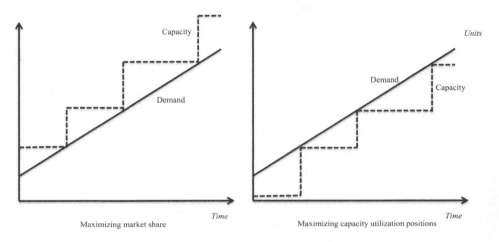

Figure 15.6 Expansion strategies (a) maximizing market share or (b) maximizing capacity utilization positions

The benefits of maximizing market share are: (1) First to market: This is often referred to as "first mover advantage" in that the first to market can command a higher price until the competition arrives. This can be critical in high technology sectors where margins (difference between revenues and costs) decrease drastically with competition, and where product life cycles are short. First to market also often results in a market share advantage that persists long-after competitors enter the market. Finally, larger volumes generally lead to declines in cost due to the learning curve. That cost advantage allows the first to market to have larger net revenues from the market price and that cost advantage can be a barrier to entry by competitors. (2) Maximizing revenue: Capacity is available to meet all potential demand.

The disadvantages to maximizing market share are (1) Early investment: in order to be able to meet demand, investments must be made earlier (than other strategies) due to lead times of bringing products to market. Given the time value of money and positive interest rates, this carries a present worth cost. (2) Greater risk exposure: investing for demand that is expected to materialize is risky as it may never materialize or conditions may change while capacity is being added.

Maximizing utilization takes on opposite attributes. The benefits are: (1) Delayed investment: investments can be made later, leading to decreased investment costs with respect to present worth; (2) Reduced exposure: as demand has already materialized, the risk associated with demand has diminished; and (3) Maximizes capacity utilization: factories are operating at maximum utilization which generally leads to greater efficiencies. The disadvantages are: (1) Last to market: as opposed to being first to market, those last to market cannot command higher prices; (2) Reduced flexibility: additional capacity provides flexibility for a company in that they can take on extra demand or convert it for other purposes. This is not possible if capacity has not yet been procured; and (3) Lower revenues: in addition to receiving lower prices, revenues are limited by capacity, not demand (as by definition, not enough capacity is available to meet all demand).

Again, these strategies represent the extremes in that demand is either always met or capacity is always utilized. Many strategies can be defined "in between" these extremes. We illustrate how we can put these strategies into practice in the next section. Figure 15.5 illustrated an expected demand curve with possible capacity expansions over time. These curves provide a basis for our approach of defining expansion decisions over some horizon. Consider a similar graph in Figure 15.7.

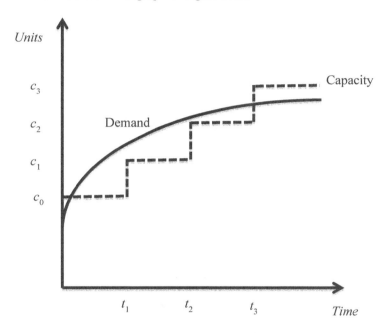

Figure 15.7 Expected demand curve and capacity levels

The graph illustrates demand growing according to some bounded geometric function over some time horizon N. Additionally, three expansions occur at times t_1, t_2, and t_3. The amount of capacity added is equal to the incremental increase over the previous level of capacity.

The actual revenues earned are defined by the minimum of the capacity and demand curves. The cash flows associated with this area include revenues from sales and costs that can be attributed to per unit sales. The capacity carries fixed costs that are incurred regardless of demand levels, such as overhead costs. Each expansion carries the cost to add capacity, which generally depends on the amount of capacity added. For areas where demand exceeds capacity, there may be an opportunity cost due to lost sales—although this value is difficult to capture in a cash flow.

If we can assign a value to each of the revenues and costs defined by the curves, then it may be possible to determine the optimal expansion policy given an expected demand curve. Let us first consider this without concern for the mathematics. Our goal is to maximize present worth, as defined here:

NPV = Expected Sales (Revenue per unit − Cost per unit) − Fixed Capacity Costs
− Expansion Costs

The profit from expected sales is merely defined by the area under the demand curve that intersects the area under the capacity curve in Figure 15.6. The fixed capacity costs are defined by the area under the capacity curve and the expansion costs are defined by the timing and amount of capacity expansions. Thus, we could use differential calculus to maximize the present worth subject to the costs defined by the difference between the curves.

Unfortunately, optimizing this function can be very difficult. Consider the number of decision variables: timing of expansions t_1, t_2, and t_3 and the amount of capacity added resulting from each expansion: c_1, c_2, and c_3.

In order to make the problem tractable (solvable), we can make some simplifying assumptions. These may not exactly match reality, but they will provide some insight when planning for the future. The assumptions are:

1. *Expansions occur at regular intervals*. This reduces the timing decisions t_1, t_2, etc., to one variable t.
2. *An expansion strategy (maximizing market share or capacity utilization) is to be followed*.
3. *A plant remains in operation through the horizon*, which may be infinite.
4. *The demand function is known* and allows for a solution.

We explicitly analyze the case of linearly increasing demand in the following two sections.

15.3.1 Maximizing market share strategy. Mathematically, maximizing market share is straightforward: minimize the difference between the demand curve and the capacity curve while not allowing the capacity curve to dip below demand. Because demand generally increases continuously or in small increments, while capacity increases in larger increments—these curves will not match.

Consider the curve in Figure 15.6 which illustrates linearly increasing demand over time. Assume the slope is defined as G. If we are to always have enough capacity to meet this demand, which defines our strategy, then we must increase capacity by G each t periods of time.

Figure 15.8 provides our traditional graph with demand and capacity over time, assuming a maximizing market share strategy is being followed. Below the graph, a cash flow diagram for these decisions is provided.

The cash flow diagram has three components. (1) Each time an expansion occurs, an investment cost P is incurred. (2) Net revenues from sales, defined as the difference between per unit revenues and per unit production costs, are defined by the demand curve, as demand is completely met in this strategy. The cash flows from net revenues are defined by an arithmetic gradient for each period t. The gradient increases at the rate G per unit of time. (3) Finally, a fixed cost for each expansion (for items such as overhead costs). These appear as steps in cash flow diagram as they are related to the capacity of the system. We do not include the effects of inflation or taxes, but the simplifications are required in order to make the model tractable (solvable). (Note that while these costs are occurring over time, we approximate them as discrete flows in the cash flow diagrams.)

15.3 Capacity Planning Strategies

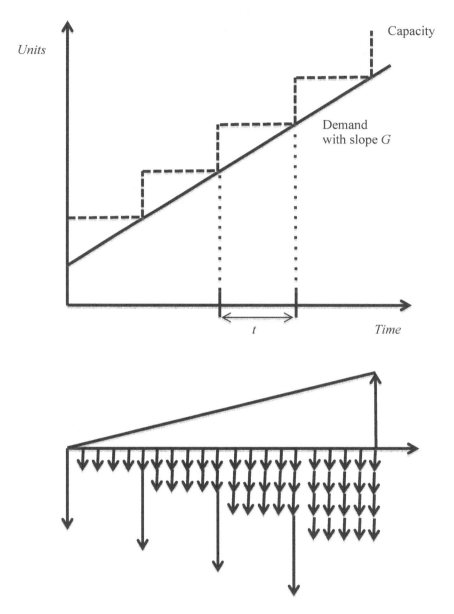

Figure 15.8 Expansion strategy of maximizing market share

To analyze the cash flow diagram, we must make a few assumptions and observations. First, it must be noted that because we are following a market share maximizing strategy, the revenues from operations are the same over time, regardless of the timing of the expansions because capacity always exceeds demand. *This greatly simplifies the problem as we must only minimize costs.*

Another simplification comes from our choice of horizon and length of time that a facility remains operational. We assume continuous compounding and infinite horizons which can often greatly simplify analysis—as will be shown soon. Furthermore, we assume that capacity, once installed, remains through our horizon.

The investment cost P is some function of Gt. We assume a power law and sizing model (see Chapter 3) with some economies of scale for greater sized expansions, such that:

$$P_p = p_0(Gt)^m, \quad 0 < m < 1$$

and p_0 is our reference cost per unit of capacity. This is a single cash flow that occurs at times $0, t, 2t, \ldots$.

Under the assumption of capacity remaining available indefinitely, the associated annual fixed charges also carry on forever. Assume the fixed operating and maintenance charges are a function of the expansion size, or:

$$A = c_0(Gt).$$

As this value is repeated without end, we can write the cost's equivalent present value as:

$$P = \lim A[(e^{rN} - 1)/(re^{rN})] = A/r.$$

Thus, the equivalent value associated with the operating costs is equal to:

$$P_c = c_0(Gt)/r,$$

at the time of the expansion. Thus, the total cost at time t for an expansion at time t is:

$$P = P_p + P_c = p_0(Gt)^m + c_0(Gt)/r$$

This value of P occurs at time period zero, time t, time $2t$, and so on. Each of these can be discounted to time zero. Thus, the total net present value cost for all expansions over the infinite horizon is:

$$PW(r,t) = P[1 + e^{-rt} + e^{-2rt} + \ldots] = [p_0(Gt)^m + c_0(Gt)/r\,][1/(1 - e^{-rt})]. \quad (15.1)$$

As r is defined, the only decision variable is t. Our goal is to find the value of t that minimizes $PW(r,t)$.

Using traditional methods from calculus, we could take the derivative of $PW(r,t)$ with respect to t and set it equal to zero. Assuming that the second derivative is negative, we would find the value of t that maximizes of $PW(r,t)$.

Examining our present worth equation, this might be a bit tedious. However, we could easily set up a spreadsheet to search for the appropriate value. Additionally, setting up a spreadsheet will allow us to perform sensitivity analysis on the critical parameters. We illustrate in the following example.

Example 15.3 Maximizing market share expansion strategy

Production capacity for a high-tech plastic is to be built incrementally using a maximizing market share strategy. Market demand is growing at the rate of 50,000 tons per year. The cost to add capacity follows a power law and sizing model with $m = 0.8$ and it costs \$150M to build 50,000 tons of annual capacity. Furthermore, operating costs are estimated at \$15M per year per 50,000 tons of annual capacity. If the nominal interest rate is 16% compounded continuously, how often should expansions take place and how large should they be?

Solution
The data translates into the following parameters:
G = 50,000 tons per year
p_0 = \$150M/50,000 tons = \$3000/ton
c_0 = \$15M/50,000 tons = \$300/ton
and r = 16%.

Plugging these values into Equation (15.1) results in:

$$PW(r,t) = [\$3000(50{,}000 t^{0.8})] + [\$300(50{,}000)t/0.16][1/(1 - e^{-0.16t})].$$

We can either search for a solution which maximizes $PW(r,t)$ (using SOLVER in EXCEL) or we can take the derivative and find t such that the derivative is zero. We use SOLVER and find that $t = 1.078$ years, or about every 13 months. This defines the amount of capacity expansion as 50,000 tons/year times 1.078 years = 53,900 tons of annual capacity.

Table 15.3 Expansion times and costs given different economies of scale

m	t (years)	$PW(r,t)$
0.9	0.67	\$1314.0
0.8	1.078	1097.0
0.7	1.231	954.8
0.6	1.22	844.3
0.5	1.131	776.8
0.4	1.015	731.1
0.3	0.899	699.2
0.2	0.79	676.5

Table 15.3 provides the solutions for a number of different economies of scale parameters. Remember that $m = 1$ is no economies of scale, and the smaller m is; the smaller is the cost of a larger facility. Note that there is a tradeoff in the choice of t. As t gets larger, greater savings can be achieved through economies of scale. However, as t gets larger, the influence of discounting becomes greater—especially as there is an investment at time zero which has the greatest impact on the decision. Thus, while m decreases, the present worth cost of the decision decreases but the time between expansions does not strictly increase.

It should be noted that the total present worth of the revenues can be found by estimating the growth in revenues according to the demand via an arithmetic gradient. For example, after one expansion, the revenues grow to size Gt at time t. Let us assume that this captures the demand for the period. After the second expansion, they grow to $2Gt$ and so on. Thus, the present value of total revenues over the infinite horizon is:

$$PW(r,t) = Gt[e^{-rt} + 2e^{-2rt} + 3e^{-3rt} + \ldots] = Gt[e^{-rt}/(1-e^{-rt})^2]$$

It is clear that we have made some simplifying assumptions in pursuing an optimal expansion strategy, but this is an excellent start for a company wishing to plan capacity decisions out into the foreseeable future. We examine our other extreme strategy next.

15.3.2 Maximizing utilization of capacity strategy. Maximizing capacity utilization is similar to the market leader strategy in that we want to minimize the difference between the demand curve and the capacity curve, but here we do not allow the capacity curve to ever exceed demand.

Analyzing this problem with cash flows follows directly from the previous analysis for the market share strategy. In the market share strategy, as soon as demand equals capacity, then more capacity is added—at the beginning of period t. When maximizing utilization, more capacity is not added until it can be fully used—at the end of period t. The difference is that investment is delayed for one period t. This results in a shift of the cash flows.

Additionally, net revenues from sales do not increase in between capacity expansions, but rather mimic the periodic fixed costs with each expansion in the market share strategy. This is because revenues are limited by capacity.

The demand and expansion curves for this strategy are given in Figure 15.6 along with the corresponding cash flows. The converted discrete cash flows are given in Figure 15.9.

When compared to the market share strategy, the investment costs occur at the end of period t rather than the beginning of period t. We could ignore revenues in the previous strategy, but that is not possible here as we do not fully meet demand. Thus, revenues depend on the expansion size. As we have previously defined costs according to capacity, we can define a similar present worth equation for costs over the horizon as:

$$PW(r,t) = P[1 + e^{-rt} + e^{-2rt} + \ldots] = [-p_0(Gt)^m + (r_0 - c_0)(Gt)/r][e^{-rt}/(1-e^{-rt})]. \quad (15.2)$$

Our goal now is to maximize the net present value as revenues are included. This is actually quite different from our previous strategy. Here, the tradeoff is between the economies of scale, such that the investment cost decreases if the time between expansions increases, and revenues, which increase if the time between expansions decreases. We illustrate by reexamining the previous example.

15.3 Capacity Planning Strategies

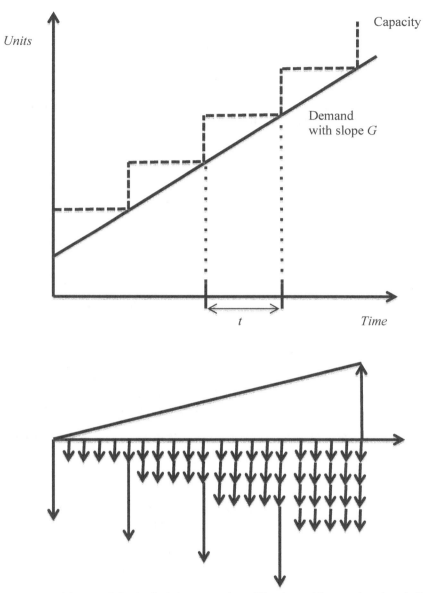

Figure 15.9 Maximizing capacity utilization with associated cash flows

Example 15.4 Maximizing capacity utilization expansion strategy
We resolve our previous example assuming *net* revenues of $200 per unit ton. All other data is as previously.

Solution
Plugging our data into Equation (15.2) results in:

$$PW(r,t) = [-\$3000(50{,}000t^{0.8}) + \$200(50{,}000)t/0.16][e^{-0.16t}/(1-e^{-0.16t})].$$

As before, we examined a number of values of m. The results from maximizing $PW(r,t)$ are given in Table 15.4.

Table 15.4 Expansion times and costs given different economies of scale

m	t (years)	$PW(r,t)$
0.8	6.54	$54.5
0.7	3.54	142.9
0.6	2.50	207.2
0.5	1.93	252.0
0.4	1.55	283.5
0.3	1.17	458.6
0.2	1.07	322.5

Note that the revenues have more influence on the decision with these parameter values as a decrease in the value of m leads to a lower value of t, which is contrary to economies of scale. Thus, while the investment costs go down with economies of scale, it is still better to have shorter times between expansions in order increase revenues.

While these two strategies simplify what a company can do over time, they provide a method for examining future expansions. Intermediate strategies between these extremes, more complicated demand functions, or a finite horizon with a finite number of expansions can also be analyzed to provide more insight at the cost of computational complexity.

15.4 SUMMARY

Capacity expansion decisions can be modeled with decision networks for deterministic data or with decision trees for stochastic data. The decision to expand capacity is difficult because it entails two decisions, including when and how much capacity to add.

Two strategies for exploring expansion planning are (1) maximizing market share, by ensuring that capacity is always ahead of the demand curve, or (2) maximizing capacity utilization, in which capacity lags demand. The first strategy is characterized by a company having first mover advantage while the second strategy is characterized by fully utilizing capacity. Economies of scale generally promote larger expansions while discounting assures a tradeoff of when capacity is made available.

The literature on capacity expansion and planning is vast and continuously growing. A significant amount of that literature is dedicated to certain scenarios, such as tooling considerations or semiconductor manufacturing (which relies heavily on re-entrant flow, complicating capacity issues). Wu et al. (2005) provides an overview for high-tech industries. For more general modeling, research has sought to expand upon the deterministic work (Smith, 1981) to include risk and uncertainty [see, for example, Birge (2000), Eppen et al. (1989) and Ryan (2004)].

REFERENCES

BIRGE, JOHN R., "Option Methods for Incorporating Risk into Linear Capacity Planning Models," *Manufacturing and Service Operations Management*, **2**(1) (2000).

EPPEN, GARY D., R. KIPP MARTIN, and LINUS SCHRAGE (1989), "A Scenario Approach to Capacity Planning," *Operations Research*, **37**(4) (1989), pp. 517–527.

MANNE, A. S. (Ed.), *Investments for Capacity Expansion: Size, Location, and Time Phasing*, (MIT Press, 1967).

NAHMIAS, S., *Production and Operations Analysis*, Fourth Edition, (McGraw-Hill Irwin, 2001).

PARASKEVOPOULOS, DIMITRIS, ELIAS KARAKITSOS, and BERC RUSTEM (1991), "Robust Capacity Planning under Uncertainty," *Management Science*, **37**(7) (1991), pp. 787–800.

RYAN, SARAH, "Capacity Expansion for Random Exponential Demand Growth with Lead Times," *Management Science*, **50**(6) (2004), pp. 740–748.

SMITH, R. L., "Optimal Expansion Policies for the Deterministic Capacity Problem," *The Engineering Economist*, **25**(3) (1980), pp.149–160.

WU, S. DAVID, MURAT ERKOC, and SULEYMAN KARABUK, "Managing Capacity in the High-Tech Industry: A Review of Literature," *The Engineering Economist*, **50**(2) (2005), pp. 125–158.

PROBLEMS

15-1. Why is the decision to expand difficult to analyze?

15-2. What are the risks associated with expansion?

15-3. What are two extreme strategies for expanding capacity? Contrast the risks associated with each.

15-4. Should firms ever follow the extreme strategies of capacity expansion? Explain.

15-5. A semiconductor manufacturing firm specializing in flash memory is considering an expansion from 9000 to 18,000 wafers per quarter. Assume that one wafer generates revenues near $500 against costs of $200. Further, assume periodic costs (quarterly) of $1M before the expansion and $1.25M after. Assume the plant runs at full production for another three years. Evaluate expanding during 2004. Regardless of when the expansion occurs, production is expected to continue at capacity through the end of 2006 with a salvage value of the additional capacity of $5M at that time. The cost of the expansion is $25M. The MARR is 5% per quarter.

(a) Assume $500 revenues per wafer in each quarter in years 2 and beyond. Revenues of $525, $520, $515, and $510 are available in the four consecutive quarters of Year 1 as competition also increases capacity. When should the expansion occur?

(b) Assume that the revenues are highly dependent on the competition, such that they have a 0.60 probability of following the expected price path in (a) or a 0.40 probability of being $500 over time. Does this information change the decision?

15-6. A microchip production firm is spending $9M to expand its design center with 10,000 additional square feet and 30 additional workers researching memory in cell phones and other digital devices. If the annual cost per employee is $150,000 (salary plus burden) and all other costs are $10 per square foot per year, what must the annual contribution of the group (in dollars) be in order to facilitate the expansion, assuming a 15% annual interest rate and five year horizon with no salvage value?

15-7. Assume each employee contributes about $250,000 in revenue in the previous problem. Using this number and the previous data, what is the present worth of adding 10 people in each of the first three periods of the horizon? Note that the $9M of the building is still spent at time zero. What is the new breakeven revenue? When might this be a viable strategy to consider?

15-8. A pulp and paper producer is looking to expand production. Assume the time zero capacity is 120,000 tons of cellulose and 80,000 tons of paper per year. Further assume that capacity expansions must be in increments of 20,000 tons and that cellulose and paper expansions occur at the same time (and in the same amount).

(a) Assume a five year horizon and that the expansion(s) can occur in any of the three periods. If capacity cannot exceed 180,000 tons per year for cellulose and 140,000 tons per year for paper, what are the total number of possible expansion plans over the five years?

(b) Restrict the options to expanding small (20,000 tons), medium (40,000 tons), or large (60,000 tons) at either time zero or time one. How many possible expansions are there now? Draw the decision network to verify your solution.

(c) Consider the previous problem. Assume the cost is $12M to install the small capacity, $22M the medium, and $30M for the large. If revenues are expected to be $400 per ton for cellulose and $600 per ton for paper, against per unit costs of $375 and $425, respectively, in addition to annual costs of $1M per year per 20,000 tons of expansion, what expansion strategy (if any) should be chosen? Assume no salvage value after five years and an interest rate of 11%.

(d) Assume the revenue per ton of paper and cellulose is $600 and $400, respectively, during the first period, and grows by $20 per pound over the horizon time. Does this change the solution to the previous problem?

(e) Restrict the decisions to be studied to either a large expansion or small expansion at time zero and, if the small expansion is taken, a medium expansion at time one. Further assume that the revenue per ton of paper and cellulose is $600 and $400, respectively, during the first period, and grows by either $20 per pound (probability of 0.50) or stays the same in each ensuing period. What decision maximizes expected present worth?

15-9. The demand for power is growing at a continuous rate of 100MW per year. An electricity producer is examining its expansion policy to meet demand. The cost to expand can be approximated by the power law and sizing model with an exponent of 0.75. Assume the investment cost of a 500 MW unit is $230M and the cost to operate is $5M per 100MW. Further assume a continuously compounded interest rate of 13% per year.

(a) Determine the optimal time between expansions assuming a market share maximization strategy is being followed.
(b) Perform a sensitivity analysis on the exponent m for the above analysis.
(c) If the revenue per MW is $700,000, determine the optimal capacity utilization strategy.
(d) Perform a sensitivity analysis on the per unit revenue for the previous problem.

16
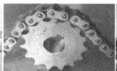

Project Selection Using Capital Asset Pricing Theory

16.1 INTRODUCTION

This chapter introduces the fourth fundamental approach to the project selection problem. The first approach, presented in Chapters 7–10, maximizes the firm's *expected return* (net present value), using an assumed known interest rate to discount future cash flows to present equivalents. The second approach, outlined in Chapter 9, assumes a known budget rather than a known interest rate. Thus, this approach ranks on internal rate of return and determines the minimum attractive rate of return as part of the model. The third approach, presented in Chapters 11–15, maximizes the *expected utility* of a set of future returns at an assumed known discount rate. This approach recognizes the *variability* of project cash flows, which is inherent in every project selection problem.

The fourth approach, originally conceived by Sharpe (1963) and Lintner (1965), is the *capital asset pricing model* (CAPM). The CAPM links the capital supply and allocation problems and develops a security market line for stocks and financial markets. While the derivation of this model is presented in this chapter, it was used earlier in Chapter 6 as one of the models for the cost of equity.

A multi-factor model called *arbitrage pricing theory* (APT) was developed by Mossin (1966) and Ross (1976), but relies more heavily than the CAPM on actions, such as selling short, that are not possible in the project selection problem. For financial markets the CAPM approach has been generalized to include multiple factors, such as firm size (Levy, 1978). However, multiple factors models that consider the state of the market (depression, rising, etc.) have not yet been extended to the project selection problem. This is an active research area, but as described in Section 16.5 extensions from financial markets to industrial projects are challenging at best. Even so, the CAPM and portfolio theory do provide useful suggestions about the selection of industrial projects.

The CAPM model differs from the cardinal utility model in two respects. First it assumes a generalized risk-avoidance posture for the owner instead of a specific, quantifiable utility function, such as the von Neumann–Morgenstern model of Chapter 11. The CAPM's generalized risk-avoidance posture merely requires that the owner require a proportionate increase/decrease in the expected return of a project, if the firm's risk exposure would be increased/decreased by accepting the project.

Secondly, the CAPM assumes that the firm's marginal cost of capital is a *random variable* instead of a known numerical constant, as does the utilitarian approach. In the CAPM, the mean and variance of the firm's random cost of capital are established by (1) the randomness of the securities market, and (2) the corelationships between the firm's dividend policies, fluctuations in the market price of the firm's shares, and the randomness of the financial market. These corelationships are given effect by and through the free market pricing of the financial markets.

The capital asset pricing approach when applied to financial assets has several practical advantages, which can be visualized by contrasting some of the assumptions and features of the CAPM to those of the deterministic and utilitarian methods. The deterministic approach obviously suffers from the defective assumption that all future events are known; including complete knowledge of what the projects will be (in all details), the depreciation models and income tax rates, the market rate of interest, and so forth. While deterministic models are useful in the firm's investment processes, they simply do not match reality.

Moreover, both the utilitarian and deterministic models depend on the assumption of a known discount rate. This discount rate is usually based on the firm's cost of capital, which implicitly allows for risk and the time value of money. Risk allowance is part of the total rate because of how the cost of capital is based on the market price of the firm's equity shares and an estimated stream of future dividends.

The dividend-to-price ratio is fixed by the securities market, and that includes an implicit collective evaluation of the riskiness of the firm's shares. While future cash flows should be discounted to their present values with a compounded *time-value-of-money* interest rate (Fisher's theory of interest), there are serious questions about compounding the risk portion of the cost-of-capital rate. That compounding may seriously distort the actual riskiness of future cash flows. Furthermore, in the utility models, the discount factors for calculating the variance of NPV are of the form $(1+i)^{-2t}$. Questions were raised in Chapter 12 about counting risk twice: once in the interest rate (i) itself in the cost-of-capital calculation, and then again in the second-order exponent ($2t$) of the variance calculation.

Other defects of the utilitarian approach are the assumptions that the utility function (1) is of known (or assumed known) form, (2) is derived from a single decision-marker (group utility functions can be shown to be the result of collective bargaining or gamesmanship rather than a preference ordering), and (3) is unchanging in the time dimension (which is contrary to observed empirical results).

Thus, for financial assets, another normative model is required—capital asset pricing theory. When we look at industrial projects, the weaknesses of each approach mentioned above—are still true. However, CAPM has an even larger weakness. For financial assets there are substantial databases of past performance, which are used to estimate the co-relationships between the different assets and the markets. Since industrial projects have only estimates of future cash flows and no history, it is not possible to calculate the corelationships which are fundamental to the CAPM. Nevertheless, we will find that there are useful lessons for industrial project selection that can be based on the CAPM applied to financial assets.

Markowitz (1959) first suggested that the return rates of pairs of securities, or of portfolios of securities, are correlated and that calculating these correlative effects could be considerably simplified by assuming that each security is correlated solely with a *marketwide* portfolio. This shortened calculation of an empirical, market-based cost of capital effectively integrates the firm's long-run project investment and dividend policies, thereby providing a criterion for selecting future investment projects.

The starting point is Markowitz' concept of a securities *portfolio* and how the variance of the portfolio return is affected by *covariation* among its securities. These relationships are used to develop *efficient* portfolios that dominate other portfolios. Next, we consider how portfolios of arbitrary composition can be coupled with borrowing or lending at a risk-free interest rate to establish the *security market line*—the locus of choices that maximize the investor's expected utility between the risky, arbitrary portfolio and the riskless security. This is Sharpe's CAPM which establishes the risk trade-off relationship between a given security and an arbitrary portfolio. Finally, recommendations for project selection based on the capital asset pricing model are made.

16.2 PORTFOLIO THEORY

16.2.1 Securities and portfolios. A *security* is generally a written, legal instrument (such as a bond or share of stock) that exchanges invested funds for a right to receive future cash flows, if and when produced. It may vest ownership in a share of the assets owned by the security's issuer. Some risk comes with owning a security because there is no guarantee that the issuer (except an implicit reputation or perhaps the worth of pledged assets) will either pay income (interest or dividends) on the security or repay its purchase price, or its future equivalent. Thus, the riskiness of securities to investors depends largely on who the issuer is. Shares of stock in Flybynight Commuter Airlines are more risky than shares of Microsoft common stock because of differences in the general uncertainty attending the *dividends* and *earnings prospects* of both. On the other hand, bonds of the United States Government are considered to be nearly riskless—or, to use better terminology, virtually default-free. The reason is obvious: The Government can tax and issue money. If inflation makes that repayment of little value to the security's owners, other securities are likely to have little value as well.

The set of securities owned by an investor is a *portfolio*, and its present value is simply its total present market price. The ratio of future income (interest, dividends, or other benefits) to present price is the *yield* or return of the portfolio.

Securities prices fluctuate moment by moment, hour by hour, day by day, and over longer intervals of time. Thus, securities prices form a time series and the prices of different securities are correlated *random variables*. The CAPM assumes that the random yield of either a single security or a portfolio can be completely characterized by specifying (or determining) the mean and variance of its yield.

The capital asset pricing model either explicitly or implicitly incorporates these other assumptions:
1. All investors are single-period expected utility-of-wealth maximizers, who choose among securities or portfolios on the basis of mean and variance (or standard deviation) of return. The dollar return increases (decreases) net present value.
2. All investors can borrow or lend an unlimited amount in the financial market, at an exogenously fixed (constant) default-free rate of interest, R_f, and there are no restrictions on the buying or short selling of any security.[1]
3. All investors have identical estimates of the means, variances, and covariances of

---

1. Borrowing or lending money can be viewed bringing cash in or sending it out. Buying a security is a cash outflow in return for future cash inflows. Selling a security *short* brings cash in now for the promise to deliver the security later—which requires a cash outflow to purchase the security for delivery.

return among all securities and portfolios. This precludes differing subjective evaluations of the securities because of a diversity of ownership.
4. All securities and portfolios are perfectly divisible (fractional shares are permissible) and perfectly liquid—all securities are marketable and there are no transaction costs.
5. There are no income taxes.
6. All investors are price takers (*no* single investor buys or sells enough to change the market price).
7. The quantities of all securities holdings are given.

16.2.2 Mean and variance of a portfolio. Consider a portfolio consisting of n securities, where n is a finite integer. Since securities are considered to be perfectly divisible and any desired amount of money can be invested in each security, any particular portfolio can be described by the *relative* amount of money or weight, w_i, invested in each security. For example, consider a portfolio consisting of $n = 5$ securities.

Security #, i	Relative Amount of Funds Invested, w_i
1	0.10
2	0.40
3	0.10
4	0.25
5	0.15
	1.00

The proportion of funds invested in security 1 is $w_1 = 0.10$; in security 2, $w_2 = 0.40$; and so forth. The relative amounts invested must sum to one, or:

$$\sum_{i=1}^{n} w_i = 1 \qquad (16.1)$$

If any $w_i = 0$, the portfolio contains none of security i. If $w_i > 0$, then security i is held in the portfolio; but if $w_i < 0$, then the portfolio's holder has sold the security short or issued security i by agreeing to pay someone else a portion of the portfolio's earnings. A portfolio is the set of n securities, held and issued in proportions w_i, such that Equation (16.1) is satisfied.

The portfolio's mean return is simply the weighted average return of the composite securities. Let $\bar{R}_P$ denote the mean return rate on the portfolio and $E(R_i)$ the mean return rate on security i. Then the mean portfolio return rate is given by:

$$E(\bar{R}_P) = \sum_{i=1}^{n} w_i E(R_i) \qquad (16.2)$$

The variance of the portfolio's return rate is based on the very realistic assumption (originally proposed by Markowitz) that the return rates of marketed securities are all *pairwise* correlated. Hence, this variance is not the simple sum of the variances of individual securities—true only if the returns on the securities were independent. Suppose that R_i and R_j are the return rates from two securities, i and j, that are held in proportions w_i

and w_j, respectively. Then, from the definition of variance for sums of random variables (derivation can be found in any good statistics text) we have:

$$\text{Var}(R_P) = \text{Var}\sum_{i=1}^{n} w_i R_i = \sum_{i=1}^{n} w_i^2 \text{Var}(R_i) + 2\sum_{i=1}^{n-1}\sum_{j>i}^{n} w_i w_j \text{Cov}(R_i, R_j) \quad (16.3)$$

16.2.3 Dominance among securities and portfolios. Since each security or portfolio can be measured by the mean and standard deviation of its return, each one can be described as a point in ($\overline{R}_P$, σ_P) coordinates, such as Figure 16.1. In this figure, to determine which of three possible securities (*A, B,* or *C*) would be preferred by a typical risk-avoiding decision-maker, the following rules can be invoked.

1. If two securities have the same standard deviation of return rate, such as *A* and *B*, the one with the larger expected return rate (*B*) is preferred.
2. If two securities have the same expected return rate, such as *B* and *C*, the one with the smaller standard deviation of return rate (*C*) will be preferred.

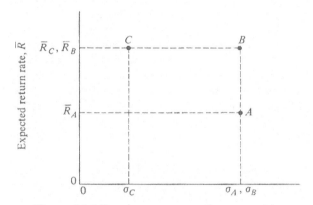

Figure 16.1 Dominance among three securities

Thus, in Figure 16.1, *B* is said to *dominate A*; likewise, *C* dominates *B*. Furthermore, it should be noted that dominance is commutative, so that if *B* dominates *A* and *C* dominates *B*, then *C* dominates *A* also. In general, securities and portfolios lying upward and to the left on ($\overline{R}_P$, σ_P) coordinates will dominate those lying downward and to the right.

16.2.4 Efficient portfolios. These portfolios in the ($\overline{R}_P$, σ) plane dominate all other possible portfolios. To illustrate, let us start with a portfolio composed of only two securities with respective proportions of w_1 and $w_2 = 1 - w_1$ (note that $\Sigma w_i = 1$). We also use the subscripts to identify the respective return rates and variances. Then the expected return rate for the portfolio is:

$$\overline{R}_P = \Sigma_i w_i E(R_i) = w_1 E(R_1) + w_2 E(R_2). \quad (16.4)$$

The Cov $(R_1, R_2) = \rho_{12}\sigma_1\sigma_2$, where ρ_{12} = the Pearson correlation coefficient ($-1 \leq \rho_{ij} \leq$

+1). Then Equation (16.5) is the two-variable version of Equation (16.3).

$$\text{Var}(R_P) = \sigma_p^2 = w_1^2\sigma_1^2 + 2w_1w_2\rho_{12}\sigma_1\sigma_2 + w_2^2\sigma_2^2 \qquad (16.5)$$

Several interesting results flow from this fundamental equation. First, we note that if $\rho_{12} = +1$, then the perfect positive correlation between R_1 and R_2 reduces Equation (16.5) to:

$$\sigma_p^2 = w_1^2\sigma_1^2 + 2w_1w_2\sigma_1\sigma_2 + w_2^2\sigma_2^2 = (w_1\sigma_1 + w_2\sigma_2)^2 \qquad (16.6)$$

In Figure 16.2, Equations (16.4) and (16.6) define the straight-line that connects the points for securities 1 and 2. If w_1 equals 1, then our portfolio consists solely of security 1 with its return and standard deviation. If w_2 equals 1, then our portfolio consists solely of security 2 with its return and standard deviation. If securities 1 and 2 are perfectly correlated, then as shown by the straight-line in Figure 16.2 and Equation (16.6), our portfolio's return and standard deviation are weighted averages of the values for the securities.

As the correlation coefficient, ρ_{12}, decreases from +1, the relationship between $\overline{R}_P$ and σ_P is nonlinear (see Figure 16.2) until finally *two* values of expected return rate can occur for the same portfolio standard deviation. If two possible values occur, investors will always prefer the portfolio with the higher expected return rate, the *efficient portfolio* (point *A* in Figure 16.2), which dominates the *inefficient* portfolio *B*.

If two perfectly negatively correlated securities could be identified, then it would be possible to combine them to produce a risk-free rate of return. In Figure 16.2 the line segments from the vertical axis to point 1 and to point 2 intersect at the resulting risk-free rate of return. Substituting −1 into Equation (16.5) leads to Equation (16.7).

$$\sigma_p^2 = w_1^2\sigma_1^2 - 2w_1w_2\sigma_1\sigma_2 + w_2^2\sigma_2^2 = (w_1\sigma_1 - w_2\sigma_2)^2 \qquad (16.7)$$

For a variance and a standard deviation of zero, we set Equation (16.7) equal to zero and solve. Thus, if securities 1 and 2 are perfectly negatively correlated then setting $w_1\sigma_1 = w_2\sigma_2$ results in a risk-free portfolio with the return given by Equation (16.4), and the line segment from the vertical axis to the coordinates of security 1 is the efficient portfolio.

A third security could be added to the set, and portfolios could be formed by combining securities 1 and 3 or securities 2 and 3, which the solid lines indicate in Figure 16.3. Furthermore, intermediate portfolios combining all three securities represent the shaded area in Figure 16.3.

Now, the dominance effect and indifference curves eliminate all but a few portfolios from the feasible set. The remaining efficient portfolios lie on the arc between points *D* and *E* in Figure (16.3) (all others are inefficient). To reach this conclusion, we recognize two limiting indifference curves. A decision-maker who cares only about risk will minimize the risk in the portfolio by choosing point E. A decision-maker who cares only about return will maximize the return in the portfolio by choosing point D. All other decision-makers will choose a portfolio along the arc that maximizes their utility. Thus, in Figure 16.4, decision-maker *G* maximizes expected utility by choosing portfolio G_{max}, while H maximizes expected utility by choosing portfolio H_{max}.

16.2 Portfolio Theory

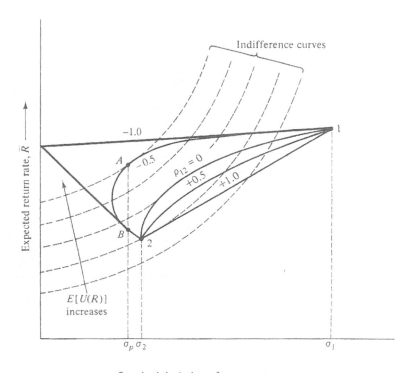

Figure 16.2 Dominance in a two-security portfolio

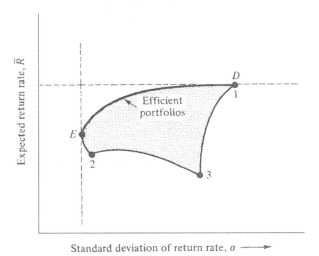

Figure 16.3 Efficient portfolios lie on D to E

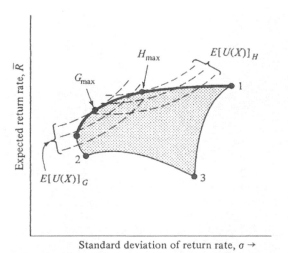

Figure 16.4 Efficient portfolios and indifference curves for utility

16.2.5 The risk in a portfolio. The previous sections have shown that *diversification* in a portfolio reduces risk—as long as the securities are *not* perfectly correlated. The reduction is substantial as the number of securities increases and as the correlations approach zero, and even more substantial if there are negative correlations.

Individual securities will have their own variances, and each pair will have their own covariance. However, as the number of securities in the portfolio increases, these differences will tend to average out. Thus it is useful to examine Equation (16.3) under the assumption that the portfolio contains n equally weighted securities all of which have an average variance and an average covariance with other securities. Under these assumptions each weight is $1/n$, and each standard deviation, σ_{avg}, and covariance, $\sigma_{ij,\,avg}$, is the same, so they are constants that can be moved outside the summation. Equation (16.8) can be derived from as Equation (16.3) as follows:

$$\sigma_p^2 = \sum_{i=1}^{n} \frac{1}{n^2} \sigma_{avg}^2 + 2 \sum_{i=1}^{n-1} \sum_{j>i}^{n} \frac{1}{n}\frac{1}{n} \sigma_{ij,avg}$$

$$\sigma_p^2 = n\left(\frac{1}{n^2}\right)\sigma_{avg}^2 + 2\left(\frac{n(n-1)}{2}\right)\left(\frac{1}{n^2}\right)\sigma_{ij,avg}^2$$

$$\sigma_p^2 = \left(\frac{\sigma_{avg}^2}{n}\right) + \left(\frac{(n-1)}{n}\right)\sigma_{ij,avg}^2 \tag{16.8}$$

As n increases the first term in Equation (16.8) goes to zero, which means the basic variance of each security matters little in a well-diversified (large) portfolio. The coefficient of the second term goes to 1. Thus, as the size of the portfolio increases, the portfolio variance is largely determined by the average covariance of the securities.

Equation (16.8) is what underlies the distinction between *unsystematic* risk and *systematic* risk. The basic random variability of the return on an individual security represents *unsystematic* risk. Since intelligent investors will be diversified, the market does not compensate investors for assuming unsystematic risk. *There is no compensation for being stupid.* However, the market is modeled as compensating investors for assuming more *systematic* risk (measured by covariance with the market and other securities).

16.3 SECURITY MARKET LINE and CAPITAL ASSET PRICING MODEL (CAPM)

16.3.1 Combinations of risky and riskless assets. The second major assumption of capital market theory permits all investors to borrow or lend unlimited funds at a given risk-free interest rate. In the real world, the term *risk free* is not really correct, because no asset is riskless. A preferable term is *virtually default-free*. For example, short-term U.S. Treasury bills (less than 1-year maturity) are considered virtually default-free, since the government can raise money by taxes and can print currency. But T-bills are not *risk free*—due to inflation. Hence, when risk-free is used, the term virtually default-free is implied. The risk-free interest rate, R_f, is assumed to have zero risk, and it is plotted as a point on the y-axis.

Fama (1968) pointed out that borrowing or lending at the risk-free, R_f, can be coupled with investment in any risky security or portfolio. Consider for Figure 16.5, the return and risk for portfolio C, which combines a default-free asset, and security A, with expected return $\overline{R}_A$ and standard deviation of return σ_A, where w_{rf} is the fraction invested in the risk-free asset.

$$\overline{R}_C = w_{rf}R_f + (1 - w_{rf})\overline{R}_A \tag{16.9}$$

$$\sigma_C = w_{rf}\sigma_f + (1 - w_{rf})\sigma_A = (1 - w_{rf})\sigma_A \quad \text{since } \sigma_f = 0 \tag{16.10}$$

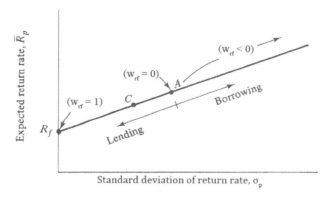

Figure 16.5 Combination of any portfolio and the riskless asset

The linear form of this equation implies that all combinations of the riskless asset and the risky security A are linear trade-offs between the risk, σ_C, of the portfolio and its expected return, $\overline{R}_C$. Thus, in Figure 16.5, portfolios such as C lie on a straight line through R_f and A. Portfolios lying between R_f and A involve the *lending* of the firm's funds at rate R_f

(since $0 \leq w_{rf} \leq 1$). That is, the firm invests a portion of its money in security A and lends the rest at the risk-free rate. Points lying beyond A involve the borrowing of funds ($w_{rf} < 0$), where all available funds are invested in security A, including those borrowed at the risk-free rate. For $w_{rf} = 0$, all available funds are invested in security A, without borrowing or lending. For $w_{rf} = 1$, all of the investor's funds are invested at the risk-free rate R_f.

16.3.2 The security market line. Borrowing or lending at the risk-free rate can be combined with any arbitrary portfolio, but consider what happens when it is combined with the locus of efficient market portfolios in Figure 16.6. The point $(0, R_f)$ is connected with the efficient frontier of risky portfolios with a tangent straight line. The point of tangency, M, describes an optimal portfolio of risky securities. Some combination of the risky portfolio, M, plus borrowing or lending at the riskless rate, R_f, dominates each other possible efficient portfolio. Different efficient portfolios will be dominated by different combinations of M and R_f.

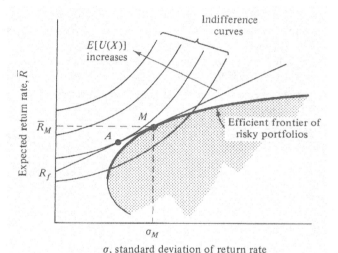

Figure 16.6 A capital market line

The point M has a special interpretation. It is the *market portfolio*, and it is composed of all securities in the market. The line of possible portfolios connecting the risk-free rate, R_f, and the market portfolio, $(\overline{R}_M, \sigma_M)$, is called the *security market line*.

Equation (16.11) is most easily understood by recognizing that the *risk premium* for the market is the difference between the expected market return and the market risk. This is the *rise* between the risk-free and market points. The *run* between the risk-free and market points is the standard deviation of the market, R_M. Thus, the slope of the line is the *rise* over the *run*. Thus the relationship between risk and return for efficient portfolios uses the risk-free rate and is the *security market line*, Equation (16.11).

$$E(R_p) = R_f + \left\{ \frac{[E(R_M) - R_f]}{\sigma_M} \right\} \sigma_p \qquad (16.11)$$

16.3.3 The capital asset pricing model (CAPM). When the security market line models how the market sets prices for individual securities using the relationship between risk and return for financial securities, it becomes the *capital asset pricing model* (CAPM). In this setting, the security market line is normally written using β as the x-axis. This beta is a measure of a security's volatility relative to the market. Since *unsystematic* risk is reduced to insignificance by diversification, beta measures the security's *systematic* risk. Beta is the security's covariance, σ_{im}, relative to the market's variance [see Equation (16.12)]. This equation is written for security i and the market m:

$$\beta_i = \frac{\sigma_{im}}{\sigma_m^2} \tag{16.12}$$

Note that by definition the beta for the market equals 1. This becomes the *run* between the risk-free and market points. The rise between the risk-free and market points is still the *market risk premium*, $R_M - R_f$. Since the *run* = 1, the slope also equals the market risk premium. For financial securities Equation (16.13) expresses the relationship between risk and return.

$$E(R_i) = R_f + \beta_i[E(R_M) - R_f] \tag{16.13}$$

This model is the foundation for teaching the relationship between risk and return in courses on finance, and it was one of the models for the cost of equity capital that was presented in Chapter 6. As our interest is in industrial projects, *not* financial securities, we are not going to go further into models for the pricing of securities.

However, we must note that there have been decades of work following development of the model by Sharpe (1963), Lintner (1965), and Mossin (1966). Much of this work has questioned the validity of the model when applied to real world data. For example, see Fama and French (1992). Yet in spite of three decades of research spurred by those results, no model has been developed and generally accepted as better. Also betas remain one of the most common measures reported for financial securities.

16.4 FIRM'S SECURITY MARKET LINE AND PROJECT ACCEPTANCE

16.4.1 Projects and the capital asset pricing model (CAPM). The final sections of this text's second edition applied the CAPM to the project selection problem. That material has been substantially revised for this edition, because we do *not* believe that the modeling of uncertainty described in Chapters 13 and 14 can be extended to estimating covariances of projects with the market.

Much of the decision-making about projects happens when the data is a mix of good quality estimates, very rough estimates, some near guesses, and sometimes some contractual numbers. Earlier chapters have covered how to convert these into useful economic measures such as the expected value and standard deviation of a project's return. As one example, distributions of first costs, annual revenues, and horizons were applied to simulate a project's economic return. Past projects may have implied where and when uniform, triangular, normal, and other distributions should be applied. Then analysis would have supplied estimates of the distribution parameters, such as each mean, standard deviation, minimum, mode, maximum, and so forth.

This modeling is radically different from the analysis of financial securities that underlies the CAPM. In that case decades of data are analyzed to reveal the correlations between the observed values for each security and the market as a whole. The required covariances can be calculated—they are not intuitively obvious quantities that can be estimated.

For financial securities, we can apply the CAPM after calculating β values that measure the security's volatility relative to the market. This also separates the security's risk into systematic risk and non-systematic risk. These steps are equivalent to calculating covariances between the security and the market. Unfortunately, neither the data nor the judgment of the decision-maker is adequate for calculating/estimating these covariance relationships for projects.

A second challenge is that for financial securities it is easy to diversify since the cost of each transaction is low and each security can be purchased or sold in small or large quantities. In contrast the typical project is a yes/no decision where all or none of the costs are incurred—and these costs are large. The ability of a firm to diversify its project portfolio is also limited by the precept of sticking to a firm's core competencies.

16.4.2 Risk/return trade-offs and the firm's security market line. Figure 16.6 describes a model for the trade-off between risk and return that can be applied for project selection. This model uses the standard deviation of return rather than β as the measure of risk. This is total risk, not just the systematic risk.

Figure 16.7 is a development of the model which emphasizes that the firm's security market line (measured with total risk) can differ from the market's corresponding line. It also shows three projects, where project A would be acceptable to the firm, project C would be acceptable to the *average* firm but not to this firm, and project B is not acceptable.

As illustrated in Figure 16.7, a project will be accepted if the project's coordinates $(\overline{R}_i, \sigma_i)$ lie above the firm's capital market line. Equivalently, the slope of a line through project$_i$ coordinates and the risk free y-intercept can be calculated and compared with the slope of firm O's capital market line in Equation (16.14).

$$\frac{R_i - R_f}{\sigma_i} > \frac{R_o - R_f}{\sigma_o} \qquad (16.14)$$

These slopes can be compared and used as a decision criterion, as developed by Sharpe (1970). These relationships are known as *reward-to-variability* ratios. A reward above the riskless rate is required for accepting the risk of such investments, that is, an acceptable project's reward-to-variability ratio must equal or exceed the firm's ratio from its existing projects.

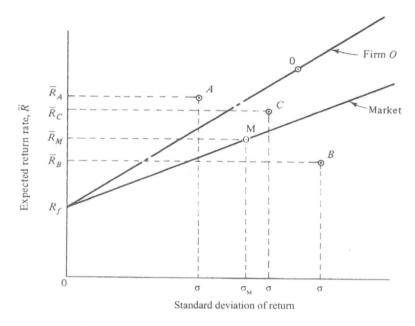

Figure 16.7 Project risk/return vs. firm risk/return

However, Sharpe argued that the reward-to-variability ratio should be replaced by measures based on volatility for application to a *single* security. He stated:

> The reward-to-variability ratio is designed to measure the performance of a portfolio. The investor is presumed to have placed a substantial portion of his wealth in the portfolio in question. Variability is thus the relevant measure of the amount of risk actually borne. To evaluate the performance of a single security, or that of a portfolio constituting only part of an investor's holdings, a different measure is needed. Variability will not adequately represent the risk actually borne. A more appropriate choice is volatility.

Note that in this context the term volatility, unlike Chapter 14, refers to a security's volatility relative to the market, β.

The shift to volatility is a laudable goal—but in most cases it is an unattainable goal for projects. However, the reward to variability ratio remains a useful measure to evaluate whether a project has a favorable return/risk ratio.

16.5 THE FIRM'S PORTFOLIO OF PROJECTS

16.5.1 Why do firms use project portfolios? When a firm uses the term *project portfolios*, this implies that a firm is looking at the interactions among projects. This may be in a strategic or marketing context with a focus on the firm's core competencies and market segments in an evolving competitive environment. The language of a project portfolio is also used within project management where the focus is on the use and scheduling of resources, personnel, and facilities. Firms also use the term with a financial focus—analogous to the portfolios of financial securities. Obviously these contexts overlap and interact.

The models that are used in these different contexts tend to differ; however, when we look at *why* firms use project portfolios in strategic, marketing, and financial contexts we find commonality. The driving reason is to maximize the firm's long-term financial performance. However, in the project management context, the focus tends to be more short-term—minimizing conflicts between projects and maximizing the efficiency and effectiveness of the resources, personnel, and facilities.

Conceptually, there is a strong link between the driving reason behind portfolios of financial securities and project portfolios—diversification reduces risk. This can be expressed as simply as an old proverb, "don't put all your eggs in one basket."

16.5.2 Can security portfolio theory be extended to project portfolios? Earlier chapters have emphasized that good economic analysis includes the estimation of the expected value and standard deviation of expected returns for potential projects. However, these economic measures are often calculated using estimated values for future revenues, costs, and horizons that may be better classified as *guesstimates* than *estimates*.

Even when high quality estimates are available it is unrealistic to expect this to include estimated covariances between projects, between projects and the firm, and between projects and the market. In a few cases, these can be ballparked with low, medium, or high numeric values. For example, oil well projects may be highly correlated with each other and the price of oil.

The problem is that covariance is not an intuitive quantity and many firms do not really audit past decisions. Even if they did, the focus would be on what projects the firm *chose* to undertake and not the ones the firm chose *not* to undertake, let alone the states of nature that did not occur and how the projects would have performed if they had occurred. Knowledge of all of the above is needed to estimate covariances. Thus, we do not believe that covariances can be estimated with sufficient accuracy to model project portfolios as portfolios of financial securities were modeled.

Another component in modeling portfolios of financial securities was the weighting of each security's returns and variability by its proportionate share of the total amount invested. With project portfolios it is often not possible to identify the proportionate share for each project—because many projects require investments over multiple years and because *portfolio* decisions on prospective additions to the project portfolio must consider existing projects as well as the candidate set. With financial securities there are clear market values for securities that are in the existing portfolio and candidate additions. Analogous values for projects often do not exist.

Thus, two required parts of security portfolio theory—covariances and weights—are not available for project portfolios. Thus we do not find that project portfolio models mirror the earlier sections of this chapter.

Instead when mathematical models are built for project portfolios the focus is typically the fit of potential projects with the set of existing projects considering the constraints on budgets, people, facilities, and resources. These models are not focused on variances, covariances, and weights; but rather are linear, integer, or mixed integer programming models—that are typically deterministic.

16.5.3 Reasonable inferences from security portfolio theory to project portfolios. Nevertheless, we believe that there are lessons for project portfolios that can be drawn from the results for security portfolios.

Even though estimating project covariances is generally not possible, we suggest that it is often possible to identify projects whose returns are likely to be highly correlated (positively or negatively) or relatively independent. One set of projects may all depend on the developmental path for the same technology. Another set may focus on supporting a certain industrial segment. Another pair may offer radically different approaches to the same problem. Some will depend on the health of the overall economy, and others will be primarily driven by other uncertainties.

With these estimates there are useful projections from securities portfolios to project portfolios. These do not rely on detailed mathematical proofs, but rather rest on logical inference. For financial portfolios we found that if the number of securities were small, then the variance of each of the securities was important. We also found that as the number of securities increased the relative importance of the covariances increased.

Thus, when we look at project portfolios we suggest that:
- High risk projects with low correlations to other projects are more valuable than they appear—since they reduce the risk of the project portfolio.
- Low risk projects with high correlations to other projects are less valuable than they appear—since they do *not* reduce the risk of the project portfolio to the same extent as projects with low or negative correlations. These correlations often arise because of links with the same commodity, product group, market segment, or technology.

We also suggest that the relative importance of project variances and covariances can be linked to the size of the firm and the size of the project portfolios.
- For larger firms with large project portfolios, the risk of the portfolio depends mainly on the covariances between the projects, the firm, and the market.
- For smaller firms with small project portfolios, the risk of the portfolio depends more on the variances of the projects.

Now the question is, "How do we link these inferences to typical firms and projects?" Here we rely principally on logic, personal experience in industry, and conversations with many part-time students with full-time jobs.
- Most firms tend to focus on projects that extend current products and market segments—which will have larger correlations and covariances and larger project portfolio risks.
- Projects which are not in the firm's "mainstream" will have lower correlations and covariances with the firm and other projects—thus they will lower the risk of the project portfolio. However, this effect is reduced by the fact that diversification from the core competencies of the firm can increase the risks of the projects and the project portfolio.
- Larger firms will have larger project portfolios, which will tend to reduce the risk of the portfolios. The divisions of those firms will inherently have smaller portfolios and larger risks than the firm as a whole.
- Smaller firms will have smaller project portfolios, which increases the risks faced by this size firm.

- Firms and managers tend to be risk averse, which is generally implemented by choosing projects which are extensions of current practice. Since this does not reduce the risk of the project portfolio, this may have the contradictory effect of increasing the firm's risk.

In closing this section, we suggest that *managing* the risks of the project portfolio is likely to be more productive than *estimating/calculating* the variance of that portfolio. That management is typically exercised through a stage-gate process where the project may be terminated at any of multiple stages that represent more knowledge about the project's costs, risks, and potential payoffs.

There is substantial anecdotal evidence that one of the most common errors made by firms is terminating projects later than the projects should have been terminated.

16.5.4 Can the capital asset pricing model for securities be extended to projects? One of the fundamental results from the CAPM is that the assumption of unsystematic risk is not compensated by the market. Instead diversification is assumed to eliminate unsystematic risk, and the market return is only adjusted for the level of systematic risk of that security. This depends on the fundamental fact that diversification is easy for financial securities. Multiple stocks, mutual funds, index funds, real estate investment trusts (REITS), and other securities can easily be purchased or sold to form a diversified portfolio.

In contrast the ability to diversify project portfolios tends to be limited at best. Firms can seek out new products and new markets, but often the best possibilities are close to the firm's current core competencies. The firm cannot easily buy or sell pieces of its projects or the projects of other firms.

In addition, with financial securities there tends to be substantial data on the systematic and unsystematic risks of each security. However, with projects the maximum level of information is often the firm's expected volumes, revenues, and costs for the next decade or so without the ability to subdivide the project's risks into systematic and unsystematic risks. For projects it is often only possible to estimate the *total* risks for the projects without the subdivision that is central to the CAPM.

Another complication is that the impact on project managers and employees is often much more closely linked to the success of the project. Managers of financial securities typically manage many securities, where a single project may involve employees and managers for six months to years. This implies that asymmetric information and asymmetric risks are even more important for projects than for financial securities. This also means that *agency theory* with its focus on the impact of differences between the incentives for managers and owners is even more important for projects than for financial securities.

The CAPM is essentially a model of returns for a *single*-period. This is an issue for financial portfolios where assets may be held for years. This is a huge issue for projects where the project only makes sense if held for its lifetime—which is measured in years and often decades. During a project's life, it is likely that sources of systematic risk, such as the state of the economy will take on multiple values.

We have identified serious difficulties in extending the CAPM to projects. In addition, there are serious questions about the CAPM's validity for securities. Finally, we have not found significant empirical work validating the extension of the CAPM to projects. Thus, we conclude here as we did earlier that the CAPM model *cannot* in general be validly extended to projects—the data simply does not support this.

However, we do want to note that the previous edition of this text presented an approach where the mathematics could be applied—if enough were assumed about the returns of the projects, the firm, and the market for a set of states of nature that define a complete model of the future. Such a model defines the states, estimates the probability for each state, and then estimates project returns for each state. It is likely that somewhere there is a project with a short life where such a model could be built, but the model cannot be realistic.

Arguably the largest challenge to this approach is that virtually every project has at least one uncertainty unique to it—R&D, drug trials, or new products from leading competitors. Only for systematic risks, such as the economy as a whole, is it possible to identify states that affect all projects. But this ignores *all* project-specific risks. Thus, we have so many doubts about the validity of this approach that we have chosen to omit it from this edition.

16.6. SUMMARY

We have examined capital market theory and its application to project selection. The fundamental proof that diversification in a portfolio reduces risk for financial securities does have implications for project selection. However, the mathematical approach relies on data that is available for financial securities but not for projects.

It was demonstrated that the best available selection criterion is the firm's reward-to-return ratio since projects that provide ratios exceeding the firm's will increase the firm's present net worth. Thus, the capital market method of project selection solves both the capital availability and capital usage aspects of the resource allocation problem simultaneously, using the market as the mediating mechanism.

At the firm level, the limitations of CAPM lie in the area of model inadequacy (the Sharpe—Lintner straight-line, single-variable model may not suffice), and in its fundamental assumption of risk avoidance by the firm. At the project level, there is the practical difficulty of estimating the volatility of the project (and hence its covariance). This is much more difficult than defining a project's variability. Thus, we have concluded that the CAPM cannot in general be validly extended to projects.

However, there are useful managerial insights from portfolio theory for financial securities that can be applied to consider how risk and return calculations for individual projects can/should be adjusted when those projects are considered as a portfolio.

REFERENCES

FAMA, EUGENE, "Risk, Return and Equilibrium: Some Clarifying Comments," *Journal of Finance,* **23**(1) (March 1968), pp. 29–40.

FAMA, EUGENE F. AND KENNETH R. FRENCH, "The Cross-Section of Expected Stock Returns," *Journal of Finance,* **47**(2) (June 1992), pp. 427–465.

LEVY, HAIM, "Equilibrium in an Inperfect Market: A Constraint on the Number of Securities in the Portfolio," *American Economic Review*, **68**(4) (September 1978), pp. 643–658.

LINTNER, JOHN, "Security Prices, Risk, and Maximal Gains from Diversification," *Journal of Finance*, **20**(4) (December 1965), pp. 587–615.

MARKOWITZ, HARRY, *Portfolio Selection: Efficient Diversification of Investments* (Wiley, 1959).

MOSSIN, JAN, "Equilibrium in a Capital Asset Market," *Econometrica*, **34**(4) (October 1966) pp. 768–783.

ROSS, STEPHEN A., "The Arbitrage Theory of Capital Asset Pricing," *The Journal of Economic Theory* **13** (December 1976), pp. 341–360.

SHARPE, WILLIAM F., "A Simplified Model for Portfolio Analysis," *Management Science*, **9**(2) (January 1963), pp. 277–293.

SHARPE, WILLIAM F., *Portfolio Theory and Capital Markets* (McGraw-Hill, 1970).

PROBLEMS

16-1. Calculate the mean and standard deviation for a portfolio made up of two independently distributed securities. Security A has an expected return of 8% with a standard deviation of 3% and security B has an expected return of 12% with a standard deviation of 5%. Assume that the portfolio has the following mix of A and B: (a) 100% A, (b) 75% A and 25% B, (c) 50% A and 50% B, (d) 25% A and 75% B, and (e) 100% B. Include a graph.

16-2. Redo problem 16-1 assuming that the two securities are not independent, but rather have the following coefficient of correlation: (a) $\rho = -1.0$, (b) $\rho = -0.5$, (c) $\rho = 0.5$, and (d) $\rho = 1.0$. Graph each.

16-3. Calculate the mean and standard deviation for a portfolio made up of two independently distributed securities. Assume that the portfolio has the following mix of D and E: (a) 100% D, (b) 75% D and 25% E, (c) 50% D and 50% E, (d) 25% D and 75% E, and (e) 100% E. Include a graph.

Security D			Security E	
Probability	Return		Probability	Return
0.3	−50%		0.4	20%
0.5	100%		0.6	90%
0.2	250%			

16-4. Redo problem 16-3 assuming that the two securities are not independent, but rather have the following coefficient of correlation: (a) $\rho = -1.0$, (b) $\rho = -0.5$, (c) $\rho = 0.5$, and (d) $\rho = 1.0$. Graph each.

16-5. (a) For the data of problem 16-2 and $\rho = -0.5$ identify two portfolio mixes that have the same standard deviation but different expected returns. Which one is efficient and which inefficient? (b) Do similar results occur for $\rho = 0$? (c) For $\rho = 0.5$?

16-6. Equation 16.8 presented the portfolio variability with a portfolio of "average" variances and covariances. Analyze the portfolio's standard deviation from $n = 1$ to 100 if the average standard deviation is 8% for average correlations ranging from 1 to −0.2.

16-7. Add a third security, C, to the data of problem 16-1. Security C has an expected return of 5% with a standard deviation of 0%. (a) What is the mean and standard deviation of a portfolio consisting of equal parts A, B, and C? (b) If $3 million must be invested in blocks of $1M, identify the 10 possible portfolios. (c) Calculate the mean and standard deviation of each portfolio. Plot and identify dominated and efficient portfolios.

16-8. For problem 16-7 contrast the envelopes of the efficient portfolios with one, two, or three securities allowed.

16-9. A firm is considering a project that has an expected rate of return of 15% with a standard deviation of 4%. The firm's rate of return on equity is 18% with a standard deviation of 6%. The risk-free lending rate is 5%. Would this project increase the firm's value to its owners?

16-10. A firm is considering a project that has an expected rate of return of 25% with a standard deviation of 12%. The firm's rate of return on equity is 18% with a standard deviation of 6%. The risk-free lending rate is 5%. Would this project increase the firm's value to its owners?

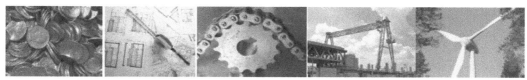

Appendix: Engineering Economy Factors ¼%

0.25% INTEREST FACTORS FOR DISCRETE COMPOUNDING PERIODS

	SINGLE PAYMENT		UNIFORM PAYMENT SERIES				GRADIENT SERIES	
	Compound Amount Factor	Present Worth Factor	Present Worth Factor	Capital Recovery Factor	Compound Amount Factor	Sinking Fund Factor	Gradient Uniform Series	Gradient Present Worth
N	Find F Given P F/P	Find P Given F P/F	Find P Given A P/A	Find A Given P A/P	Find F Given A F/A	Find A Given F A/F	Find A Given G A/G	Find P Given G P/G
1	1.003	0.9975	0.9975	1.0025	1.000	1.0000	0.0000	0.000
2	1.005	0.9950	1.9925	0.5019	2.002	0.4994	0.4994	0.995
3	1.008	0.9925	2.9851	0.3350	3.008	0.3325	0.9983	2.980
4	1.010	0.9901	3.9751	0.2516	4.015	0.2491	1.4969	5.950
5	1.013	0.9876	4.9627	0.2015	5.025	0.1990	1.9950	9.901
6	1.015	0.9851	5.9478	0.1681	6.038	0.1656	2.4927	14.826
7	1.018	0.9827	6.9305	0.1443	7.053	0.1418	2.9900	20.722
8	1.020	0.9802	7.9107	0.1264	8.070	0.1239	3.4869	27.584
9	1.023	0.9778	8.8885	0.1125	9.091	0.1100	3.9834	35.406
10	1.025	0.9753	9.8639	0.1014	10.113	0.0989	4.4794	44.184
11	1.028	0.9729	10.8368	0.0923	11.139	0.0898	4.9750	53.913
12	1.030	0.9705	11.8073	0.0847	12.166	0.0822	5.4702	64.589
13	1.033	0.9681	12.7753	0.0783	13.197	0.0758	5.9650	76.205
14	1.036	0.9656	13.7410	0.0728	14.230	0.0703	6.4594	88.759
15	1.038	0.9632	14.7042	0.0680	15.265	0.0655	6.9534	102.244
16	1.041	0.9608	15.6650	0.0638	16.304	0.0613	7.4469	116.657
17	1.043	0.9584	16.6235	0.0602	17.344	0.0577	7.9401	131.992
18	1.046	0.9561	17.5795	0.0569	18.388	0.0544	8.4328	148.245
19	1.049	0.9537	18.5332	0.0540	19.434	0.0515	8.9251	165.411
20	1.051	0.9513	19.4845	0.0513	20.482	0.0488	9.4170	183.485
21	1.054	0.9489	20.4334	0.0489	21.533	0.0464	9.9085	202.463
22	1.056	0.9466	21.3800	0.0468	22.587	0.0443	10.3995	222.341
23	1.059	0.9442	22.3241	0.0448	23.644	0.0423	10.8901	243.113
24	1.062	0.9418	23.2660	0.0430	24.703	0.0405	11.3804	264.775
25	1.064	0.9395	24.2055	0.0413	25.765	0.0388	11.8702	287.323
26	1.067	0.9371	25.1426	0.0398	26.829	0.0373	12.3596	310.752
27	1.070	0.9348	26.0774	0.0383	27.896	0.0358	12.8485	335.057
28	1.072	0.9325	27.0099	0.0370	28.966	0.0345	13.3371	360.233
29	1.075	0.9301	27.9400	0.0358	30.038	0.0333	13.8252	386.278
30	1.078	0.9278	28.8679	0.0346	31.113	0.0321	14.3130	413.185
31	1.080	0.9255	29.7934	0.0336	32.191	0.0311	14.8003	440.950
32	1.083	0.9232	30.7166	0.0326	33.272	0.0301	15.2872	469.570
33	1.086	0.9209	31.6375	0.0316	34.355	0.0291	15.7736	499.039
34	1.089	0.9186	32.5561	0.0307	35.441	0.0282	16.2597	529.353
35	1.091	0.9163	33.4724	0.0299	36.529	0.0274	16.7454	560.508
36	1.094	0.9140	34.3865	0.0291	37.621	0.0266	17.2306	592.499
37	1.097	0.9118	35.2982	0.0283	38.715	0.0258	17.7154	625.322
38	1.100	0.9095	36.2077	0.0276	39.811	0.0251	18.1998	658.973
39	1.102	0.9072	37.1149	0.0269	40.911	0.0244	18.6838	693.447
40	1.105	0.9050	38.0199	0.0263	42.013	0.0238	19.1673	728.740

½%

0.5% INTEREST FACTORS FOR DISCRETE COMPOUNDING PERIODS

	SINGLE PAYMENT		UNIFORM PAYMENT SERIES				GRADIENT SERIES	
	Compound Amount Factor	Present Worth Factor	Present Worth Factor	Capital Recovery Factor	Compound Amount Factor	Sinking Fund Factor	Gradient Uniform Series	Gradient Present Worth
N	Find F Given P F/P	Find P Given F P/F	Find P Given A P/A	Find A Given P A/P	Find F Given A F/A	Find A Given F A/F	Find A Given G A/G	Find P Given G P/G
1	1.005	0.9950	0.9950	1.0050	1.000	1.0000	0.0000	0.000
2	1.010	0.9901	1.9851	0.5038	2.005	0.4988	0.4988	0.990
3	1.015	0.9851	2.9702	0.3367	3.015	0.3317	0.9967	2.960
4	1.020	0.9802	3.9505	0.2531	4.030	0.2481	1.4938	5.901
5	1.025	0.9754	4.9259	0.2030	5.050	0.1980	1.9900	9.803
6	1.030	0.9705	5.8964	0.1696	6.076	0.1646	2.4855	14.655
7	1.036	0.9657	6.8621	0.1457	7.106	0.1407	2.9801	20.449
8	1.041	0.9609	7.8230	0.1278	8.141	0.1228	3.4738	27.176
9	1.046	0.9561	8.7791	0.1139	9.182	0.1089	3.9668	34.824
10	1.051	0.9513	9.7304	0.1028	10.228	0.0978	4.4589	43.386
11	1.056	0.9466	10.6770	0.0937	11.279	0.0887	4.9501	52.853
12	1.062	0.9419	11.6189	0.0861	12.336	0.0811	5.4406	63.214
13	1.067	0.9372	12.5562	0.0796	13.397	0.0746	5.9302	74.460
14	1.072	0.9326	13.4887	0.0741	14.464	0.0691	6.4190	86.583
15	1.078	0.9279	14.4166	0.0694	15.537	0.0644	6.9069	99.574
16	1.083	0.9233	15.3399	0.0652	16.614	0.0602	7.3940	113.424
17	1.088	0.9187	16.2586	0.0615	17.697	0.0565	7.8803	128.123
18	1.094	0.9141	17.1728	0.0582	18.786	0.0532	8.3658	143.663
19	1.099	0.9096	18.0824	0.0553	19.880	0.0503	8.8504	160.036
20	1.105	0.9051	18.9874	0.0527	20.979	0.0477	9.3342	177.232
21	1.110	0.9006	19.8880	0.0503	22.084	0.0453	9.8172	195.243
22	1.116	0.8961	20.7841	0.0481	23.194	0.0431	10.2993	214.061
23	1.122	0.8916	21.6757	0.0461	24.310	0.0411	10.7806	233.677
24	1.127	0.8872	22.5629	0.0443	25.432	0.0393	11.2611	254.082
25	1.133	0.8828	23.4456	0.0427	26.559	0.0377	11.7407	275.269
26	1.138	0.8784	24.3240	0.0411	27.692	0.0361	12.2195	297.228
27	1.144	0.8740	25.1980	0.0397	28.830	0.0347	12.6975	319.952
28	1.150	0.8697	26.0677	0.0384	29.975	0.0334	13.1747	343.433
29	1.156	0.8653	26.9330	0.0371	31.124	0.0321	13.6510	367.663
30	1.161	0.8610	27.7941	0.0360	32.280	0.0310	14.1265	392.632
31	1.167	0.8567	28.6508	0.0349	33.441	0.0299	14.6012	418.335
32	1.173	0.8525	29.5033	0.0339	34.609	0.0289	15.0750	444.762
33	1.179	0.8482	30.3515	0.0329	35.782	0.0279	15.5480	471.906
34	1.185	0.8440	31.1955	0.0321	36.961	0.0271	16.0202	499.758
35	1.191	0.8398	32.0354	0.0312	38.145	0.0262	16.4915	528.312
36	1.197	0.8356	32.8710	0.0304	39.336	0.0254	16.9621	557.560
37	1.203	0.8315	33.7025	0.0297	40.533	0.0247	17.4317	587.493
38	1.209	0.8274	34.5299	0.0290	41.735	0.0240	17.9006	618.105
39	1.215	0.8232	35.3531	0.0283	42.944	0.0233	18.3686	649.388
40	1.221	0.8191	36.1722	0.0276	44.159	0.0226	18.8359	681.335

1%

1% INTEREST FACTORS FOR DISCRETE COMPOUNDING PERIODS

	SINGLE PAYMENT		UNIFORM PAYMENT SERIES				GRADIENT SERIES	
	Compound Amount Factor	Present Worth Factor	Present Worth Factor	Capital Recovery Factor	Compound Amount Factor	Sinking Fund Factor	Gradient Uniform Series	Gradient Present Worth
N	Find F Given P F/P	Find P Given F P/F	Find P Given A P/A	Find A Given P A/P	Find F Given A F/A	Find A Given F A/F	Find A Given G A/G	Find P Given G P/G
1	1.010	0.9901	0.9901	1.0100	1.000	1.0000	0.0000	0.000
2	1.020	0.9803	1.9704	0.5075	2.010	0.4975	0.4975	0.980
3	1.030	0.9706	2.9410	0.3400	3.030	0.3300	0.9934	2.921
4	1.041	0.9610	3.9020	0.2563	4.060	0.2463	1.4876	5.804
5	1.051	0.9515	4.8534	0.2060	5.101	0.1960	1.9801	9.610
6	1.062	0.9420	5.7955	0.1725	6.152	0.1625	2.4710	14.321
7	1.072	0.9327	6.7282	0.1486	7.214	0.1386	2.9602	19.917
8	1.083	0.9235	7.6517	0.1307	8.286	0.1207	3.4478	26.381
9	1.094	0.9143	8.5660	0.1167	9.369	0.1067	3.9337	33.696
10	1.105	0.9053	9.4713	0.1056	10.462	0.0956	4.4179	41.843
11	1.116	0.8963	10.3676	0.0965	11.567	0.0865	4.9005	50.807
12	1.127	0.8874	11.2551	0.0888	12.683	0.0788	5.3815	60.569
13	1.138	0.8787	12.1337	0.0824	13.809	0.0724	5.8607	71.113
14	1.149	0.8700	13.0037	0.0769	14.947	0.0669	6.3384	82.422
15	1.161	0.8613	13.8651	0.0721	16.097	0.0621	6.8143	94.481
16	1.173	0.8528	14.7179	0.0679	17.258	0.0579	7.2886	107.273
17	1.184	0.8444	15.5623	0.0643	18.430	0.0543	7.7613	120.783
18	1.196	0.8360	16.3983	0.0610	19.615	0.0510	8.2323	134.996
19	1.208	0.8277	17.2260	0.0581	20.811	0.0481	8.7017	149.895
20	1.220	0.8195	18.0456	0.0554	22.019	0.0454	9.1694	165.466
21	1.232	0.8114	18.8570	0.0530	23.239	0.0430	9.6354	181.695
22	1.245	0.8034	19.6604	0.0509	24.472	0.0409	10.0998	198.566
23	1.257	0.7954	20.4558	0.0489	25.716	0.0389	10.5626	216.066
24	1.270	0.7876	21.2434	0.0471	26.973	0.0371	11.0237	234.180
25	1.282	0.7798	22.0232	0.0454	28.243	0.0354	11.4831	252.894
26	1.295	0.7720	22.7952	0.0439	29.526	0.0339	11.9409	272.196
27	1.308	0.7644	23.5596	0.0424	30.821	0.0324	12.3971	292.070
28	1.321	0.7568	24.3164	0.0411	32.129	0.0311	12.8516	312.505
29	1.335	0.7493	25.0658	0.0399	33.450	0.0299	13.3044	333.486
30	1.348	0.7419	25.8077	0.0387	34.785	0.0287	13.7557	355.002
31	1.361	0.7346	26.5423	0.0377	36.133	0.0277	14.2052	377.039
32	1.375	0.7273	27.2696	0.0367	37.494	0.0267	14.6532	399.586
33	1.389	0.7201	27.9897	0.0357	38.869	0.0257	15.0995	422.629
34	1.403	0.7130	28.7027	0.0348	40.258	0.0248	15.5441	446.157
35	1.417	0.7059	29.4086	0.0340	41.660	0.0240	15.9871	470.158
36	1.431	0.6989	30.1075	0.0332	43.077	0.0232	16.4285	494.621
37	1.445	0.6920	30.7995	0.0325	44.508	0.0225	16.8682	519.533
38	1.460	0.6852	31.4847	0.0318	45.953	0.0218	17.3063	544.884
39	1.474	0.6784	32.1630	0.0311	47.412	0.0211	17.7428	570.662
40	1.489	0.6717	32.8347	0.0305	48.886	0.0205	18.1776	596.856

3%

3% INTEREST FACTORS FOR DISCRETE COMPOUNDING PERIODS

	SINGLE PAYMENT		UNIFORM PAYMENT SERIES				GRADIENT SERIES	
	Compound Amount Factor	Present Worth Factor	Present Worth Factor	Capital Recovery Factor	Compound Amount Factor	Sinking Fund Factor	Gradient Uniform Series	Gradient Present Worth
N	Find F Given P F/P	Find P Given F P/F	Find P Given A P/A	Find A Given P A/P	Find F Given A F/A	Find A Given F A/F	Find A Given G A/G	Find P Given G P/G
1	1.030	0.9709	0.9709	1.0300	1.000	1.0000	0.0000	0.000
2	1.061	0.9426	1.9135	0.5226	2.030	0.4926	0.4926	0.943
3	1.093	0.9151	2.8286	0.3535	3.091	0.3235	0.9803	2.773
4	1.126	0.8885	3.7171	0.2690	4.184	0.2390	1.4631	5.438
5	1.159	0.8626	4.5797	0.2184	5.309	0.1884	1.9409	8.889
6	1.194	0.8375	5.4172	0.1846	6.468	0.1546	2.4138	13.076
7	1.230	0.8131	6.2303	0.1605	7.662	0.1305	2.8819	17.955
8	1.267	0.7894	7.0197	0.1425	8.892	0.1125	3.3450	23.481
9	1.305	0.7664	7.7861	0.1284	10.159	0.0984	3.8032	29.612
10	1.344	0.7441	8.5302	0.1172	11.464	0.0872	4.2565	36.309
11	1.384	0.7224	9.2526	0.1081	12.808	0.0781	4.7049	43.533
12	1.426	0.7014	9.9540	0.1005	14.192	0.0705	5.1485	51.248
13	1.469	0.6810	10.6350	0.0940	15.618	0.0640	5.5872	59.420
14	1.513	0.6611	11.2961	0.0885	17.086	0.0585	6.0210	68.014
15	1.558	0.6419	11.9379	0.0838	18.599	0.0538	6.4500	77.000
16	1.605	0.6232	12.5611	0.0796	20.157	0.0496	6.8742	86.348
17	1.653	0.6050	13.1661	0.0760	21.762	0.0460	7.2936	96.028
18	1.702	0.5874	13.7535	0.0727	23.414	0.0427	7.7081	106.014
19	1.754	0.5703	14.3238	0.0698	25.117	0.0398	8.1179	116.279
20	1.806	0.5537	14.8775	0.0672	26.870	0.0372	8.5229	126.799
21	1.860	0.5375	15.4150	0.0649	28.676	0.0349	8.9231	137.550
22	1.916	0.5219	15.9369	0.0627	30.537	0.0327	9.3186	148.509
23	1.974	0.5067	16.4436	0.0608	32.453	0.0308	9.7093	159.657
24	2.033	0.4919	16.9355	0.0590	34.426	0.0290	10.0954	170.971
25	2.094	0.4776	17.4131	0.0574	36.459	0.0274	10.4768	182.434
26	2.157	0.4637	17.8768	0.0559	38.553	0.0259	10.8535	194.026
27	2.221	0.4502	18.3270	0.0546	40.710	0.0246	11.2255	205.731
28	2.288	0.4371	18.7641	0.0533	42.931	0.0233	11.5930	217.532
29	2.357	0.4243	19.1885	0.0521	45.219	0.0221	11.9558	229.414
30	2.427	0.4120	19.6004	0.0510	47.575	0.0210	12.3141	241.361
31	2.500	0.4000	20.0004	0.0500	50.003	0.0200	12.6678	253.361
32	2.575	0.3883	20.3888	0.0490	52.503	0.0190	13.0169	265.399
33	2.652	0.3770	20.7658	0.0482	55.078	0.0182	13.3616	277.464
34	2.732	0.3660	21.1318	0.0473	57.730	0.0173	13.7018	289.544
35	2.814	0.3554	21.4872	0.0465	60.462	0.0165	14.0375	301.627
36	2.898	0.3450	21.8323	0.0458	63.276	0.0158	14.3688	313.703
37	2.985	0.3350	22.1672	0.0451	66.174	0.0151	14.6957	325.762
38	3.075	0.3252	22.4925	0.0445	69.159	0.0145	15.0182	337.796
39	3.167	0.3158	22.8082	0.0438	72.234	0.0138	15.3363	349.794
40	3.262	0.3066	23.1148	0.0433	75.401	0.0133	15.6502	361.750

5% INTEREST FACTORS FOR DISCRETE COMPOUNDING PERIODS

5%

	SINGLE PAYMENT		UNIFORM PAYMENT SERIES				GRADIENT SERIES	
	Compound Amount Factor	Present Worth Factor	Present Worth Factor	Capital Recovery Factor	Compound Amount Factor	Sinking Fund Factor	Gradient Uniform Series	Gradient Present Worth
N	Find F Given P F/P	Find P Given F P/F	Find P Given A P/A	Find A Given P A/P	Find F Given A F/A	Find A Given F A/F	Find A Given G A/G	Find P Given G P/G
1	1.050	0.9524	0.9524	1.0500	1.000	1.0000	0.0000	0.000
2	1.103	0.9070	1.8594	0.5378	2.050	0.4878	0.4878	0.907
3	1.158	0.8638	2.7232	0.3672	3.152	0.3172	0.9675	2.635
4	1.216	0.8227	3.5460	0.2820	4.310	0.2320	1.4391	5.103
5	1.276	0.7835	4.3295	0.2310	5.526	0.1810	1.9025	8.237
6	1.340	0.7462	5.0757	0.1970	6.802	0.1470	2.3579	11.968
7	1.407	0.7107	5.7864	0.1728	8.142	0.1228	2.8052	16.232
8	1.477	0.6768	6.4632	0.1547	9.549	0.1047	3.2445	20.970
9	1.551	0.6446	7.1078	0.1407	11.027	0.0907	3.6758	26.127
10	1.629	0.6139	7.7217	0.1295	12.578	0.0795	4.0991	31.652
11	1.710	0.5847	8.3064	0.1204	14.207	0.0704	4.5144	37.499
12	1.796	0.5568	8.8633	0.1128	15.917	0.0628	4.9219	43.624
13	1.886	0.5303	9.3936	0.1065	17.713	0.0565	5.3215	49.988
14	1.980	0.5051	9.8986	0.1010	19.599	0.0510	5.7133	56.554
15	2.079	0.4810	10.3797	0.0963	21.579	0.0463	6.0973	63.288
16	2.183	0.4581	10.8378	0.0923	23.657	0.0423	6.4736	70.160
17	2.292	0.4363	11.2741	0.0887	25.840	0.0387	6.8423	77.140
18	2.407	0.4155	11.6896	0.0855	28.132	0.0355	7.2034	84.204
19	2.527	0.3957	12.0853	0.0827	30.539	0.0327	7.5569	91.328
20	2.653	0.3769	12.4622	0.0802	33.066	0.0302	7.9030	98.488
21	2.786	0.3589	12.8212	0.0780	35.719	0.0280	8.2416	105.667
22	2.925	0.3418	13.1630	0.0760	38.505	0.0260	8.5730	112.846
23	3.072	0.3256	13.4886	0.0741	41.430	0.0241	8.8971	120.009
24	3.225	0.3101	13.7986	0.0725	44.502	0.0225	9.2140	127.140
25	3.386	0.2953	14.0939	0.0710	47.727	0.0210	9.5238	134.228
26	3.556	0.2812	14.3752	0.0696	51.113	0.0196	9.8266	141.259
27	3.733	0.2678	14.6430	0.0683	54.669	0.0183	10.1224	148.223
28	3.920	0.2551	14.8981	0.0671	58.403	0.0171	10.4114	155.110
29	4.116	0.2429	15.1411	0.0660	62.323	0.0160	10.6936	161.913
30	4.322	0.2314	15.3725	0.0651	66.439	0.0151	10.9691	168.623
31	4.538	0.2204	15.5928	0.0641	70.761	0.0141	11.2381	175.233
32	4.765	0.2099	15.8027	0.0633	75.299	0.0133	11.5005	181.739
33	5.003	0.1999	16.0025	0.0625	80.064	0.0125	11.7566	188.135
34	5.253	0.1904	16.1929	0.0618	85.067	0.0118	12.0063	194.417
35	5.516	0.1813	16.3742	0.0611	90.320	0.0111	12.2498	200.581
36	5.792	0.1727	16.5469	0.0604	95.836	0.0104	12.4872	206.624
37	6.081	0.1644	16.7113	0.0598	101.628	0.0098	12.7186	212.543
38	6.385	0.1566	16.8679	0.0593	107.710	0.0093	12.9440	218.338
39	6.705	0.1491	17.0170	0.0588	114.095	0.0088	13.1636	224.005
40	7.040	0.1420	17.1591	0.0583	120.800	0.0083	13.3775	229.545

8%

8% INTEREST FACTORS FOR DISCRETE COMPOUNDING PERIODS

	SINGLE PAYMENT		UNIFORM PAYMENT SERIES				GRADIENT SERIES	
	Compound Amount Factor	Present Worth Factor	Present Worth Factor	Capital Recovery Factor	Compound Amount Factor	Sinking Fund Factor	Gradient Uniform Series	Gradient Present Worth
N	Find F Given P F/P	Find P Given F P/F	Find P Given A P/A	Find A Given P A/P	Find F Given A F/A	Find A Given F A/F	Find A Given G A/G	Find P Given G P/G
1	1.080	0.9259	0.9259	1.0800	1.000	1.0000	0.0000	0.000
2	1.166	0.8573	1.7833	0.5608	2.080	0.4808	0.4808	0.857
3	1.260	0.7938	2.5771	0.3880	3.246	0.3080	0.9487	2.445
4	1.360	0.7350	3.3121	0.3019	4.506	0.2219	1.4040	4.650
5	1.469	0.6806	3.9927	0.2505	5.867	0.1705	1.8465	7.372
6	1.587	0.6302	4.6229	0.2163	7.336	0.1363	2.2763	10.523
7	1.714	0.5835	5.2064	0.1921	8.923	0.1121	2.6937	14.024
8	1.851	0.5403	5.7466	0.1740	10.637	0.0940	3.0985	17.806
9	1.999	0.5002	6.2469	0.1601	12.488	0.0801	3.4910	21.808
10	2.159	0.4632	6.7101	0.1490	14.487	0.0690	3.8713	25.977
11	2.332	0.4289	7.1390	0.1401	16.645	0.0601	4.2395	30.266
12	2.518	0.3971	7.5361	0.1327	18.977	0.0527	4.5957	34.634
13	2.720	0.3677	7.9038	0.1265	21.495	0.0465	4.9402	39.046
14	2.937	0.3405	8.2442	0.1213	24.215	0.0413	5.2731	43.472
15	3.172	0.3152	8.5595	0.1168	27.152	0.0368	5.5945	47.886
16	3.426	0.2919	8.8514	0.1130	30.324	0.0330	5.9046	52.264
17	3.700	0.2703	9.1216	0.1096	33.750	0.0296	6.2037	56.588
18	3.996	0.2502	9.3719	0.1067	37.450	0.0267	6.4920	60.843
19	4.316	0.2317	9.6036	0.1041	41.446	0.0241	6.7697	65.013
20	4.661	0.2145	9.8181	0.1019	45.762	0.0219	7.0369	69.090
21	5.034	0.1987	10.0168	0.0998	50.423	0.0198	7.2940	73.063
22	5.437	0.1839	10.2007	0.0980	55.457	0.0180	7.5412	76.926
23	5.871	0.1703	10.3711	0.0964	60.893	0.0164	7.7786	80.673
24	6.341	0.1577	10.5288	0.0950	66.765	0.0150	8.0066	84.300
25	6.848	0.1460	10.6748	0.0937	73.106	0.0137	8.2254	87.804
26	7.396	0.1352	10.8100	0.0925	79.954	0.0125	8.4352	91.184
27	7.988	0.1252	10.9352	0.0914	87.351	0.0114	8.6363	94.439
28	8.627	0.1159	11.0511	0.0905	95.339	0.0105	8.8289	97.569
29	9.317	0.1073	11.1584	0.0896	103.966	0.0096	9.0133	100.574
30	10.063	0.0994	11.2578	0.0888	113.283	0.0088	9.1897	103.456
31	10.868	0.0920	11.3498	0.0881	123.346	0.0081	9.3584	106.216
32	11.737	0.0852	11.4350	0.0875	134.214	0.0075	9.5197	108.857
33	12.676	0.0789	11.5139	0.0869	145.951	0.0069	9.6737	111.382
34	13.690	0.0730	11.5869	0.0863	158.627	0.0063	9.8208	113.792
35	14.785	0.0676	11.6546	0.0858	172.317	0.0058	9.9611	116.092
36	15.968	0.0626	11.7172	0.0853	187.102	0.0053	10.0949	118.284
37	17.246	0.0580	11.7752	0.0849	203.070	0.0049	10.2225	120.371
38	18.625	0.0537	11.8289	0.0845	220.316	0.0045	10.3440	122.358
39	20.115	0.0497	11.8786	0.0842	238.941	0.0042	10.4597	124.247
40	21.725	0.0460	11.9246	0.0839	259.057	0.0039	10.5699	126.042

10%

10% INTEREST FACTORS FOR DISCRETE COMPOUNDING PERIODS

	SINGLE PAYMENT		UNIFORM PAYMENT SERIES				GRADIENT SERIES	
	Compound Amount Factor	Present Worth Factor	Present Worth Factor	Capital Recovery Factor	Compound Amount Factor	Sinking Fund Factor	Gradient Uniform Series	Gradient Present Worth
N	Find F Given P F/P	Find P Given F P/F	Find P Given A P/A	Find A Given P A/P	Find F Given A F/A	Find A Given F A/F	Find A Given G A/G	Find P Given G P/G
1	1.100	0.9091	0.9091	1.1000	1.000	1.0000	0.0000	0.000
2	1.210	0.8264	1.7355	0.5762	2.100	0.4762	0.4762	0.826
3	1.331	0.7513	2.4869	0.4021	3.310	0.3021	0.9366	2.329
4	1.464	0.6830	3.1699	0.3155	4.641	0.2155	1.3812	4.378
5	1.611	0.6209	3.7908	0.2638	6.105	0.1638	1.8101	6.862
6	1.772	0.5645	4.3553	0.2296	7.716	0.1296	2.2236	9.684
7	1.949	0.5132	4.8684	0.2054	9.487	0.1054	2.6216	12.763
8	2.144	0.4665	5.3349	0.1874	11.436	0.0874	3.0045	16.029
9	2.358	0.4241	5.7590	0.1736	13.579	0.0736	3.3724	19.421
10	2.594	0.3855	6.1446	0.1627	15.937	0.0627	3.7255	22.891
11	2.853	0.3505	6.4951	0.1540	18.531	0.0540	4.0641	26.396
12	3.138	0.3186	6.8137	0.1468	21.384	0.0468	4.3884	29.901
13	3.452	0.2897	7.1034	0.1408	24.523	0.0408	4.6988	33.377
14	3.797	0.2633	7.3667	0.1357	27.975	0.0357	4.9955	36.800
15	4.177	0.2394	7.6061	0.1315	31.772	0.0315	5.2789	40.152
16	4.595	0.2176	7.8237	0.1278	35.950	0.0278	5.5493	43.416
17	5.054	0.1978	8.0216	0.1247	40.545	0.0247	5.8071	46.582
18	5.560	0.1799	8.2014	0.1219	45.599	0.0219	6.0526	49.640
19	6.116	0.1635	8.3649	0.1195	51.159	0.0195	6.2861	52.583
20	6.727	0.1486	8.5136	0.1175	57.275	0.0175	6.5081	55.407
21	7.400	0.1351	8.6487	0.1156	64.002	0.0156	6.7189	58.110
22	8.140	0.1228	8.7715	0.1140	71.403	0.0140	6.9189	60.689
23	8.954	0.1117	8.8832	0.1126	79.543	0.0126	7.1085	63.146
24	9.850	0.1015	8.9847	0.1113	88.497	0.0113	7.2881	65.481
25	10.835	0.0923	9.0770	0.1102	98.347	0.0102	7.4580	67.696
26	11.918	0.0839	9.1609	0.1092	109.182	0.0092	7.6186	69.794
27	13.110	0.0763	9.2372	0.1083	121.100	0.0083	7.7704	71.777
28	14.421	0.0693	9.3066	0.1075	134.210	0.0075	7.9137	73.650
29	15.863	0.0630	9.3696	0.1067	148.631	0.0067	8.0489	75.415
30	17.449	0.0573	9.4269	0.1061	164.494	0.0061	8.1762	77.077
31	19.194	0.0521	9.4790	0.1055	181.943	0.0055	8.2962	78.640
32	21.114	0.0474	9.5264	0.1050	201.138	0.0050	8.4091	80.108
33	23.225	0.0431	9.5694	0.1045	222.252	0.0045	8.5152	81.486
34	25.548	0.0391	9.6086	0.1041	245.477	0.0041	8.6149	82.777
35	28.102	0.0356	9.6442	0.1037	271.024	0.0037	8.7086	83.987
36	30.913	0.0323	9.6765	0.1033	299.127	0.0033	8.7965	85.119
37	34.004	0.0294	9.7059	0.1030	330.039	0.0030	8.8789	86.178
38	37.404	0.0267	9.7327	0.1027	364.043	0.0027	8.9562	87.167
39	41.145	0.0243	9.7570	0.1025	401.448	0.0025	9.0285	88.091
40	45.259	0.0221	9.7791	0.1023	442.593	0.0023	9.0962	88.953

12%

12% INTEREST FACTORS FOR DISCRETE COMPOUNDING PERIODS

	SINGLE PAYMENT		UNIFORM PAYMENT SERIES				GRADIENT SERIES	
	Compound Amount Factor	Present Worth Factor	Present Worth Factor	Capital Recovery Factor	Compound Amount Factor	Sinking Fund Factor	Gradient Uniform Series	Gradient Present Worth
N	Find F Given P F/P	Find P Given F P/F	Find P Given A P/A	Find A Given P A/P	Find F Given A F/A	Find A Given F A/F	Find A Given G A/G	Find P Given G P/G
1	1.120	0.8929	0.8929	1.1200	1.000	1.0000	0.0000	0.000
2	1.254	0.7972	1.6901	0.5917	2.120	0.4717	0.4717	0.797
3	1.405	0.7118	2.4018	0.4163	3.374	0.2963	0.9246	2.221
4	1.574	0.6355	3.0373	0.3292	4.779	0.2092	1.3589	4.127
5	1.762	0.5674	3.6048	0.2774	6.353	0.1574	1.7746	6.397
6	1.974	0.5066	4.1114	0.2432	8.115	0.1232	2.1720	8.930
7	2.211	0.4523	4.5638	0.2191	10.089	0.0991	2.5515	11.644
8	2.476	0.4039	4.9676	0.2013	12.300	0.0813	2.9131	14.471
9	2.773	0.3606	5.3282	0.1877	14.776	0.0677	3.2574	17.356
10	3.106	0.3220	5.6502	0.1770	17.549	0.0570	3.5847	20.254
11	3.479	0.2875	5.9377	0.1684	20.655	0.0484	3.8953	23.129
12	3.896	0.2567	6.1944	0.1614	24.133	0.0414	4.1897	25.952
13	4.363	0.2292	6.4235	0.1557	28.029	0.0357	4.4683	28.702
14	4.887	0.2046	6.6282	0.1509	32.393	0.0309	4.7317	31.362
15	5.474	0.1827	6.8109	0.1468	37.280	0.0268	4.9803	33.920
16	6.130	0.1631	6.9740	0.1434	42.753	0.0234	5.2147	36.367
17	6.866	0.1456	7.1196	0.1405	48.884	0.0205	5.4353	38.697
18	7.690	0.1300	7.2497	0.1379	55.750	0.0179	5.6427	40.908
19	8.613	0.1161	7.3658	0.1358	63.440	0.0158	5.8375	42.998
20	9.646	0.1037	7.4694	0.1339	72.052	0.0139	6.0202	44.968
21	10.804	0.0926	7.5620	0.1322	81.699	0.0122	6.1913	46.819
22	12.100	0.0826	7.6446	0.1308	92.503	0.0108	6.3514	48.554
23	13.552	0.0738	7.7184	0.1296	104.603	0.0096	6.5010	50.178
24	15.179	0.0659	7.7843	0.1285	118.155	0.0085	6.6406	51.693
25	17.000	0.0588	7.8431	0.1275	133.334	0.0075	6.7708	53.105
26	19.040	0.0525	7.8957	0.1267	150.334	0.0067	6.8921	54.418
27	21.325	0.0469	7.9426	0.1259	169.374	0.0059	7.0049	55.637
28	23.884	0.0419	7.9844	0.1252	190.699	0.0052	7.1098	56.767
29	26.750	0.0374	8.0218	0.1247	214.583	0.0047	7.2071	57.814
30	29.960	0.0334	8.0552	0.1241	241.333	0.0041	7.2974	58.782
31	33.555	0.0298	8.0850	0.1237	271.293	0.0037	7.3811	59.676
32	37.582	0.0266	8.1116	0.1233	304.848	0.0033	7.4586	60.501
33	42.092	0.0238	8.1354	0.1229	342.429	0.0029	7.5302	61.261
34	47.143	0.0212	8.1566	0.1226	384.521	0.0026	7.5965	61.961
35	52.800	0.0189	8.1755	0.1223	431.663	0.0023	7.6577	62.605
36	59.136	0.0169	8.1924	0.1221	484.463	0.0021	7.7141	63.197
37	66.232	0.0151	8.2075	0.1218	543.599	0.0018	7.7661	63.741
38	74.180	0.0135	8.2210	0.1216	609.831	0.0016	7.8141	64.239
39	83.081	0.0120	8.2330	0.1215	684.010	0.0015	7.8582	64.697
40	93.051	0.0107	8.2438	0.1213	767.091	0.0013	7.8988	65.116

15%

15% INTEREST FACTORS FOR DISCRETE COMPOUNDING PERIODS

	SINGLE PAYMENT		UNIFORM PAYMENT SERIES				GRADIENT SERIES	
	Compound Amount Factor	Present Worth Factor	Present Worth Factor	Capital Recovery Factor	Compound Amount Factor	Sinking Fund Factor	Gradient Uniform Series	Gradient Present Worth
N	Find F Given P F/P	Find P Given F P/F	Find P Given A P/A	Find A Given P A/P	Find F Given A F/A	Find A Given F A/F	Find A Given G A/G	Find P Given G P/G
1	1.150	0.8696	0.8696	1.1500	1.000	1.0000	0.0000	0.000
2	1.323	0.7561	1.6257	0.6151	2.150	0.4651	0.4651	0.756
3	1.521	0.6575	2.2832	0.4380	3.472	0.2880	0.9071	2.071
4	1.749	0.5718	2.8550	0.3503	4.993	0.2003	1.3263	3.786
5	2.011	0.4972	3.3522	0.2983	6.742	0.1483	1.7228	5.775
6	2.313	0.4323	3.7845	0.2642	8.754	0.1142	2.0972	7.937
7	2.660	0.3759	4.1604	0.2404	11.067	0.0904	2.4498	10.192
8	3.059	0.3269	4.4873	0.2229	13.727	0.0729	2.7813	12.481
9	3.518	0.2843	4.7716	0.2096	16.786	0.0596	3.0922	14.755
10	4.046	0.2472	5.0188	0.1993	20.304	0.0493	3.3832	16.979
11	4.652	0.2149	5.2337	0.1911	24.349	0.0411	3.6549	19.129
12	5.350	0.1869	5.4206	0.1845	29.002	0.0345	3.9082	21.185
13	6.153	0.1625	5.5831	0.1791	34.352	0.0291	4.1438	23.135
14	7.076	0.1413	5.7245	0.1747	40.505	0.0247	4.3624	24.972
15	8.137	0.1229	5.8474	0.1710	47.580	0.0210	4.5650	26.693
16	9.358	0.1069	5.9542	0.1679	55.717	0.0179	4.7522	28.296
17	10.761	0.0929	6.0472	0.1654	65.075	0.0154	4.9251	29.783
18	12.375	0.0808	6.1280	0.1632	75.836	0.0132	5.0843	31.156
19	14.232	0.0703	6.1982	0.1613	88.212	0.0113	5.2307	32.421
20	16.367	0.0611	6.2593	0.1598	102.444	0.0098	5.3651	33.582
21	18.822	0.0531	6.3125	0.1584	118.810	0.0084	5.4883	34.645
22	21.645	0.0462	6.3587	0.1573	137.632	0.0073	5.6010	35.615
23	24.891	0.0402	6.3988	0.1563	159.276	0.0063	5.7040	36.499
24	28.625	0.0349	6.4338	0.1554	184.168	0.0054	5.7979	37.302
25	32.919	0.0304	6.4641	0.1547	212.793	0.0047	5.8834	38.031
26	37.857	0.0264	6.4906	0.1541	245.712	0.0041	5.9612	38.692
27	43.535	0.0230	6.5135	0.1535	283.569	0.0035	6.0319	39.289
28	50.066	0.0200	6.5335	0.1531	327.104	0.0031	6.0960	39.828
29	57.575	0.0174	6.5509	0.1527	377.170	0.0027	6.1541	40.315
30	66.212	0.0151	6.5660	0.1523	434.745	0.0023	6.2066	40.753
31	76.144	0.0131	6.5791	0.1520	500.957	0.0020	6.2541	41.147
32	87.565	0.0114	6.5905	0.1517	577.100	0.0017	6.2970	41.501
33	100.700	0.0099	6.6005	0.1515	664.666	0.0015	6.3357	41.818
34	115.805	0.0086	6.6091	0.1513	765.365	0.0013	6.3705	42.103
35	133.176	0.0075	6.6166	0.1511	881.170	0.0011	6.4019	42.359
36	153.152	0.0065	6.6231	0.1510	1014.346	0.0010	6.4301	42.587
37	176.125	0.0057	6.6288	0.1509	1167.498	0.0009	6.4554	42.792
38	202.543	0.0049	6.6338	0.1507	1343.622	0.0007	6.4781	42.974
39	232.925	0.0043	6.6380	0.1506	1546.165	0.0006	6.4985	43.137
40	267.864	0.0037	6.6418	0.1506	1779.090	0.0006	6.5168	43.283

20%

20% INTEREST FACTORS FOR DISCRETE COMPOUNDING PERIODS

	SINGLE PAYMENT		UNIFORM PAYMENT SERIES				GRADIENT SERIES	
	Compound Amount Factor	Present Worth Factor	Present Worth Factor	Capital Recovery Factor	Compound Amount Factor	Sinking Fund Factor	Gradient Uniform Series	Gradient Present Worth
N	Find F Given P F/P	Find P Given F P/F	Find P Given A P/A	Find A Given P A/P	Find F Given A F/A	Find A Given F A/F	Find A Given G A/G	Find P Given G P/G
1	1.200	0.8333	0.8333	1.2000	1.000	1.0000	0.0000	0.000
2	1.440	0.6944	1.5278	0.6545	2.200	0.4545	0.4545	0.694
3	1.728	0.5787	2.1065	0.4747	3.640	0.2747	0.8791	1.852
4	2.074	0.4823	2.5887	0.3863	5.368	0.1863	1.2742	3.299
5	2.488	0.4019	2.9906	0.3344	7.442	0.1344	1.6405	4.906
6	2.986	0.3349	3.3255	0.3007	9.930	0.1007	1.9788	6.581
7	3.583	0.2791	3.6046	0.2774	12.916	0.0774	2.2902	8.255
8	4.300	0.2326	3.8372	0.2606	16.499	0.0606	2.5756	9.883
9	5.160	0.1938	4.0310	0.2481	20.799	0.0481	2.8364	11.434
10	6.192	0.1615	4.1925	0.2385	25.959	0.0385	3.0739	12.887
11	7.430	0.1346	4.3271	0.2311	32.150	0.0311	3.2893	14.233
12	8.916	0.1122	4.4392	0.2253	39.581	0.0253	3.4841	15.467
13	10.699	0.0935	4.5327	0.2206	48.497	0.0206	3.6597	16.588
14	12.839	0.0779	4.6106	0.2169	59.196	0.0169	3.8175	17.601
15	15.407	0.0649	4.6755	0.2139	72.035	0.0139	3.9588	18.509
16	18.488	0.0541	4.7296	0.2114	87.442	0.0114	4.0851	19.321
17	22.186	0.0451	4.7746	0.2094	105.931	0.0094	4.1976	20.042
18	26.623	0.0376	4.8122	0.2078	128.117	0.0078	4.2975	20.680
19	31.948	0.0313	4.8435	0.2065	154.740	0.0065	4.3861	21.244
20	38.338	0.0261	4.8696	0.2054	186.688	0.0054	4.4643	21.739
21	46.005	0.0217	4.8913	0.2044	225.026	0.0044	4.5334	22.174
22	55.206	0.0181	4.9094	0.2037	271.031	0.0037	4.5941	22.555
23	66.247	0.0151	4.9245	0.2031	326.237	0.0031	4.6475	22.887
24	79.497	0.0126	4.9371	0.2025	392.484	0.0025	4.6943	23.176
25	95.396	0.0105	4.9476	0.2021	471.981	0.0021	4.7352	23.428
26	114.475	0.0087	4.9563	0.2018	567.377	0.0018	4.7709	23.646
27	137.371	0.0073	4.9636	0.2015	681.853	0.0015	4.8020	23.835
28	164.845	0.0061	4.9697	0.2012	819.223	0.0012	4.8291	23.999
29	197.814	0.0051	4.9747	0.2010	984.068	0.0010	4.8527	24.141
30	237.376	0.0042	4.9789	0.2008	1181.882	0.0008	4.8731	24.263
31	284.852	0.0035	4.9824	0.2007	1419.258	0.0007	4.8908	24.368
32	341.822	0.0029	4.9854	0.2006	1704.109	0.0006	4.9061	24.459
33	410.186	0.0024	4.9878	0.2005	2045.931	0.0005	4.9194	24.537
34	492.224	0.0020	4.9898	0.2004	2456.118	0.0004	4.9308	24.604
35	590.668	0.0017	4.9915	0.2003	2948.341	0.0003	4.9406	24.661
36	708.802	0.0014	4.9929	0.2003	3539.009	0.0003	4.9491	24.711
37	850.562	0.0012	4.9941	0.2002	4247.811	0.0002	4.9564	24.753
38	1020.675	0.0010	4.9951	0.2002	5098.373	0.0002	4.9627	24.789
39	1224.810	0.0008	4.9959	0.2002	6119.048	0.0002	4.9681	24.820
40	1469.772	0.0007	4.9966	0.2001	7343.858	0.0001	4.9728	24.847

25%

25% INTEREST FACTORS FOR DISCRETE COMPOUNDING PERIODS

	SINGLE PAYMENT		UNIFORM PAYMENT SERIES				GRADIENT SERIES	
	Compound Amount Factor	Present Worth Factor	Present Worth Factor	Capital Recovery Factor	Compound Amount Factor	Sinking Fund Factor	Gradient Uniform Series	Gradient Present Worth
N	Find F Given P F/P	Find P Given F P/F	Find P Given A P/A	Find A Given P A/P	Find F Given A F/A	Find A Given F A/F	Find A Given G A/G	Find P Given G P/G
1	1.250	0.8000	0.8000	1.2500	1.000	1.0000	0.0000	0.000
2	1.563	0.6400	1.4400	0.6944	2.250	0.4444	0.4444	0.640
3	1.953	0.5120	1.9520	0.5123	3.812	0.2623	0.8525	1.664
4	2.441	0.4096	2.3616	0.4234	5.766	0.1734	1.2249	2.893
5	3.052	0.3277	2.6893	0.3718	8.207	0.1218	1.5631	4.204
6	3.815	0.2621	2.9514	0.3388	11.259	0.0888	1.8683	5.514
7	4.768	0.2097	3.1611	0.3163	15.073	0.0663	2.1424	6.773
8	5.960	0.1678	3.3289	0.3004	19.842	0.0504	2.3872	7.947
9	7.451	0.1342	3.4631	0.2888	25.802	0.0388	2.6048	9.021
10	9.313	0.1074	3.5705	0.2801	33.253	0.0301	2.7971	9.987
11	11.642	0.0859	3.6564	0.2735	42.566	0.0235	2.9663	10.846
12	14.552	0.0687	3.7251	0.2684	54.208	0.0184	3.1145	11.602
13	18.190	0.0550	3.7801	0.2645	68.760	0.0145	3.2437	12.262
14	22.737	0.0440	3.8241	0.2615	86.949	0.0115	3.3559	12.833
15	28.422	0.0352	3.8593	0.2591	109.687	0.0091	3.4530	13.326
16	35.527	0.0281	3.8874	0.2572	138.109	0.0072	3.5366	13.748
17	44.409	0.0225	3.9099	0.2558	173.636	0.0058	3.6084	14.108
18	55.511	0.0180	3.9279	0.2546	218.045	0.0046	3.6698	14.415
19	69.389	0.0144	3.9424	0.2537	273.556	0.0037	3.7222	14.674
20	86.736	0.0115	3.9539	0.2529	342.945	0.0029	3.7667	14.893
21	108.420	0.0092	3.9631	0.2523	429.681	0.0023	3.8045	15.078
22	135.525	0.0074	3.9705	0.2519	538.101	0.0019	3.8365	15.233
23	169.407	0.0059	3.9764	0.2515	673.626	0.0015	3.8634	15.362
24	211.758	0.0047	3.9811	0.2512	843.033	0.0012	3.8861	15.471
25	264.698	0.0038	3.9849	0.2509	1054.791	0.0009	3.9052	15.562
26	330.872	0.0030	3.9879	0.2508	1319.489	0.0008	3.9212	15.637
27	413.590	0.0024	3.9903	0.2506	1650.361	0.0006	3.9346	15.700
28	516.988	0.0019	3.9923	0.2505	2063.952	0.0005	3.9457	15.752
29	646.235	0.0015	3.9938	0.2504	2580.939	0.0004	3.9551	15.796
30	807.794	0.0012	3.9950	0.2503	3227.174	0.0003	3.9628	15.832
31	1009.742	0.0010	3.9960	0.2502	4034.968	0.0002	3.9693	15.861
32	1262.177	0.0008	3.9968	0.2502	5044.710	0.0002	3.9746	15.886
33	1577.722	0.0006	3.9975	0.2502	6306.887	0.0002	3.9791	15.906
34	1972.152	0.0005	3.9980	0.2501	7884.609	0.0001	3.9828	15.923
35	2465.190	0.0004	3.9984	0.2501	9856.761	0.0001	3.9858	15.937

30%

30% INTEREST FACTORS FOR DISCRETE COMPOUNDING PERIODS

	SINGLE PAYMENT		UNIFORM PAYMENT SERIES				GRADIENT SERIES	
	Compound Amount Factor	Present Worth Factor	Present Worth Factor	Capital Recovery Factor	Compound Amount Factor	Sinking Fund Factor	Gradient Uniform Series	Gradient Present Worth
N	Find F Given P F/P	Find P Given F P/F	Find P Given A P/A	Find A Given P A/P	Find F Given A F/A	Find A Given F A/F	Find A Given G A/G	Find P Given G P/G
1	1.300	0.7692	0.7692	1.3000	1.000	1.0000	0.0000	0.000
2	1.690	0.5917	1.3609	0.7348	2.300	0.4348	0.4348	0.592
3	2.197	0.4552	1.8161	0.5506	3.990	0.2506	0.8271	1.502
4	2.856	0.3501	2.1662	0.4616	6.187	0.1616	1.1783	2.552
5	3.713	0.2693	2.4356	0.4106	9.043	0.1106	1.4903	3.630
6	4.827	0.2072	2.6427	0.3784	12.756	0.0784	1.7654	4.666
7	6.275	0.1594	2.8021	0.3569	17.583	0.0569	2.0063	5.622
8	8.157	0.1226	2.9247	0.3419	23.858	0.0419	2.2156	6.480
9	10.604	0.0943	3.0190	0.3312	32.015	0.0312	2.3963	7.234
10	13.786	0.0725	3.0915	0.3235	42.619	0.0235	2.5512	7.887
11	17.922	0.0558	3.1473	0.3177	56.405	0.0177	2.6833	8.445
12	23.298	0.0429	3.1903	0.3135	74.327	0.0135	2.7952	8.917
13	30.288	0.0330	3.2233	0.3102	97.625	0.0102	2.8895	9.314
14	39.374	0.0254	3.2487	0.3078	127.913	0.0078	2.9685	9.644
15	51.186	0.0195	3.2682	0.3060	167.286	0.0060	3.0344	9.917
16	66.542	0.0150	3.2832	0.3046	218.472	0.0046	3.0892	10.143
17	86.504	0.0116	3.2948	0.3035	285.014	0.0035	3.1345	10.328
18	112.455	0.0089	3.3037	0.3027	371.518	0.0027	3.1718	10.479
19	146.192	0.0068	3.3105	0.3021	483.973	0.0021	3.2025	10.602
20	190.050	0.0053	3.3158	0.3016	630.165	0.0016	3.2275	10.702
21	247.065	0.0040	3.3198	0.3012	820.215	0.0012	3.2480	10.783
22	321.184	0.0031	3.3230	0.3009	1067.280	0.0009	3.2646	10.848
23	417.539	0.0024	3.3254	0.3007	1388.464	0.0007	3.2781	10.901
24	542.801	0.0018	3.3272	0.3006	1806.003	0.0006	3.2890	10.943
25	705.641	0.0014	3.3286	0.3004	2348.803	0.0004	3.2979	10.977
26	917.333	0.0011	3.3297	0.3003	3054.444	0.0003	3.3050	11.005
27	1192.533	0.0008	3.3305	0.3003	3971.778	0.0003	3.3107	11.026
28	1550.293	0.0006	3.3312	0.3002	5164.311	0.0002	3.3153	11.044
29	2015.381	0.0005	3.3317	0.3001	6714.604	0.0001	3.3189	11.058
30	2619.996	0.0004	3.3321	0.3001	8729.985	0.0001	3.3219	11.069
31	3405.994	0.0003	3.3324	0.3001	11349.981	0.0001	3.3242	11.078
32	4427.793	0.0002	3.3326	0.3001	14755.975	0.0001	3.3261	11.085
33	5756.130	0.0002	3.3328	0.3001	19183.768	0.0001	3.3276	11.090
34	7482.970	0.0001	3.3329	0.3000	24939.899	0.0000	3.3288	11.094
35	9727.860	0.0001	3.3330	0.3000	32422.868	0.0000	3.3297	11.098

40%

40% INTEREST FACTORS FOR DISCRETE COMPOUNDING PERIODS

	SINGLE PAYMENT		UNIFORM PAYMENT SERIES				GRADIENT SERIES	
	Compound Amount Factor	Present Worth Factor	Present Worth Factor	Capital Recovery Factor	Compound Amount Factor	Sinking Fund Factor	Gradient Uniform Series	Gradient Present Worth
N	Find F Given P F/P	Find P Given F P/F	Find P Given A P/A	Find A Given P A/P	Find F Given A F/A	Find A Given F A/F	Find A Given G A/G	Find P Given G P/G
1	1.400	0.7143	0.7143	1.4000	1.000	1.0000	0.0000	0.000
2	1.960	0.5102	1.2245	0.8167	2.400	0.4167	0.4167	0.510
3	2.744	0.3644	1.5889	0.6294	4.360	0.2294	0.7798	1.239
4	3.842	0.2603	1.8492	0.5408	7.104	0.1408	1.0923	2.020
5	5.378	0.1859	2.0352	0.4914	10.946	0.0914	1.3580	2.764
6	7.530	0.1328	2.1680	0.4613	16.324	0.0613	1.5811	3.428
7	10.541	0.0949	2.2628	0.4419	23.853	0.0419	1.7664	3.997
8	14.758	0.0678	2.3306	0.4291	34.395	0.0291	1.9185	4.471
9	20.661	0.0484	2.3790	0.4203	49.153	0.0203	2.0422	4.858
10	28.925	0.0346	2.4136	0.4143	69.814	0.0143	2.1419	5.170
11	40.496	0.0247	2.4383	0.4101	98.739	0.0101	2.2215	5.417
12	56.694	0.0176	2.4559	0.4072	139.235	0.0072	2.2845	5.611
13	79.371	0.0126	2.4685	0.4051	195.929	0.0051	2.3341	5.762
14	111.120	0.0090	2.4775	0.4036	275.300	0.0036	2.3729	5.879
15	155.568	0.0064	2.4839	0.4026	386.420	0.0026	2.4030	5.969
16	217.795	0.0046	2.4885	0.4018	541.988	0.0018	2.4262	6.038
17	304.913	0.0033	2.4918	0.4013	759.784	0.0013	2.4441	6.090
18	426.879	0.0023	2.4941	0.4009	1064.697	0.0009	2.4577	6.130
19	597.630	0.0017	2.4958	0.4007	1491.576	0.0007	2.4682	6.160
20	836.683	0.0012	2.4970	0.4005	2089.206	0.0005	2.4761	6.183
21	1171.356	0.0009	2.4979	0.4003	2925.889	0.0003	2.4821	6.200
22	1639.898	0.0006	2.4985	0.4002	4097.245	0.0002	2.4866	6.213
23	2295.857	0.0004	2.4989	0.4002	5737.142	0.0002	2.4900	6.222
24	3214.200	0.0003	2.4992	0.4001	8032.999	0.0001	2.4925	6.229
25	4499.880	0.0002	2.4994	0.4001	11247.199	0.0001	2.4944	6.235
26	6299.831	0.0002	2.4996	0.4001	15747.079	0.0001	2.4959	6.239
27	8819.764	0.0001	2.4997	0.4000	22046.910	0.0000	2.4969	6.242
28	12347.670	0.0001	2.4998	0.4000	30866.674	0.0000	2.4977	6.244
29	17286.737	0.0001	2.4999	0.4000	43214.343	0.0000	2.4983	6.245
30	24201.432	0.0000	2.4999	0.4000	60501.081	0.0000	2.4988	6.247

50%

50% INTEREST FACTORS FOR DISCRETE COMPOUNDING PERIODS

	SINGLE PAYMENT		UNIFORM PAYMENT SERIES				GRADIENT SERIES	
	Compound Amount Factor	Present Worth Factor	Present Worth Factor	Capital Recovery Factor	Compound Amount Factor	Sinking Fund Factor	Gradient Uniform Series	Gradient Present Worth
N	Find F Given P F/P	Find P Given F P/F	Find P Given A P/A	Find A Given P A/P	Find F Given A F/A	Find A Given F A/F	Find A Given G A/G	Find P Given G P/G
1	1.500	0.6667	0.6667	1.5000	1.000	1.0000	0.0000	0.000
2	2.250	0.4444	1.1111	0.9000	2.500	0.4000	0.4000	0.444
3	3.375	0.2963	1.4074	0.7105	4.750	0.2105	0.7368	1.037
4	5.063	0.1975	1.6049	0.6231	8.125	0.1231	1.0154	1.630
5	7.594	0.1317	1.7366	0.5758	13.188	0.0758	1.2417	2.156
6	11.391	0.0878	1.8244	0.5481	20.781	0.0481	1.4226	2.595
7	17.086	0.0585	1.8829	0.5311	32.172	0.0311	1.5648	2.947
8	25.629	0.0390	1.9220	0.5203	49.258	0.0203	1.6752	3.220
9	38.443	0.0260	1.9480	0.5134	74.887	0.0134	1.7596	3.428
10	57.665	0.0173	1.9653	0.5088	113.330	0.0088	1.8235	3.584
11	86.498	0.0116	1.9769	0.5058	170.995	0.0058	1.8713	3.699
12	129.746	0.0077	1.9846	0.5039	257.493	0.0039	1.9068	3.784
13	194.620	0.0051	1.9897	0.5026	387.239	0.0026	1.9329	3.846
14	291.929	0.0034	1.9931	0.5017	581.859	0.0017	1.9519	3.890
15	437.894	0.0023	1.9544	0.5011	873.788	0.0011	1.9657	3.922
16	656.841	0.0015	1.9970	0.5008	1311.682	0.0008	1.9756	3.945
17	985.261	0.0010	1.9980	0.5005	1968.523	0.0005	1.9827	3.961
18	1477.892	0.0007	1.9986	0.5003	2953.784	0.0003	1.9878	3.973
19	2216.838	0.0005	1.9991	0.5002	4431.676	0.0002	1.9914	3.981
20	3325.257	0.0003	1.9994	0.5002	6648.513	0.0002	1.9940	3.987
21	4987.885	0.0002	1.9996	0.5001	9973.770	0.0001	1.9958	3.991
22	7481.828	0.0001	1.9997	0.5001	14961.655	0.0001	1.9971	3.994
23	11222.741	0.0001	1.9998	0.5000	22443.483	0.0000	1.9980	3.996
24	16834.112	0.0001	1.9999	0.5000	33666.224	0.0000	1.9986	3.997
25	25251.168	0.0000	1.9999	0.5000	50500.337	0.0000	1.9990	3.998

Index

Abandonment, 221
  option, 428–430
Actionable volatility, 407–408, 411, 416–417, 418, 425, 437
Activity based costs, 80–82
After–tax, 116
  income, 122
  replacement, 132
  replacement analysis, 238–241
Agency theory, projects, 486
Amortization, 89, 111
Annuity functions, 43
Arithmetic gradient, 29–33
Autocorrelated cash flows, 356–357

Basis, 91, 92–93
Before–tax, 116
Benefit–cost ratio, 190
Benefit estimating, 55–82, 75–77
Bernhard's generalized horizon model, 293–302
Bernoulli principle, 333–335
Beta, 151
Beta distribution, 352–353
Binomial lattices, 411–412
  abandonment option, 429–430
  compound option, 432–436
  dementia example, 418–420
  oil well example, 426–428
Black–Scholes equation, 402–403
  derivation, 439–444
Bonds, 143–145
Book value, 92–93
  of stock, 150
Borrowing, 479
Budget allocation, chance–constrained
  dynamic programming, 289
  goal programming, 289
  integer programming, 289
  linear programming, 289
  programming, 289
Budget constraints, 278

Capacity expansion, 453–468
Capacity functions, 69–70
Capacity planning, 453–468
  dynamic deterministic, 454–456
  dynamic probabilistic, 456–468
  strategies, 458–468
Capital, 22
Capital asset pricing model (CAPM), 151–153, 471–473
  assumptions, 473–474
  project selection, 481–482, 486–487
  security market line, 479–481
  systematic risk, 481
  unsystematic risk, 481
Capital asset pricing theory, 471–487
Capital expenditure, 1
  plans, 453
Capital gains, 117
Capital rationing, 252–254
  external, 252
  internal, 252
Cash flows, autocorrelated, 356–357
  stochastic, 351–364
  variance of cross–correlated, 360–362
Certainty equivalents, 351
Chance–constrained programming, 289
Choices under uncertainty, St. Petersburg paradox, 331–333
Common reinvestment assumption, 257
Common stock, 146–154
Competitive projects, 252
Complementary projects, 252

505

506 Index

Compound interest factors, 25–33
Compound option, 431–437
Constrained project selection, 276–281, 289
   mutually exclusive bundles, 266–271
Continuous interest, 40, 41–42
Conventional investment, 196–197
Cost estimating, 55–82
Cost of capital, 141, 142–167
   marginal, 157–166, marginal, 254
   opportunity cost, 277, 279
   weighted average, 142
Cost of waiting, real options, 406–407
Costs: direct, 77–78
   indirect, 78–82
   relationships (CERs), 63–75
Covariance, 482

Debt capital, 142–145
Decision networks, 371–372
   expansion options, 454
Decision trees, 371, 372–377
   outcome variability, 380–383
   sequential, 377–380
Declining balance depreciation, 96–97
Default free interest rate, 479
Deferral option: see delay option
Delay cost, real options, 406–407
Delay option
   dementia example, 412–420
   oil well example, 420–428
Dependent projects, 251–252
Depletion, 89, 109–111
Depreciation, 89–109
   amortization, 111
   declining balance, 96–97
   depletion, 109–111
   MACRS, 99–101
   MACRS recaptured, 125
   NPV, 103–105
   recaptured, 125
   property, 90
   straight–line, 94–95
   strategies, 105–108
   sum–of–the–years' digits, 97–98
   units of production, 99
Derivative securities, financial options, 402

Deterioration, 221
Diminishing marginal utility, 333
Direct costs, 77–78
Discounted payback period, 179–180
Diversification, 478
   limits in project portfolio, 482
   risk reduction, 484
Dividend policy, 154
Dominance, securities and portfolios, 475
Dynamic capacity planning
   deterministic, 454–456
   probabilistic, 456–468
Dynamic programming, 230–237
   budget allocation, 289

Effective interest rates, 38–40
Efficient portfolios, 475–478
Equity capital, 146–154
Equivalent, 3
Equivalent annual, 46–47
Estimating
   capacity functions, 69–70
   classification, 57–59
   cost estimation relationships (CERs), 63–75
   costs & benefits, 55–82
   factor, 62–63
   growth curves, 75–77
   Lang functions, 69–70
   learning curves, 70–75
   regression, 65–68
   unit, 61–63
Excise taxes, 115
Expected utility, 333–337
Expected value, 341–342
   decision–making challenges, 344
   perfect information of, 383–384
External capital rationing, 252

Factor estimates, 62–63
Financial options, 402–403
   financial vs. real, 405
Financing, 4
   curve, 277
   function, 141–166
Finite horizon replacement, 222

Firm, 1–11
Firm objectives, 6–8
Fisher's intersection, 262

Gain on sale, 125
Geometric gradients, 36–38
Goal programming
    budget allocation, 289
    project selection, 302–316, 317–320
Gordon–Shapiro growth model, 147–149
Gradient
    arithmetic, 29–33
    geometric, 36–38
Growth curves, 75–77

Horizon model
    Bernhard's generalized, 293–302
    Weingartner's, 291–293

Imperfect market assumption, 289
Income taxes, 115, 117–120
    investment tax credit, 121–122
    Section 1231, 124
Incremental rates of return, 264–271
Indexes, 60–61
Indirect costs, 78–82
    activity based, 80–82
Infinite horizon replacement, 222
Input–output firm, 3–4
Integer programming, 256, 270
    budget allocation, 289
    project portfolio, 484
Interest, 15–23
    continuous, 40, 41–42
Interest factors, 23, 25–33
    extended, 42–44
Interest rates
    nominal and effective, 38–40
    real options, 405–406
Internal capital rationing, 252
Internal rate of return, 175, 191–198
    conventional investment, 197
    NPV, 280
    ranking on, 276–281
    real option volatility, 410–411
    reinvestment assumption, 262–263
    WACC, assumptions, 279
Investing, 4
Investment balance, 192–195
Investment rate, marginal, 257
Investment tax credit, 121–122
Investment, conventional and nonconventional, 196–197
IRR. *See* internal rate of return

Kuhn–Tucker conditions, 293, 297–301

Lang functions, 69–70
Learning curves, 70–75
Lending, 479
Life cycle costs committed, 59
Life cycle estimation, 56–57
Like–for–like replacement, 93
Linear programming,
    budget allocation, 289
    project portfolio, 484
    replacement, 237–238
Loans, 196–197
Logarithmic cash flow, real option volatility, 408
Lorie–Savage problem, 204, 206–212, 251
    present worth, 212
Loss on sale, 125

MACRS, 99–101
    depreciation recaptured, 125
Marginal cost of capital, 157–166, 254
Marginal cost to extend, 224
Marginal investment rate, 257
Market portfolio, 151, 480
Market rate, 151
Market risk, 151
Market risk premium, 152, 481
Market share strategy, 459–460, 462–466
MARR, 166–167, 277, 280
Mathematical programming, Bernhard's generalized horizon model, 293–302
Maximize capacity utilization, 459–460

Index

Maximize capacity utilization, 466–468
Minimum (acceptable/attractive) rate of return. *See* MARR
Minimum cost replacement, 228
Mixed investment, 199, 206, 209–212
Modified accelerated cost recovery system. *See* MACRS
Multiple project selection, 251–281
Multiple roots, 205–206
   positive roots, 206–209
Mutually exclusive, 221
   alternatives, 173
   bundles, 266–271

Net present value, 173, 175, 181–189
   depreciation, 103–105
   expectation and variance, 354–356
   and IRR, 280
   probability statements, 357–359
   reinvestment assumption, 258–263
Neumann–Morgenstern:
   preference theory, 335–337
   utility function, 338–340
Nominal interest rates, 38–40
Nonconventional investment, 196–197, 198–204
Nonstationary replacement, 230–237
NPV. *See* net present value

Objectives firm, 6–8
Obsolescence, 221
Opportunity cost, 9
Opportunity cost of capital, 277, 279
Options, financial vs. real, 405

Parallel replacement, 222, 241–245
Payback period, 175–180
   discounted, 179–180
   rate of return, 177–179
Pecking order model, WACC, 167
Perfect information, expected value of, 383–384
Perfect market, 15, 174, 271, 277
   assumption, 289
Planning horizon, 236

Portfolio:
   efficient, 475–478
   mean and variance, 474–475
   projects, 483–487
   risk, 478–479
   securities and projects, 485–486
   theory, 473–479
   See also project portfolios
Preference theory
   Neumann–Morgenstern, 335–337
Preferred stock, 145–146
Present value. *See* net present value
Producing, 4
Production–consumption opportunities, 182–186
Profitability index, 175
Programming:
   dynamic, 230–237
   integer, 256, 270
   linear and replacement, 237–238
Project balance, 198
Project dependence, 251–252
Project portfolio, 483–487. *See also* portfolios
   limits to diversification, 482
   linear and integer programming, 484
   vs. securities, 485–486
Project selection
   agency theory, 486
   capital asset pricing theory, 471–487
   CAPM, 481–482, 486–487
   constrained, 289
   goal programming, 302–316, 317–320
   mutually exclusive bundles, 266–271
   semivariance, 354
   utility theory, 329–345
Property taxes, 115

Ranking on IRR, 276–281
Rate of return, 173
   incremental, 264–271
Rationing of capital, 252–254
Real options, 401–438
   abandonment, 428–430
   actionable volatility, 411, 416–417, 418, 425, 437

basis, 402
compound, 431–437
current issues, 437–438
delay option, dementia example, 412–420
delay option, oil well example, 420–428
financial vs. real, 405
historical development, 403–405
interest rates, 405–406
Real option volatility
internal rate of return, 410–411
logarithmic cash flow, 408
logarithmic present value returns, 409–410
management estimate, 409
standard deviation of cash flows, 410
stock proxy, 408
Recaptured depreciation, 125
Recovery period, 91
Regression: cost estimating, 65–68
Reinvestment assumption, 181
IRR, 262–263
NPV, 258–263
Reinvestment rate, 256–263
problem, 257
Replacement, 221–245
after–tax, 132, 238–241
like–for–like, 93
linear programming, 237–238
nonstationary, 230–237
parallel, 241–245
planning horizon, 236
Retained earnings, 153–154, 9, 90
Return on invested capital, 209–212
Return on investment, 175
Revenue estimation, 75–77
Reward–to–variability ratio, 482–483
Risk, 330
aversion, utility function, 340–344
decision making under, 371–393
multiple projects, 359–363
portfolio, 478–479
premium, 341
reduce by diversification, 484
Risk–free rate, 151, 479
Risk/return trade–off, security market line of firm, 482–483
Root space for P, A, and F, 204–208

Sales taxes, 115
Salvage value, 91
Satisficing, 7
Section 179, 92–93
Section 1231, 124
Security market line, CAPM, 479–481
of firm, risk/return trade–off, 482–483
Semivariance, project selection, 354
Sensitivity analysis, 82
Separation theorem, invalidation, 290–291
Serial replacement, 222
Simulation, 384–393
Solomon growth model, 149–151
Spreadsheet
annuity functions, 43
cash flow tables, 44–46
St. Petersburg paradox, choices under uncertainty, 331–333
Stationary replacement, 222–229
Stochastic cash flows, 351–364
Stochastic dominance, 354
Straight–line depreciation, 94–95
Strategies, capacity planning, 458–468
Subscription/membership problem, 213–214
Sum–of–the–years' digits depreciation, 97–98
Synergistic projects, 252
Systematic risk, 479, 481

Taxes, 115–133
Tax value–added, 133
Technological change and replacement, 228–229
Terminal wealth, 296–297
Time value of money examples, 33–36
Treasury stock, 154
Triangular distribution, 388, 389

Uncertain future states, 363–364
Uncertainty, 330
Unconstrained project selection, 173
assumptions, 174–175
Uniform series factor, 27–29
Unit estimates, 61–63

Units of production depreciation, 99
Unrecovered investment balance, 192–195, 198
Unsystematic risk, 479, 481
Utility
　expected, 333–337
　　firm's and behavior by employees and managers, 344
　theory, 329–345
　function
　　linear approximation, 342
　　Neumann–Morgenstern, 338–340
　　risk aversion, 340–344

Value–added tax, 115, 133

Volatility, 401, 403, 483
　actionable, 407–408
　issues in multi–stage, 437
　real options volatility, 407–411

WACC. *See* weighted average cost of capital
Warranty, 227
Weighted average cost of capital (WACC), 142, 154–156
　IRR, assumptions, 279
　pecking order model, 167
Weingartner formulation, 271–276
　horizon model, 291–293